Springer-Verlag

Geschäftsbibliothek - Heidelberg

Titel: Springer Series in Solid-State Sciences, Vol. 76 (by Kuzmany, Mehring, Roth)

Aufl.-Aufst.: 1. Auflage

Drucker: Druckhaus Beltz, Hemsbach

Buchbinder: Fa. Schäffer, Grünstadt

Auflage: 1200   Bindequote: 1200

Schutzkarton/Schuber: nein

Satzart: Filmsatz

Filme vorhanden: ja

Reproabzüge vorhanden: ja

Preis: DM 110,--

Fertiggestellt: 22.12.87

Sonderdrucke: ja

Bemerkungen: keine

Berichtigungszettel: nein

Hersteller: R. Michels   Datum: 09.02.88

76 Springer Series in Solid-State Sciences
Edited by Hans-Joachim Queisser

Electronic Properties of Conjugated Polymers

Proceedings of an International Winter School,
Kirchberg, Tirol, March 14–21, 1987

Editors: H. Kuzmany, M. Mehring,
and S. Roth

With 265 Figures

Springer-Verlag Berlin Heidelberg New York
London Paris Tokyo

Professor Dr. Hans Kuzmany

Institut für Festkörperphysik, Universität Wien, Strudlhofgasse 4, A-1090 Wien, Austria and
Ludwig Boltzmann Institut für Festkörperphysik, Wien,
Kopernikusgasse 15, A-1090 Wien, Austria

Professor Dr. Michael Mehring

Physikalisches Institut II, Universität Stuttgart,
Pfaffenwaldring 57, D-7000 Stuttgart 80, Fed. Rep. of Germany

Dr. Siegmar Roth

Max-Planck-Institut für Festkörperforschung,
Heisenbergstrasse 1, D-7000 Stuttgart 80, Fed. Rep. of Germany

Series Editors:

Professor Dr., Dres. h. c. Manuel Cardona
Professor Dr., Dr. h. c. Peter Fulde
Professor Dr. Klaus von Klitzing
Professor Dr. Hans-Joachim Queisser

Max-Planck-Institut für Festkörperforschung, Heisenbergstrasse 1
D-7000 Stuttgart 80, Fed. Rep. of Germany

ISBN-13: 978-3-642-83286-4   e-ISBN-13: 978-3-642-83284-0
DOI: 10.1007/978-3-642-83284-0

Printing: Druckhaus Beltz, 6944 Hemsbach/Bergstr.
Binding: J. Schäffer GmbH & Co. KG., 6718 Grünstadt
2153/3150-543210

Preface

The International Winter School on Electronic Properties of Conjugated Polymers held March 14–21, 1987, in Kirchberg (Austria) was a sequel to a meeting held in Kirchberg two years before on a similar subject. The 1987 winter school was organized in cooperation with the "Bundesministerium für Wissenschaft und Forschung" in Austria and the "Bundesministerium für Forschung und Technologie" in the Federal Republic of Germany. The basic idea of the meeting was to provide an opportunity for experienced scientists from universities and industry to discuss their most recent results and for students and young scientists to inform themselves about the present state of the research in this field.

As in 1985, the scientific interest was concentrated on the electronic structure of various conjugated polymers and related compounds. The focus of interest in the field now appears to have broadened and covers not only conductivity and relaxation phenomena of polyacetylene but also nonlinear optical properties, highly oriented and single-crystal polymers, and electrochemical and opto-electrochemical properties of special materials such as polypyrrole and polyaniline. Exciting results on conductivity – the mass specific conductivity (i.e., the conductivity divided by the density) of polyacetylene is more than twice that of copper (!) – and a detailed interpretation of the meaning of conjugation length are reported. In spite of the high degree of orientation in several polymers, the mechanism for the conductivity was confirmed to be similar to the mechanism in amorphous systems. Theoretical and experimental results have proved the importance of electron-electron correlation. Possible applications such as electrochemical cells, electrode materials, processable conducting polymers, nonlinear optics devices, etc., are presented and now appear to be much more realistic than in previous reports.

Discussion meetings were devoted to the conduction mechanism and possible limitations of the conductivity of conjugated polymers in general, and to the physical and chemical properties of polyaniline. In addition, a spontaneous meeting on the new high T_c oxidic superconductors and their relation to the conjugated polymers was held.

This book summarizes the tutorial and research papers presented at the winter school. We thank all the authors for their contributions and all the discussion speakers at the winter school for their stimulating remarks, which played an essential role in making the winter school an exciting and informative event.

We acknowledge in particular the "Bundesministerium für Wissenschaft und Forschung" (Austria) and the "Bundesministerium für Forschung und Technologie" (FRG), as well as the sponsors from industry, for their financial support. This support was not only a great help but was, in fact, indispensable for the goal of the meeting to be attained.

Finally, we thank the manager of the Hotel Sonnalp, Herr J.R. Jurgeith, for his continuous support and for his patience with the many special arrangements required during the meeting.

Wien, Stuttgart   *H. Kuzmany*
June, 1987   *M. Mehring*
                                        S. Roth

Contents

Part III   **Electron Energy Loss, Optical and Raman Spectroscopy**

Part IV   **Magnetic Resonance**

Part VII   **Polypyrrole, Polythiophene and Polyparaphenylene**

Electrical Conducting Molecular Crystals in a Polymer Matrix
By G. Heywang (With 1 Figure) 363

Part X   **Applications**

Part I

Conductivity

Electronic Transport in Low-Conductivity Metals and Comparison with Highly Conducting Polymers

A.B. Kaiser

Alexander von Humboldt Fellow,
Max-Planck-Institut für Festkörperforschung,
Postfach 800665, D-7000 Stuttgart 80, Fed. Rep. of Germany

We review the observed conductivity and thermopower behaviour in low-conductivity metals, and mention possible analogies with electronic transport in highly conducting polymers. As the electronic mean free path in metals becomes comparable to the interatomic spacing, precursor effects of localization are seen. For low-conductivity metals, incipient localization usually causes an <u>increase</u> of conductivity with temperature (i.e. opposite to typical metallic behaviour). This sign of temperature dependence is also seen in polymers, but the size of the temperature dependence suggests the presence of some thermally activated conductivity contribution even in the most highly conducting polymers. Another consequence of weak localization is a relatively large effect of magnetic field upon conductivity at low temperatures; qualitatively similar effects have been seen in conducting polymers.

As the conductivity of amorphous metals decreases, the thermopower increases in size but remains metallic in temperature dependence. We point out that metallic thermopower is almost never linear in temperature, but shows a change in slope due to electron-phonon enhancement at low temperatures. The resulting thermopower shape is the same as that seen in some highly conducting polymers, although the separation of effects due to electron-phonon enhancement and a hopping contribution to conductivity is difficult. For highly conducting polymers, the more metallic temperature dependence of thermopower compared to conductivity supports a picture of relatively long conducting regions separated by thin heat-conducting barriers.

1. Introduction

With the advent of highly conducting polymers, the concepts used for the interpretation of the electronic transport properties of metals, particularly disordered metals, are beginning to be used also by polymer physicists. EPSTEIN et al. [1] suggested that disorder localization of the electron wave functions played an important role in their samples of doped polyacetylene with conductivity of a few S/cm. For samples of much higher conductivity (~3000 S/cm) GOULD et al. [2] raised the possibility that localization and electron-electron interaction models could account for the observed logarithmic temperature dependence of conductivity at very low temperatures. We review here the main features of the conductivity, magnetoconductivity and thermoelectric power of low-conductivity metals, and make some comparisons with corresponding polymer properties to help clarify the conductivity mechanism in highly conducting polymers.

2. Conductivity

The conductivities of pure metals and lightly doped polymers are each strongly temperature dependent, but in opposite senses, as illustrated in Fig. 1. Between these extreme cases, the conductivity of the latest highly conducting polymers overlaps that of amorphous metal alloys, and temperature dependences are much smaller; it is therefore of particular interest to compare electron transport in these two classes of system.

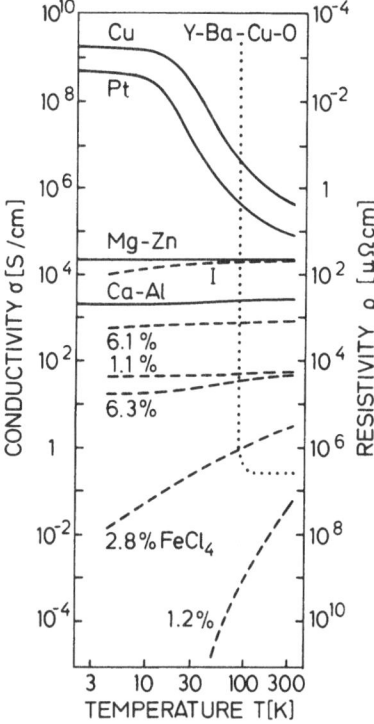

Fig. 1: Comparison of the conductivities of metals (full lines) and polymers (dashed lines). The data are for $FeCl_3$-doped polyacetylene (labelled by the doping concentration in %) [3], with the 6.1% sample from [4]; I-doped polyacetylene [5] (label I); $Mg_{70}Zn_{30}$ and $Ca_{60}Al_{40}$ glassy metals [6]; high-purity Cu and Pt; and the new high-T_c superconductor Y-Ba-Cu-O [7]. The conductivity of Y-Ba-Cu-O above its transition temperature is low but varies considerably between samples

2.1 Metal-semiconductor transition

The conductivities of amorphous alloys that are stable at room temperature and comprise only metallic elements typically occupy a relatively narrow band in Fig. 1, from about 25000 S/cm ($Mg_{70}Zn_{30}$) to 2500 S/cm ($Ca_{60}Al_{40}$) and show only a very small temperature variation (not more than about 13% change between 1K and 300K). However, in special cases where metallic elements are alloyed with non-metallic elements, it is possible by varying the concentration to drive the system through the metal-semiconductor transition. An example, shown in Fig. 2, is Fe-Sb alloys (amorphous Sb being a semiconductor). In the Sb-rich alloys, the low-temperature conductivity follows Mott's variable-range-hopping law, as in semiconducting polymers. For Fe concentrations of 15% and greater, however, there is a qualitative change in the low-temperature conductivity σ, which tends to a finite value rather than zero as the temperature goes to zero. This signals the metal-semiconductor transition, usually discussed in terms of Anderson (i.e. disorder-induced) localization, although as we shall see there are also aspects of Mott localization (induced by correlations amongst

3

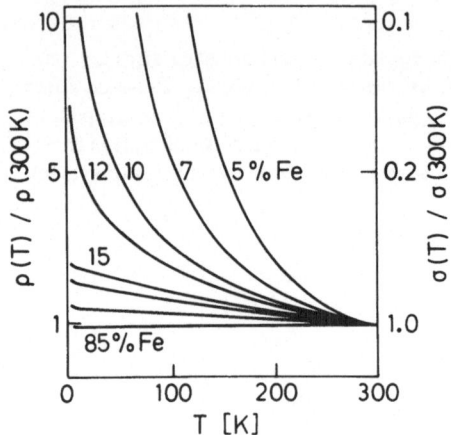

Fig. 2: Sketch of the temperature dependence of resistivity $\rho(T)$ for amorphous Fe-Sb alloys as measured by XIAO and CHIEN [8], showing the metal-semiconductor transition as the Fe concentration is varied. Between 12% and 15% Fe, the temperature dependence of the conductivity at low temperatures changes from the Mott $\exp(-T^{-1/4})$ law for an amorphous semiconductor to the $T^{1/2}$ law characteristic of disordered metals

the electrons that ultimately confine each to one site). Note that on the metallic side of the transition the temperature dependence of conductivity is still opposite in sign to the usual metallic behaviour for all but the 85% Fe sample. The reason for this is precursor effects of the localization transition that remove the discontinuity in the transition. Mott originally suggested that there was a minimum electron mean-free-path l in metals, given by the Ioffe-Regel criterion $k_F l > \pi$ (where k_F is the Fermi wavevector); this implies a minimum metallic conductivity of order 10^3 S/cm (depending on the density of carriers), and a discontinuous jump to zero conductivity going through the metal-semiconductor transition at zero temperature. However, in the scaling theory of Anderson localization the 3-D conductivity goes smoothly to zero as the Fermi energy is driven past a mobility edge [9]. Experiments tend to agree with the scaling theory.

We note that for conduction in 1-D or 2-D, the scaling theory predicts that $\sigma \to 0$ as $T \to 0$, i.e. there is no transition to metallic behaviour at zero temperature.

2.2 Weak Localization

We show in Fig. 3b the typical temperature dependence of conductivity in disordered metals. For metals of conductivity of order 10^4 S/cm (e.g. $Fe_{85}Sb_{15}$), we see the usual metallic sign of temperature dependence (although the size of the T dependence is very small). For most metals of conductivity less than 7000 S/cm, on the other hand, the precursor effects to Anderson localization (called 'weak' or 'incipient' localization) reverse the sign of the conductivity T dependence (this effect is known as the 'Mooij correlation' [10,11]). The change of sign is expected at lower conductivities for systems with a lower density of conduction electrons. An example of this sign reversal is shown in Fig. 3b for $Ta_{44}Ir_{56}$ ($\sigma = 3600$ S/cm), which has one of the largest temperature dependences for amorphous metal-metal alloys.

We can give a brief explanation of weak localization with the help of Fig. 3a. The curve labelled σ_B is the usual 'classical' Boltzmann metallic conductivity, which decreases as T increases. Consider an electron that is scattered so that its direction of propagation is reversed. This result can occur by means of two or more scatterings in various ways, and because scattering by the disordered ions is elastic, the phase

4

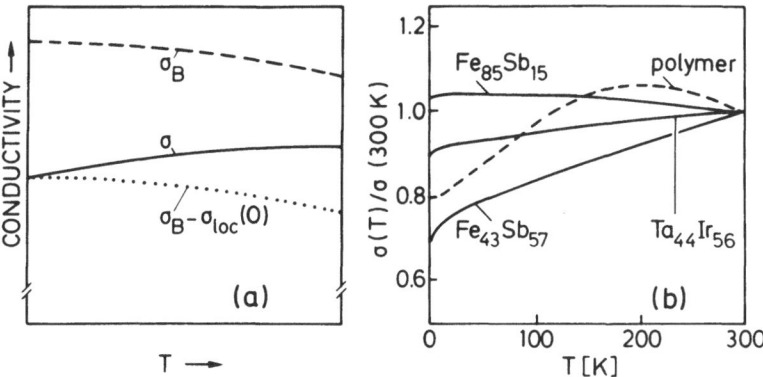

Fig. 3: (a) Effect of weak localization on conductivity (see text); (b) observed conductivity temperature dependence in Fe-Sb [8], Ta-Ir [12] and the polymer $[CH(FeCl_4)_{0.061})]_x$ [4], which has an unusually small T dependence

of the electron wave functions will be coherent in the backscattered direction. These waves will constructively interfere, strengthening the backscattering [9] and so reducing the conductivity by an amount $\sigma_{loc}(0)$, as shown by the dotted line in Fig. 3a. The reversal of the sign of conductivity T dependence arises because the presence of inelastic scattering as T increases destroys the coherence of the waves and so the interference, and the conductivity tends to revert to its classical value. This is shown by the full line in Fig. 3a, which represents the measured conductivity. The size of this weak localization effect increases as electron mean-free-path decreases. This description is valid only when the weak localization correction is small, but qualitatively similar behaviour (with larger T dependence of conductivity) is seen up to the metal-semiconductor transition, as shown for Fe-Sb in Figs. 2 and 3b.

Turning now to highly conducting polymers, it might be thought that weak localization can account for the rather puzzling lack of 'metallic' conductivity T dependence at all temperatures, as shown by the polymer data in Fig. 3b. (Weak localization may lead to a reversion to metallic behaviour at higher temperatures.) The latest data [5] for I-doped polyacetylene are of particular importance, since the room-temperature conductivity is similar to that of Mg-Zn, one of the highest conductivity amorphous metal alloys. In Mg-Zn, the electron mean-free-path is of order 12 Å and localization effects are insignificant except possibly at very low temperatures ($T < 10K$) [11]. The mean-free-path in I-doped $(CH)_x$ is much larger, owing to the lower density of conduction electrons [5], which should mean even smaller localization effects. Hence it would be surprising if weak localization could produce the observed increase in conductivity from 4K to 300K of more than a factor of two. Our comparison therefore supports the conclusion [5] that the conductivity T dependence in I-doped $(CH)_x$ is probably caused by a thermally activated contribution to conductivity, possibly across interfibril contacts. Localization effects in this temperature range would need to be enhanced by orders of magnitude owing to 1-D or 2-D conduction to change this conclusion.

2.3 Electron-electron interactions

A low-temperature conductivity anomaly is seen almost universally in amorphous metals, namely the onset of $T^{1/2}$ conductivity behaviour below about 10K. Examples can be seen in Figs. 2 and 3b, and in more detail in Fig. 4. The size of the anomaly is usually within a factor of two or three of the 'universal' size $\Delta\sigma = 6\,T^{1/2}$ S/cm K$^{1/2}$ [13]. This anomaly is a precursor of Mott localization, and arises from the loss of dynamic screening of the long-range Coulomb interaction as the electron mean-free-path becomes very short, i.e. from electron-electron interactions [9].

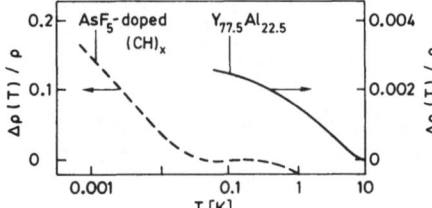

Fig. 4: Low-temperature $T^{1/2}$ conductivity anomaly in a Y-Al glassy metal ($\sigma = 3600$ S/cm) [14], compared to the anomaly in a polymer of similar conductivity ($\sigma = 3180$ S/cm) [2]

A logarithmic low-temperature conductivity anomaly has been seen in highly doped polyacetylene (Fig. 4), possibly due to localization and electron-electron interaction effects [2] (a logarithmic anomaly arises in 2-D conduction from either mechanism). The occurrence of the anomaly only at such low temperatures supports our conclusion above that the mean-free-path is too long in the most highly conducting polymers for localization to have an important effect at temperatures up to 100K or more. The polymer anomaly in Fig. 4 is more than two orders of magnitude larger, but at temperatures more than two orders of magnitude smaller than that in metal-metal alloys of similar conductivity magnitude. This difference is possibly related to the 3-D nature of conduction in amorphous metals (with σ finite as $T \to 0$, compared to the $\sigma \to 0$ behaviour in reduced dimensions).

3. Magnetoconductivity

Weak localization is sensitive to magnetic field, since the field will alter the relative phase of the backscattered wavefunctions and destroy the constructive interference. This causes conductivity to increase with field H, first as H^2 and then as $H^{1/2}$, the effect reducing strongly as temperature increases [10]. Such behaviour is seen in Ca-Al alloys (Fig. 5). However, the presence of the spin-orbit interaction, especially

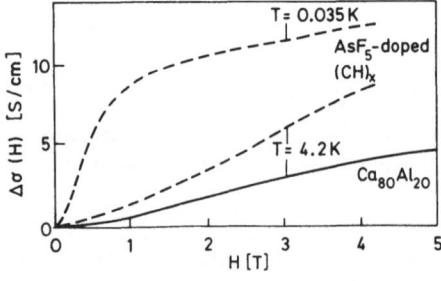

Fig. 5: Change in conductivity $\Delta\sigma(H)$ in a magnetic field for an amorphous Ca-Al alloy ($\sigma = 7700$ S/cm) [16], compared to that in a highly conducting polymer ($\sigma \sim 1000$ S/cm) [17]

for heavier nuclei, also affects the electron's phase and can reverse the sign of the magnetoconductivity. Many amorphous metal alloys therefore show negative magnetoconductance [15]. The effect of a magnetic field on the electron-electron interaction is usually small in disordered metals, so although the temperature dependence of conductivity at low temperatures is dominated by interactions, the magnetoconductivity is dominated by weak localization. We see from Fig. 5 that behaviour qualitatively similar to that in Ca-Al in seen in AsF$_5$-doped (CH)$_x$, but the effect is larger in the polymer.

4. Thermopower

The thermoelectric power is a property that has often been measured in polymers, because it signals rather dramatically the metal-semiconductor transition; the change in its temperature dependence is more striking than that seen in conductivity. Thermopower, however, is a rather subtle property, so we begin with its definition (Fig. 6) and a brief explanation.

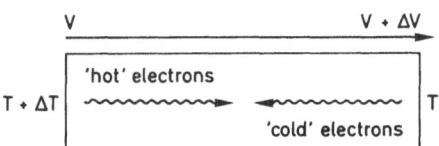

Fig. 6: Definition of thermopower: when a temperature difference ΔT is applied across the conductor, a charge accumulation, and so potential difference ΔV, occurs. Thermopower S is defined as $S = \Delta V / \Delta T$

Physically thermopower arises as follows (we take as an example the free electron case): when a temperature gradient is established, electrons at the hot end diffuse faster than those at the cold end, producing an accumulation of electrons at the cold end and so a negative thermopower S. Thermopower therefore depends on the differences of conductivity above and below the Fermi surface, as reflected by the derivative in the standard formula

$$S = \frac{\pi^2 k_B^2 T}{3e} \frac{\partial \ln \sigma(E)}{\partial E}\bigg|_{E_F} \tag{1}$$

where k_B is Boltzmann's constant, e is the electronic charge and $\sigma(E)$ the conductivity as a function of electron energy. Note that this formula is valid only in homogeneous media with $\sigma(E)$ a slowly varying function of E. Thermopower is an intrinsic property, independent of the physical dimensions of the system, and so is not affected as much as conductance (from which conductivity is calculated) by partial breaks in electrical continuity in poor samples.

4.1 Effect of localization

For amorphous metal-metal alloys, weak localization has only a small effect on the temperature dependence of thermopower [18], although S appears to level off at higher temperatures near the metal-semiconductor transition [19]. As the conductivity decreases, thermopower magnitude tends to increase in amorphous metals [18] - this effect is a precursor of the huge increase in S seen across the metal-semiconductor transition, as illustrated in Fig. 7. In contrast to the monotonic increase with T

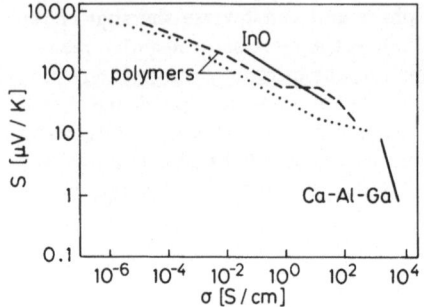

Fig. 7: Increase of thermopower magnitude as conductivity decreases in amorphous Ca-Al-Ga alloys [20], amorphous InO films [19], in FeCl$_3$-doped (CH)$_x$ (top dashed line) [21], and in MoCl$_5$-doped (CH)$_x$ (lower dashed line) [22]

seen in metals, S usually decreases with temperature for semiconductors, besides being orders of magnitude larger. The small size of S in highly conducting polymers suggests that localization effects are not large, in agreement with our conclusions for conductivity.

4.2 Electron-phonon interaction

In crystals, two effects produce highly non-linear thermopowers which are very difficult to interpret. Firstly, strongly energy and temperature dependent scattering from magnetic impurities affects the energy dependence of the conductivity in (1), giving large thermopower peaks near the characteristic 'Kondo' temperature T_K for the magnetic impurity. In amorphous metals, however, disorder scattering is so much larger than scattering from impurities that magnetic impurity peaks in thermopower are washed out [23]. Secondly, phonons transport heat down the thermal gradient in Fig. 6, 'dragging' electrons with them by imparting momentum in collisions and causing a 'phonon-drag' thermopower. This adds to the usual electron diffusion thermopower given by (1) and typically causes a large peak at about $T_D/5$, where T_D is the Debye temperature. This source of non-linearity is also absent in amorphous metals, because the phonon current and the importance of electron-phonon collisions are both greatly reduced, eliminating the phonon drag peak in S [24].

The thermopower is however <u>still</u> non-linear, because it is enhanced at low temperatures by the electron-phonon interaction [25], as shown by the example in Fig. 8a.

The origin of this enhancement effect can be understood from (1), although a more general formulation is required for the complete calculation [28]. At low temperatures the bare electron energy ϵ (measured from the Fermi level) is changed by the response of the positive ions as the electron moves past to a renormalized energy $E = \epsilon/(1+\lambda)$, where λ is the appropriate electron-phonon coupling parameter (not necessarily the same as that which determines the Peierls gap). Hence in (1) the derivative with respect to E is $(1 + \lambda)$ times the derivative with respect to ϵ, and the thermopower is therefore enhanced by the factor $(1 + \lambda)$ due to the electron-phonon interaction (the conductivity for each electron state is unchanged by the electron-phonon interaction). As the temperature increases, the enhancement λ decays, so the slope of the thermopower decreases, giving a characteristic 'knee' at about 50K. This behaviour is a general feature of metallic thermopower, although occasionally cancellation with other small effects does lead to a linear thermopower [28]. It occurs even for the highest conductivity amorphous alloys and so is not associated with localization.

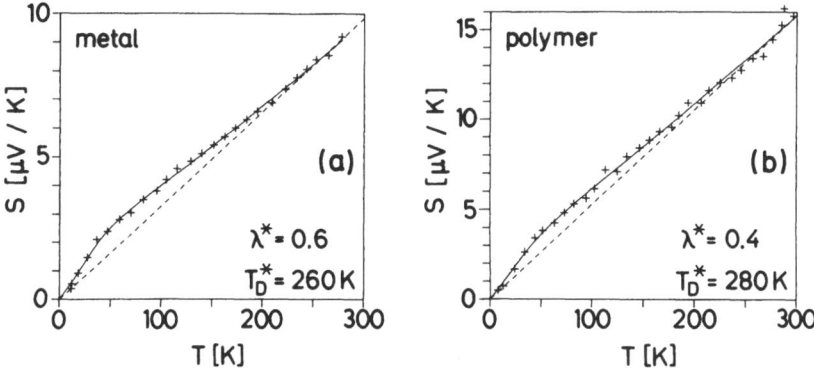

Fig. 8: Measured temperature dependence of the thermopower (crosses) in (a) the Chevrel compound $Cu_{1.8}Mo_6Se_5Te_3$ [26] and (b) the highly conducting polymer $[CH(FeCl_4)_{0.061}]_x$ [27]. The full lines are fits to the theoretical electron-phonon enhancement [23] with fitting parameters (the effective electron-phonon coupling parameter λ^* and Debye temperature T_D^*) as indicated. The dotted lines show linear behaviour for comparison

Comparing the thermopower of highly conducting polymers and amorphous metals, we notice a striking similarity in both magnitude and temperature dependence. The temperature dependence of polymer thermopower is also very rarely linear through the origin (AsF_5-doped polyacetylene [27] is an exception). In some cases, as in I-doped polyacetylene, the slope decreases continuously as T increases, ascribed to the presence of a $T^{1/2}$ hopping conduction term [27].

In polymers such as $FeCl_3$-doped polyacetylene [27], the thermopower is approximately linear above 100K but does not extrapolate through the origin, having an enhanced slope at low temperatures. This is just the behaviour seen in disordered metals due to the electron-phonon enhancement: comparing Figs. 8a and 8b suggests that the thermopower knee for the doped polyacetylene could be simply the usual electron-phonon enhancement effect. If this were to be confirmed, some information could be gained regarding the effective electron-phonon coupling parameter λ^* (which may differ from the real value λ owing to additional complications [28]). A fuller discussion of thermopower in highly conducting polymers will be given in a later publication.

Conclusions

We can briefly summarize the main features of the transport properties of disordered metals as follows:

i) Weak localization reverses the sign of the conductivity temperature dependence for most metals with conductivity less than 7000 S/cm, produces a relatively large magnetoconductivity (of either sign) at low temperatures, and causes a tendency for the thermopower to increase as the metal-semiconductor transition is approached.

ii) Electron-electron interactions produce a $T^{1/2}$ conductivity term below a temperature of about 10K.

iii) The electron-phonon interaction produces a characteristic knee in thermopower at a temperature of about 50K, but other deviations from linearity due to phonon drag and magnetic impurities are absent.

Our comparison with transport in highly conducting polymers leads to the following general conclusions:

i) Although weak localization due to disorder is likely to play some role in polymers, the small size of thermopower, the extremely low temperature of the apparent interaction-localization conductivity anomaly, and the long electron mean-free-paths in the most highly conducting polymers [5], suggest that for these materials localization effects are not large enough to cause the observed increase of conductivity with temperature above helium temperatures.

ii) The thermopower in some highly conducting polymers is in good agreement with the characteristic knee of metallic thermopower caused by electron-phonon enhancement, although similar behaviour may also arise in models involving a hopping contribution to conductivity.

iii) The thermopowers of highly conducting polymers and amorphous metals are more similar than their conductivities. This observation is consistent with a model [3] of long metallic conducting regions (e.g. fibrils) separated by thin electrical but not thermal barriers (e.g. interfibril contacts), as indicated by the following reasoning. Most of an electric potential difference applied across the sample would appear across the barriers, while most of an applied temperature difference would appear across the highly conducting regions. Thus resistivity would be mostly due to the barriers, and the temperature dependence of conductivity would be dominated by some kind of thermally activated hopping across the barriers [3]. On the other hand, the thermopower would tend to be that characteristic of the highly conducting regions where the largest temperature difference appears, unless the ratio of the characteristic thermopowers of the barriers to that of the conducting regions is even larger than the corresponding resistivity ratio.

Acknowledgements

I am very grateful to D. Baeriswyl, D. Moses, S. Roth, D. Schäfer-Siebert and Y.Q.Shen for helpful comments, and to the Alexander von Humboldt Stiftung for the award of a Fellowship.

References

1. A.J. Epstein, H.W. Gibson, P.M. Chaikin, W.G. Clark, G. Grüner: Phys. Rev. Lett. 45, 1730 (1980)
2. C.M. Gould, D.M. Bates, H.M. Bozler, A.J. Heeger, M.A. Druy, A.G. MacDiarmid: Phys. Rev. B 23, 6820 (1981)
3. K. Ehinger, S. Roth: Phil. Mag. B 53, 301 (1986)
4. Y.W. Park, J.C. Woo, K.H. Yoo, W.K. Han, C.H. Choi, T. Kabayashi, H. Shirakawa: Solid State Commun. 46, 731 (1983)
5. N. Basescu, Z.-X. Liu, D. Moses, A.J. Heeger, H. Naarmann: this volume
6. D.G. Naugle: J. Phys. Chem. Solids 45, 367 (1984)
7. J.M. Tarascon, L.H. Greene, W.R. McKinnon, G.W. Hull: to be published
8. G. Xiao, C.L. Chien: Phys. Rev. B 34, 8430 (1986)
9. P.A. Lee, T.V. Ramakrishnan: Rev. Mod. Phys. 57, 287 (1985)

10. C.C. Tsuei: Phys. Rev. Lett. 57, 1943 (1986)
11. A.B. Kaiser: Phys. Rev. Lett. 58, 1384 (1987)
12. K.D.D. Rathnayaka, H.J. Trodahl, A.B. Kaiser: Solid State Commun. 57, 207 (1986)
13. R.W. Cochrane, J.O. Strom-Olsen: Phys. Rev. B 29, 1088 (1984)
14. M. Olivier, J.O. Strom-Olsen, Z. Altounian, R.W. Cochrane, M. Trudeau: Phys. Rev. B 33, 2799 (1986)
15. D. Greig, B.J. Hickey, M.A. Howson, M.J. Walker: Z. Physikalische Chem., to appear
16. C.L. Tsai, F.C. Lu: J. Non-cryst. Solids 61-62, 1403 (1984)
17. J.F. Kwak, T.C. Clarke, R.L. Greene, G.B. Street: Solid State Commun. 31, 355 (1979)
18. A.B. Kaiser: Phys. Rev. B 35, 2480 (1987)
19. Z. Ovadyahu: J. Phys. C 19, 5187 (1986)
20. D.G. Naugle, R. Delgado, H. Armbrüster, C.L. Tsai, T.O. Callaway, D. Reynolds, V.L. Moruzzi: Phys. Rev. B 34, 8279 (1986)
21. M. Przybylski, B.R. Bulka, I. Kulszewicz, A. Pron: Solid State Commun. 48, 893 (1983)
22. M. Galtier, J.M. Gay, A. Montaner, J.L. Ribet: J. de Phys. Suppl. 44, C3-107 (1983)
23. A.B. Kaiser: Phys. Rev. B 35, 3677 (1987)
24. J. Jäckle: J. Phys. F 10, L43 (1980)
25. B.L. Gallagher: J. Phys. F 11, L207 (1981)
26. V. Vasudeva Rao, G. Rangarajan, R. Srinivasan: J. Phys. F 14, 973 (1984)
27. Y.W. Park, W.K. Han, C.H. Choi, H. Shirakawa: Phys. Rev. B 30, 5847 (1984)
28. A.B. Kaiser: Phys. Rev. B 29, 7088 (1984)

Synthesis of New Conductive Electronic Polymers

H. Naarmann

BASF Plastics Research Laboratory,
D-6700 Ludwigshafen, Fed. Rep. of Germany

"Who can find a wise and original idea which hasn't been thought before?"
Goethe, Faust

Introduction

Conductive materials widen the horizon in science and technology, and it is not only chemists to whom they present a challenge. This paper is an attempt to demonstrate the infinite variety of scientific thought required to explore the various stimulating areas of organic chemistry involved. The scope is far too wide to be covered by the one contribution.

Basic systems and principles

Many routes are available for the synthesis of the polymers, conductive backbones. They include Wittig, Horner and Grignard reactions; polycondensation processes, and metal-catalyzed polymerization techniques. Oxidative coupling with oxidizing Lewis acid catalysts generally leads to polymers with aromatic or heterocyclic building blocks.

The Ziegler-Natta process has been widely adopted for the production of film-forming polyacetylenes, but it is only one of the many feasible reactions in which the catalysts are combinations of reducing agents and transition metals. The polymers thus obtained are made conductive by doping with n-type materials by reduction and with p-type materials by oxidation.

Direct polymerization of acetylene or aromatics in which the dope is incorporated or which proceeds via anodic oxidation, e. g. of pyrrole, leads to highly conductive materials.

No plastics or other organic polymers display the beneficial combination of properties peculiar to metals, i. e.

- great mechanical strength
- ductility
- thermal and electrical conductivity.

Conductive metals have a high specific gravity, whereas organic polymers are lighter and are typical dielectrics. It is the combination of low weight and high conductivity that makes the new class of synthetic materials particularly interesting from the economic aspect and assures them of a wide field of application.

In the last two years, astonishing progress has been made in attaining greater conductivity and stability. If metals are compared with polyacetylene on a basis of the conductivity/weight ratio, it can be seen that the

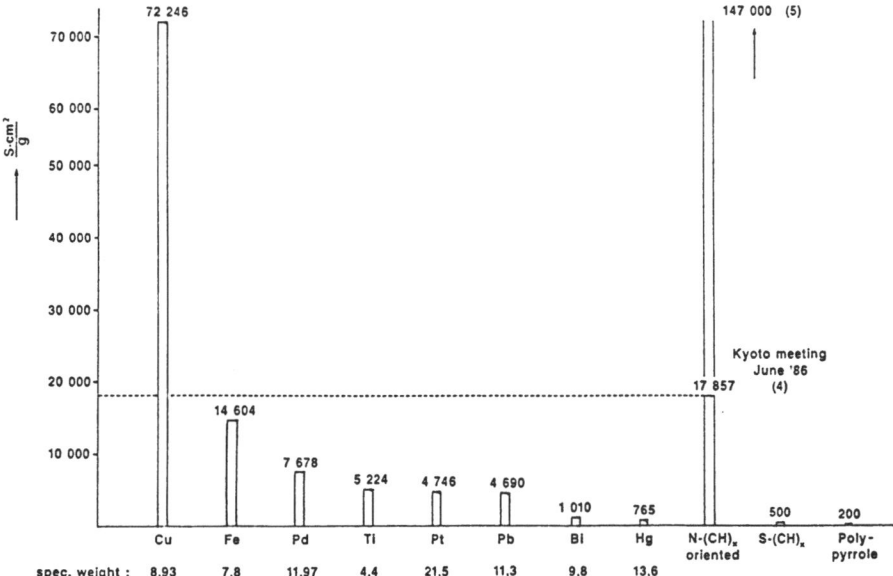

Fig. 1: The conductivities ($S \cdot cm^2 \cdot g^{-1}$) of different metals compared with organic polymers

polymers have already reached the same level as that of many metals (Fig. 1).

A review of recent trends has shown that scientific progress and practical application depend on the following.

1. Reproducible preparation of specimens of extensive area.

2. Determination of the conditions for synthesis and the laws relating them to properties.

New syntheses - conditions and methods

1. (CH)$_x$-polymerisation in silicone oil - a room temperature process

A decisive step towards the production of polyacetylene was the determination of the direct relationship between conductivity and crystallinity and the indirect relationship to the number of sp^3 orbitals (1).

If the polymerization conditions are modified, e. g. by silicone oil, a new N-(CH)$_x$ polyacetylene can be obtained at room temperature. Its quality is at least just as good as that of the standard S-(CH)$_x$ prepared by the Shirakawa technique at - 78 °C. The results are presented in Table 1 (2).

Aging of the standard catalyst is responsible for another, surprising improvement in the properties of the (CH)$_x$.The reduction in the number of sp^3 orbitals, i. e. the production of a system free from defects, is a great advantage.

13

Table 1: Properties of the different $(CH)_x$ types

	crystallinity [a] [%]	conductivity [b] [S/cm] undoped	doped with I_2	sp^3 content [c] [rel. %]	surface [d] area [m²/g]	configuration [e] cis content [%]
S-$(CH)_x$	70	10^{-6}	200	4	300	50
N-$(CH)_x$	65	10^{-8}	2000	0	100	80

a) Phillips diffractometer Cu K_α radiation b) four probe measurement c) determined by ^{13}C-NMR spectroscopy d) BET-method e) determined by IR spectroscopy

2. $(CH)_x$-orientation by mechanical stretching

Orientation of thin $(CH)_x$ types leads to surprising values for the conductivity, as is shown in Figure 1.

Figure 2 shows the differences in morphology between the S$(CH)_x$ and the oriented N$(CH)_x$.

3. $(CH)_x$-highly conductive and transparent

Transparent $(CH)_x$ film with conductivities higher than 5000 S/cm can be prepared by the method mentioned above (5). The polyacetylene is produced on a plastic film and stretched jointly with the supporting material. Afterwards, it can be peeled off to yield a self-supporting film and a coordination compound, e. g. with iodine, under standard conditions.

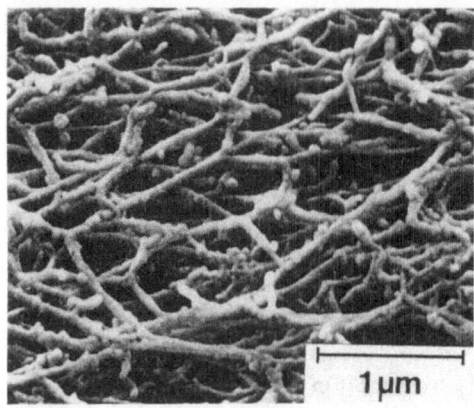

1 µm

Fig. 2a. Polyacetylene old - S$(CH)_x$: conductivity (iodine-doped) 200 S/cm (electron microscope photograph)

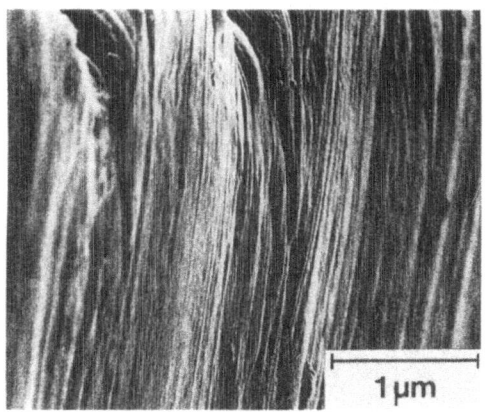

Fig. 2b: Polyacetylene new - N(CH)$_x$: conductivity (iodine-doped)
> 10 000 S/cm (electron microscope photograph)

4. Continuous preparation of polypyrrole

Polypyrrole and other polyheterocycles are more stable than polyacetylene
and can be produced quite easily by electrochemical techniques (3) (Fig.
3).

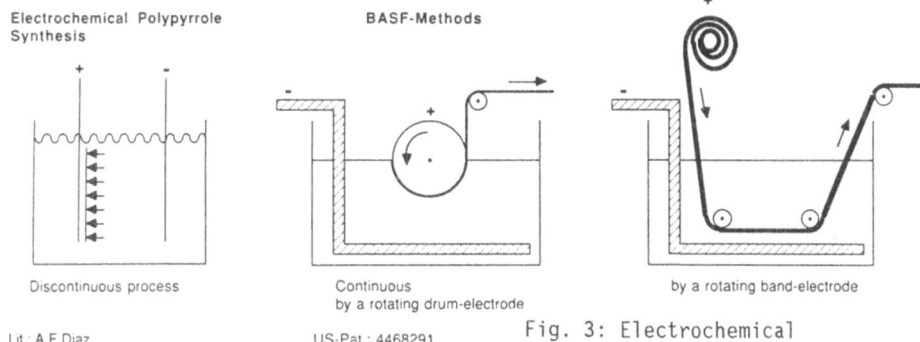

Electrochemical Polypyrrole
Synthesis

BASF-Methods

Discontinuous process

Continuous
by a rotating drum-electrode

by a rotating band-electrode

Lit.: A.F Diaz
J.C.S.Chem.Comm. 1979.635

US-Pat.: 4468291
June 27, 1983

Fig. 3: Electrochemical
polypyrrole processes

The continuous process allows homogeneous material to be obtained in
the form of self-supporting film of 20 um to 500 um gauge. Its properties
depend on the production conditions, e. g. the current density, the pH,
the electrolyte, counterions and monomers.

Applications

1. Applications in which the conductivity is exploited

The conductivity of polypyrrole film directly suggests some potential
applications, e. g. flexible conductive paths, heating film, and - owing
to its resilience - film keyboards.

Since polypyrrole film has even higher conductivity than plastics extended with conductive fillers, it is only logical to use it for electromagnetic interference shields. The results presented in Figure 4 demonstrate the good shielding effect of polypyrrole film over a wide range of frequencies in a radiation field. The attenuation required in this case is about 40 dB, i. e. the amplitude of the eletromagnetic radiation should be reduced by a factor of about 100.

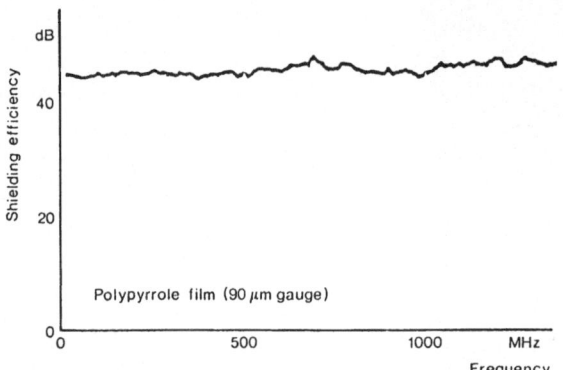

Fig. 4: Shielding efficiency as a function of frequency (ASTM ES 7-83)

2. Applications in which the electrochemical reversibility is exploited

Polypyrrole is a suitable electrode material for rechargeable electro-chemical cells. The advantage of polymer electrodes is that they can be easily shaped. As a result, they allow new battery types to be designed, e. g. for the electronics sector, and new production methods involving lower costs to be adopted.

Consideration

It can be concluded that we are on the right track. Resounding successes have been achieved with the multiplicity of components. Multifarious applications have been realized to quicken the already exciting ECP field of activity.

Acknowledgements

I am indebted to H. Heckmann for REM, H. Haberkorn for X-ray, P. Simak and M. Passlack for IR and R. Voelkel for C^{13}-measurements. Thanks to N. Theophilou and G. Köhler for $(CH)_x$ preparations.

This work is part of the projekt No. 03-C1-340 sponsored by BMFT. The author thanks the Bundesministerium für Forschung und Technologie for supporting this work.

References

(1) H. Haberkorn et al.: Structure and conductivity of $(CH)_x$, Synthetic Metals, 5 (1982) 51

(2) (CH)$_x$ synthesis in silicone oil at room temperature, EP 88301,
     Mar. 5, 1982/Feb. 25, 1983; BASF Germany

(3) US Pat. 4468291, Jun. 27, 1983/May 28, 1984, Continuous production
     of polypyrrole films, BASF Germany

(4) H. Naarmann, ICSM 86, Kyoto, 4A-07, "Synthesis of new conductive
     polymers"

(5) N. Theophilou et al.: "Highly conducting polyacetylene", submitted
     to "Synthetic Metals", presented on the ACS-Meeting, Denver,
     April 10, 1987

Long Mean Free Path Coherent Transport in Doped Polyacetylene

N. Basescu[1], Z.-X. Liu[1], D. Moses[1], A.J. Heeger[1], H. Naarmann[2], and N. Theophilou[2]

[1]Department of Physics and Institute for Polymers and Organic Solids, University of California, Santa Barbara, CA 93106, USA
[2]BASF Laboratories, D-6700 Ludwigshafen, Fed. Rep. of Germany

A modified technique has been used to synthesize stretch oriented polyacetylene with fewer sp^3 defects than in material prepared by other methods. The higher quality material exhibits substantially higher electrical conductivity. The data indicate a long electronic mean free path consistent with coherent metallic transport.

I. Introduction

The electrical conductivity of conducting polymers results from mobile charge carriers introduced into the π-electronic system through doping [1]. Because of the large intrachain transfer integrals, the transport of charge is believed to be principally along the conjugated chains with interchain hopping as a necessary secondary step.

In conducting polymers, as in all metals and semiconductors, charge transport is limited by a combination of intrinsic electron-phonon scattering and sample imperfection. Although relatively high conductivities ($\sigma \approx 10^3$ S/cm) have been reported for partially oriented and heavily doped polyacetylene [1,2], the absence of a metal-like temperature dependence implies that the observed values are not intrinsic. In doped polyacetylene, $(CH)_x$, electrical transport can be limited both by microscopic defects (leading to scattering and localization) and by the more macroscopic complex fibrillar morphology [1] and associated interfibrillar contacts. Thus, with improvements in material quality, one might anticipate corresponding improvements in the electrical conductivity.

II. Synthesis and Electrical Measurements

Free-standing polyacetylene films (thickness $\approx 30\mu m$) were synthesized using the Zeigler-Natta catalyst technique, initially developed by Shirakawa and colleagues [3], with important modifications introduced by Naarmann [4]. The catalyst (tetrabutoxytitanium and triethylaluminum suspended in silicone oil) [5] was stirred for two hours at 120° C, followed by stirring and degassing (pumping) while cooling slowly to room temperature. The polymerization reaction was carried out at room temperature (rather than using [3] toluene with polymerization at -78° C) in a controlled atmosphere drybox. The resulting films (which are roughly a 50/50 mixture of cis- and trans-polyacetylene [6]) were thoroughly washed [7] to remove all the catalyst. Isomerization to the all-trans material can be achieved by subsequent heating or by doping. The material can be stretch-oriented with maximum elongation ratio of about 6:1.

The polymer synthesized by this technique has been thoroughly characterized [4]. In most respects it is essentially identical to polyacetylene prepared by the Shirakawa technique. The material has the characteristic fibrillar morphology (approximately the same fibril diameter as Shirakawa material); the oriented material has a density of about 1 gm/cm^3. X-ray scattering studies show features which are indistinguishable from Shirakawa material. The principal difference is a major reduction in the number of sp^3 defects to a level which is not detectable by high resolution ^{13}C nmr. No bands corresponding to -CH$_3$- or -CH$_2$- could be detected at wavelengths of 2960, 2910, and 2830 μm in the infrared spectrum. The low density of sp^3 defects implies a higher degree of chain perfection in this material.

The (CH)$_x$ films were doped to saturation by immersion for one hour in a saturated solution of iodine in CCl$_4$, followed by rinsing three times in pure solvent for two minutes each. Since (CCl$_3$)$^-$ is produced in the photon catalyzed reaction, (I)$^-$ + CCl$_4$ --→ (CCl$_3$)$^-$ + ICl, and attacks the double bonds on the polyacetylene chain, care must be taken not to expose the solution to light. Weight uptake analysis indicates a doping concentration of y ≈ 0.06 assuming a composition of [(CH)$^{+y}$(I$_3^-$)$_y$]$_x$; we find that most of the doping occurs during the first 15 minutes of immersion. Immersion in the iodine solution for more than 70-80 minutes yields films with somewhat lower conductivity, possibly due to degradation of the polymer.

Immediately after rinsing, four copper wires were attached to the doped film (under an inert atmosphere) using electrodag. The samples, mounted with the stretch direction either parallel ($\sigma_{\parallel}$) or perpendicular ($\sigma_{\perp}$) to the current flow, were placed in a teflon cell for high pressure measurements (1 atm. to 15 kbar). The pressure cell was designed for interchangeable use in a closed cycle Displex refrigerator for measurements from 20 K to 300 K and in a He^3 cryostat for measurements down to 0.48 K.

The most striking feature of the Naarmann polyacetylene is the high electrical conductivity. Figure 1 shows $\sigma_{\parallel}$ as a function of temperature for a four-fold stretched sample at 10 kbar. At room temperature, the conductivity is greater than 20,000 S/cm; at 0.5 K the conductivity is still above 9000 S/cm. The inset shows the temperature dependence below 4 K in more detail. The results shown in Fig. 1 are typical, with good reproducibility from sample to sample. For a single sample, the conductivity was observed to decrease monotonically with time (down by about 30% over a period of a week), indicative of the irreversible degradation known for iodine doping [1].

Figure 2 shows the temperature dependence of $\sigma_{\parallel}$ at several different pressures; high pressure suppresses the decrease in $\sigma_{\parallel}$(T) at low temperatures. This effect is emphasized in the inset to Fig. 2 where we plot the ratio [$\sigma_{\parallel}$(7 kbar)/$\sigma_{\parallel}$(1 atm)] as a function of temperature.

The anisotropy of the conductivity is relatively large compared to that obtained from previous measurements on material prepared by the Shirakawa method; we find ($\sigma_{\parallel}/\sigma_{\perp}$) ≈ 80 at 7 kbar. The anisotropy is somewhat higher at high pressure, since $\sigma_{\parallel}$ increases with pressure while $\sigma_{\perp}$ remains essentially constant. The temperature dependence of ($\sigma_{\parallel}/\sigma_{\perp}$) is shown in Fig. 4 for pressures of 1 atm. and 7 kbar. Note that at high pressure, the ratio is constant.

Further improvements in the preparation of oriented polyacetylene were recently announced [8] leading to a room temperature electrical conductivity of σ ≈ 1.5x10^{+5} S/cm with an anisotropy of ≈ 10^3 (at one atmosphere pressure).

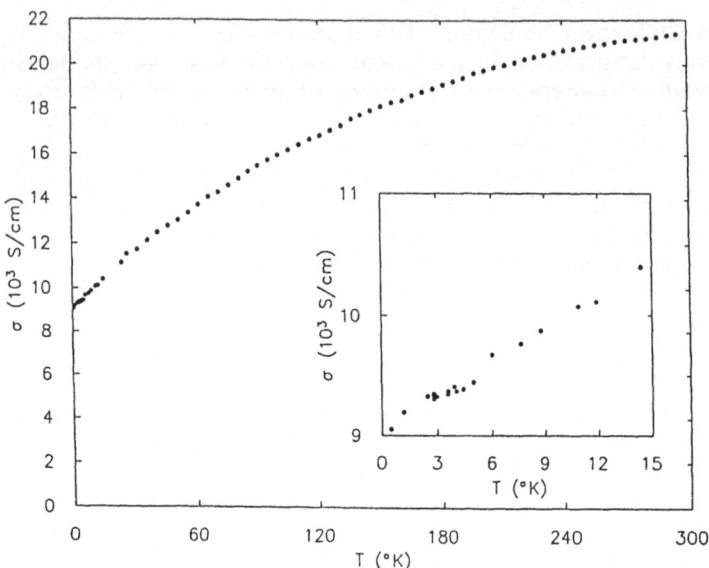

Fig. 1. $\sigma_{\|}(T)$ for a four-fold stretched sample at 10 kbar; the inset shows the temperature dependence below 4K in more detail. At room temperature, the conductivity is greater than 20,000 S/cm; at 0.5K the conductivity is still above 9000 S/cm.

III. Mobility and Mean Free Path

The electrical conductivity of a metal results from the existence of free carriers (N per unit volume) with a mobility μ;

$$\sigma = Ne\mu. \tag{1}$$

Alternatively, this can be writtten as

$$\sigma = \omega_p^2 \tau / 4\pi \tag{2}$$

where ω_p is the plasma frequency, $\omega_p^2 = 4\pi Ne^2/m^*$, and m^* is the electronic effective mass.

For metallic (doped) polyacetylene, there is at present an ambiguity concerning the free carrier density. Assuming a single carrier per dopant (i.e., per I_3^- with y = 0.06) and a material density of about 1 gm-cm^{-3}, one obtains a minimum number, $N_{min} = 3\times10^{21}$ cm^{-3}. If, on the other hand, one assumes that in the metallic doped polymer the energy gap has been reduced to zero, then all the π-electrons can contribute to the electrical conductivity. This gives an upper limit for the number of carriers; $N_{max}=N_\pi=5\times10^{22}$ cm^{-3}. Thus,

$$3\times10^{21} \text{ cm}^{-3} < N < 5\times10^{22} \text{ cm}^{-3};$$

N_{min} would be consistent with the concept of a polaronic metal [9], whereas N_{max} would be consistent with a conventional metallic state in which the

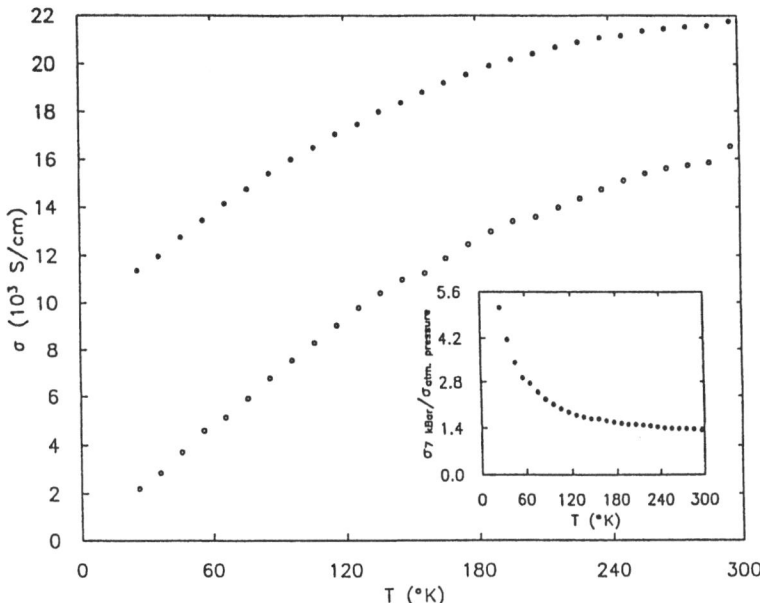

Fig.2. Temperature dependence of $\sigma_{\parallel}$ at 7 kbar (solid dots) and at 1 atmosphere (open cirlces); stretching ratio 6:1.   Pressure suppresses the decrease in $\sigma_{\parallel}(T)$ at   low temperatures. Inset: [$\sigma_{\parallel}$(7 kbar)/$\sigma_{\parallel}$ (1 atm)] as a function of temperature.

Peierls' gap had been reduced to zero by doping.   Polarized reflectance measurements [10] on oriented metallic Durham polyacetylene indicate an experimental value for the plasma frequency of $\hbar\omega_p \sim 3.5$ eV. Assuming a free electron mass, this corresponds to N ~ $9 \times 10^{21} cm^{-3}$, a value in the middle of the anticipated range.

Keeping in mind the uncertainty in the carrier density, we   use N = $9 \times 10^{21} cm^{-3}$ (equivalent to a plasma frequency of 3.5 ev with a free electron mass) to estimate the mobility and the electronic mean free path. Taking the experimental value of $\sigma = 1.5 \times 10^5$ S/cm, the implied mobility (Eq. 1) is $\mu \sim 10^2$ cm^2/V-s, or alternatively, $\tau \sim 6 \times 10^{-14}$ s (Eq. 2). Since the π-electron transfer integral is $t_0 = 3$ eV[11,12], the Fermi velocity is given by $v_F = 2t_0 a/\hbar \sim 10^8$ cm/s, implying a mean free path of $\lambda = v_F \tau \sim$ 600 Å.   For comparison, the corresponding values for copper (at room temperature) are $\mu \approx 40$ cm^2/V-s and $\lambda \approx 300$ Å. Since the intrinsic conductivity of doped polyacetylene is higher than the measured value (the experimental values remain limited by a combination of microscopic defects and the more macroscopic complex fibrillar morphology and associated interfibrillar contacts), the actual values for metallic polyacetylene may be considerably higher.

IV. Conclusion

Although improvement of solid state properties through higher quality materials is a general feature of materials science, there has been little optimism that this rule would be applicable to conducting polymers. Perhaps the reason for this was the argument that the high level of impurities in doped conducting polymers would negate any improvements toward molecular chain perfection. The recent studies demonstrate that this is not the case in polyacetylene or (by implication) in other conducting polymers: the achievement of still higher quality material from improved synthesis and processing can be expected to lead to correspondingly better electronic properties.

From the estimates of the mobility and mean free path, it is clear that coherent electronic transport is implied for doped metallic polyacetylene. The large mean free path (at least several hundred lattice constants) rules out a picture of polyacetylene as a "dirty" metal in which the transport would result from hopping between localized states. On the contrary, this large electronic mean free path is somewhat difficult to understand in the context of a quasi-one dimensional polymer doped with $\sim 10^{22}$ charged impurities, and in which the structural coherence length (from x-ray diffraction) is well below 100 Å.

Acknowledgment: The research at UCSB was supported by the Office of Naval Research.

References
1.	For reviews, see the following:
	a) Handbook of Conducting Polymers, ed. Skotheim, T.J. (Marcel Dekker, Inc., NY, 1986).
	b) Etemad, S., Heeger, A.J. and MacDiarmid, A.G., Ann. Rev. Phys. Chem. 33, 443 (1982).
2.	Gould, C.M., Bates, D.M., Bozler, H.M., Heeger, A.J., Druy, M.A., and MacDiarmid, A.G.; Phys. Rev B23, 6820 (1981).
3.	Shirakawa, H. and Ikeda, S.; Synth. Met. 1, 175 (1980).
4.	Naarmann, H., Proc. of ICSM 86, Synth. Met. (to be published).
5.	Catalyst: 31 ml $Al(C_2H_5)_3$ and 41 ml $Ti(C_4H_9O)_4$ in 50 ml silicone oil (Wacker).
6.	Elemental analysis: C, 91.7-92.2%; H, 7.6-7.7%; O, <0.5%; Al/Ti. 0.01%.
7.	Three hours in toluene followed by sixteen hours in methanol containing 6% HCl and finally two periods of 1-1.5 hours each with methanol.
8.	H. Naarmann, Symposium on "Conducting Polymers: Their Emergence and Future"; American Chemical Society Meeting, Denver, Colorado; April 8 and 9, 1987.
9.	S. Kivelson and A.J. Heeger, Phys. Rev. Lett. 55, 308 (1985).
10.	E. J. Mele and M.J. Rice, Phys. Rev. B 23. 5397 (1981).
11.	P. M. Grant and I. Batra, J. Phys. (Paris) Colloq. 44, C3-4377, (1983).
12.	G. Fink and G. Leising, Phys. Rev. B 34 (1986).

Frequency- and Temperature-Dependent Dielectric Losses in Lightly Doped Conducting Polymers

J.P. Parneix and M. El Kadiri

Centre Hyperfréquences et Semiconducteurs, UA CNRS 287,
Université de Lille Flandres Artois,
F-59655 Villeneuve d'Ascq Cedex, France

1 - Introduction

In this paper, we report the results of an extensive series of measurements concerning the frequency and temperature dependence of the complex conductivity $\sigma^* = \sigma' + j\sigma''$ of conducting polymers. Numerous data were obtained from a very wide range of compounds such as polyacetylene, polyparaphenylene, polypyrrole, polyaniline and polythiophene.

The general trends of the data are quite similar for all these compounds and appear to be comparable to the well-known behavior of disordered systems such as amorphous semiconductors. Since such a response is common to various materials, the data must be analyzed very carefully, and the ranges of frequency and temperature investigated as large as possible.

2 - Experimental techniques and results

The complex conductivity measurements were made using the reflectometric technique from 5 Hz to 26×10^9 Hz with more than 20 experimental points per decade. At frequencies up to 10^9 Hz, data were obtained by two impedance analyzers (HP 4192 and 4191) controlled by a HP 9826 computer. Above 10^9 Hz, a vectorial network analyzer (HP 8510) piloted by a HP 300 type computer was employed.

Our experiments were performed on samples with a sandwich structure (pressed powder pellets or cylinders). The evaporated gold contacts were ohmic in the temperature range investigated. To check the contacts, the following curves were systematically plotted : the current-voltage characteristic I (V) ; the variations of the capacitance against bias-voltage C(V) up to the microwave region ; the variations with frequency of the impedance in the complex plane. Concerning contact effects, another point must be noted. With this technique, the conductance and also the capacitance of the sandwich were obtained. These quantities are not independent, being related by the Kramers-Krönig relationships. Hence, by measuring both parameters, one has an important check on the accuracy of the experimental data.

The temperature range studied goes from 77 K to 400 K. In some cases measurements were made in an inert atmosphere (nitrogen) to prevent sample degradation.

Typical experimental results for σ' are shown in Fig. 1 for different doping levels.

23

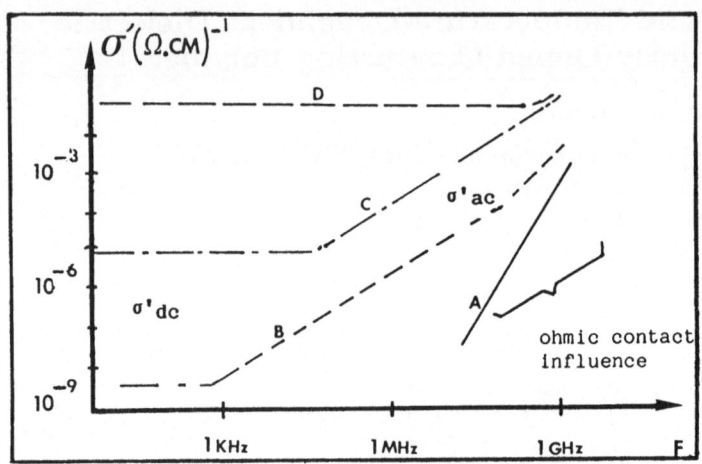

Figure 1 : Variation of $\sigma'(f)$ at constant temperature for different
doping levels : A undoped, B lightly doped, C moderately
doped, D heavily doped.

As usual, at light doping levels, the total conductivity σ' may be
separated into a frequency-independent contribution σ'_{dc}, and a
frequency-dependent contribution σ'_{ac}. σ'_{dc} changes significantly with
temperature. This term is mainly influenced by contact effects. At high
frequencies, σ'_{ac} increases as w^s ($s \lesssim 1$). The temperature dependence of
σ'_{ac} is weak and the exponent s always increases as temperature
decreases. σ'_{ac} is nearly contact independent, except at very high
frequencies where the influence of the series resistance of the contacts
becomes apparent. Nevertheless, this effect can be easily deduced from
the data.
At high doping levels, the dc behavior becomes predominant and then the
total conductivity σ' remains frequency independent.

3 - Discussion

Various theoretical models have been applied and discussed to account
for the experimental results /1-4/. Qualitatively, it seems obvious that
transport processes are probably different whether the carriers have to
cross the sample from one end to the other, or have to relax between
localized sites inside the bulk. At low frequencies σ'_{dc} is predominant
because the carriers have enough time to cross the sample following a
percolation path. In contrast, at high frequencies, the investigated
regions decrease, and the frequency behavior becomes predominant.

dc - conductivity : σ'_{dc} (T)

The T dependence of σ'_{dc} is shown in Fig. 2 for poly 3 methylthiophene. Over the limited temperature range studied, a linear behavior of
$\log (\sigma'_{dc})$ vs $(T)^{-1/4}$ is found. This behavior is in good agreement with
the three-dimensional variable range hopping model which leads to the
well-known Mott power law $\sigma = \sigma_o \exp (- T_o/T)^{-1/4}$.   (1)

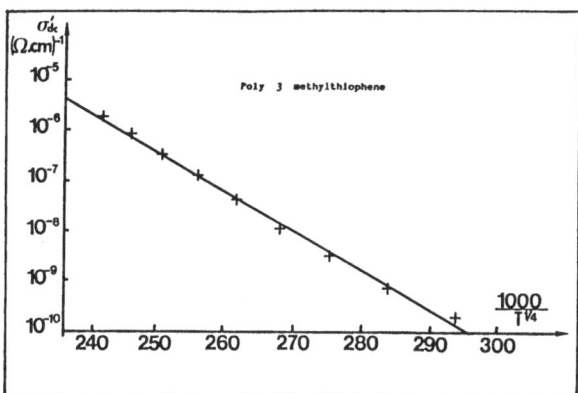

Figure 2 : σ'_{dc} conductivity vs $(T)^{-1/4}$ for poly 3 methylthiophene.

The Mott parameters σ_0, T_0 are then adjusted and give reasonable values of the density of localized states $N(E_F)$ at the Fermi level and the wave function decay α^{-1} /1/.

ac - conductivity : σ'_{ac} (ω, T)

To account for the ac data, the hopping model derived by ELLIOT/5/ has been used. The model is based on a hopping of carriers over barriers. It predicts a frequency dependence of σ'_{ac} in the form : ω^S where the exponent s is given by the following relationship :

$$s = 1 - 6 \; kT \; / \; W_M \; ; \qquad\qquad\qquad (2)$$

W_M is the well depth for an isolated well.

Figure 3 shows that the fit is quite good for poly 3 methylthiophene. The W_M value deduced is $W_M \sim 1.1eV$, which corresponds approximately to the midgap of poly 3 methylthiophene /6/.

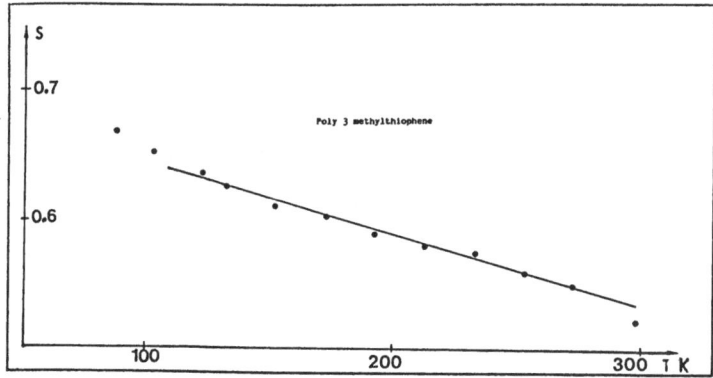

Figure 3 : T dependence of the exponent s for poly 3 methylthiophene.

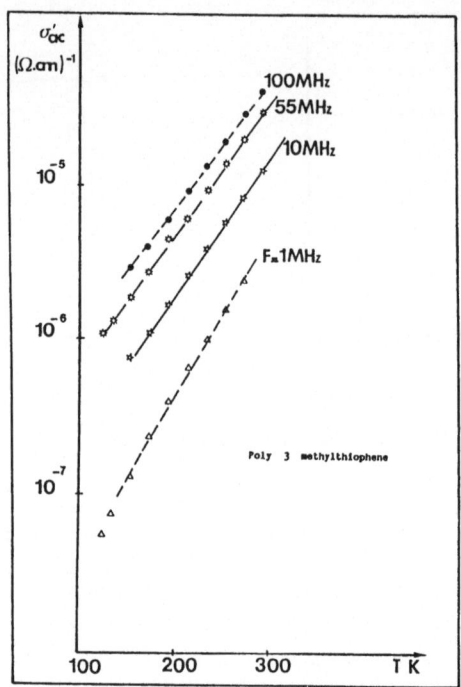

Figure 4 : T dependence of σ'_{ac} magnitude at various frequencies for poly 3 methylthiophene.

Moreover, it is possible to plot at a given frequency the variation of the σ'_{ac} magnitude vs temperature. It is shown in Fig. 4 that log (σ'_{ac}) is linear with T which is in agreement with the hopping model. The slope p decreases as frequency increases, which also agrees with the model. From the variation of p with w, one can again deduce the W_M parameter, which is found to be 1.1 eV in agreement with the preceding value.

References

1. M. El Kadiri, Thesis, Lille (1987)
2. K. Ehinger, W. Bauhofer, K. Menke, S. Roth, J. de Physique, C3, 44, 115 (1983)
3. A.J. Epstein, H. Rommelmann, H.W. Gibson, Synth. Met., 9, 103 (1984)
4. H. Stubb, H. Isotalo, P. Kuivalainen, J.J. Lindberg, J. de Physique, C3, 44, 737 (1983)
5. For a discussion of the hopping models see A.R. Long, N. Balkan, Philosophical Mag. B, 41, 3, 287 (1980)
6. G. Tourillon, F. Garnier, J. Electroanal. Chem., 135, 173 (1982).

Variable-Range Hopping in Polymers Prepared from Iodo(2-nitrophenyl)-acetylene

*M. Rotti, H. Krikor, and P. Nagels**

Rijksuniversitair Centrum, University of Antwerp,
B-2020 Antwerpen, Belgium

Iodo-(2-nitrophenyl)acetylene ($o-NO_2C_6H_4C{\equiv}CI$) was thermally polymerized at 400°C for 4 hours. The temperature dependence of the dc conductivity of samples having a different iodine content can be best described by $\sigma_{dc} = \sigma_0[\exp(-T_0/T)^{1/4}]$, typical of variable-range hopping. The thermopower is negative for the sample with the highest iodine content and positive for the other ones. The ac conductivity shows a frequency dependence of the form $\sigma_{ac} = A\omega^s$ with s close to 0.6.

1. INTRODUCTION

Doping of polyphenylacetylene with certain electron acceptors such as iodine yields a material with poor semiconductor characteristics. The aim of our study was to find out in how far the electrical properties are affected by a direct substitution of these acceptors into the monomer. For this purpose, we prepared iodo-(2-nitrophenyl)acetylene ($o-NO_2C_6H_4C{\equiv}CI$) which was thermally polymerized at 400°C.

2. EXPERIMENTAL PROCEDURES

The compound o-nitrophenylacetylene was synthesized from commercially available o-nitrocinnamic acid by bromine addition at the double bond, followed by HBr elimination and decarboxylation. The monomer iodo-(2-nitrophenyl)acetylene was readily prepared by adding a solution of iodine in liquid ammonia to the acetylene derivative. It was thermally polymerized at 400°C for 4 hours in a quartz ampoule filled with argon. Four fractions of the material were washed with methanol only once or up to four times. The iodine content of these fractions was lower than expected (10, 7, 5 and 2 wt %), indicating that molecular iodine is split off during polymerization. As reported in a previous paper [1], this might result from poly-condensation and cross-linking reactions.

3. RESULTS AND DISCUSSION

The dc conductivity σ_{dc} of the samples with different iodine content is plotted versus $10^3/T$ in Fig. 1. At room temperature σ_{dc} is of the order of 10^{-4} ohm^{-1}cm^{-1}, which exceeds that of polyphenylacetylene by many orders of magnitude. The reason for this may be found in an "in-situ" iodine doping of the polymers during their preparation. In the log σ vs 1/T plot the lines

---

*Also at Physics Department, S.C.K./C.E.N., B-2400 Mol (Belgium)

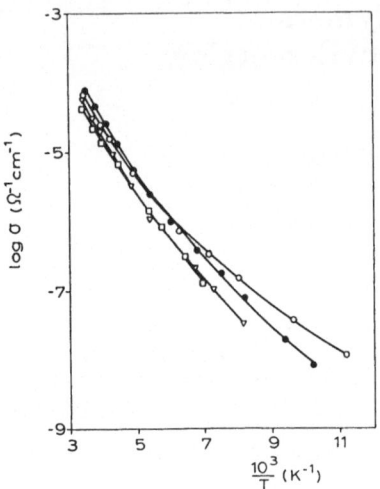 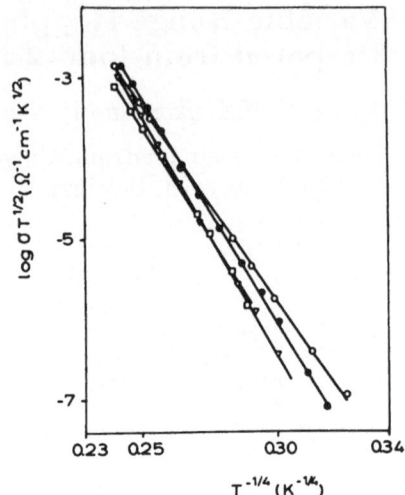

FIG. 1. Log σ_{dc} vs $10^3/T$ FIG. 2. Log $\sigma T^{1/2}$ vs $T^{-1/4}$
of polymers prepared from iodo-(2-nitrophenyl)acetylene by heating at
400°C. Remaining iodine content in wt %: ○ 10; ▽ 7; ● 5; □ 2

are continuously curved and, hence, the temperature dependence of σ_{dc} does
not follow an Arrhenius-type behaviour. A conduction process often
encountered in disordered materials at low temperature is variable range
hopping (VRH). In Fig. 2 the dc conductivity data are plotted as $\sigma T^{1/2}$ vs
$T^{-1/4}$ according to Mott's formula for hopping in three dimensions:

$$\sigma_{dc} = \frac{9e^2 v_0}{64\alpha^2} (T_0/T)^{1/2} N(E_F) \exp [-(T_0/T)^{1/4}], \qquad (1)$$

where $T_0 = 18.1 \, \alpha^3/k \, N(E_F)$.
Here, v_0 is an attempt frequency, $N(E_F)$ the density of states at the Fermi
energy and α describes the spatial extent of the localized wavefunction.
The dc conductivity data obey very well the $T^{-1/4}$ relationship.

The thermopower (Fig. 3) is negative for the sample with the highest
iodine content (10 wt %) and positive for the other ones. The observation of
low S values is consistent with our explanation in terms of VRH. Many
authors have treated the temperature dependence of the thermopower in VRH.
Nagels et al. [2] have demonstrated that, under special conditions of a slow
variation of the density of states at E_F and a low hopping energy ΔE, the
thermopower exhibits a simple temperature dependence of the form
$S/T \propto T^{-1/n+1}$, where n is the dimension of the system. In Fig. 4 the thermopower of the samples with a positive sign is plotted according to the
S/T vs $T^{-1/4}$ relationship valid in three dimensions. In Fig. 5 is
represented the ac conductivity $\sigma(\omega)$ of the sample containing 10 wt % I
measured in the frequency range from 5 to 500 kHz and plotted as a function
of reciprocal temperature. As can be seen from this figure, $\sigma(\omega)$ is strongly
frequency dependent over the entire temperature interval. In Fig. 6 the ac

28

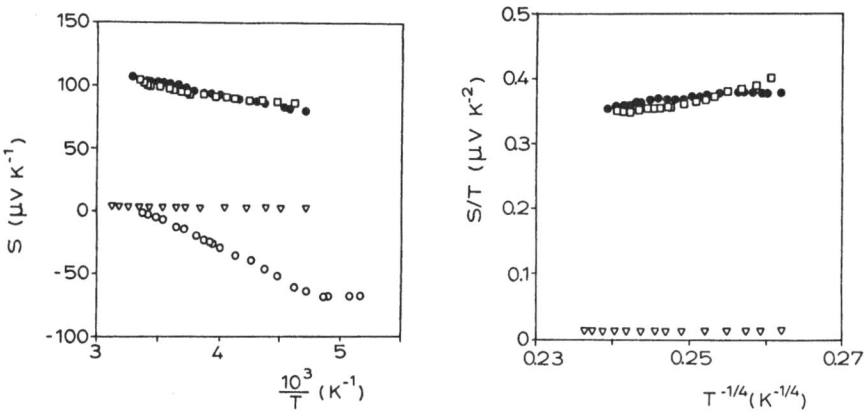

FIG. 3. Thermopower S vs $10^3/T$ FIG. 4. Plot of S/T vs $T^{-1/4}$
of polymers prepared from iodo-(2-nitrophenyl)acetylene by heating
at 400°C. Remaining iodine content in wt %: O 10; ▽ 7; ● 5; ☐ 2

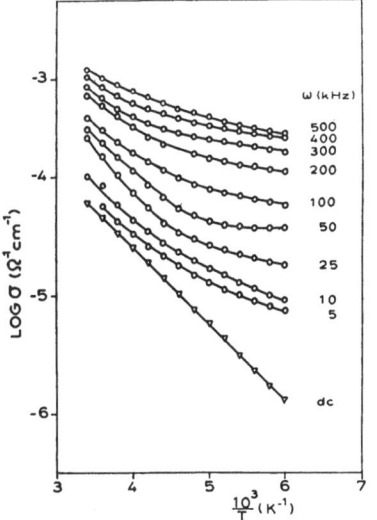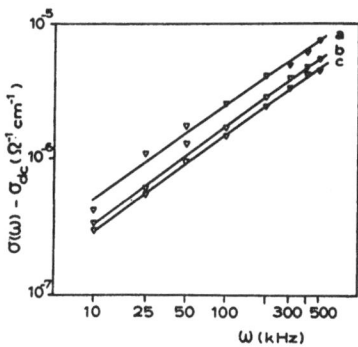

FIG. 5. σ_{ac} at frequencies 5 - 500 kHz   FIG. 6. Frequency dependence of
as a function of $10^3/T$ for a polymer σ_{ac} at 250 K (a), 200 K (b) and
with 10 wt % I   167 K (c) for a polymer with
   10 wt % I

conductivity, $\sigma_{ac} = \sigma(\omega) - \sigma_{dc}$, is represented versus ω for three different
temperatures. The frequency dependence obeys the empirical relationship
$\sigma \propto \omega^s$ with s = 0.61. Mott's formula describing the ac conductivity for
three-dimensional variable-range hopping is expressed by

$$\sigma_{ac} = \frac{\pi^2}{48} [N(E_F)]^2 kT \frac{e^2}{\alpha^5} \omega \ln^4 \left(\frac{\nu_0}{\omega}\right); \quad\quad\quad (2)$$

29

$\omega \ln^4 (\nu_0/\omega)$ can be written as ω^s with $s = 1 - 4/\ln (\nu_0/\omega)$. The comparison of our dc and ac data with the predictions of the variable-range hopping theory shows that this type of charge transfer mechanism is most adequate to explain our results. The hopping parameters involved in this hopping process were evaluated from the dc data using eq. (1) (Table I). The hopping distance R and the hopping energy ΔE were calculated from the expressions derived by Mott in his VRH theory.

The values of ν_0 (10^{20} to 7.10^{22} s^{-1}) may appear surprisingly high. In many other disordered materials where VRH applies, values exceeding the optical phonon frequency ($\cong 10^{13}$ s^{-1}) by many orders of magnitude have also been observed. A solution to this discrepancy has been proposed by Colson and Nagels [3] who showed that the original expression of ν_0 by Miller and Abrahams used to calculate the transition probability yields very high values for most materials. When incorporating the calculated ν_0 values into the frequency exponent s, one finds a value of the order of 0.9 with $\omega_0 = 10^4$ s^{-1}. The problem related to the deviations from the well-known $\omega^{0.8}$ law, in principle obtained for $\nu_0 = 10^{13}$ s^{-1} and $\omega = 10^4$ s^{-1}, has been the subject of many discussions, see e.g. [4,5]. However, the reason for the low s values (0.61), observed in our polymeric materials, is not clear to us.

Table I. Parameters at 300 K for three-dimensional VRH deduced from dc conductivity and assuming $\alpha^{-1} = 5Å$

I content	T_0	ν_0	$N(E_F)$	R	ΔE
(wt %)	(K)	(s^{-1})	(eV^{-1} cm^{-3})	(Å)	(eV)
10	1.6×10^8	1.1×10^{20}	1.0×10^{19}	51	0.17
7	3.7×10^8	7.3×10^{22}	4.6×10^{18}	62	0.21
5	2.8×10^8	1.1×10^{22}	6.0×10^{18}	58	0.20
2	2.5×10^8	2.3×10^{21}	6.6×10^{18}	57	0.19

References

1. M. Rotti, H. Krikor and P. Nagels, J. Synthetic Metals, in print
2. P. Nagels, M. Rotti and R. Gevers, J. Non-Cryst. Solids 59&60, 65 (1983)
3. R. Colson and P. Nagels, J. Non-Cryst. Solids 35&36, 129 (1980)
4. S.R. Eliott, Phil. Mag. 36, 1291 (1977)
5. A.R. Long, Adv. Phys. 31, 553 (1982)

Transport Studies on Polypyrrole Films Prepared from Aqueous TsONa Solutions of Different Concentrations

Shen Yueqiang[1;*], *K. Carneiro*[1], *Wang Ping*[2], *and Qian Renyuan*[2]

[1]Physics Laboratory, University of Copenhagen,
  Universitetsparken 5, DK-2100 Copenhagen Ø, Denmark
[2]Institute of Chemistry, Academia Sinica, Beijing, China

Recently we have prepared a group of polypyrrole films from aqueous sodium toluene sulphonate (TsONa) solutions of different TsONa concentrations. Here we report the DC- conductivity, thermoelectric power and Voltage Shorted Compaction (VSC) investigations on this group of polypyrrole films.

The results of our transport studies indicate that the electronic transport process in the polypyrrole films consists of two parts: one part is the charge transport by mobile charge carriers (such as polarons and bipolarons), the other part is the hopping transport between the mobile carrier regions. In the polypyrrole films prepared from highly concentrated TsONa solutions, the mobile charge carriers give a large contribution to the transport process, while in the polypyrrole films prepared from low concentrated TsONa solutions, the transport is totally dominated by hopping. For the polypyrrole films prepared from middle concentrated TsONa solutions, the data of the DC-conductivity and thermoelectric power measurements can be well fitted by a model which describes the hopping process between mobile localized states.

A very interesting fact in our study is that a kind of phase transition at ca. 100K is observed in one of the polypyrrole films.

1 Introduction

Since the first preparation by Kanazawa, Diaz and coworkers of a flexible film from pyrrole by electrochemical oxidation [1-3], a lot of attention has been paid to polypyrrole. The polypyrrole film is an interesting conducting polymer material not only because of its high conductivity [4] and stable characteristics, but also because of its possible applications [5, 6].

The electronic transport process in polypyrrole is believed to be dominated by hopping [7, 8]. Our earlier investigation shows that in the homogeneous polypyrrole films the transport processes are the transport of the mobile charge carriers (i.e. polarons or bipolarons) along the chain and hopping transport between these mobile charge carriers [9, 10]. For the highly conductive polypyrroles the charge transport by mobile charge carriers plays a important role.

In order to study the transport properties of polypyrrole films containing different amounts of counter ions and thereby having different room-temperature conductivities, we have prepared a group of polypyrrole films from aqueous Sodium Toluene Sulphonate (TsONa) solutions of different TsONa concentrations. This group of polypyrrole films contains different amounts of TsO$^-$ counter ions and the differences between their room-temperature conductivities are very large. Our investigations show also that they have different morphologies. The preparation details, results from the elemental analyses and the structure study of this group of polypyrrole films are reported in [11].

Here we report the transport study of this group of polypyrrole films by DC four-probe conductivity, Voltage-Shorted Compaction(VSC) and Thermoelectric Power (TEP) investigations.

2 Experimental Details

This group of six polypyrrole films are prepared electrochemically from aqueous TsONa solutions. The only difference between the preparations of the films is in the concentration of the electrolyte TsONa, which leads to differences in the amounts of the counter ions contained in the film. However, the structure study by Scanning Electron Microscopy (SEM) showed that the morphologies of these films were also different [11]. These six samples are listed in Table 1, together with their room temperature conductivities, the amount of TsO$^-$ counter ions, and the TsONa concentrations used in the preparation.

Table 1. The results of the elemental analyses and the room temperature conductivities of the polypyrrole films prepared from aqueous TsONa solution.

Samples	TsONa [M]	Mole ratios of C_4H_3N:TsO$^-$	σ(300K) [$\Omega^{-1}cm^{-1}$]
PP-TsONa-1	0.0	3.21:1	85.7
PP-TsONa-2	0.05	3.42:1	73.4
PP-TsONa-3	0.02	5.91:1	8.5
PP-TsONa-4	0.01	8.10:1	1.82
PP-TsONa-5	0.005	5.77:1	0.05
PP-TsONa-6	0.002	7.18:1	0.013

The DC-conductivity is measured by using the conventional four-probe method. The contacts of the sample were formed by gold wires and silver paste. The Voltage Shorted Compaction (VSC) technique was introduced by Coleman [12] for the qualitative investigation of temperature dependences of conductivity in compact samples of polycrystalline charge-transfer complexes. The application of this technique to conducting polymer films is described and discussed in [13-15]. The VSC measurements may enable us to study the intrachain electronic properties of the conducting polypyrrole films. The experimental set-up for VSC is the same as for four-probe measurement. On the VSC sample the area between the two contacts in the middle is shorted by a thin layer of silver paste. The thermoelectric power of a material is basically measured by constructing a thermocouple arrangement with a well known thermoelectric power S_0. During each measurement of the TEP of the sample, S, a small temperature difference, ΔT, between the junction is induced, holding the average sample temperature of T, and measuring the small emf., ΔV. Then the S can be obtained by means of $\Delta V = (S - S_0) \Delta T$. The apparatus used is almost the same as that described by Mortensen in [16].

For all the measurements the whole system is located on a cold head of a CTI closed cycle cryogenic refrigerator. The cold head is isolated by vacuum, and the temperature is electronically controlled from 300K to 10K.

3 Experimental Results and Discussions

The results of DC conductivities, VSC and TEP are well correlated to the preparation of the samples, and they are shown in Fig.1. For samples PP-TsONa-1,2 which were prepared from concentrated TsONa solutions, the DC-conductivities are high and less temperature dependent, and the VSC results show a metallic behavior of the polypyrrole chains. These agree with the results in the literature, and may be due to the large number of mobile bipolarons. For PP-TsONa-4 which was prepared from less concentrated TsONa solution, the DC-conductivity is much lower and its temperature dependence is much higher, and the VSC result shows a non-metallic property of the intrachain transport. For PP-TsONa-5,6 which were prepared from very low concentration TsONa solutions, have very low DC-conductivities and thermo-conductivities. The conductivities are so low that the VSC and TEP measurement could not be done.

The behavior of PP-TsONa-3 is similar to PP-TsONa-4, except below 100K, where it behaves completely differently from other samples. In the VSC measurement the relative conductivity changes at ca. 100K from increasing with decreasing temperature to decreasing with decreasing temperature. In the TEP measurement the TEP value dropped very drastically to zero at ca. 100K. This strange behavior of PP-TsONa-3 may indicate that there is a phase transition at ca. 100K.

The transport process in PP-TsONa-3 above 120K and PP-TsONa-4 can be interpreted as charge hopping between mobile charge carriers. We have fit the DC and TEP measurements by the extended Variable Range Hopping model [9]. From the fitting (see Fig.2) we found that for PP-TsONa-3 above 120K the localization length is 10 monomers and density of states at the Fermi level is $4 \times 10^{-4} eV^{-1}$ per monomer, and the corresponding value for PP-TsONa-4 are $10^{-4} eV^{-1}$ per monomer. In Table 1 we see that the amounts of TsO^- counter ions contained in samples PP-TsONa-3,4,5,6 are almost the same, and their morphologies are also similar [11]. But their electronic properties are very different. The differences in property must be due to the differences in their chain arrangement, the conjugation length of the chain and the position of the counter ions with respect to the chains. All these factors seems to be strongly dependent on the concentration of the electrolyte in the preparation. From our investigations we conclude that the electronic tranport in polypyrrole films is carried on by hopping charges and free charge carriers in the form of polarons and bipolarons. For highly conducting films the free charge carriers give a great contribution, for the less conducting films the contribution from mobile polarons and bipolarons is very small, and for very low conductivity films there may be no free charge carriers at all. The electronic properties of the films are not simply determined by 'doping concentrations' but also by the conjugation length of the chain, the chain arrangement and the position of the counter ions relative to the chains. In some case, like sample PP-TsONa-3, there may be

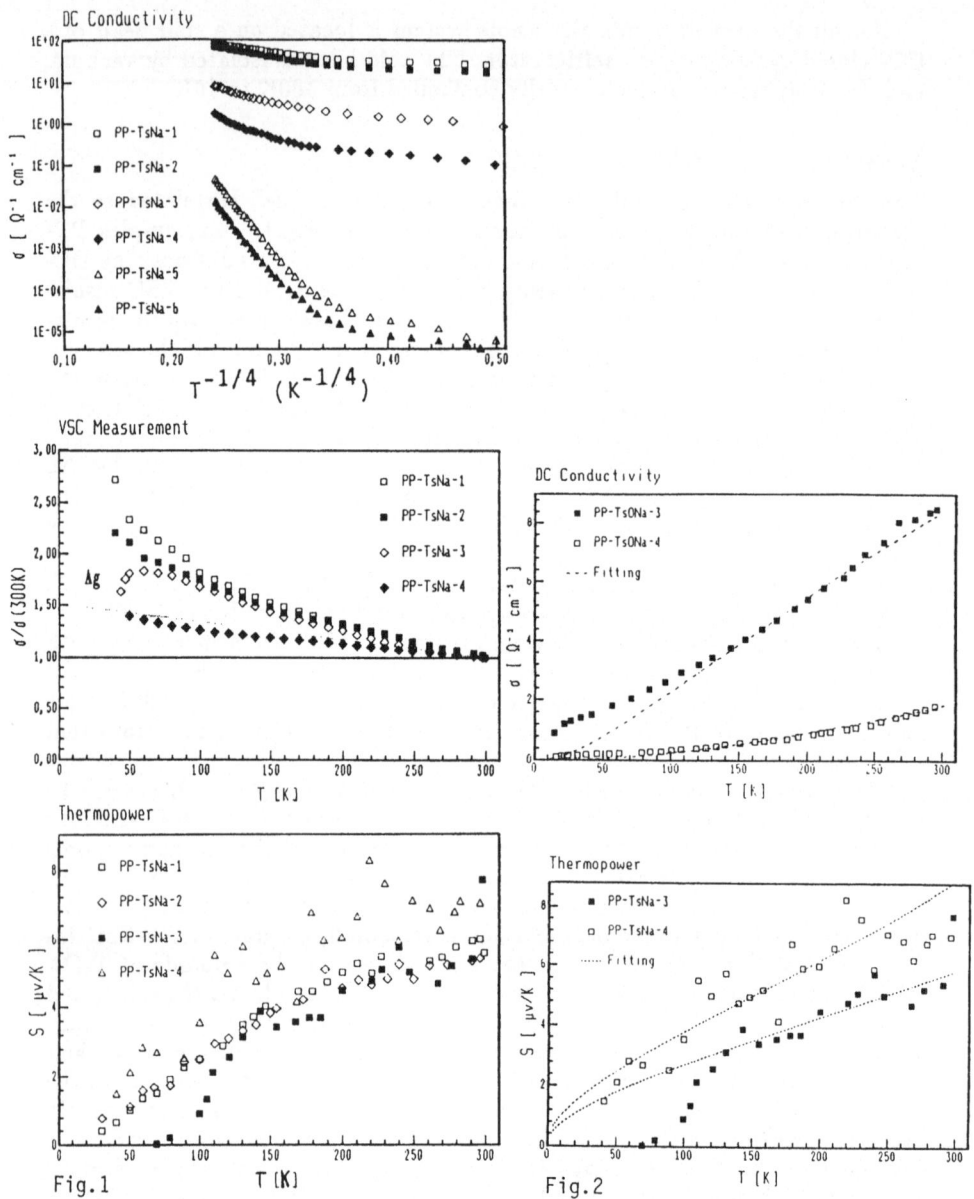

Fig.1 The results of DC-conductivity, VSC and TEP measurements on samples PP-TsONa-1,2,3,4,5,6.

Fig.2 The temperature dependence of DC-conductivities and TEP of samples PP-TsONa-3,4 and the fitting curves in terms of the extended variable range hopping model.

a phase transition. The properties of PP-TsONa-3 at ca. 100K are now under further investigation.

4 Acknowledgments

One of the authors would like to acknowledge financial support from the Danish Ministry of Education. We thank the staff members of the organic conductor group in the Institute of Chemistry, Academia Sinica, for their help in various ways during this investigation.

*. Present address: Max-Planck-Institut für Festkörperforschung, Heisenbergstrasse 1, D-7000 Stuttgart 80, West Germany.

5 References

1. A. F. Diaz, K. K. Kanazawa and G. P. Gardini: J. Chem. Soc., Chem. Commun., 635(1979)
2. K. K. Kanazawa, A. F. Diaz, R. H. Geiss, W. D. Gill, J. F. Kwak, J. A. Logan, J. F. Rabolt and G. B. Street: J. Chem. Soc., Chem. Commun., 854(1979)
3. K. K. Kanazawa, A. F. Diaz, W. D. Gill, P. M. Gran, G. B. Street, G. P. Gardini and J. F. Kwak: Synth. Mat. $\underline{1}$, 329(1979/80)
4. M. Satoh, K. Kaneto and K. Yoshino: Jpn. J. Appl. Phys. $\underline{24}$, L423(1985)
5. H. Münstedt: In Electronic Properties of Polymers and Related Compounds, ed. by H. Kuzmany, M. Mehring and S. Roth, Springer Ser. of Solid-state Sci., Vol.63 (Springer, Berlin, Heidelberg, 1985) p.8
6. H. Münstedt, G. Köler, H. Möhwald, D. Naegele, R. Bitthin, G. Ely and E. Meissner: Syn. Metals $\underline{18}$, 259(1987)
7. J. P. Travers, P. Audebert and G. Bidan: Mol. Cryst. Liq. Cryst. $\underline{118}$(B), 149(1985)
8. Y. Q. Shen, K. Carneiro, C. Jacobsen, T. Freltoft, R. Y. Qian and X. T. Bi: In Electronic Properties of Polymers and Related Compounds, ed. by H. Kuzmany, M. Mehring and S. Roth, Springer Ser. of Solid-state Sci., Vol.63 (Springer, Berlin, Heidelberg, 1985) p.187
9. Y. Q. Shen, K. Carneiro, C. Jacobsen, R. Y. Qian and J. J. Qiu: Syn. Metals $\underline{18}$, 77(1987)
10 Y. Q. Shen, K. Carneiro, T. Freltoft and R. Y. Qian: Syn. Metals $\underline{18}$, 65(1987)
11. Y. Q. Shen, K. Carneiro, J. J. Qiu and R. Y. Qian: Makromol. Chem., $\underline{188}$, 2041(1987)
12. L. B. Coleman: Rev. Sci. Instrum. $\underline{49}$, 58(1978)
13. M. Wan, P. Wang, Y. Cao and R. Y. Qian: Solid State Commun. $\underline{47}$, 759(1983)
14. M. Wan and P. Wang: Acta Physica Sinica $\underline{35}$(1), 82(1986)
15. J. Ulanski, D. T. Glatzhofer and G. Wegner: Makromol. Chem., Rapid Commun. $\underline{7}$, 361(1986)
16. K. Mortensen: Solid State Commun. $\underline{44}$, 643(1982)

Part II

Defects and Conjugation Length

Influence of the Conjugation Length of Polyacetylene Chains on the DC-Conductivity

D. Schäfer-Siebert[1], C. Budrowski[2], H. Kuzmany[3], and S. Roth[1]

[1]Max-Planck-Institut für Festkörperforschung,
Heisenbergstr. 1, D-7000 Stuttgart 80, Fed. Rep. of Germany
[2]Institute of Inorganic Chemistry, TU, PL-Warsaw
[3]Institut für Festkörperforschung der Universität Wien,
A-1090 Wien, Austria

Though the $(CH)_x$ chains are mainly responsible for the charge transport in polyacetylene, there is only a little knowledge about the influence of the length of these $(CH)_x$ chains on the transport properties of polyacetylene. Experimentally, it is nearly impossible to control the length of the $(CH)_x$ chains during sample preparation, but it is possible to shorten the length of the π-systems of the $(CH)_x$ chains (= conjugation length) by introducing sp^3 defects by chemical methods.

We investigated the effect of the shortened conjugation length on the transport properties of polyacetylene by varying both the sp^3 defect- and the doping concentration. The changes of the conjugation lengths were monitored by resonant Raman scattering. A strong dependence of the DC-conductivity on the conjugation length could be found. The general correlation between the room temperature conductivity and the temperature coefficient previously reported for standard polyacetylene also holds for these samples with reduced conjugation length.

1. Introduction

Electrically conducting polymers on the basis of extended systems of conjugated double bonds have attracted significant scientific and technical interest in recent years. The conductivity can be well described in terms of a hopping model [1]. Till now mainly the influence of the doping species and concentration on the transport properties has been studied, but some recent publications also deal with the effect of defects, which break the conjugation of double bonds in polyacetylene, and hence shorten the conjugation length [2]-[7],[15]. From the knowledge of this influence a deeper understanding of the transport properties is expected, because it finally should allow to distinguish between the resistivity along the chains and the resistivity caused by hops between the chains.

2. Sample preparation

Trans-polyacetylene was prepared by the Shirakawa method [8]. Artificial defects were created by a reaction of water vapour with the carbocations of $(AlCl_4)^-$ doped samples [9], as shown in fig. 1 in a very simplified manner. From IR studies with samples prepared with D_2O, Budrowski [9] found that always two carbocations react in a pinacoline type rearrangement with two $(AlCl_4)^-$ ions to form one paired conjugational defect, which consists of an sp^3 defect and a neighbouring carbonyl group. The number of these paired defects is just half of the initial number of $(AlCl_4)^-$ ions, and was determined by mass uptake of the $(AlCl_4)^-$-doped samples. It is in good agreement with the results obtained by elemental analysis of the hydrolized samples [9]. Since the maximum concentration of $(AlCl_4)^-$ ions in polyacetylene cannot exceed 7 mol %, the whole procedure of $(AlCl_4)^-$-doping and hydrolysis was repeated several times in order to create higher concentrations of defects.

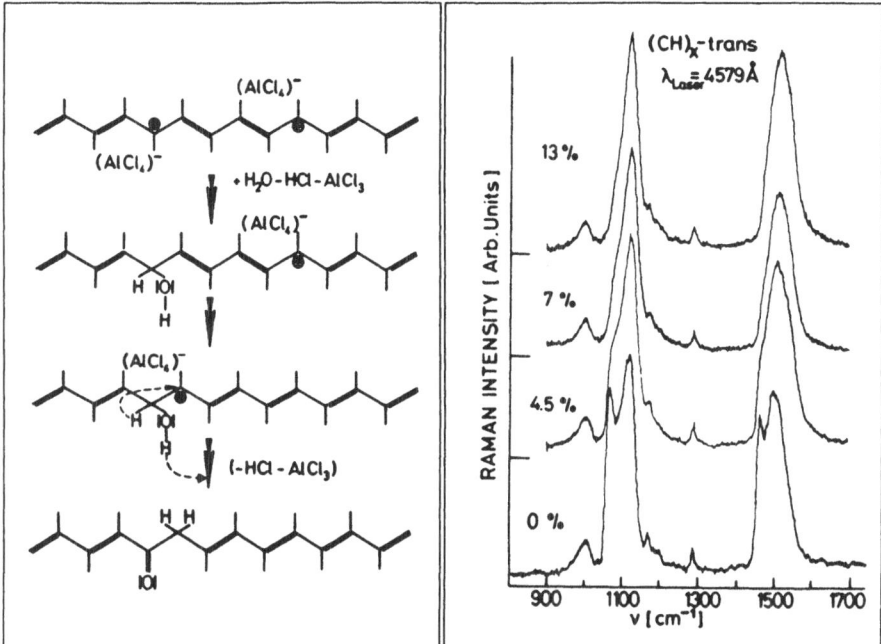

Fig. 1 : Creation of artificial sp³ defects in polyacetylene

Fig. 2 : Raman spectra of polyacetylene with artificial sp³ defects

Fig. 3 : Equilibrium reaction between conjugational defect and non-conjugational defect

This paired defect is in equilibrium with an OH-group, which results from a shift of a proton from the sp³ defect to the carbonyl group (Fig. 3). This OH-group does not interrupt the conjugation. IR-measurements suggest, that this equilibrium is far shifted to the side of the conjugational defect, so that we can neglect this for the calculation of the defect concentration. Subsequent to the introduction of defects, the polyacetylene films were doped with iodine at different doping levels, which were also measured by mass uptake. To insure comparability, all samples of one series were carefully prepared at the same time and in the same vessel.

3. Resonant Raman scattering

To determine the influence of these sp³ defects, we made a resonant Raman scattering experiment. Each conjugated segment of the polymer has a specific electronic structure of the π-electronic system which is dependent on the conjugation length and a characteristic vibrational frequency for modes which couple strongly to this system [10],[12]-[14]. A fit of a corresponding calculation to the experimentally observed lineshapes and linepositions allows to determine the distribution function of the conjugation lengths.

Figure 2 shows experimental results of Raman spectra for trans polyacetylene samples of various concentrations of defects. The blue laser line at 457.9 nm was used to excite the spectra. The high frequency part of each peak represents scattering from the short segments, whereas the low frequency part (at 1070 cm^{-1} and 1460 cm^{-1} ,respectively) represents scattering from long (nominally infinitely long) segments. The spectrum at the bottom corresponds to material with no artificial defects. The double peak structure of the modes around 1100 cm^{-1} and 1500 cm^{-1} is characteristic for a material with a large fraction of long chains.

In order to characterize the distribution of the conjugation lengths with one 'conjugation length parameter', the ratio of the Raman scattering intensity corresponding to short conjugated segments, to the Raman scattering intensity corresponding to long conjugated segments, may be considered. We introduce the ratio $I(8)/I(\infty)$ as a 'conjugation length parameter', which corresponds to 8 double bonds and nominally infinite long chains. Kürti [14] has calculated that the sp^3 defect does not fully interrupt the π-electron system, as a total cut off would do, but that it interrupts the π-electron system only at approx. 78%.

4. Conductivity

4.1 Room-Temperature Conductivity

The DC-conductivity was measured using a standard 4-probe technique with platinum pressure contacts. We have measured the conductivity of two series of polyacetylene samples which were doped with 3.5% and 15% of iodine, respectively.

Figure 4 shows the dependence of the room-temperature conductivity on the defect concentration. A fit leads to the empirical formula

$$\sigma(d) = \sigma(0) * e^{-55d}, \tag{1}$$

where d is the defect concentration. For both doping concentrations the same decrease of the conductivity with the defect concentration is observed, so that (1) is independent of the doping concentration. Some authors [2]-[4] suggested a fit of (log σ) versus (log d) which would also work fairly well with our data, but we found the fit in fig. 4 to work significantly better, especially in the regime of low-defect concentrations.

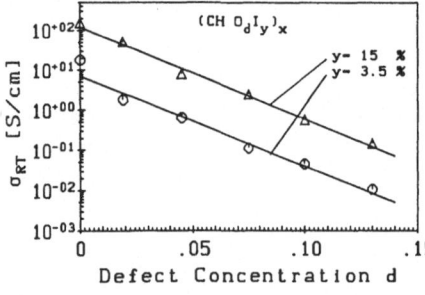

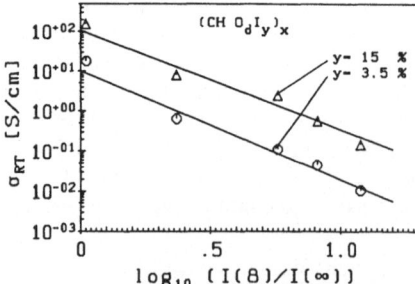

Fig. 4 : Room-temperature conductivity of polyacetylene depending on the defect concentration

Fig. 5 : Room-temperature conductivity depending on the 'conjugation length parameter' $I(8)/I(\infty)$

Combining the conductivity and the Raman data we can now plot the room temperature conductivity versus the 'conjugation length parameter' defined above. This is shown in fig. 5. As can be seen from the comparison of fig. 4 and fig. 5 the conductivity is nearly as well correlated with the 'conjugation length parameter' as it is with the defect concentration.

4.2 Temperature Dependence of the Conductivity

The temperature dependence of the conductivity of all segmented samples follows strongly the behaviour

$$\sigma = \sigma_0 * e^{-\frac{T}{T_0}^{-\gamma}} \tag{2}$$

with $\frac{1}{2} \geq \gamma \geq \frac{1}{4}$. This temperature dependence has also been observed for ordinary doped polyacetylene without artificial defects and is quite common for conductive polymers [11]. It formally corresponds to Mott's law of variable-range hopping in $\left(\frac{1}{\gamma} - 1\right)$ dimensions. The value γ can not be determined uniquely by fitting because all fits with γ in the mentioned regime lead to comparably good fits.

It has been shown [11] that there is a universal correlation between the room-temperature conductivity and T_0 in (2) in the case of polyacetylene. The analysis of conductivity data of other conducting polymers (our measurements and data from [16],[17]) and of amorphous silicon as well, show that this relation also holds for these materials (fig. 6).

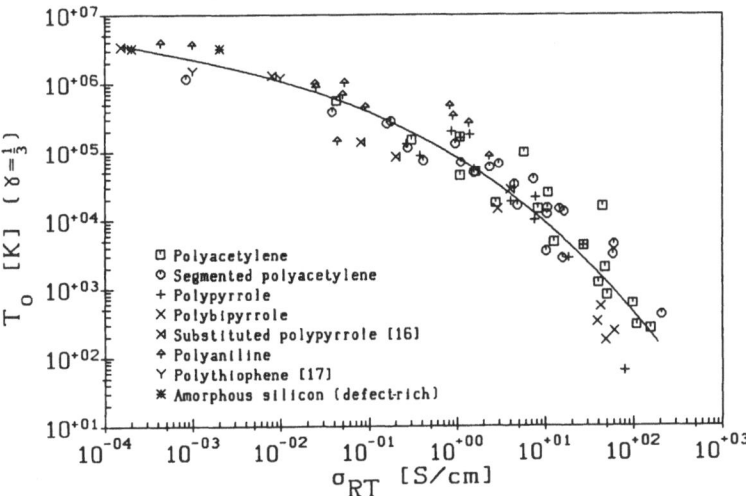

Fig. 6 : Universal correlation between the parameters T_0 and σ_{RT} for some conducting polymers and amorphous silicon

Therefore one should not be surprised that also the parameters of segmented polyacetylene follow this universal correlation, as shown in fig. 6. This correlation seems to be universal for all conductive polymers, for which hopping between conducting molecules is limiting the conductivity.

The consequence of this universal relation is that the temperature dependence T_0 is insensitive to the chemical composition and morphology of conductive polymers. It only depends on the room temperature conductivity σ_{RT} , which can be adjusted by varying the doping level of the conductive polymer. Of course, these doping levels are different for samples with different chemical composition and morphology to get the same σ_{RT} .

5. Summary

The room temperature conductivity σ_{RT} of polyacetylene is strongly related to the defect concentration d. This relation is independent of the doping level. The conjugation length plays a dominant role for the transport process in polyacetylene. The shortening of the conjugation length does not break the universal correlation between T_0 and σ_{RT} so that for all conductive polymers which we investigated so far the knowledge of σ_{RT} allows to predict the temperature dependence T_0 of the DC-conductivity. This is independent of the chemical composition and morphology of the conductive polymer.

References
1. K.Ehinger, S.Roth: Phil. Mag. B 53, 301 (1986)
2. J.C.W. Chien, G.N. Babu: J.Chem.Phys. 82, 441 (1985)
3. K.Soga, M.Nakamaru: J.Chem.Soc., Chem.Commun., 1495 (1983)
4. S.I.Yaniger, M.J.Kletter, A.G.MacDiarmid: Polymer Preprints 25, 264 (1984)
5. R.H.Baughman, L.W.Shacklette: Synthetic Metals, 17,173 (1987)
6. F.Zuo, A.J.Epstein, X.Q. Yang, D.B.Tanner, G.Arbuckle, A.G.MacDiarmid: Synthetic Metals, 17,433 (1987)
7. X.Q.Yang, D.B.Tanner, G.Arbuckle, A.G.MacDiarmid, A.J.Epstein: Synthetic Metals, 17,277 (1987)
8. T.Ito, H.Shirakawa, S.Ikeda: J.Polym. Sci., Polym. Chem. Ed. 12,11 (1974)
9. C.Budrowski, S.Roth, H.Kuzmany, J.Przyluski: Synthetic Metals, 16,291 (1986)
10. H.Kuzmany: Pure and Applied Chem. 57,235 (1985)
11. C.Budrowski, A.Pron, J.Przyluski, K.Ehinger, S.Roth: Synthetic Metals, 16,117 (1986)
12. D.Schäfer-Siebert, C.Budrowski, H.Kuzmany, S.Roth: submitted to Synthetic Metals, 21 (1987)
13. P.Knoll, H.Kuzmany, G.Leising: this volume and private communication
14. J.Kürti, H.Kuzmany: this volume and private communication
15. J.L.Brédas, J.M. Toussaint, J.M.André: this volume and private communication
16. Ch.Kröhnke, J.Rühe, G.Wegner: this volume and private communication
17. H.Isolato, H.Stubb, J.Saarilahti: this volume and private communication

Conjugation Length and Localization in Conjugated Polymers

J. Kürti and H. Kuzmany

Institut für Festkörperphysik, Universität Wien and
Ludwig Boltzmann Institut für Festkörperphysik,
Strudlhofgasse 4, A-1090 Wien, Austria

1. Dispersion of Raman lines

Raman spectra of trans-polyacetylene are well known to show a characteristic double-peak structure at around 1500 cm^{-1} (double bond stretch mode) and 1100 cm^{-1} if excited with a blue laser [1]. The frequency shift between main and **satellite** peak increases with increasing excitation energy (dispersion effect). We obtained similar results for polydiacetylene single crystals of toluene sulfonate type (PDA-TS). Fig.1 shows the shift of the **satellite** peak for the C=C stretch mode and for the C≡C stretch mode as a function of the exciting laser energy. Both lines extrapolate to about 1.9 eV which is the exciton transition energy for **crystalline PDA-TS**. This value for the extrapolation and the fact that we observe **the satellite line for the C=C stretch mode and for the C≡C stretch** mode is good evidence for its origin from disordered areas of the polymer rather than from an oxygen defect [2].

The Raman-lineshape for excitation in the visible spectral region can be very well described by the conjugation length model (CLM) [3-5]. This model assumes a distribution of the length of undisturbed segments of conjugation on the chain and takes into account the dependence of the energy gap and the vibrational frequencies on segment length. The CLM is useful for interpreting other physical properties like e.g. the change of the conductivity upon defects as well [6].

In spite of the good agreement between model calculation and experiments the definition of conjugation length as the length of the undisturbed segments is vague, since defects on the chain may only partly interrupt the conjugation. On the other hand the physically relevant quantity is not the chain or segment length but the excitation energy. For a chain with defect(s) we can define an effective conjugation length using the relationship between first excitation energy and the (undisturbed) chain length. This is of course only meaningful in the case when the energy gap of a chain increases due to an incorporated defect, which can be expected for a defect with strong interruption. However, as is known, disorder can lead to bandtailing and finally to a closure of the gap. Therefore it is very important to investigate whether a defect in effect increases or decreases the bandgap. On the other hand, disorder in solids leads to localization of the electrons. This raises the question whether one should consider a chain with defects as equivalent to a shorter chain, or rather equivalent to the original chain with respect to its length but with the electronic states becoming localized. Which is the more relevant quantity for characterising the effect of defects: the (effective) conjugation length or the localization length?

2. The Influence of Disorder on the Band Gap

To investigate the role of disorder in a linear chain we performed quantum mechanical calculations for trans-polyacetylene on the Hückel-level and on the CNDO level. For the Hückel calculations we used a tight-binding Hamiltonian with σ-core potential after Longuett-Higgins and Salem /7,8/:

$$H = \sum_{n,s} \alpha_{ns} a_{ns}^+ a_{ns} + \sum_{n,s} \beta_n^\circ(r_n)(\cos\phi_n)(a_{ns}^+ a_{n+1,s} + a_{ns} a_{n+1,s}^+) + \sum_n f_n(r_n). \quad (1)$$

The resonance integral has the form

$$\beta_n^\circ(r_n) = -A \exp(-r_n/B), \quad (2)$$

the σ-core potential

$$f_n(r_n) = C \cdot \beta_n^\circ(r_n) \cdot (r_n - r_0 + B). \quad (3)$$

The α_n and $\cos\phi_n$ terms describe the diagonal and off-diagonal disorder, respectively. The latter are considered as rotations around the carbon-carbon bond. The program optimizes the geometry so that **Coulson's linear relationship between bond length and bond order** is satisfied:

$$r_n = r_0 - K*P_n. \quad (4)$$

We have chosen the parameters so that for a long chain (200 carbon atoms) we obtain 1.4 eV energy gap , 10 eV total band width, and a bondlength alternation with 1.47 Å / 1.33 Å . Furthermore we required that the total electronic energy should have a minimum after relaxation. The optimized parameters were:
$A = 42.32$ eV, $B = 0.493$ Å, $r_0 = 1.61$ Å, $K = 0.33$ Å and $C = 6.06$ Å^{-1}.
From the calculations we found a great difference between random **weak** disorder at every site and strong disorder at few randomly selected places. Fig. 2a shows the energy levels for a linear chain of 100 carbon atoms if we have a statistical distribution of the α_n . Each column corresponds to the same statistical distribution but to different (increasing) amplitudes of the disorder. As

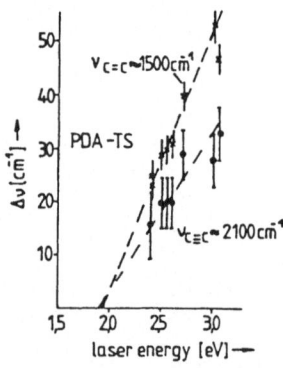

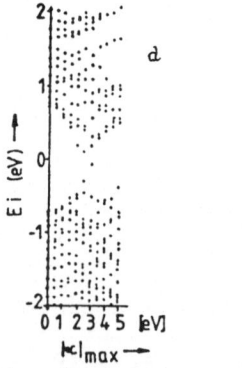

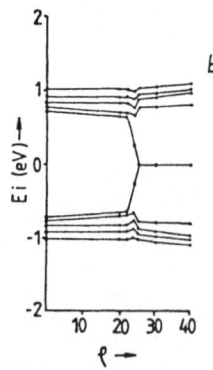

Fig. 1 Dispersion of Raman lines ("gap-phonon relation") in PDA-TS

Fig. 2 Energy levels E_i vs amplitude of diagonal disorder $|\alpha|_{max}$ (a). Energy levels vs rotation angle ϕ for 10% discrete rotations (b)

44

can be seen, disorder in this case leads to bandtailing and finally
to a closure of the gap. We obtained the same tendency for the
off-diagonal disorder. Random rotation around each bond with increasing amplitude leads again to decreasing of the energy gap.
Similar results have been reported previously /9-10/. A different
behaviour is observed for strong discrete disorder. Fig. 2b shows
results where 10 % of the bonds are rotated for a chain of 100 C
atoms. The rotated bonds are randomly selected and the rotation
angles are the same with increasing amplitude. For small angles
there is again a bandtailing effect. For stronger rotations there
is an increase of the gap which is accompanied by a creation of
midgap states. From the evaluated lattice distortion this state is
identified as a soliton state. Very similarly, carbonyl groups on
the chain lead to an increase of the band gap and sometimes to
midgap states depending on the sites of the carbonyl groups.

3. Correlation Between Conjugation Length and Localization Length

From the Hückel-calculation we obtain the $C_{i,j}$ linear combination
factors for the j-th molecular orbital on the i-th carbon atom.
This allows to study the localization of the wave functions quantitatively. To characterize the localization for each case we use
as a localization length parameter the participation ratio P_j /11/

$$P_j = \{ \sum_i |C_{i,j}|^4 \}^{-1} \tag{5}$$

for the j-th molecular orbital. P equals 1 for complete localization and P equals N for complete delocalization where N is the
number of C-atoms on the chain. It should be mentioned that for an
unperturbed chain without any disorder P/N is only about 0,7 because of the bond alternation. Fig. 3 shows the relation between
effective conjugation length (L) and localization length (P) for a
40 carbon chain with 4 rotations with the same angle around 2. 14.
16. and 32. bond. P is the localization length of the HOMO and L
is calculated from the energy gap-chain length relation we obtained with the same Hückel calculation for unperturbed chains. Fig. 3
shows that at least for certain kind of defects and defect distribution there is a linear correlation between effective conjugation
length and localization length.

4. The Role of Chemical Segmentation

Trans-polyacetylene samples with a controlled number and a specific
type of defects were prepared. This preparation was carried out by
reaction of $AlCl_4$ doped samples with water. The created defect is
an sp^3 defect with a carbonyl group attached to the neighbouring C
atom. For details see Ref. /12/. We carried out Raman and ESR
measurements on these samples. Resistivity measurements are reported in Ref. /13/. Fig 4. shows the increase of the Raman-dispersi-
on, the ESR-linewidth and the resistivity with increasing defect
concentration. These results are in agreement with the conjugation
length model.
    We investigated the effect of sp^3, carbonyl and combined de-
fects on the electronic energy spectrum for a short transpolyace-
tylene chain. We performed all-valence-electron calculations using

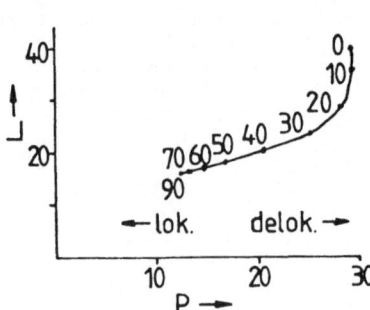

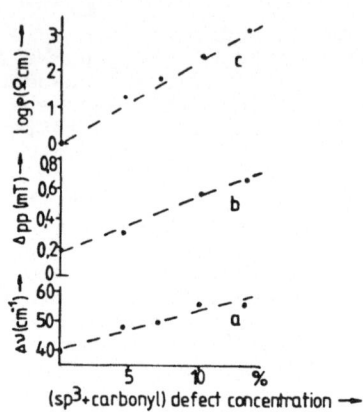

Fig. 3 Correlation between localization length (P) and effective
conjugation length (L) for a polyacetylene chain of 40 carbon atoms
with 4 rotations around single
bonds. The rotations are at randomly selected places and have equal
amplitudes

Fig. 4 The increase of the
shift between main and satellite Raman line (a), ESR
linewidth (b) and resistivity (c) with increasing
(sp^3 + carbonyl) defect
concentration. The lines
are guidelines for the eye

the spectroscopic CNDO/S-CI method. For more details of the method
see /8/. To characterize the strength of interruption of the conjugation we introduced an interruption parameter (η) :

$$\eta(y,N,def) := \frac{\Delta E_1(N,def) - \Delta E_1(N)}{\Delta E_1(y) - \Delta E_1(N)} \quad ; \tag{6}$$

$N = n_1 + n_2 + n_{def}$ is the total number of carbon atoms of the chain.
n_{def} is the number of carbon atoms "in contact" with the defect.
(n_{def} =1 for an sp^3 or a carbonyl defect and =2 for the combined
defect.) n_1 and n_2 are the numbers of carbon atoms left and right
from the defect, respectively, $y = \max(n_1,n_2)$. ΔE_1 is the first
transition energy (energy gap) for the chain. Thus, the interruption parameter is the ratio between increase of the gap due to the
defect and the highest possible increase where the defect would be
a real "cut". $\eta \approx 0$ for weak interruption and $\eta \approx 1$ for strong interruption. We calculated the interruption parameter for a chain of 9
carbon atoms with an sp^3 or a carbonyl defect in the middle and
for "combined defects". In the latter case the chain contained 10
carbon atoms and two sp^3 or two carbonyl or one (sp^3+carbonyl) defect in the center. Before the all-valence-electron calculation we
performed geometry optimization with the Hückel-method. In this
case however we could calculate only the effect of one or two carbonyl groups. As an approximation we choose the same geometry for
the sp^3 containing defects. The results are shown in Table 1. We
want to point out two facts. First: the interruption effect of the
sp^3+carbonyl is the same as that of the sp^3 alone. This is in agreement with the observations of S. LEFRANT et al /14/ that an e-
qual concentration of sp^3 defects and sp^3 + carbonyl defects leads
to equal changes in the Raman line shape for the C=C stretch mode.

46

Second: the interruption effect of the sp^3+carbonyl defect is large enough so we can use the conjugation length model for interpreting the experimental results shown on Fig. 4 .

Table 1. Interruption parameter for defects in polyacetylene

chain	sp^3	carb.	sp^3+sp^3	carb.+carb.	sp^3+carb.
(4, 9)	71%	46%			
(4,10)			88%	62%	71%

5. Conclusions

In summary, we demonstrated that for strong discrete defects the concept of conjugation length is physically meaningful and well defined. Weak, smeared-out distribution of disorder leads to bandtailing. Some kinds of disorder create midgap states (solitons). Furthermore we showed for the first time an example for a linear correlation between conjugation length and localization length.

Acknowledgements

The authors acknowledge the submission of the computer programs by P. Surjan and many valuable discussions with him.
This work was supported by the Stiftung Volkswagenwerk.

References

1 L.S. Lichtmann, A. Sarhangi and D.B. Fitchen:
   Chemica Scripta 17, 149 (1981)
2 N.J. Poole and D.N. Batchelder:
   Mol. Cryst. Liq. Cryst. 105, 55 (1984)
3 H. Kuzmany: phys. stat. solidi b97, 521 (1980)
4 H. Kuzmany, E.A. Imhoff, D.B. Fitchen and A. Sarhangi:
   Phys. Rev. B26, 7109 (1982)
5 E. Mulazzi, G.P. Brivio, E. Faulques and S. Lefrant:
   Solid State Commun. 46, 851 (1983)
6 R.H. Baughman and L.W. Shacklette:
   Synth. Metals 17, 173, (1987)
7 H.C. Longuett-Higgins and L. Salem:
   Proc.R.Soc. London, Ser.A251, 172 (1959)
8 P. Surjan, H. Kuzmany: Phys. Rev. B33, 2615 (1986)
9 C.T. White, M.L. Elert and J.W. Mintmire:
   Journal de Phys. C3, 481 (1983)
10 D. Vanderbilt and E.J. Mele: Phys. Rev. B22, 3939 (1980)
11 K. Tanaka, M. Nagaoka and T. Yamabe:
   Int. J. of Quantum Chem.23, 1101 (1983)
12 C. Budrowski, S. Roth, H. Kuzmany and J. Przyluski:
   Synth. Met. to be published
13 D. Schäfer-Siebert, C. Budrowski, H. Kuzmany and S. Roth:
   Synth. Met. to be published
14 S. Lefrant et al. this volume

Theoretical Investigations
of Segmented Polyacetylene

J.L. Brédas[1], J.M. Toussaint[1], G. Hennico[1], J. Delhalle[1], J.M. André[1], A.J. Epstein[2], and A.G. MacDiarmid[3]

[1]Laboratoire de Chimie Théorique Appliquée, Centre de Recherches sur les Matériaux Avancés, Facultés Universitaires Notre-Dame de la Paix, B-5000 Namur (Belgium)
[2]Department of Chemistry and Department of Physics, Ohio State University, Columbus, OH 43210, USA
[3]Department of Chemistry, University of Pennsylvania, Philadelphia, PA 19104, USA

Segmented polyacetylene is formed by the introduction of sp^3 defects through deuteration along the chains of ordinary trans-polyacetylene. A number of experimental data have recently been collected on segmented polyacetylene, $(CHD_y)_x$, containing concentrations (y) in deuterium atoms as high as 17%, or even about 30% [1-3].The experimental results have been previously interpreted in terms implying that the presence of sp^3-bonded carbons has only a moderate influence on the conjugation of the polyacetylene π bands (due to hyperconjugation effects) [1-3].

In this paper, after a presentation of the main experimental findings on segmented polyacetylene, we describe preliminary theoretical investigations on this compound. Our major purpose is to assess whether the interpretation that sp^3 defects only moderately affect the conjugation along polyacetylene chains is entirely valid. Our results indicate that conjugation may be strongly interrupted and we discuss an alternative model that is in accord with much of the experimental and theoretical data.

I. Summary of experimental findings on segmented polyacetylene

The synthesis of segmented polyacetylene containing sp^3 defects is a two-step process starting with ordinary Shirakawa-type trans-polyacetylene [4, 5]. First, trans-polyacetylene is doped with sodium naphthalide to produce n-doped polyacetylene:

$$t\text{-}(CH)_x + (xy)\, Na^+ Napht^- \rightarrow [\, Na^+_y (CH)^{-y}\,]_x + (xy)\, Napht$$

Second, the n-doped compound is treated with deuterated methanol to lead to undoped deuterated (segmented) trans-polyacetylene and sodium methoxide:

$$[\, Na^+_y (CH)^{-y}\,]_x + (xy)\, CH_3OD \rightarrow (CHD_y)_x + (xy)\, CH_3ONa$$

48

In this way, concentrations (y) in deuteriums can be achieved up to y=0.17, i.e. one sp^3 defect per three double bonds. More recently, y values up to 0.30 have been obtained by starting from $(CHD_{0.17})_x$ and performing again the two-step process described above. Segmented polyacetylene can be doped with iodine up to dopant concentrations on the order of 3% to form $[CHD_y(I_3)_{0.03}]_x$. The maximum iodine concentration achieved here is about half that attainable in ordinary trans-polyacetylene.

Optical absorption data on $(CHD_y)_x$ have been reported by Yang et al. [2]. With respect to ordinary polyacetylene, the effects of deuteration on the π-π^* interband transition are threefold: (i) the peak of absorption shifts a small amount to higher energies (about 0.4-0.5 eV from $\approx$ 2.0 eV up to $\approx$ 2.5 eV for $(CHD_{0.17})_x$); (ii) there is a substantial broadening of the absorption to the high frequency side; and (iii) the overall intensity of the oscillator strength is reduced by a factor of 2. The magnetic data of Zuo et al. indicate that the number of Curie spins is around 440 ppm, very similar to that in ordinary trans-polyacetylene [1].

EPR linewidths measurements on segmented polyacetylene have been performed by Arbuckle et al. [3]. For $(CHD_{0.17})_x$, the linewidth is on the order of 3.5 Gauss at 120 K and gets motionally narrowed as temperature increases. At room temperature, the linewidth is about 2 Gauss. A comparable behavior is found in normal trans-polyacetylene (the EPR linewidth decreasing from $\approx$ 3 Gauss to $\approx$ 1 Gauss as the temperature evolves from 120 to 300 K) whereas in cis-polyacetylene, the linewidth is about 8 Gauss and temperature-independent, indicative of fixed spins. The comparison of all these results shows that in $(CHD_y)_x$, there is a reduction but not a complete blockage of the motion of the spins (neutral solitons) along the chains.

Upon iodine doping, the conductivity increases by six orders of magnitude (much as in ordinary iodine-doped trans-polyacetylene) but the maximum conductivity is always about 10^3 to 10^4 times smaller than in iodine-doped polyacetylene. For instance, the maximum conductivity at room temperature for $[CHD_{0.17}(I_3)_{0.03}]_x$ is $\approx 2\times10^{-2}$ S/cm [1]. Magnetic measurements indicate that the charges that appear on the chains upon doping go into spinless states, i.e. into charged soliton states [1]. These observations are further supported by the evolution of the optical absorption data and infra-red data upon doping [2]. The former show the appearance of midgap soliton electronic absorptions, and the latter, the presence of infra-red doping-induced vibrational modes characteristic of charged solitons at $\approx$ 900, 1240, and 1400 wavenumbers. These modes are observed in $(CHD_y)_x$ even for sp^3 concentrations as high as 17% but their integrated oscillator strengths is 5 to 6 times smaller than in doped ordinary trans-polyacetylene [2]. Following Yang et al., this feature is most likely due to an increase in the effective mass of the solitons confined by sp^3 defects [2].

Since all available experimental data are totally incompatible with a complete segregation between a pure polyacetylene phase and a pure deuterated (polyethylene) phase, the experimental data described above have been interpreted at first as implying that sp^3 defects are only a small perturbation of conjugation (and a model of hyperconjugation through the sp^3 defects has been put forth) [1, 3]. This idea has been further nurtured namely by theoretical results of Surjan and Kuzmany [6], indicating that interruption of conjugation by sp^3 defects is far from being complete. In order to investigate theoretically this interpretation, we have performed a number of Hartree-Fock *ab initio*, Modified Neglect of Differential Overlap (MNDO), and Valence Effective Hamiltonian (VEH) calculations on polyacetylene chains containing sp^3 defects.

II. Theoretical calculations on segmented polyacetylene

We have first examined at the *ab initio* Hartree-Fock minimal basis set (STO-3G) and MNDO levels, the influence of an sp^3 defect on the geometry of the neighboring conjugated segments. We have considered a $C_{17}H_{20}$ chain containing an sp^3 defect in the middle and, for the sake of comparison, a fully conjugated $C_{16}H_{18}$ chain. Geometries have been fully optimized in both cases. Bond-lengths results are presented in Figure 1. As can be observed, the influence of the sp^3 defect on the geometry of the conjugated segments is very weak (the nearest double bond shortens by less than 0.01 Å) and does not extend beyond $\approx$ 2-3 carbons.

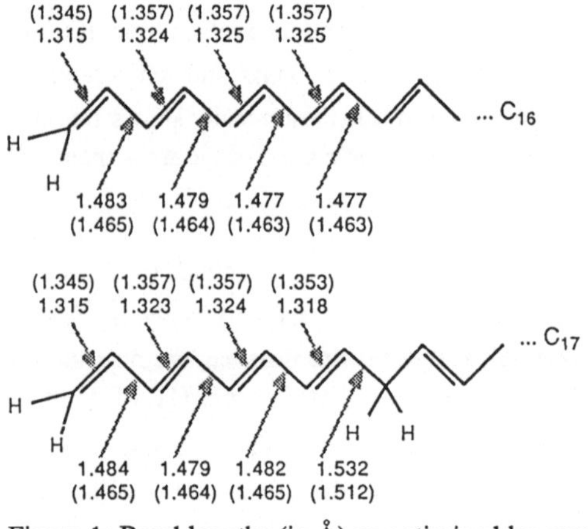

Figure 1: Bond lengths (in Å) as optimized by means of an Hartree-Fock *ab initio* STO-3G technique and MNDO technique fot a fully conjugated $C_{16}H_{18}$ chain and a $C_{17}H_{20}$ chain containing an sp^3 carbon in the middle. The bond lengths presented between parentheses refer to the MNDO results.

An interesting aspect is to consider how the electronic energy levels of two octatetraene (C_8H_{10}) moieties evolve when they are connected, on the one hand to form a fully conjugated system ($C_{16}H_{18}$) or, on the other hand, through an sp^3 carbon ($C_{17}H_{20}$). Calculating the Hartree-Fock *ab initio* evolution of the octatetraene highest occupied (HOMO) electronic states, we find that these states split by 1.17 eV in the first case but only by 0.16 eV in the second case. The HOMO level splitting is thus seven times smaller when the two conjugated segments are separated by an sp^3 carbon. This result is a first indication that the sp^3 defect constitutes a strong interruption of conjugation, the hyperconjugation effect being very small, as might have been expected.

We have also compared from Hartree-Fock *ab initio* calculations, the energetics of two configurations of a $C_{19}H_{24}$ chain containing three sp^3 carbons, see Figure 2. In configuration A, the three defects are clustered in the middle of the chain and separate two conjugated segments formed by four double bonds each. In configuration B, the sp^3 defects are regularly spaced and connect four conjugated segments formed by two double bonds each. We find that configuration A is significantly favored over configuration B. (At the minimal basis set STO-3G level, this difference is about 20 kcal/mol. Using a larger basis set, we expect this difference to be somewhat affected but not qualitatively modified.) These results indicate that some clustering of the sp^3 carbons should occur, in agreement with intuitive ideas from organic chemistry. Note that the most stable conformation for polyacetylene chains containing sp^3 carbons remains all-trans coplanar, a behavior which is normal since, in both pure polyacetylene and polyethylene, that conformation is the most stable.

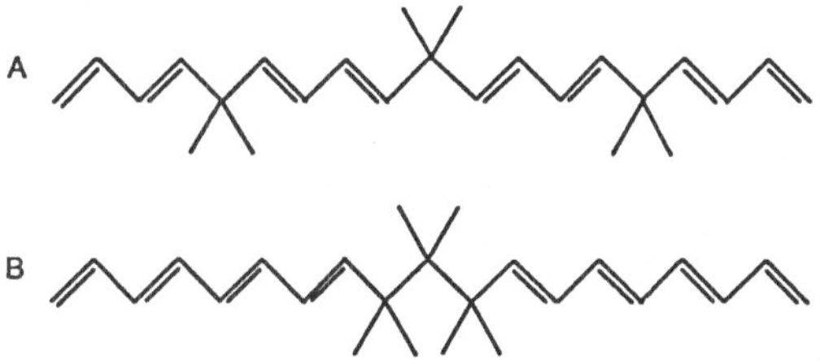

Figure 2: Illustration of the two configurations chosen for the $C_{19}H_{24}$ chain.

We have performed VEH band structure calculations on a perfectly regular $(CHD_{0.17})_x$ chain, Figure 3. In this context, regular means that three double bonds are always followed by an sp^3 carbon in a regular way. The input geometries are taken from the MNDO optimizations described above. We calculate for $(CHD_{0.17})_x$ a bandgap of 3.64 eV, to be contrasted with 1.44 eV in the case of trans-polyacetylene. The π bands are found to be flat, the width of the highest occupied π band being only 0.42 eV, instead of 6.5 eV for trans-polyacetylene. These model results are also indicative of strong conjugation interruption due to sp^3 defects. The bandgap value calculated for a regular distribution of sp^3 carbons is thus much larger than the bandgap observed experimentally. Furthermore, the solid state ionization potential of $(CHD_{0.17})_x$ is here theoretically estimated to be ≈ 5.9 eV, which constitutes too large a value to allow for any significant iodine doping.

An agreement with experiment can be obtained if we take account of a random distribution of sp^3 defects. Indeed, previous works in Namur have shown that in polyacetylene chains interrupted by C=O defects, the bandgap decreases by about 40% when

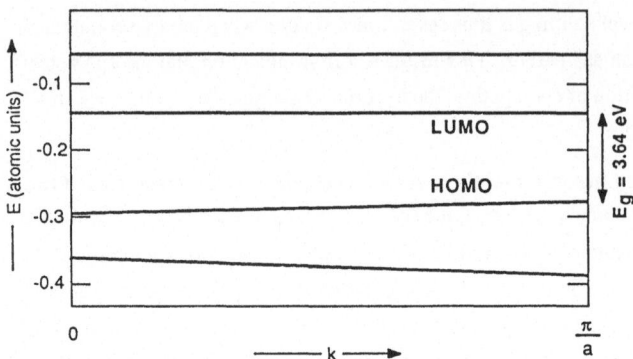

Figure 3: VEH band structure for a regular chain of $(CHD_{0.17})_x$.

switching from a regular to a random distribution of the defects [7]. We can expect at least a similar decrease in the $(CHD_{0.17})_x$ case — work is in progress to obtain precise values — and therefore a bandgap value on the order of 2 eV, in good agreement with experimental data. Such a decrease in bandgap value in the C=O situation has been correlated with the presence of a kind of bimodal distribution of conjugation lengths: some fraction of the chains are substantially longer than the average while the rest are much shorter. Resonant Raman Scattering measurements for $(CHD_y)_x$ have been interpreted by Lefrant and Mulazzi [8] to indicate a bimodal distribution of chain lengths. Preliminary analysis of the same data using the amplitude mode formalism [9] indicates [10] some deviation from the Peierls relation for $(CHD_{0.16})_x$. A bimodal distribution has also been suggested by Yang et al. [2] in order to rationalize the bandgap evolution as a function of increasing sp^3 carbon concentration.

III. Conclusions

In summary, the experimental observations reported on segmented polyacetylene, $(CHD_y)_x$:
(i) cannot be explained on the basis that sp^3 defects constitute only a small perturbation of the electronic structure. Our theoretical calculations indeed indicate that sp^3 defects provoke a strong interruption of conjugation;
(ii) can be rationalized by taking account of:
$\rightarrow$ some clusterization of the sp^3 carbons. We find that thermodynamically, aggregation of the sp^3 defects lead to a more stable configuration, in agreement with an organic chemist's intuition;
$\rightarrow$ a random distribution of the sp^3 carbons. Disorder effects can be very important in leading to gap values similar to that in pure trans-polyacetylene despite the presence of a large number of sp^3 defects.

Acknowledgements

This work has been supported by the Belgian National Fund for Scientific Research (FNRS). We are indebted to FNRS, IBM Belgium, and the Facultés Universitaires Notre-Dame de la Paix for the use of the Namur Scientific Computing Facility (SCF). One of us (AJE) acknowledges partial support from grant US-NSF/INT/8514202.

References

a. Chercheur Qualifié of the Belgian National Fund for Scientific Research (FNRS).

1. F. Zuo, A.J. Epstein, X.Q. Yang, D.B. Tanner, G. Arbuckle, A.G. MacDiarmid: Synth. Met. 17, 433 (1987).

2. X.Q. Yang, D.B. Tanner, G. Arbuckle, A.G. MacDiarmid, A.J. Epstein: Synth. Met. 17, 277 (1987).

3. G. Arbuckle, M.X. Wan, A.G. MacDiarmid, F. Zuo, A.J. Epstein: Bull. Am. Phys. Soc. 32, 456 (1987).

4. S.I. Yaniger, M.J. Kletter, A.G. MacDiarmid: Polym. Prepr. 25, 264 (1984).

5. K. Soga, S. Kawakami, H. Shirakawa: Makromol. Chem. Rapid Commun. 1, 643 (1980).

6. P.R. Surjan, H. Kuzmany: Phys. Rev. B 33, 2615 (1986).

7. G. Hennico, J. Delhalle, J.M. André: to be published.

8. S. Lefrant, E. Mulazzi: these Proceedings.

9. Z. Vardeny, E. Ehrenfreund, O. Brafman, and B. Horowitz: Phys. Rev. Lett. 51, 2326 (1983).

10. J.M. Ginder et al.: to be published.

Analysis of the Raman Spectra
of Modified trans-(CH)$_x$

S. Lefrant[1], *G. Arbuckle*[2], *E. Faulques*[1], *E. Perrin*[1], *A. Pron*[3],
and E. Mulazzi[4]

[1]Université de Nantes, 2 rue de la Houssinière,
 F-44072 Nantes Cedex 03, France
[2]Department of Chemistry, University of Pennsylvania,
 Philadelphia, PA 19104, USA
[3]Technical University of Warsaw, PL-00664 Warsaw, Poland
[4]Dipartimento di Fisica dell'Università di Milano,
 I-20133 Milano, Itay

1. Introduction

The relationship between the conjugation length in polyacetylene and the
maximum d.c. conductivity obtained after doping can be very important in
understanding transport properties in this material. In this context,
samples with additional defects have recently been prepared and their
electrical and optical properties studied. The addition of hydrogen atoms
can be achieved in a controlled manner as shown by Soga and Nakamaru /1/.
The treatment can be followed even more accurately by using deuterium atoms
since the C-D stretching vibration is detected by infrared spectroscopy. As
a result, a breaking of the conjugation length of the carbon chain is
expected by the creation of carbon atoms in Sp^3 orbital configuration.

Vibrational spectroscopy and in particular resonance Raman scattering is
very sensitive to the conjugation length in polyacetylene and can therefore
be used as a probe for the chain modifications provided by this kind of
treatment. A similar study has been reported by Furukawa et al. /2/ and we
present in this paper Raman results obtained on deuterated samples of
different compositions. All our data are then theoretically investigated by
using the bimodal distribution of conjugated segments elaborated by Brivio
and Mulazzi /3/. Finally, a comparison is made with samples in which the
disruption of the conjugation is achieved by introducing more complicated
defects consisting of a carbonyl defect adjacent to an Sp^3 one. Results are
discussed in view of the recent theoretical results presented by Brédas et
al. /4/.

2. Experimental Data and Discussion

In Fig. 1, we report RRS spectra of $(CHD_y)_x$ with y = 0, y = 0.069, y = 0.11
and y = 0.17 respectively, for the two excitation wavelengths λ_L = 676.4 nm
and λ_L = 457.9 nm. For λ_L = 676.4 nm, the two Raman bands peaked at 1073
and 1462 cm^{-1} for pristine $(CH)_x$ (y = 0, Fig. 1a) exhibit small changes
when y goes from 0 to 0.17 consisting in a shift in frequency to 1087 and
1472 cm^{-1} respectively (Fig. 1d). Also, for this concentration, the bands
become more symmetrical.

For λ_L = 457.9 nm, the spectrum of the pristine polymer does not exhibit
a well resolved double peak structure and the satellite components are
peaked at 1119 cm^{-1} and 1503 cm^{-1} (Fig. 1e). By deuteration, significant
changes are observed in the profile of the bands leading to peaks at 1131

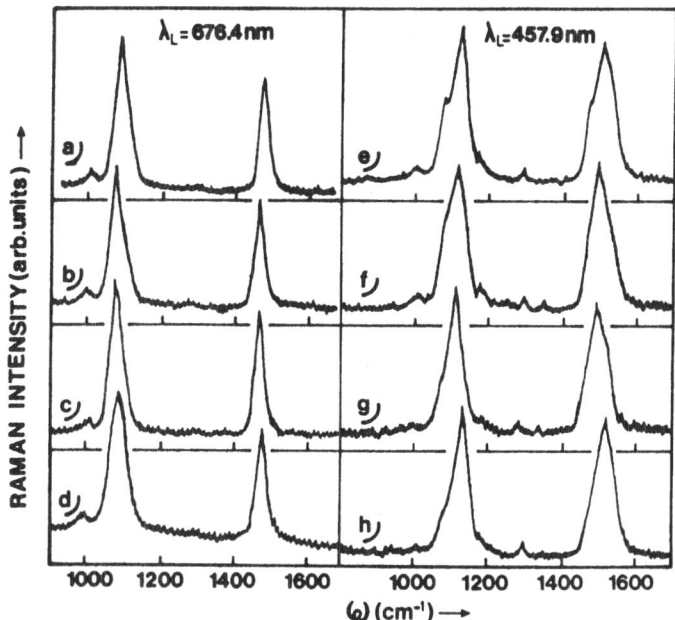

Fig. 1 : Raman spectra of $(CHD_y)_x$; T = 20°C
λ_L = 676.4 nm : a) y = 0, b) y = 0.069, c) y = 0.11, d) y = 0.17
λ_L = 457.9 nm : e) y = 0, f) y = 0.069, g) y = 0.11, h) y = 0.17

and 1514 cm^{-1} for $(CHD_{0.17})_x$ (Fig. 1h). Also, it can be mentioned that the low-frequency components due to the contribution from the long conjugated segments can hardly be seen. The spectra are qualitatively identical to those presented in /5/ in the case of hydrolyzed p-doped films. Moreover, it should be noticed that these Raman spectra also show some similarities with those of doped $(CH)_x$ at low doping levels /6/.

The different features of the Raman spectra given above can be described in terms of a double distribution of short and long conjugated segments respectively. In this model /3,7/ the electronic and vibrational properties and the electron vibration interaction in the excited electronic states of the conjugated segments are studied as a function of the number of double bonds (N). The Hückel model is used for 3 < N < 30 whereas for N ≫ 30, calculations are made with a tight binding model. Details of the parameters are given in /3,7/ and applications of this theoretical approach have been successfully used to fit Raman spectra in many different cases /8/.

In the present case, we have determined the parameters which give a good fit for all the series of Raman spectra. They are reported in Table 1 in which N_1 and N_2 are the two distribution centers, σ_1 and σ_2 their standard deviations and G the relative weight of the long segments' distribution.

In this study, it appears that the first stage of deuteration leads to a significant decrease of the conjugation length in the polymer since the two distribution centers are shifted downwards from 60 to 40 and from 20 to 15 double bonds respectively. This is an important effect if we keep in mind that the starting material was not of very good quality with respect to the

Table 1 - Values of the parameters of the two distributions calculated to fit Raman spectra of $(CHD_y)_x$.

Sample	N_1	N_2	σ_1	σ_2	G
$(CH)_x$-Trans	60	20	30	7	0.55
$(CHD_{0.069})_x$	40	15	20	7	0.5
$(CHD_{0.11})_x$	40	15	20	7	0.4
$(CHD_{0.17})_x$	30	10	15	5	0.4

conjugation length of the segments in the chain. When the deuteration increases up to 17 %, the shortening of the conjugated segments gradually increases to end up with a sample in which the long chain contribution is small, but nevertheless present. These calculations which evidence the predominance of short segments in the 17 % deuterated film, are in agreement with the optical absorption spectra which show peaks shifted to the violet and with the infrared absorption bands recorded after doping with iodine of such samples /9/. In fact, by following the formalism developed in Refs /10,11/, it is possible to explain the peak position of the doped induced infrared modes reported in /9/ by considering a perturbation due to the dopant which turns out to be more important on short segments. Similar experimental results have been obtained by Furukawa et al. /2/ on a sample hydrogenated at a similar level and doped with iodine.

One relevant question is to what extent the breaking of the conjugation is achieved by the introduction of Sp^3 defects. As already mentioned, another group of defects consisting of a carbonyl defect adjacent to an Sp^3 one can be created by hydrolysis of $(CH(AlCl_4)_y)_x$ /5/. The result is a partial hydrogenation of the unsaturated polymer backbone but the treatment can be repeated many times (three times in the study reported in /5/, leading to a concentration of y = 0.009, 0.067 and 0.12 respectively). Raman spectra, although the calculations of the distributions have not been made, are qualitatively similar to those reported here. For λ_L = 457.9 nm, the decrease of the long conjugated segment contribution is observed, as well as a shift of the bands towards higher frequencies. In a parallel study, the EPR line width was measured at room temperature and was found to vary from 0.90 Gauss to 3.4 Gauss going from y = 0.009 to y = 0.12, showing that the mobility of the unpaired spins is lowered in the segments of trans-$(CH)_x$. This corroborates the hypothesis of a shortening of the conjugated segments. It is worth noting that, in recent calculations performed by Kürti and Kuzmany /12/, the degree of breaking of the conjugation on a polyacetylene chain was estimated. It was found that an Sp^3 defect is supposed to break the conjugation by 71 % and in particular, that a carbonyl defect with a CH_2 group adjacent gives exactly the same result. This is in very good agreement with what is reported here.

Finally, it must appear surprising that in $(CHD_{0.17})_x$, long segments are still present even if their concentration is low. If deuterium atoms were equally distributed along the chains, that would give conjugated segments with only three C=C bonds. Therefore, one has to postulate the existence of Sp^3 clusters in order to understand the behaviour of the Raman spectra. Such an hypothesis recently received strong support from energy calculations performed by J.L. Brédas et al. /4/ who showed that aggregates of defects are indeed energetically more favorable.

In conclusion, we have carried out Raman experiments on so-called segmented polyacetylene. We have shown that the bimodal distribution model can give a comprehensive and realistic interpretation of our results. Quantitative numbers are derived for the short and long segment distributions which are both gradually shortened as a function of the deuteration.

Acknowledgments :

We thank Dr. G.P. Brivio for helping with the computer programm and Dr. S.I. Yaniger for assistance in the preparation of the samples.

References

1 K. Soga and M. Nakamaru: J. Chem. Soc., Chem. Commun. 1495 (1983)
2 Y. Furukawa, T. Arakawa, H. Takeuchi, I. Harada and H. Shirakawa: J. Chem. Phys. 81, 2907 (1984)
3 G.P. Brivio and E. Mulazzi: Phys. Rev. B 30, 676 (1984)
4 J.L. Brédas, J.M. Toussaint, J.M. André, A.J. Epstein and A.G. MacDiarmid: Proceedings of this conference
5 A. Pron, E. Faulques and S. Lefrant: Polymer Commun. 28, 27 (1987)
6 E. Faulques and S. Lefrant: J. de Phys. (Paris) 44, C3-337 (1983)
7 R. Tiziani, G.P. Brivio, E. Mulazzi: Phys. Rev. B 31, 4015 (1985)
8 S. Lefrant, E. Faulques, G.P. Brivio and E. Mulazzi: Solid St. Commun. 53, 583 (1985)
9 X.Q. Yang, D.B. Tanner, G. A. Arbuckle, A.G. MacDiarmid and A.J. Epstein: Synth. Metals 17, 277 (1987)
10 G.P. Brivio and E. Mulazzi: Solid St. Commun. 60, 203 (1986)
11 P. Piaggo, G. Dellepiane, E. Mulazzi and R. Tubino: Polymer 28, 563 (1987)
12 J. Kürti and H. Kuzmany: Proceedings of this conference.

Conformation of Conjugated Polymers and Their Relation to Electron Delocalisation

J.P. Aimé, M. Rawiso*, and M. Schott

Groupe de Physique des Solides de l'E.N.S., Université Paris VII, 2, place Jussieu, F-75251 Paris Cedex 05, France

The notion of "conjugation length" has been commonly used in the interpretation of Resonance Raman Scattering spectra from solid polyacetylene [1] or conjugated polymer solutions [2,3]. This "length" ξ is usually of the order of 30 to 200 Å. A simple way of interrupting electronic delocalization is by introducing a strong defect : a chemical impurity like a $-CH_2-$ group into the $CH)_x$ chain, or a physical defect of unspecified nature. Several papers in this volume consider the influence of such defects theoretically and experimentally.

A good approximation to a 1-D disordered system would be a single conjugated macromolecule in solutions. It has been known since 1978 that suitably substituted high molecular weight polydiacetylenes (and not only oligomers) are soluble in several solvents [4]. The absorption spectrum suggests strong electron localization on the chain, which cannot be due to chemical defects since it can be modified reversibly by changes in solvent quality [5]. Polymer scientists have developed powerful theoretical and experimental methods for studying the conformation of macromolecules in solution [6,7]. Specifically, small angle scattering methods allow to study polymer in the spatial range of the "conjugation lengths". Thus, conjugated polymer solutions are good systems for studying the interplay of geometrical conformation and electronic conjugation.

Here we shall report on small angle scattering (SANS) studies of solutions of two polydiacetylenes : poly-3BCMU and poly-4BCMU having side groups with formula $(CH_2)_nOCONHCH_2C_4H_9$ with n = 3 or 4. More detailed discussions will appear elsewhere [8-10]. The relation between ξ and a length characterizing the chain stiffness will be briefly discussed.

Recently, it has been found that solutions of doped conjugated polymers can be prepared [11,12]. Such solutions can in principle be studied by the same SANS method. We recently obtained the first such results, which will be briefly presented here, mainly to show the potentialities of the SANS method.

SANS Study of Polydiacetylene Solutions

In a SANS experiment, a sample is irradiated with a parallel beam of low energy neutrons, approximately monochromatic with wavelength λ in the present case.

* I.L.L., Grenoble - Present address : Laboratoire de Physique des Solides - 91405 Orsay - France

Neutrons are elastically scattered within the sample at an angle θ defining a scattering vector

$$q = \frac{4\pi \sin \theta/2}{\lambda} \quad .$$

Suitably chosen experimental conditions allowed us to study the range

$$5.10^{-3} \leq q \leq 2.10^{-1} \text{ Å}^{-1}$$

corresponding in real space to distances of the order of 20 to 1000 Å. The outcome of such measurements is the structure factor $S(q)$ of an isolated macromolecule, which is the Fourier Transform of $<\rho(r)\rho(o)>$, the monomer density autocorrelation function. At large enough q, $S(q)$ reflects directly intrachain monomer-monomer correlations, that is the chain conformation [13]. $S(q)$ is then usually compared to theoretically computed $S(q)$ from reasonable models.

Two models have been proposed for poly-3 and 4BCMU in solution. In the first one, the chain and its side-groups form platelets, stabilized by H-bonds between adjacent side-groups. The conjugation length of a platelet is its actual length. Successive platelets are therefore separated by physical defects where inter-side-group H-bonds are broken, plate planes are tilted and misoriented [5,14]. In the second one, the conjugated chain is approximated by a continuously curved line [15]. This is the Porod-Kratky, or "worm-like-chain" model [16].

As we shall see below, the experimental SANS results are in complete agreement with the Porod-Kratky model. Neglecting first the fact that side-groups give to the chain a finite lateral extension, $S(q)$ measured in absolute values depends on only two parameters, the mass per unit length M_L and the persistence length b, which is a measure of the decay of orientational correlations along the chain. Specifically, if s_i is the curvilinear coordinate along the chain and $r(s_i)$ the direction of the tangent at s_i, then

$$<r(s_1)\, r(s_2)> = \exp - \frac{2(s_1 - s_2)}{b}$$

defines b. Thus, the larger b is, the more rigid the chain. It can be shown that, in our experimental conditions, the total length of the chain L (the "contour length"), is irrelevant provided $L/b > 10$. In this model, the asymptotic behavior of the structure factor for $L,q \to \infty$, is given by

$$\lim \frac{L}{b} P(q) = \frac{\pi}{qb} + \frac{4}{3} (\frac{1}{qb})^2 \quad ,$$

where $P(q) = S(q)/S(o)$ is normalized [8]. That is, the chain behaves like a stiff rod at small enough scale.

As an example of a polymer in good solvent, poly 3-BCMU in deuterated DMF was studied at 295 K. The results are shown in Fig. 1, compared to the predictions of the Porod-Kratky model. The agreement is excellent up to q

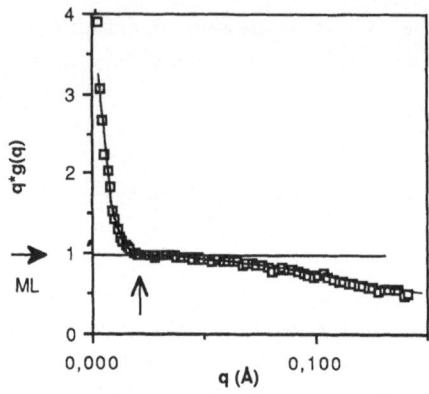

Figure 1 : Normalized structure factor, plotted as $qg(q) = q(M_W/m)[S(q)/S(o)]$ versus q, where m is the monomer mass. Absolute measurements are reported. Squares are experiment, line is the fit described in text.

~0.04 Å^{-1} but at larger $qP(q)$ decays faster than q^{-1} ($qP(q)$ is not constant). This is due to the finite lateral extension of the side groups. Taking that into account by approximating the polymer by a thin semi-rigid ribbon, the asymptotic behavior becomes $2\pi/Sq^2$, depending on the mass per unit surface M_S, a third parameter. Fig. 1 shows that a perfect fit is obtained with b = 310 Å, which means that the polymer is fairly rigid ; M_L = 94 g/Å mole, which shows that the polymer is in the trans configuration of the double bonds ; and M_S = 3.5 g/Å^2 mole, corresponding to an effective ribbon width of 27Å, showing that the side groups are not completely extended [8].

This shows that a fairly detailed description of a conjugated chain in solution can be obtained by SANS. However, one might think that a continuously curved ribbon might look very much like a succession of rigid platelets, provided they are not too long and that their relative misorientation is not too large. In fact, assuming the chain made of identical rigid elements of length a, b is related to the angle θ between successive elements by

$$b = \frac{a}{1 - <\cos\theta>} \quad ,$$

which means that, if the rigid unit is a monomer a = 5 Å and θ ~ 10-11 degrees, whereas if a = 30 Å, θ ~ 25 degrees. We have shown in [8] that the data are consistent with the worm-like chain model, but not with the values expected from a chain made of rods, either of constant length, or with a statistical distribution of lengths. This would not seem compatible with the first model described above : the rigid unit of the chain might be the monomer itself. Indeed, assuming a perfect polydiacetylene chain and taking into account only the vibrations and rotations of bond with known force constants, Allegra et al. [17] calculate b ~ 300 Å. The agreement is so good it must be at least in part accidental.

In conclusion, "strong" physical defects are too spare to really influence the polydiacetylene chain conformation and rigidity. One may therefore doubt that they are the dominant factor in electron localization. If localization is affected by the same numerous "weak" defects which determine b, there must be a relation between b and $<\xi>$. A simple measure of $<\xi>$ is the wavenumber λ_{max} of maximum absorption in the solution, since the broad, structureless, absorption band in the blue is in fact inhomogeneously broadened [2].

We have now several measurements of b and the optical absorption of 3- and 4-BCMU samples in good solvents. Fig. 2 shows clearly that b and λ_{max} are indeed correlated, a smaller rigidity corresponding to greater localization. System a) is poly-3BCMU in $CDBr_3$ at 30°C, where incipient H-bonding increases chain rigidity [9] but not the point of incipient aggregation [8]. Systems b) and c) correspond to the same polymer in DMF at different temperatures. Systems d) and e) correspond to high and low M_W which dependence is not yet well understood. This relation between b and λ_{max} is strongly suggestive. One should not neglect however the influence of other factors, like different van der Waals interactions with different solvents. The conclusion at present is only qualitative. Further work is in progress on this subject.

Generalizing these conclusions to all neutral conjugated polymers in solution is certainly bold. They suggest, however, that polyacetylene is not *per se* a very rigid molecule, and that chain extension in the solid state is rather the consequence of interchain interactions than a single chain property. In other conditions, $CH)_x$ might assume other conformations. It also suggests that localization on the "long chain" segments invoked in RRS [1] might in part be due to the effects discussed above.

Solutions of Polythiophenes in the Doped and Undoped States [18]

We are presently investigating by SANS the systems described by Elsenbaumer et al. in this volume : poly-nButyl and nOctyl Thiophene in deuterated Nitrobenzene neat or doped with $NOSbF_6$ or $NOPF_6$. Only preliminary data are

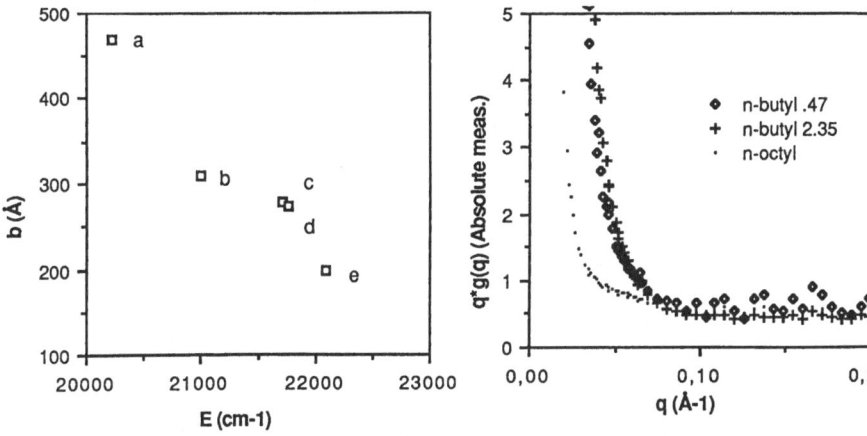

Figure 2: Relation between b and maximum optical absorption. Significance of points a) to e) is described in text.

Figure 3: Normalized structure factor of three undoped polythiophene solutions plotted as qg(q) versus q.

61

available, and will be presented to show further the usefulness of SANS. Fig. 3 shows data obtained on three solutions of the two undoped polymers, showing:

- that at least above $q > 0.06$ Å^{-1} they behave as rigid molecules, suggesting that they too are worm-like chains with $b > 80$ Å. Only a lower limit can be given, since the small-q range is complicated by strong intermolecular interactions ;

- that M_L has the value expected from the molecular formula, showing that in this q range single chain behavior is indeed observed ;

- that the side groups are far from fully extended, since no indication of a marked decrease of qg(q) is found up to $q \sim 0.2$ Å^{-1}.

Upon doping, the scattering behavior is strongly modified : a constant value of qg(q) is observed from 0.01 to 0.2 Å^{-1}, M_L being unchanged. The system is now very different, being a polyelectrolyte solution. However, we can conclude that the chains are now highly rigid and that interaction between chains is greatly reduced. Further work is in progress, particularly as a function of doping level.

Conclusion

The aim of this note was to advocate the usefulness of SANS in the study of conjugated polymers. The informations available have been described on two examples, and the role of geometrical disorder on electron localization ("conjugation length") is illustrated.

Acknowledgements

The SANS experiments were made possible by several beam time allocations at ILL. We are grateful to our coworkers mentioned in the references for their collaboration, and to ILL staff for technical assistance.

References

[1] There is a very abundant literature on that subject. Several papers can be found in the proceedings of this and the previous IWEPP "Electronic Properties of Polymers and Related Compounds", H. Kuzmany, M. Mehring and S. Roth eds., Springer Series in Solid State Sciences 63 (1985) as well as those of the conferences at Abano Terme, Les Arcs, etc... Mol. Cryst. Liq. Cryst. 117 (1985) J. Physique Coll. 44 C3 (1983).
[2] M.L. Shand, R.R. Chance, M. Le Postollec, M. Schott: Phys. Rev. B25, 4431 (1982)
[3] J. Berréhar, C. Lapersonne-Meyer, M. Schott, M. Le Postollec: Chem. Phys. 77, 11-19 (1983)
[4] G.N. Patel: J. Polym. Sci. Polym. Lett. ed. 16, 607 (1978)
[5] G.N. Patel, R.R. Chance and J.D. Witt: J. Chem. Phys. 70, 4387 (1979)
[6] P.J. Flory: Statistical Mechanics of Chain Molecules (J. Wiley and Sons, New-York, 1969)
[7] P.G. de Gennes: Scaling Concepts in Polymer Physics (Cornell U. Press, New-York, 1979)
[8] M. Rawiso, J.P. Aimé, M. Schott, M.A. Müller, M. Schmidt and G. Wegner: to be published

[9] J.P. Aimé, F. Bargain, J.L. Fave and M. Schott: to be published

[10] J.P. Aimé, F. Bargain, J.L. Fave, M. Rawiso and M. Schott: in preparation

[11] J. Frommer: Accts. Chem. Res. 19, 2 (1986)

[12] R.L. Elsenbaumer: this volume

[13] A. Guinier and R. Fournet: Small angle scattering of X-rays (J. Wiley and Sons, New-York, 1955)

[14] R.R. Chance: Macromolecules 13, 396 (1980)

[15] G. Wenz, M.A. Müller, M. Schmidt and G. Wegner: Macromolecules 17, 837 (1984)

[16] G. Porod: Monatshift Chemie 80, 251 (1949)
     O. Kratky, G. Porod: Rec. Trav. Chim. Pays-Bas 68, 1166 (1949)

[17] G. Allegra, S. Brückner, M. Schmidt and G. Wegner: Macromolecules 19, 399 (1986)

[18] J.P. Aimé, M. Schott, R.L. Elsenbaumer, H. Eckhardt and G.G. Miller: in preparation.

Polyacetylene Segments in a Polyvinylidene-Chloride Matrix: Anisotropic Optical Properties

*G. Leising, B. Ankele, H. Kahlert, and P. Knoll**

Institut für Festkörperphysik, Technische Universität Graz,
Petersgasse 16, A-8010 Graz, Austria

Thermal degradation of polyvinylidene-chloride (PVDC) under applied mechanical stress leads to the formation of oriented conjugated sequences embedded in the PVDC matrix. Polarized infrared spectra show absorptions related to C-H- vibrations with a polarization dependence typical for polyacetylene. Transmission spectra in the visible range show highly anisotropic properties even at low concentration of conjugated bonds. Distinct absorptions in the perpendicular polarization can be attributed to different conjugation lengths. Long-time storage in air results in a polarized absorption peak in the band gap region, which can be completely removed by long-time compensation with NH_3.

Introduction

Polyacetylene is the simplest linear conjugated system with a degenerate ground state, which is the basis of the theoretical model for topological solitons or moving domain walls /1/. Three-dimensional correlations or interchain coupling are not included in this model, which deals only with one single chain, but are expected to play an important role in the real (bulk) polyacetylene /2/. To study the influence of the interchain contributions it would be interesting to produce polyacetylene chains in a nonconjugated matrix. In this case interchain contributions are expected to be smaller or even negligible and the doping-induced gap states could be compared to predictions of the soliton theory /3/.

Results and Discussion

As we have already shown /4/, the thermal degradation of polyvinylidene-chloride (PVDC) and copolymers thereof under applied mechanical stress leads to the formation of oriented conjugated sequences embedded in the PVDC matrix. Besides the typical bands of the PVDC, polarized infrared spectra show also absorptions related to C-H vibrations with a polarization behaviour typical for trans-polyacetylene /5/. The position of the C-H deformation vibration with the transition moment perpendicular to the chain axis is slightly dependent on the length of the conjugated chain. However, for chains with more than about 10 double bonds the peak position already reaches

*Institut für Experimentalphysik, Karl-Franzens-Universität Graz
Universitätsplatz 5, A-8010 Graz, Austria

its asymptotic value of about 1015 cm^{-1} /6/. The peak position in our case is about 1015 cm^{-1}; so it is impossible to give even an estimate of the conjugation length (n) based on the infrared results. However, the reality in polyacetylene can not be described by a distinct conjugation length. The shape of the C-H deformation band shows a characteristic tailing to lower wavenumbers, which indicates that we have to deal with a distribution of the conjugation lengths. In our case the maximum of this conjugation length distribution is expected for n>10. Since the molar absorption coefficient for the 1015 cm^{-1}-band is known for polyacetylene /5/, the content of polyacetylene in the PVDC matrix can be estimated to be about 0.4 % . The unoriented original PVDC-film is transparent in the visible and near infrared region. The transmission spectra of a PVDC-film, stretched by a factor of 2.5 during the conversion at 190° C for 25 hours are shown in Fig. 1 together with the transmission spectrum of the original PVDC-film.

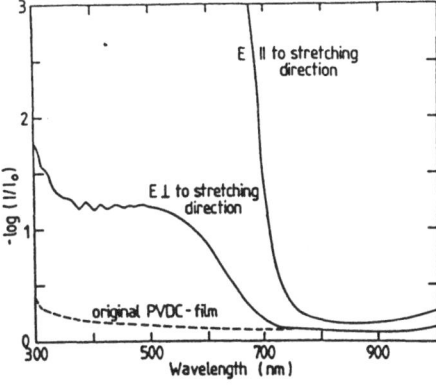

Fig.1 Optical absorption for a PVDC-film before (dashed curve) and after stretching at 190°C (full curves obtained with polarized light)

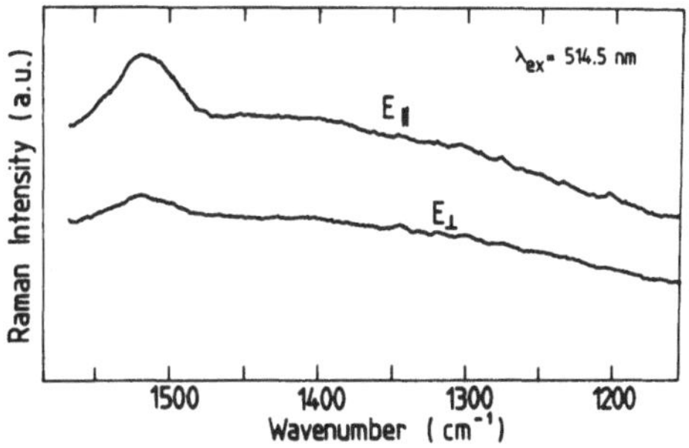

Fig.2 Polarized Raman spectrum of the stretched PVDC-film

For light polarized parallel to the stretching direction (polyacetylene chain axis), the strong interband transition of polyacetylene dominates the spectrum with an onset of the absorption at about 1.6 eV, which is correlated to the band gap. So we can conclude that the band gap of polyacetylene sequences in PVDC is somewhat higher than in the case of pure polyacetylene (1.4 eV), assuming a lower value for the conjugation length in this case. Since the thickness of the PVDC-film is about 50 μm and the absorption coefficient of trans-polyacetylene is about 8.5 10^5cm^{-1}at 2 eV for the parallel polarization /7/, the absolute height of the interband absorption in this material could not be determined. For light polarized perpendicular to the stretching direction, the absorption edge is shifted to higher energies and exhibits oscillatory behaviour, which results from distinct lengths of conjugated sequences in the PVDC matrix /4/ . The existence of oriented conjugated trans-sequences is also demonstrated by means of polarized Raman spectroscopy. In Fig.2 we present Raman spectra of the oriented conjugated sequences in PVDC for an excitation wavelength of 514.5 nm for both polarization directions in the region of the C=C double-bond stretching mode. The C=C mode has its maximum at 1520 cm^{-1} with a pronounced anisotropy. This wavenumber is higher than that observed in trans-polyacetylene for the same excitation wavelength, which can be explained by the fact that at 514.5 nm we are in resonance mainly with conjugated chains of about 11 to 12 double bonds /4,8/.

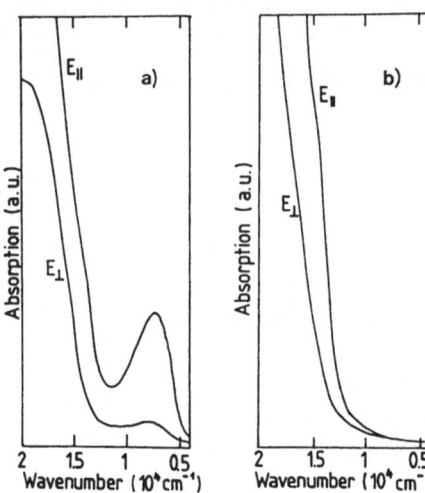

Fig.3 Polarized optical absorption spectra for the stretched PVDC-film: a) sample after 700 days in air b) sample from a) after 100 days NH$_3$ compensation

Long-time storage (700 days) of a stretched and thermally degraded PVDC sample at ambient conditions in air results in some oxidation of the material, which is clearly demonstrated by the appearence of oxidation-induced absorption in the energy region below the interband gap. In Fig.3a) we present the polarized optical absorption spectra for such a sample. At about 0.95 eV a marked peak is seen, which is polarized predominantly parallel to the polyene chain direction. If one applies the picture of soliton doping /3/ to this system, the above described behaviour would be expected. The doping process proceeds through the formation of a positively charged soliton upon doping (oxidation) and the band observed at 0.95 eV is then associated with transitions from the valence band to these soliton levels, which lay inside the gap, with the transition moment mainly in chain direction. According to this model it should be possible to compensate the charged soliton levels by exposing the oxidized polymer to NH$_3$ vapour. In Fig.3b) we show the polarized optical absorption spectra of the stretched PVDC sample from Fig.3a) after 100 days of compensation with NH$_3$ under a vapour pressure of 1000 mbar. The doping-induced band is completely removed from the spectra for both polarization directions. The possibility of this cycle shows that there is a reasonable diffusion coefficient for gas (oxygen) in PVDC, which was not expected from short time storage experiments /4/.

Acknowledgement

This work was supported by the Austrian Science Research Fund under project Nr. 6198.

References

1. W.P. Su, J.R. Schrieffer and A.J. Heeger:
   Phys.Rev.Lett. 42, 1698 (1979); Phys.Rev. B22, 2099 (1980)
2. D. Baeriswyl and K. Maki: Phys.Rev. B28, 2068 (1983)
3. N. Suzuki, M. Ozaki, S. Etemad, A.J. Heeger and
   A.G. MacDiarmid: Phys.Rev.Lett. 45, 1209 (1980)
4. B. Ankele, G. Leising and H. Kahlert:
   Solid State Commun. 62, 245 (1987)
5. R. Uitz, G. Temmel, G. Leising and H. Kahlert:
   Z.Phys.B, Condensed Matter (in print)
6. H. Shirakawa and S, Ikeda: Polymer Journal 2, 231 (1971)
7. J. Fink and G. Leising: Phys.Rev. B34, 5320 (1986)
8. H.J. Bowley, D.L. Gerrard and W.F. Maddams:
   Makromol. Chem. 188, 899 (1987)

Part III

Electron Energy Loss,
Optical and Raman Spectroscopy

Electronic Structure of Conducting Polymers by Electron Energy-Loss Spectroscopy

J. Fink, N. Nücker, B. Scheerer, W. Czerwinski, A. Litzelmann, and A. vom Felde

Kernforschungszentrum Karlsruhe, Institut für Nukleare Festkörperphysik, Postfach 3640, D-7500 Karlsruhe, Fed. Rep. of Germany

1. INTRODUCTION

The evolution of the band structure of conducting polymers upon doping is the basis for the understanding of the mechanism leading to the high conductivities observed in these materials. Experimentally, many investigations around the energy range of the fundamental gap have been performed by optical spectroscopy. In addition, the density of occupied states of many conducting polymers has been studied by photoelectron spectroscopy. Another powerful method for investigations of the electronic structure of conducting polymers is electron energy-loss spectroscopy (EELS). It covers a large energy range from 0.1 to ~2000 eV. Thus, it is possible to study not only excitations of π and σ electrons but also of core electrons similar to X-ray absorption spectroscopy (XANES, EXAFS) using synchrotron radiation. Moreover, EELS allows one to perform excitations with varying momentum transfer, thus giving information on the dispersion of bands which cannot be obtained by optical spectroscopy.

2. THE METHOD AND THE SPECTROMETER

In Fig. 1 we describe the principle of EELS in transmission. High energy electrons (E = 170 keV) are transferred through thin samples (d ~ 1000 Å). The energy loss of the electrons in the sample gives information on the frequency of possible excitations in the solid ($\Delta E = \hbar\omega$). The momentum transfer q in the scattering process is related to the scattering angle Θ. Thus, measurements as a function of Θ yield information on the wavelength ($q = 2\pi/\lambda$)

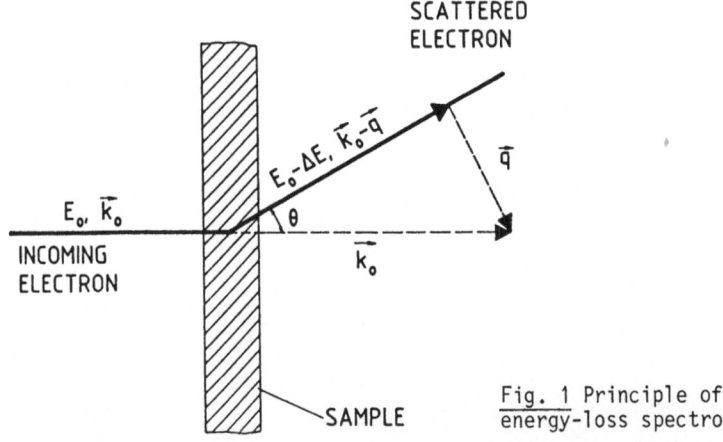

Fig. 1 Principle of electron energy-loss spectroscopy

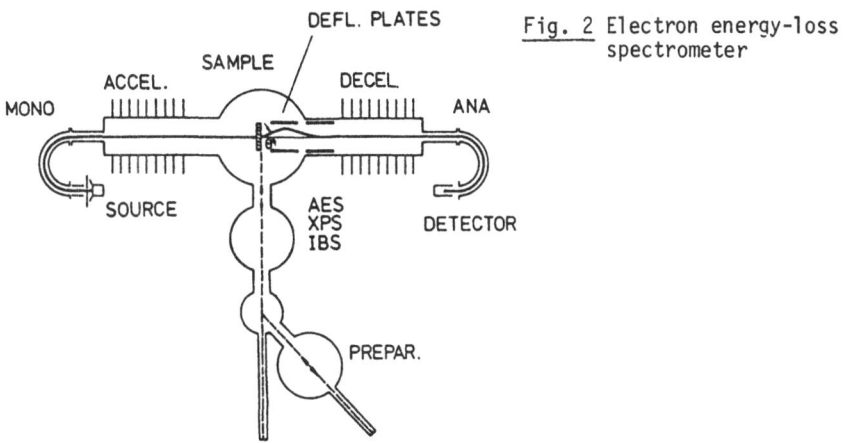

Fig. 2 Electron energy-loss
spectrometer

of excitations in the solid. In Fig. 2, we show a sketch of the spectrometer.
Electrons from a heated tungsten dispenser cathode are monochromatized in
a spherical electrostatic deflection monochromator and then accelerated to
170 keV. After the transmission through the sample the electrons are de-
celerated again and the energy-loss is measured in the analyzer. The energy
resolution can be varied between 0.08 and 0.7 eV, corresponding to currents
of 1 to 200 nA transmitted through the spectrometer. The momentum transfer
is varied by voltages on the deflection plates. Samples can be prepared or
doped in a preparation chamber attached to the spectrometer. In addition,
there is a further characterization chamber, where Auger spectroscopy (AES),
X-ray induced photoelectron spectroscopy (XPS) and ion backscattering spectroscopy (IBS) can be performed.

3. VALENCE BAND EXCITATIONS

In EELS, we measure the loss function $Im(-1/\varepsilon(q,\omega))$ where $\varepsilon(q,\omega)$ is the complex dielectric function. In Fig. 3 we show loss functions for various un-
doped conjugated polymers for $q = 0.1$ Å^{-1} which is small with respect to the
extension of the Brillouin zone (BZ). For polyacetylene (PA), we have measured the loss functions parallel and perpendicular to the chain axis on
highly oriented films [1]. In both cases, the spectra are dominated by a
broad maximum near 22 eV which is a plasmon related to all valence electrons
$(\pi+\sigma)$. At 4.9 eV a further plasmon is observed for $\vec{q}\|\vec{c}$ which is related
only to the π electrons. Knowing the refractive index at low energy from
optical spectroscopy, the absolute value of the loss function can be obtained and by a Kramers-Kronig analysis, the real and the imaginary part
of the dielectric function ε_1 and ε_2 can be calculated. For highly oriented
PA and $\vec{q}\|\vec{c}$ a strong oscillator at 1.9 eV is realized in ε_2. The same
oscillator causes a zero crossing of ε_1 at 4.9 eV. Therefore, the loss
function $Im(-1/\varepsilon) = \varepsilon_2/(\varepsilon_1^2+\varepsilon_2^2)$ shows a maximum at the energy where $\varepsilon_1=0$ and
ε_2 is small. As this maximum is related to the π oscillator (a transition
from the π band to the π^* band at the edge of the BZ, well known from opti-
cal spectroscopy) we call it a π plasmon. For $\vec{q}|\vec{c}$, the π oscillator is
strongly reduced and an anisotropy of 170 is observed, indicating a much
lower polarizability of the π electrons perpendicular to the chain axis.
This shows up also in the loss function, where for $\vec{q}|\vec{c}$ almost no π plasmon
is observed. Above ∿8 eV various $\sigma\to\sigma^*$ transitions and $\pi\to\sigma^*$ transitions are
observed which are discussed in detail in Ref. 1.

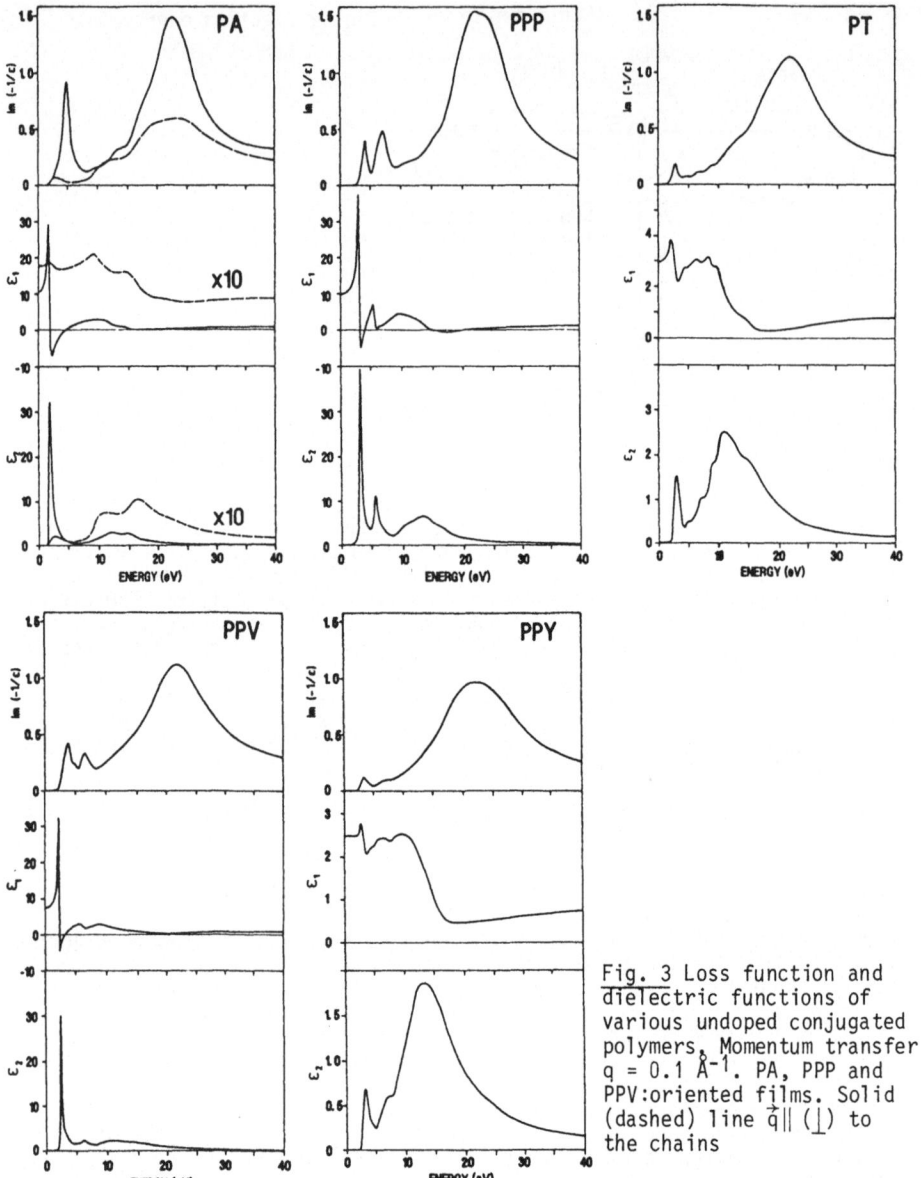

Fig. 3 Loss function and dielectric functions of various undoped conjugated polymers. Momentum transfer $q = 0.1$ Å^{-1}. PA, PPP and PPV:oriented films. Solid (dashed) line $\vec{q}\parallel$ ($\perp$) to the chains

Increasing the momentum transfer $\vec{q}\parallel\vec{c}$, the energy of the π plasmon in PA increases linear in q, i.e., the π plasmon shows a positive linear dispersion (see Fig. 4). The origin of this is shown in Fig. 5c. In PA, the valence (π) band as well as the conduction (π^*) band has a strong dispersion. For q=0 we perform, like in optical spectroscopy, a vertical transition across the fundamental gap. For q>0, we switch to non-vertical transitions. A high joint density of states due to parallel bands is achieved at higher

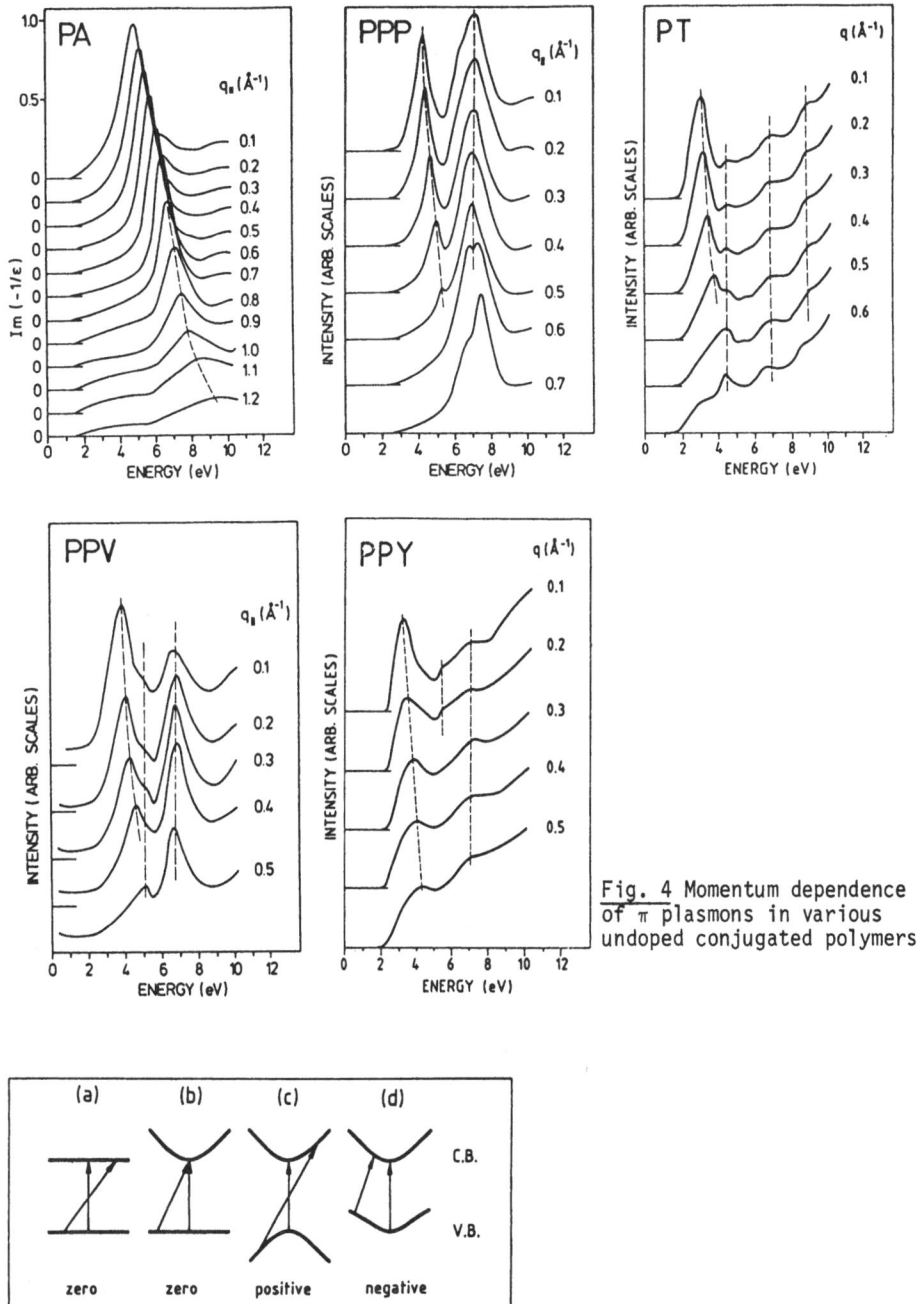

Fig. 4 Momentum dependence of π plasmons in various undoped conjugated polymers

Fig.5 Nonvertical transitions for different bands giving zero, positive or negative dispersion

transition energies. Thus with increasing q, the oscillator is shifted to higher energy and therefore also the energy of the π plasmon increases. In Fig. 6 we show for $\vec{q} \parallel \vec{c}$ the plasmon dispersion and the dispersion of the $\pi{\to}\pi^*$ oscillator (maximum of $\omega \cdot \varepsilon_2$). Due to the special form of the π bands, a splitting of the $\pi{\to}\pi^*$ transition is observed at higher momentum transfer. For comparison, we show the theoretical calculations of the plasmon dispersion and the dispersion of the maximum of the $\pi{\to}\pi^*$ transition, on the basis of the Su-Schrieffer-Heeger model including local field contributions [2]. The comparison between theory and experiment yields a gap energy E_g = 1.8 eV and a total width of the π-electron system W = 11 eV. The latter value is slightly smaller than the value W = 12.8 eV deduced without taking into account local field corrections [1]. On the other hand, it is in good agreement with theoretical results from band structure calculations W = 10-12 eV [3].

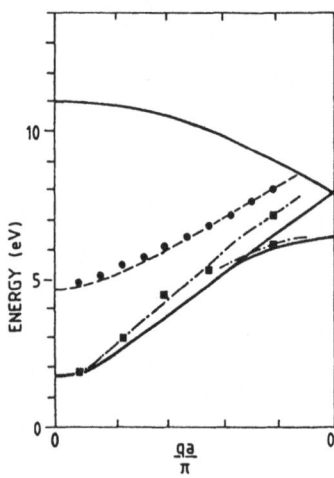

Fig. 6 Momentum dependence of the π plasmon (•) and the maximum of the $\pi{\to}\pi^*$ interband transitions (■) in PA. Dashed and chain-like curves: Calculations given in Ref. 2. Solid lines: Possible range of $\pi{\to}\pi^*$ transitions

This example illustrates the possibility of yielding information on the width of π bands by momentum-dependent measurements. The measurements also indicate that the dispersion of the $\pi{\to}\pi^*$ transition is not exactly the same as that of the π plasmon. However, in the other conjugated polymers, where the strength of the individual $\pi{\to}\pi^*$ oscillator is smaller compared to that in PA, we think that the slope of the plasmon is close to that of the interband transitions. Therefore, an estimate of the width of the π bands can be obtained directly from the dispersion of the π plasmon in the loss function.

The results on partially oriented polyparaphenylene (PPP) for $\vec{q} \parallel \vec{c}$ are shown in Figs. 3 and 4. Due to the non-perfect orientation of the films, no reliable data for $\vec{q} \perp \vec{c}$ could be obtained as such measurements suffer from contributions from the loss function for $\vec{q} \parallel \vec{c}$. Two π plasmons are realized at ∿4 eV and ∿7 eV. The lower π plasmon shows a strong linear dispersion for 0.07 Å^{-1} ≤ q ≤ 0.3 Å^{-1} with a slope of 2.9 eVÅ. From this, a width W = 2.2 eV of the highest π band or the lowest π^* band can be derived. For comparison, we give the width W = 3.4 eV derived from band structure calculations [4]. As usually the band width calculated by valence effective Hamiltonian techniques has to be reduced by a factor of 1.15-1.3, the experimental results are close to the theoretical calculations. The π plasmon at ∿7 eV shows no dispersion, typical of a transition where at least the π band or the π^* band or both have no dispersion. According to Ref. 4 this transition

is related to π electrons which are localized in the benzene rings forming
a narrow π and a narrow π* band as shown in Fig. 5a. For polyphenylenevinyl-
lene (PPV), the situation is very similar to PPP. There is, on the other
hand, a shoulder near 5 eV which probably can be assigned to a mixed transition between a wide π band and a flat π* band as shown in Fig. 5b. Such a
transition should show no dispersion, in agreement with the experimental
results.

In Figs. 3 and 4 we also show results for undoped polypyrrole (PPY) and
undoped polythiophene (PT). In both cases, the lowest π plasmon shows a
strong dispersion, indicating that the transition across the fundamental
gap is related to a wide π and to a wide π* band. The π plasmons at higher
energies show almost no dispersion, indicating that a least the π or the
π* band or both should be narrow.

Before leaving the undoped conjugated polymers, we mention another important result on the electronic structure of PA which was derived by an
evaluation of the sum rule on the optical conductivity for $\vec{q} \| \vec{c}$ which gives
information on the optical effective mass m*/m. From experimental data we
obtain m*/m = 1.7 while a calculation on the basis of the Su-Schrieffer-
Heeger (SSH) Hamiltonian including local field corrections [2] yields m*/m =
1.44. The difference between these values may be explained by electron-
electron correlations, not taken into account in the SSH Hamiltonian.
According to calculations of the sum rule in the framework of the one-dimensional Hubbard model [5] and taking into account the small influence of the
dimerization [6] our experimental value can be explained by a high on-site
correlation energy U ∿ 7 eV or U/W = 0.6. These values are again slightly
lower than previous values derived without taking into account local field
contributions [1]. On the other hand, they are close to values derived
from other experiments [7,8]. These high values for U and U/W have important consequences on the interpretation of the fundamental absorption
edge in trans-PA.

The changes of the band structure of trans-PA upon doping are described
elsewhere [9]. In this contribution we illustrate such changes on PPP doped
with Cs. The PPP films were fully doped by Cs vapour at room temperature in
the preparation chamber of the spectrometer shown in Fig. 2. Subsequently,
the films were moved to the EELS spectrometer and heated to ∿300°C. At that
temperature, Cs is evaporated out of the film and an undoping during several
hours could be achieved. During that time, the spectra shown in Fig. 7 were
measured. At low doping concentrations (lower curves), two transitions appear
in the gap which can be assigned to transitions from occupied bipolaron
levels to the empty π* band [4]. At higher Cs concentrations, the bipolaron
transitions increase at the expense of that due to the lowest π→π* transition at ∿4 eV. At ∿25% doping concentration per monomer, the π→π* transition
has disappeared and a shoulder at ∿3 eV at the upper side of the bipolaron
transition appears. At the highest Cs concentrations (upper curves) this
shoulder has disappeared and there remain two transitions, at 2-3 eV and the
π→π* transition at ∿ 7 eV due to the localized π electrons in the benzene
rings. The fact that the latter transition is not changed upon doping, indicates that the polymer chain is not destroyed upon doping. The transition at
∿2.5 eV may be explained by a π plasmon related to a π band which is closed
upon doping i.e. a free-carrier plasmon. However, the momentum dependence of
this transition, as shown in Fig. 8 indicates a much more complicated band
structure of the fully doped PPP. There appear two components at higher momentum transfer. One component shows almost no dispersion, indicating a remaining narrow bipolaron band also in the fully doped PPP. The second component shows a strong dispersion suggesting an assignment to a π→π* transition.
Probably the lowest π* band is already partially filled, a fact which allows
π→π* transitions only at higher q. In summary, the present results of the
fully doped PPP can be explained by a band structure with overlapping bi-

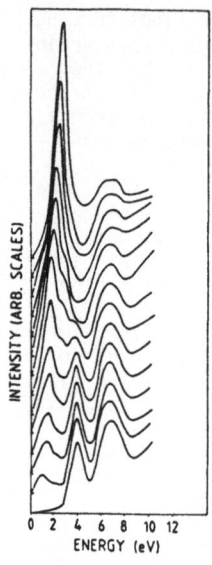

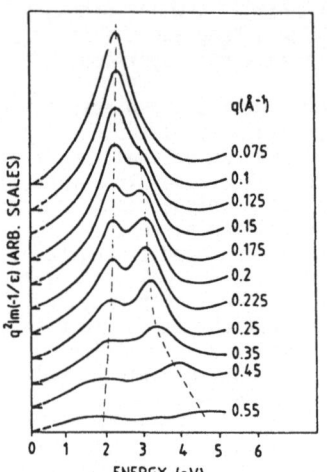

Fig. 7 Loss function of
PPP as a function of Cs
doping. Upper (lower)
curves high (low) doping
concentrations

Fig. 8 Momentum depen-
dence of the loss
function of fully Cs
doped PPP

polaron bands and π bands having a reduced gap ($\Delta E \sim 1$ eV). Details of these investigations at intermediate concentrations will be reported elsewhere [10]. Similar data on PPP doped with Li [11] and Na [12] have already been published.

4. CORE LEVEL EXCITATIONS

Core level excitations can give important information on the chemistry of carbon and nitrogen atoms in conducting polymers. This is illustrated in Fig. 9 where we show excitations of electrons from the K-shell (1s level) of carbon into unoccupied states above the Fermi level for PA and for polymethineimine (PMI). In principle, they should reflect the density of unoccupied π^* and σ^* states. However, due to the strong interaction with the core hole, a resonance-like excitation into the lower part of the density of states of π^* and σ^* states appear [13]. For PA, those maxima appear at 284 and at 292 eV, respectively. A further maximum at 287.5 eV can be probably assigned to a satellite, where in addition to the core elec-tron also a $\pi \rightarrow \pi^*$ transition is excited. In PMI, the 1s$\rightarrow \pi^*$ transition appears at 286.5 eV. The higher energy, compared to PA, can be explained by a chemical shift due to the two more electronegative N atoms bonded to C. Thus, in some cases, information on the charge of the C atoms can be obtained like in photoelectron spectroscopy. In Fig. 10, we show the C 1s edge of oxidized polyaniline (PAN). In this case, two 1s$\rightarrow \pi^*$ transitions appear at 285 and at 286.8 eV, corresponding to carbon atoms bonded only to C and H atoms and to those carbon atoms bonded to one N atom. The chemical shift of the two C atoms is only about one half of the shift ob-

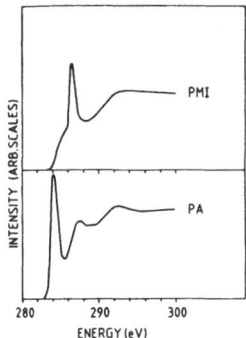

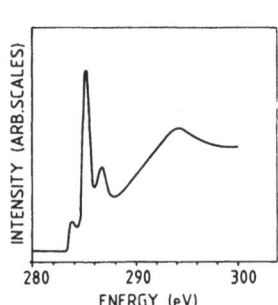

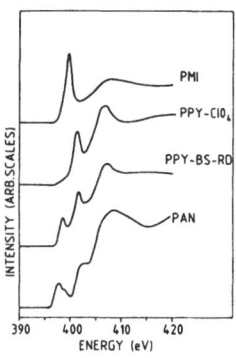

Fig. 9 Carbon K-edges of PA and PMI

Fig. 10 Carbon K-edge of PAN

Fig. 11 Nitrogen K-edges of various conducting polymers

served between PA and PMI. A shoulder below the 1s→π* transition appears at 284 eV for PAN as in various other oxidized polymers [14,15]. It is due, either to a transition into defect states in the gap or due to a negative shift which is caused by the potential of the counter ions close to the C atoms.

Finally, we show in Fig. 11 some N 1s edges of conducting polymers. Normally, the 1s→π* transition appears at ∿401 eV and the 1s→π* transition appears at ∿406 eV. In PMI, the N atoms are not bonded to an H atom. Therefore, they have probably more charge and the 1s→π* transition appears at 399 eV. Deprotonated N atoms are also observed in PPY samples which have suffered strong radiation damage (PPY-BS-RD)[14]. The spectra taken on oxidized PAN also indicate a considerable amount of deprotonized N atoms. Besides these examples which illustrate the influence of the chemical bonding on the core edge spectra, similar spectra on counterions give information on the charge on the counterions and thus on the charge transferred to the polymer [11,12,14].

ACKNOWLEDGEMENT

This work has been performed in collaboration with J. Heinze, H. Kuzmany, G. Leising, H. Lindenberger, H. Neugebauer, S. Roth, M. Stamm, B. Tieke, W. Wernet, G. Wegner, R. Weizenhöfer and D. Wöhrle.

REFERENCES

1. J. Fink and G. Leising, Phys. Rev. B 34, 5320 (1986)
2. C.-S. Neumann and R. von Baltz, Phys. Rev. B to be published
3. P.M. Grant and I.P. Batra, Synth. Met. 1, 193 (1979/80);
    Solid State Commun. 29, 225 (1979)
4. J.C. Brédas, B. Thémans, J.G. Fripiat, J.M. André and R.R. Chance,
    Phys. Rev. B 29, 6761 (1984)
5. D. Baeriswyl, J. Carmelo, and A. Luther, Phys. Rev. B 33, 7247 (1986)
6. D. Baeriswyl, personal communication
7. D. Baeriswyl, see contribution in this volume
8. A. Grupp, P. Höfer, H. Käss, and M. Mehring, see contribution in
    this volume
9. J. Fink, N. Nücker, B. Scheerer, A. vom Felde and G. Leising, see
    contribution in this volume

10. J. Fink et al., to be published
11. J. Fink, B. Scheerer, M. Stamm, B. Tieke, B. Kanellakopulos and E. Dornberger, Phys. Rev. B $\underline{30}$, 4867 (1984)
12. H. Fark, J. Fink, B. Scheerer, M. Stamm, and B. Tieke, Synth. Met. $\underline{17}$, 583 (1987)
13. E.J. Mele and J.J. Ritsko, Phys. Rev. Lett. $\underline{43}$, 68 (1979)
14. J. Fink, B. Scheerer, W. Wernet, M. Monkenbusch, G. Wegner, H.J. Freund, and H. Gonska, Phys. Rev. B $\underline{34}$, 1101 (1986)
15. J. Fink, N. Nücker, B. Scheerer, and H. Neugebauer, Synth. Met. $\underline{18}$, 163 (1987)

Electronic Structure
of Undoped and Doped Polyphenylenevinylene

J. Fink[1], *N. Nücker*[1], *B. Scheerer*[1], *A. vom Felde*[1], *H. Lindenberger*[2], *and S. Roth*[2]

[1]Kernforschungszentrum Karlsruhe GmbH,
Institut für Nukleare Festkörperphysik,
Postfach 3640, D-7500 Karlsruhe, Fed. Rep. of Germany
[2]Max-Planck-Institut für Festkörperforschung,
Heisenbergstr. 1, D-7000 Stuttgart 80, Fed. Rep. of Germany

1. Introduction

Poly(phenylenevinylene) (PPV) has a molecular structure between that of
polyacetylene and polyphenylene. It is at present the only non-degenerate
ground state polymer which can be prepared in a highly oriented form.
Therefore, it is of particular interest for studies of the changes of the
electronic structure upon doping. According to theoretical models [1] the
non-degenerate ground state implies the formation of polarons and of bipolarons at lower doping concentrations. At higher concentrations the transition to the metallic state should be observable. In this contribution, we
have studied the electronic structure of undoped and n-type doped PPV by
electron energy-loss spectroscopy (EELS).

2. Experimental

The precursor polymer poly(p-xylene-x-dimethylsulphoniumchloride) was prepared by the method of Wessling and Zimmerman [2] as detailed by Karasz et
al. [3]. Non-oriented PPV films with a thickness of about 2000 Å were obtained by casting the precursor polymer solution and subsequent heat treatment at 300°C. Highly oriented PPV films with a thickness of about 4000 Å
were prepared by stretching the precursor film at temperatures in the range
60-150°C. Stretch ratios of about 5 could be achieved for these thin films.
n-type doping of the PPV films was obtained by treating the polymers in
alkali metal vapour at room temperature or by evaporating a certain amount
of alkali metal onto the film and subsequent annealing. Electron diffraction and EELS was performed with a 170 keV spectrometer [4] with an energy
and momentum transfer resolution of 0.17 eV and 0.04 Å^{-1}, respectively.

3. Results and Discussion

a. Electronic Structure of undoped PPV
In the upper part of Fig. 1 we show the loss function $Im(-1/\varepsilon)$ of oriented
PPV in the energy range 0-40 eV for $\vec{q} = 0.1$ Å^{-1} parallel to the chain axis $\vec{c}$.
The spectrum is dominated by a broad plasmon at 22 eV due to the sum of all
valence band electrons ($\pi + \sigma$). In addition, π-plasmons at 4 and 6.8 eV, and
a shoulder at 5 eV can be realized. By a Kramers-Kronig analysis, we derive
the real and the imaginary part of the dielectric function ε_1 and ε_2, also
shown in Fig. 1. The strong maxima in ε_2 at 2.5 eV and at 6 eV and the
shoulder at 5 eV are due to transitions from occupied π-bands to unoccupied
π^*-bands. These $\pi \rightarrow \pi^*$ transitions cause the π-plasmons in the loss function.
$\sigma \rightarrow \sigma^*$ transitions appear in the energy range 8-22 eV. More information on the
π-electron band structure can be obtained by looking at the momentum dependence of the π-plasmons as shown in Fig. 2. While the π-plasmon at 4 eV
shows a strong dispersion in momentum transfer and disappears for $q > 0.5$ Å^{-1},
the energy and the intensity of the shoulder near 5 eV and the π-plasmon

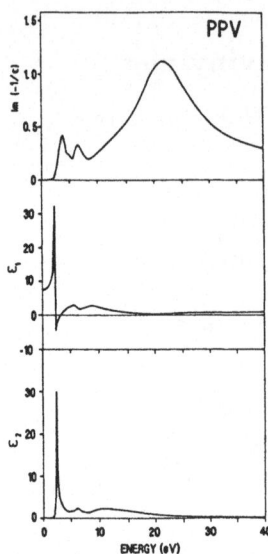

Fig. 1 Loss function and
dielectric functions for
$\vec{q} = 0.1 \text{ Å}^{-1} \parallel \vec{c}$

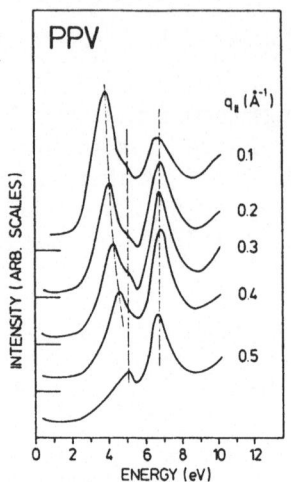

Fig. 2 Momentum dependence
of π-plasmons for $\vec{q} \parallel \vec{c}$

at 6.8 eV is almost not changed. These results are in line with recent band
structure calculations [5,6] which show that, as in polyparaphenylene (PPP)
the π^*-bands look like the π-bands with the Fermi level acting as a mirror
plane. The highest π-band and the lowest π^*-band have a width of about 2.8
eV [5]. From the observed dispersion of the lowest π-plasmon of 2.3 eV/Å^{-1},
a width of 2.2 eV can be derived. As bandwidth calculated by valence effective Hamiltonian techniques usually have to be reduced by a factor of 1.15-
1.3, the agreement between experiment and theory is almost perfect. According to the band-structure calculations [5] the gap should be 2.5 eV and a
second transition between π and π^*-bands should appear near 6 eV, which is
again in line with our experimental data of ε_2. As the latter transition is
related to flat bands (localized π-electrons in the benzene rings), no dispersion should be observed, in agreement with the data shown in Fig. 2. The
weak shoulder in the loss function and in ε_2 may be caused by a mixed transition from the wide π-band to the flat π^*-band or from the flat π-band to
the wide π^*-band. The observed transitions and the π-band structure near
the Fermi level are very similar to those in PPP. However this "mixed"
transition was not observed in PPP [7].

Up to now, we were not able to obtain reliable data on the dielectric
functions of PPV for $\vec{q}$ perpendicular to the chain axis, as due to the non-
perfect orientation of the films measurements suffer from contributions due
to Im($-1/\varepsilon_\parallel$). However, we would like to point out that such measurements
are feasible when orientations of the chains better than $\pm 2°$ can be achieved.

Having analyzed the π-band structure of fully converted PPV it is interesting to look at the evolution of the band structure during conversion from
the precursor to PPV. In Fig. 3 we show the loss function of the dimethyl-
sulphoniumchloride precursor at various stages of thermal conversion to PPV.

80

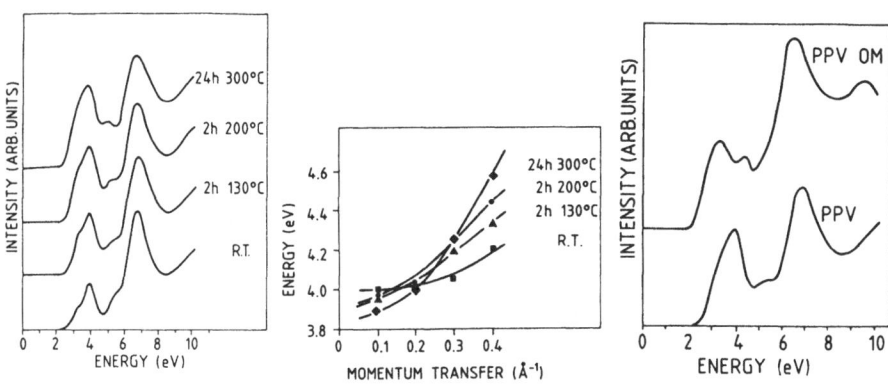

Fig. 3 Loss function during conversion from the prepolymer

Fig. 4 Momentum dependence of the π-plasmon during conversion

Fig. 5 Loss function of PPV and PPV OM

In Fig. 4 we show similar data for the dispersion of the lowest π-plasmon. The as-cast precursor shows structure in the absorption near 4 eV typical of molecular transitions in short conjugated systems. In addition, the dispersion of the lowest π→π* transitions is very small, indicating again flat molecular levels. With increasing temperature, the ratio of unsaturated to saturated units increases. Therefore, the structure in the lowest absorption peak is reduced, the gap decreases and in particular the dispersion as shown in Fig. 4 increases, indicating the transformation from molecular levels to a wide π-band with delocalized π-electrons. The present data clearly show the gradual transformation of the precursor to oligomers and finally to the polymer. Similar results were derived from optical spectroscopy [8].

We also have investigated the band structure of substituted PPV. In Fig. 5 we compare the loss function of PPV with that of poly(2,5-dimethoxy-p-phenylenevinylene)(PPV-OM). The data clearly show the shift of the lowest

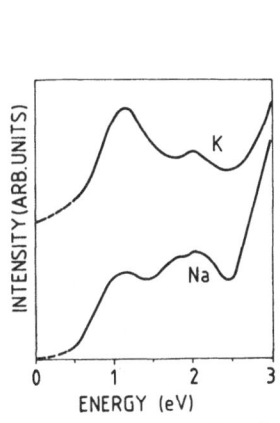

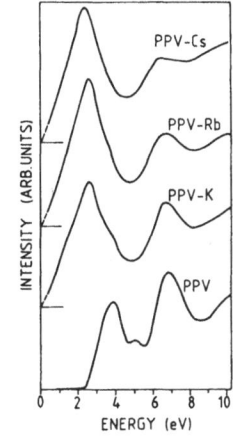

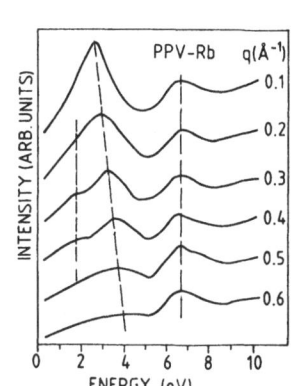

Fig. 6 Loss function for n-type doped PPV (low doping concentrations)

Fig. 7 Loss function of heavily doped PPV

Fig. 8 Momentum dependence of the loss function of PPV-Rb

π-plasmon from 4 eV to 3.4 eV. This indicates a reduction of the gap by about 0.6 eV due to the CH_3O substituent which transfers electrons into the polymer chain. The change of the colour of the film from yellow to red upon CH_3O substitution is in line with our ELS data. Probably, the reduction of the gap and the widening of the π-bands is related to the higher conductivities observed in the CH_3O-substituted doped PPV [9]. In the substituted PPV the shoulder near 5 eV is shifted to lower energy by about 0.6 eV while the higher π-plasmon at 6.8 eV is not shifted. These findings support our assignment of the 5 eV shoulder to a "mixed" transition between wide and flat π-bands.

b. n-type doped PPV

The doping by alkali metals leads to considerable changes of the π-electron band structure. At low doping concentrations two peaks appear in the gap at 1.1 eV and at 2.0 eV for Na and K doped PPV (see Fig. 6). This is consistent with the formation of bipolaron defects leading to two occupied levels in the gap. The maxima are then related to transitions from the occupied bipolaron levels to the lowest π^*-band. There is no indication of a narrow third line in the gap which should appear when single-charged polaron defects are present [10]. The intensity of the two transitions is rather different from what is expected from theoretical calculations [10]. In particular, for the Na-doped PPV, the second line has almost the same intensity as the first line. Similar results were also obtained for p-type doped PPV [5]. We would like to emphasize that in all non-degenerate ground state polymers investigated so far, the intensity ratio of the first bipolaron transition to the second is much higher. A possible explanation for this anomalous intensity ratio could be a particular high correlation energy in PPV leading to strong changes of transition probabilities [11].

At the highest doping concentrations of K, Rb and Cs, the gap is closed and the π-plasmon at 4 eV is shifted to 2.5 eV, while the second π-plasmon near 6.8 eV is not shifted (see Fig. 7). The π-plasmon at 2.5 eV may be interpreted by a free-carrier plasmon of a partially filled π-band. However, the investigation of the momentum dependence as shown in Fig. 8 for PPV doped with Rb clearly contradicts this explanation. With increasing momentum transfer, the lower π-plasmon splits into two components, one showing a strong dispersion (due to a $\pi \rightarrow \pi^*$ transition) and another one showing no dispersion. The latter one is probably related to a transition from a still existing bipolaron band to the empty π^*-band. The results indicate that the band structure of the fully doped PPV is more complicated than just a closing of the $\pi \rightarrow \pi^*$-band. There are probably still bipolaron bands in the gap, which partially overlap with the highest π- and the lowest π^*-band. The shift of the lowest π-plasmon by only 1.5 eV compared to the gap of $\sim$2.5 eV indicates that there remains a π-π^* gap of $\sim$1 eV in the fully doped PPV.

References

1. S.A. Brazovskii and N.N. Kirova, JEPT Lett. 33, 4 (1981)
2. R.A. Wessling and R.G. Zimmerman, U.S. Patent # 3, 401, 152 (1968)
3. F.E. Karasz, J.P. Capistran, D.R. Gagnon and R.W. Lenz, Mol. Cryst. Liq. Cryst. 118, 327 (1985)
4. J. Fink, Z. Phys. B 61, 463 (1985)
5. J.L. Brédas, unpublished results
6. C.W. Duke and W.K. Ford, Int. J. Quantum Chem., Quantum chemistry symposium, no. 17, 597 (1983)
   C.W. Duke, A. Paton, and W.R. Salaneck, Mol. Cryst. Liq. Cryst. 83, 1209 (1982)

7. G. Crecelius, J. Fink, J.J. Ritsko, M. Stamm, H.-J. Freund and
    H. Gonska, Phys. Rev. B 28, 1802 (1983)
8. D.D.C. Bradley, G.P. Evans and R.H. Friend, Synth. Metals 17, 651 (1987)
9. I. Murase, T. Ohnishi, T. Noguchi, and M. Hirooka, Synth. Metals 17,
    639 (1987)
10. K. Fesser, A.R. Bishop and D.K. Campbell, Phys. Rev. B 27, 4804 (1983)
11. U. Sum, K. Fesser, and H. Büttner, Solid State Commun. 61, 607 (1987)

Electronic Structure
of Doped Highly Oriented Polyacetylene

J. Fink[1], *N. Nücker*[1], *B. Scheerer*[1], *A. vom Felde*[1], *and G. Leising*[2]

[1]Kernforschungszentrum Karlsruhe GmbH,
Institut für Nukleare Festkörperphysik,
Postfach 3640, D-7500 Karlsruhe, Fed.Rep.of Germany
[2]Institut für Festkörperphysik, TU Graz,
Petersgasse 16, A-8010 Graz, Austria

1. INTRODUCTION

Despite very active investigations of the electronic structure of conducting polymers during the last 10 years, there is a continuous debate on the semi-conductor-metal transition and on the nature of the highly conducting state of the prototype polyacetylene (PA). Various models have been proposed. The metallic droplet model explained the transition by a percolation by an increasing number of small metallic regions [1]. In the soliton model, the charged defects ("solitons") form at sufficiently high concentrations a soliton band,and the metallic state is reached due to a disorder-induced quenching of the Peierls distortion [2]. Recently, a first order transition from a soliton lattice to a polaron lattice has been discussed [3]. In this contribution we give a first report on investigations of the band structure of heavily doped highly oriented PA by electron energy loss spectroscopy (EELS). This method gives more information compared to optical spectroscopy because nonvertical transitions can be excited, which give information on the dispersion of the bands.

2. EXPERIMENTAL

Free-standing highly oriented crystalline films of trans-PA with a thickness of $\sim$3000 Å were prepared as described previously [4]. p-type doping with AsF_6^- was performed from the gas phase. n-type doping with Na, K, and Cs was achieved by evaporating a small amount of alkali metal onto the PA films and subsequent annealing or by keeping the films in alkali metal vapour at room temperature. EELS measurements in transmission were performed with a 170 keV spectrometer described elsewhere [5].

3. RESULTS

In Fig. 1 we show the electron energy-loss function for PA doped with various amounts of AsF_6^-. Up to now, only relative concentrations of AsF_6^- in PA could be determined due to measurements of the relative intensity of As 3d excitations at 46 eV and the $\pi+\sigma$ plasmon at 22 eV. Because the error bars of the determination of the concentration are still rather high, we characterize the AsF_6^- content by the measured conductivity parallel to the chain axis $\sigma_\parallel$. For a uniaxial crystal, the loss function can be written $Im(-1/(\varepsilon^\parallel\cos^2\Theta + \varepsilon^\perp\sin^2\Theta))$, where $\varepsilon^\parallel$ and $\varepsilon^\perp$ are the principal components of dielectric tensor $\bar{\varepsilon}(\vec{q},\omega)$ parallel and perpendicular to the symmetry axis (for PA the chain axis $\vec{c}$). Θ is the angle between the symmetry axis and the momentum transfer $\vec{q}$. For undoped PA and $\vec{q} = 0.1$ Å$^{-1} \| \vec{c}$ a π plasmon at 4.9 eV and a plasmon width (FWHM) of 1.3 eV is observed. This plasmon is caused by a strong $\pi \rightarrow \pi^*$ transition at $\sim$1.9 eV showing up in $Re(\varepsilon)=\varepsilon_1$ by a zero crossing at 4.9 eV and in $Im(\varepsilon)=\varepsilon_2$ by a strong maximum at 1.9 eV [6]. With increasing doping concentration, a small broad maximum at 0.8 eV and a shoul-

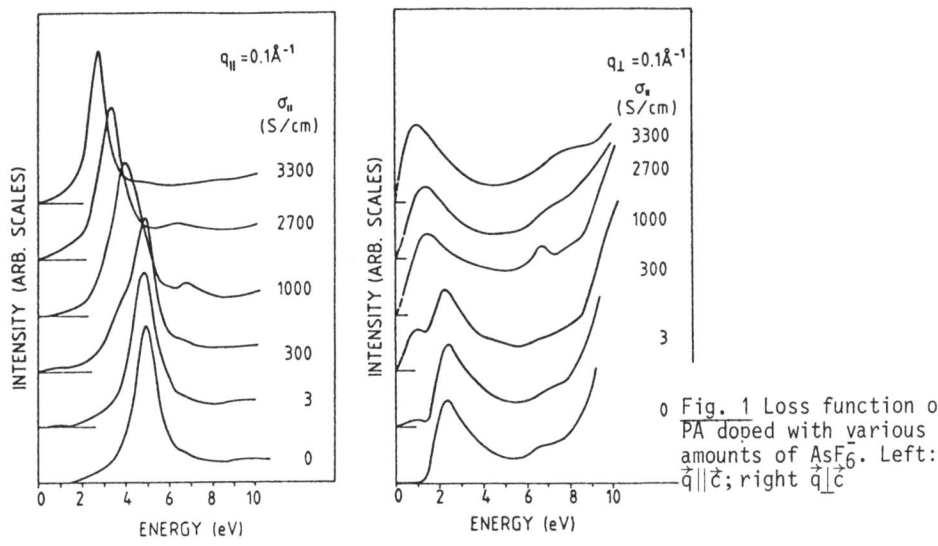

Fig. 1 Loss function of
PA doped with various
amounts of AsF$_6^-$. Left:
$\vec{q}\|\vec{c}$; right $\vec{q}\bot\vec{c}$

der at ∿4eV appear. The latter increases in intensity and leads to a broadening of the π plasmon up to 1.8 eV. Close to the transition to the metallic
state σ∿1000 S/cm, the plasmon is shifted very rapidly to lower energy
(∿2.8 eV) and the width decreases to ∿0.9 eV which is below the value of
undoped PA. The shift of the π plasmon by 2.1 eV, which is close to the gap
energy (1.8 eV), may be explained by a growing oscillator at almost zero
energy and a decreasing strength of the oscillator at 1.9 eV. This is in
line with the results for $\vec{q}$ = 0.1 Å^{-1} |$\vec{c}$, which are also shown in Fig. 1.
In the undoped PA, the maximum at 2.2 eV corresponds to a very small π→π*
oscillator. The anisotropy in ε_2 is 170 [6]. As the dielectric function due
to the π electrons is probably small compared to the background dielectric
function due to the σ electrons ε_b, the loss function is given in this case
by Im$(-1/\varepsilon)=\varepsilon_2/(\varepsilon_1^2+\varepsilon_2^2)=\varepsilon_2/\varepsilon_b^2$ and is therefore proportional to ε_2. With increasing doping concentration, a new oscillator at lower energy appears
which causes a maximum at 0.8 eV. At σ$_⊥$ = 1000 S/cm, the maximum in the
loss function at 2.2 eV has disappeared. At even higher conductivities the
lower maximum slightly moves to lower energies. The results clearly indicate the reduction of the gap to almost zero energy (less than some tenths
of an eV). However, no Drude-like increase of ε_2 is observed. Investigations
of the structure by electron diffraction on the same samples show for σ$_‖$ up
to 1000 S/cm a well-ordered crystalline structure. For higher conductivities
there remains a high anisotropy in the EELS data while the diffraction
pattern reveals a remarkable degradation of the crystal structure [7]. Therefore, we cannot exclude for high doping concentrations contributions from
Im$(-1/\varepsilon_‖)$ in the loss function taken for $\vec{q}\bot\vec{c}$.
In Fig. 2 we show for various doping concentrations the dispersion of the
π plasmon $\vec{q}\|\vec{c}$ which is related to the momentum dependence of the π→π* tran-
sitions and therefore directly reflects the band structure of the π bands
[6]. For AsF$_6^-$ doping up to σ$_‖$ = 300 S/cm, the dispersion of the π plasmon is
not changed compared to that of undoped PA. Between σ$_‖$ = 300 and ∿1800 S/cm
there is a rapid transition to a new dispersion curve with a steeper slope,
indicating a stronger dispersion of the initial band and the final band
both related to the transition. In addition, for the highest doping concentrations a step of about 0.3 eV is observed for q∿0.25 Å^{-1}.

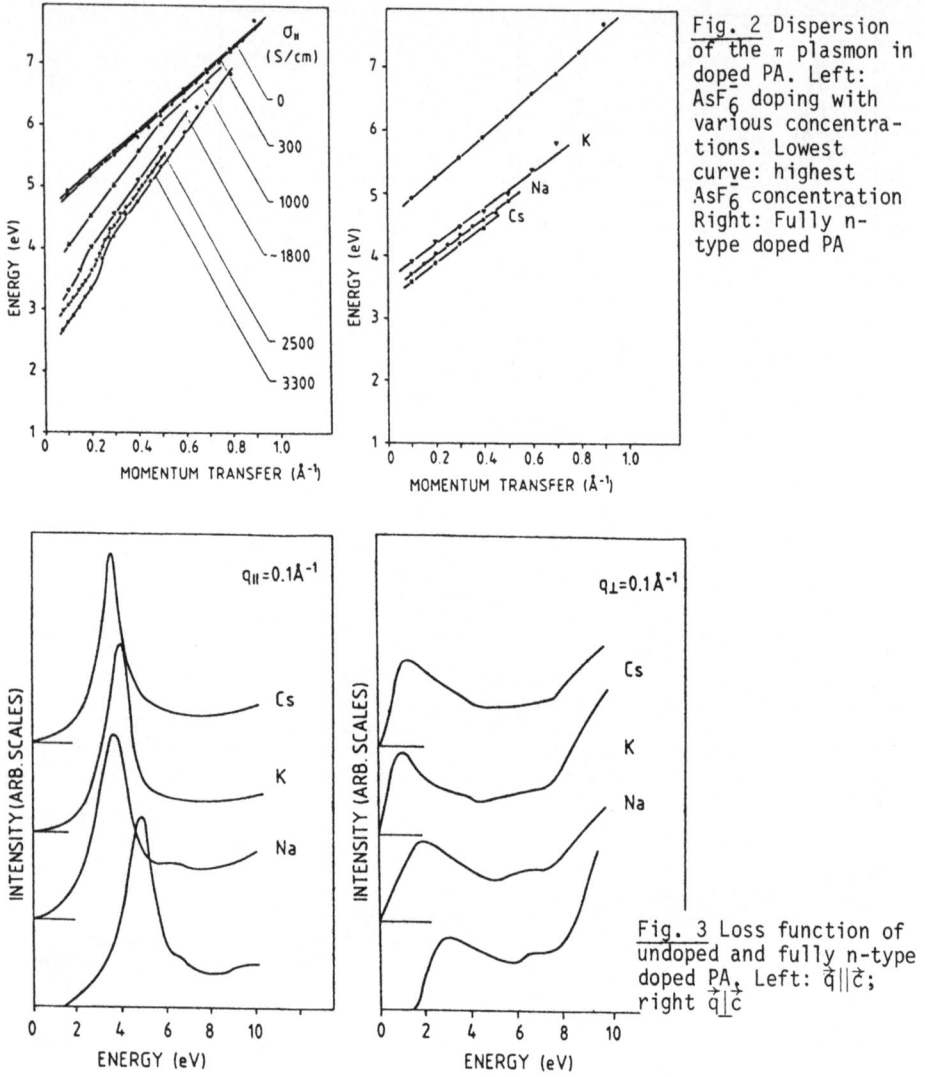

Fig. 2 Dispersion of the π plasmon in doped PA. Left: AsF$_6^-$ doping with various concentrations. Lowest curve: highest AsF$_6^-$ concentration Right: Fully n-type doped PA

Fig. 3 Loss function of undoped and fully n-type doped PA. Left: $\vec{q}\|\vec{c}$; right $\vec{q}\perp\vec{c}$

In Fig. 3 we compare the loss functions of undoped PA with that of n-type doped PA. Only the highest doping concentrations which could be achieved are shown. In all cases of n-type doping, a crystalline structure was observed by electron diffraction. For the n-type doped PA, the shift of the π plasmon to lower energy is smaller compared to that of AsF$_6^-$ doped PA. For Na, K, and Cs the lowest π plasmons were observed at 3.8, 3.9, and 3.5 eV, respectively. The dispersion of the π plasmon is not changed upon n-type doping (see Fig. 2). This indicates that at least in the initial state band or in the final state band the dispersion is not changed. Up to now, no steps in the dispersion were observed for n-type doping.

4. DISCUSSION

The results at high concentration can be explained on the basis of a soliton lattice which is formed at higher doping concentrations and which forms a soliton band in the gap. Various calculations on the optical conductivity for the soliton lattice model have been carried out [8,9,10]. In the fully AsF$_6^-$ doped PA, the plasmon at 2.8 eV is caused by a "zone-edge" transition [10] between the valence band (VB) and the empty soliton lattice band (SB) at the zone boundary of the reduced Brillouin zone (BZ). From the shift of the plasmon compared to the undoped case, this transition should be close to zero, i.e. the gap between VB and SB is less than some tenths of an eV. In the fully p-type doped PA, no indication of a "zone-center" transition [10] between VB and SB in the reduced BZ is detected,which should show up in a second plasmon at higher energy. The stronger dispersion of the π plasmon compared to that of undoped PA indicates that the gap between VB and conducting band (CB) is almost completely filled with a broad SB. The step of 0.3 eV in the dispersion of the π plasmon at q = 0.25 Å^{-1} may be explained by possible transitions from the VB to the CB at higher momentum transfer, which should appear at higher energies compared to VB → SB transitions due to the gap between SB and CB. If this interpretation is correct, this gap would be 0.3 eV. The assignment is supported by the fact that the step in the dispersion is shifted to lower momentum transfer with decreasing AsF$_6^-$ content. The data at intermediate doping concentrations do not support a two-phase model, as at no concentration the metallic π plasmon and the undoped π plasmon appear in the same spectrum. Rather, the data can be explained by a gradual decrease of the strength of the VB → CB transition and an increase of the strength of VB → SB transitions. However, more experimental data, a full evaluation of the data by a Kramers-Kronig analysis and calculations of the loss function by a Maxwell Garnet theory are necessary to exclude definitely the two-phase model.

The lower dispersion of the π plasmon in n-type doped PA suggests a smaller width of the SB. The origin of this difference of the width of the SB between n-type doped PA and p-type doped PA is not clear up to now but can be unambiguously derived from the experiment. It may be related to the observation that for n-type doping the crystallinity is preserved up to the highest concentrations which could be achieved in the present experiments. The investigation of the range of intermediate doping concentrations for n-type doped PA was started recently. Preliminary results look similar to those of p-type doped PA. In both cases, a finite polarizability of π electrons at low energy ($\sim$0.8 eV) was observed for $\vec{q} \| \vec{c}$.

Finally, we mention that the present results are not compatible with the formation of narrow polaron bands in the gap at high doping concentrations as proposed in Ref. 3. In this model, a transition without dispersion and thus a plasmon without dispersion should appear,which was not detected in our spectra.

1. J. Tomkiewicz, T.D. Schultz, H.B. Brom, T.C. Clarke, and G.B. Street, Phys. Rev. Lett. 43, 1532 (1979)
2. E.J. Mele and M.J. Rice, Phys. Rev. B 23, 5397 (1981)
3. S. Kivelson and A.J. Heeger, Phys. Rev. Lett. 55, 308 (1985)
4. G. Leising, Polymer Bulletin, 11, 401 (1984)
5. J. Fink, Z. Phys. B 61, 463 (1985)
6. J. Fink and G. Leising, Phys. Rev. B 34, 5320 (1986)
7. H. Kahlert, G. Leising, and J. Fink, see contribution in this volume.
8. J.P. Albert and C. Jouanin, Mol. Cryst. and Liq. Cryst. 117, 283 (1985)
9. B. Horovitz, Solid State Commun. 41, 593 (1982)
10. S. Jeyadev and E.M. Conwell, Phys. Rev. B 33, 2530 (1986).

Polarization Dependence of Recombination Kinetics in Stretch-Oriented trans-Polyacetylene

H. Bleier, S. Roth, and G. Leising*

Max-Planck-Institut für Festkörperforschung,
Heisenbergstr. 1, D-7000 Stuttgart 80, Fed. Rep. of Germany

The transient photoconductivity of fully oriented Durham-Graz trans-polyacetylene has been investigated. In this paper we report on a detailed series of measurements on the polarization dependence of the transient photocurrent. The influence of the external field, of the intensity and wavelength of the incident light, and of the temperature on the measured anisotropy has been studied. The results suggest interchain charge creation is the predominant mechanism of photoexcitation of carriers.

Introduction

For a better understanding of the transport and recombination mechanisms in trans-polyacetylene $(CH)_x$ we have studied the behaviour of optically excited charged carriers by transient photoconductivity (PC) experiments on fully oriented trans-$(CH)_x$. Undoped trans-$(CH)_x$ is a 1d-semiconductor with a bandgap of about 1.7 eV. Photoexcitation across this gap leads to the generation of charged carriers, which can be observed by PC experiments.

The PC in trans-$(CH)_x$ (Shirakawa-type /1,2/ as well as Durham-Graz material /3,4/) consists of a fast /1-5/ and a slow /1,6-8/ component, the former one due to free, highly mobile carriers with a "mobility relaxation time" of about 100 ps /4/. These created hot carriers (fast PC component) lose their initial mobility by falling into various traps, the further motion of these carriers (slow PC component) can be described as hopping, in analogy to the mechanism observed for the dark conductivity. We have investigated the dependence of both components

---

* Institut für Festkörperphysik, Petersgasse 16, A-8010 Graz

on temperature, light intensity, external electric field
(strength and orientation) as well as on the wavelength and the
polarization of the incident light. Some of these results have
already been published /1,3,4/ or will be presented in a subsequent publication. In this paper we concentrate on the polarization dependence of both the slow and the fast components of
the photocurrent.

Experimental

Films of trans-$(CH)_x$ have been prepared by the Durham route /9/
and oriented by applying uniaxial stress during the thermal
conversion of the precursor polymer /10/. The films (1-2 µm
thickness) were mounted onto a 150 µm quartz substrate, onto
which gold contacts of two different geometries (Fig. 1) had
been evaporated. All sample handling was carried out in an
argon glovebox without any exposure to air.

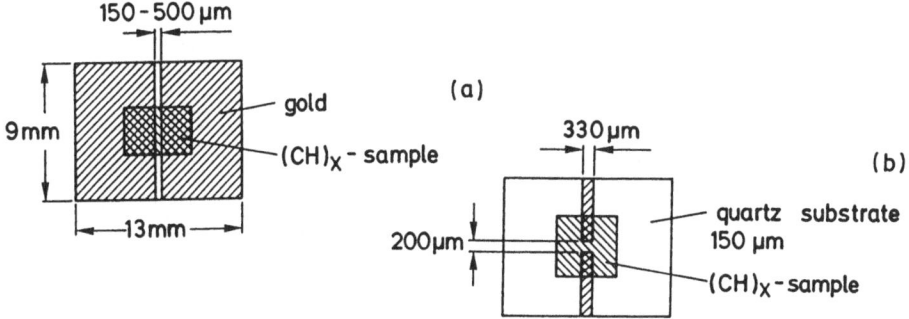

Fig. 1: Two contact geometries used for different experiments

The transient measurements were carried out with 500-ps
pulses of a nitrogen-laser pumped dye laser with a tunable
range from 1.2 to 3.2 eV as light source and a TEK 7912 AD
(bandwidth=700 MHz) or a Lecroy 9400 (125 MHz) digital oscillo-
scope as the electronic detection system. For measurements of
the slow component a chopped cw-light-source (He-Ne laser or a
mercury-xenon lamp) was used together with lock-in detection.

The influence of thermal modulation of the dark current on
the principal results of the slow component are excluded by

comparing the room temperature data with measurements at sufficiently low temperatures where the proportion of the dark current in the total signal is drastically reduced.

Results

The high degree of orientation of the polymer chains in the
Durham-Graz $(CH)_x$ (misalignment < 2°) allows us to study the
anisotropic properties of this 1d-semiconductor. The measured
PC varies significantly on turning the polarization of the
incident light with respect to the chain direction. For the
slow component of the PC and for photoinduced absorption (PA)
measurements such a behaviour was reported for the first time
by Townsend et al. in Ref. /11/. As shown in Fig. 2a this can
also be observed for the fast component of the PC when the
contact geometry shown in Fig. 1a is chosen. With the contacts
of Fig. 1b, which one has to use to achieve very high time
resolution /12/ (picosecond measurements), the polarization
effect on the photocurrent is partially smeared out by the
inhomogeneity of the external electric field.

The necessity of correction of the measured raw data (full
circles in Fig. 2) for the anisotropy of the optical absorption
and reflection /13-15/ is demonstrated in Figs. 2a and 2b. The
apparently different raw data of Fig. 2a (excitation energy=-
2.6 eV) and Fig. 2b (1.7 eV) result in two similar curves with
the same anisotropy of about 1.7 between parallel and perpendicular polarized light. With a sample thickness of 1.7 μm we
obtain that 59% of the incident light at 2.6 eV is reflected
and about $10^{-6}0\%$ (≈ 0) is transmitted for parallel polarization.
For the perpendicular direction these values are 2.4% and 3.5%
respectively, with the consequence that in the latter case 94%
is absorbed and in the former case only 41%. At an ·excitation
energy of 1.7 eV we get values of 49% and $10^{-11}\%$ for the parallel and 2.4% and 62% for the perpendicular direction (the
sample becomes semitransparent for perpendicular polarization).
Contrary to what was stated in a preliminary publication /3/
this anisotropy factor of 1.7 for the fast component of the
photocurrent does not change with light intensity and applied
electric field.

90

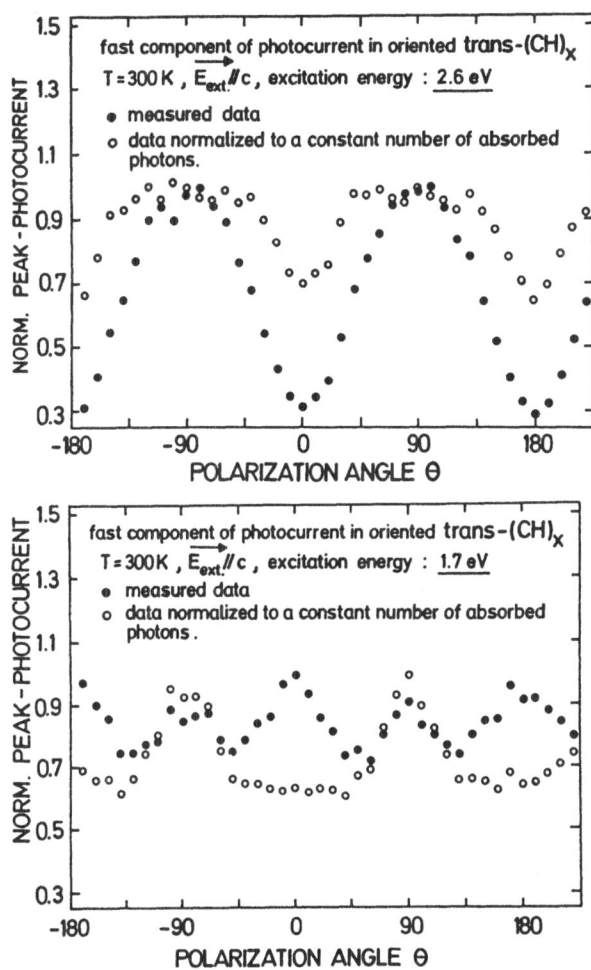

Fig. 2: Polarization anisotropy of the fast component of PC at two different excitation energies, 2.6 eV (2a, upper figure) and 1.7 eV (2b, lower picture).

For the slow component this anisotropy is larger and sample dependent, varying between 3 and 5 (Fig. 3). In the case of Fig. 3 the total number of absorbed photons has been kept constant during the measurement by changing the intensity of the incident light to compensate the anisotropy of the reflectance and absorbance. Both methods, the directly controlled intensity as well as the numerical correction of the data after the measurements, lead to the same results. The difference in the values at -90°, 90°, and 270° of about ±3% of the peak value is within the limits of the experimental error.

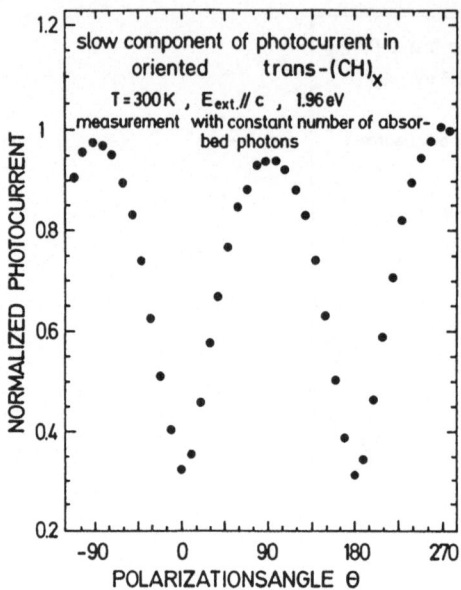

Fig. 3: Polarization anisotropy of the slow component of the photocurrent.

Discussion

Polarization anisotropy has been found not only in PC but also in PA measurements /11/ at the so-called low-energy (LE) peak at 0.45 eV, which is directly correlated to the charged excitations in trans-$(CH)_x$ /16/. The close relation between PA and PC results clearly indicates that this anisotropy is due to the number of charged carriers and not to their mobility, because PA is proportional to the number of carriers whereas PC is proportional to the product of the carrier number and their mobility.

Bimolecular recombination, as proposed by Dorsinville et al. /17/ for the explanation of a similar result obtained in stretch-oriented modified Shirakawa-$(CH)_x$, has to be excluded, since the decay rates of PC /3,4/ as well as PA experiments /18/ contradict such a mechanism. Therefore we explain this anisotropy by a higher probability for the photoexcited electron-hole pairs to separate immediately onto different polymer chains if the absorbed light is polarized perpendicular to the chain direction. Carriers on different chains escape geminate recombination more easily. Interchain carrier creation was also proposed by Townsend et al. /11/ and more recently calculated by Danielsen /19/ to explain the anisotropy in the PA experiments. In addition to the direct interchain-carrier creation, some intrachain carriers should be able to hop to a neighbouring chain before they can relax into charged solitons on a subpicosecond time scale. All other carriers, remaining on the same chain, recombine or convert into the neutral excitations (responsible for the so-called high-energy peak in PA experiments at 1.4 eV) as proposed by Kivelson and Wu /20/. These carriers, except for a few reaching the contacts, cannot contribute to PC.

Therefore we conclude that only interchain carriers give rise to PC. Even in the case of a certain separation of the intrachain carriers for a short time in the external electric field, the optically excited pairs will mainly recombine by geminate recombination (and not by recombination between different pairs). Therefore they cannot give any net contribution to the observed PC.

Due to symmetry arguments the interchain carriers must be polarons rather than solitons. Consequently they do not contribute to the LE peak in PA measurements but they are seen in the fast component of the photocurrent. By collison with neutral soliton defects they convert, however, into charged solitons. The retarded contribution (after ≈40 ps) to the LE peak in time-resolved PA experiments /21/ was recently explained by this argument.

In summary, we have established a polarization anisotropy for the fast and the slow components of the photocurrent, which are independent of excitation energy, light intensity, external electric field and temperature. This result was explained by interchain carriers as the main contribution to the photocurrent. Since these interchain carriers have to be polarons, the fast component of the photocurrent should have for the most part its origin in polarons before they recombine with neutral solitons to form charged solitons. These charged solitons will then contribute to the slow component of the photocurrent.

Financial support from the Volkswagen foundation and the Deutsche Forschungsgemeinschaft are gratefully acknowledged. We thank all colleagues from Polymer Hill for valuable discussions.

References

1. S. Roth and H. Bleier: Synth.Metals $\underline{17}$, 503 (1987)
2. M. Sinclair, D. Moses, and A.J. Heeger: Solid State
   Commun. $\underline{59}$, 343 (1986)
3. H. Bleier, G. Leising, and S. Roth: Synth.Metals $\underline{17}$, 521
   (1987)
4. H. Bleier, S. Roth, H. Lobentanzer, and G. Leising:
   to be published
5. Y. Yacoby, S. Roth, K. Menke, F. Keilmann, and J. Kuhl:
   Solid State Commun. $\underline{47}$, 869 (1983)
6. T. Tani, P.M. Grant, W.D. Gill, G.B. Street, T.C. Clarke: :
   Solid State Commun. $\underline{33}$, 499 (1980)
7. S. Etemad, T. Mitani, M. Ozaki, T.C. Chung, A.J. Heeger,
   A.G. MacDiarmid: Solid State Commun. $\underline{40}$, 75 (1981)
8. H. Kiess, R. Keller, D. Baeriswyl, and G. Harbeke: Solid
   State Commun. $\underline{44}$, 1443 (1982)
9. J.H. Edwards and W.J. Feast: Polymer $\underline{21}$, 595 (1980)
10. G. Leising: Polymer Bulletin $\underline{11}$, 401 (1984)
11. P.D. Townsend, D.D.C. Bradley, M.E. Horton, C.M. Pereira,
    R.H. Friend, N.C. Billingham, P.D. Calvert, P.J.S. Foot,
    D.C. Bott, C.K. Chai, N.S. Walker, K.P.J. Williams: Solid
    State Sciences $\underline{63}$, 50 (1985)
12. D.H. Auston: "Picosecond Optoelectronic Devices"
    $\underline{4}$, 73 (1984), Academic Press, New York
13. P.D. Townsend and R.H. Friend: Synth.Metals $\underline{17}$, 361 (1987)
14. R. Uitz, G. Temmel, G. Leising and H. Kahlert: submitted
    to Z.Phys. B
15. G. Leising: private communication
16. Z. Vardeny, J. Orenstein, and G.L. Baker: Phys.Rev.Lett.
    $\underline{50}$, 2032 (1983)
17. R. Dorsinville, S. Krimchansky, R.R. Alfano, J.L. Birman,
    R. Tubino, Dellepiane: Solid State Commun. $\underline{56}$, 857 (1985)
18. Z. Vardeny and E. Ehrenfreund: Proc. of the "Conf. on
    Transport and Relaxation Processes in Random Materials",
    Gaithersburg (1985)
19. P.L. Danielsen: J.Phys.C $\underline{19}$, L741 (1986)
20. S. Kivelson and W.-K. Wu: Phys.Rev. $\underline{B34}$, 5423 (1986)
21. L. Rothberg, T.M. Jedju, S. Etemad, and G.L. Baker:
    Phys.Rev.Lett. $\underline{57}$, 3229 (1986)

Photogeneration Mechanism and Mobility in Polydiacetylene

D. Moses, M. Sinclair, and A.J. Heeger

Department of Physics and Institute for Polymers and Organic Solids, University of California, Santa Barbara, CA 93106, USA

Abstract

Transient photoconductivity experiments have been carried out on single crystals of polydiacetylene-(bis_ p-toluene sulfonate), PDA-TS. The low electric field photocurrent decay consists of a temperature–independent fast (picosecond) initial component and a longer time (nanosecond) component with magnitude that is strongly temperature dependent. Using very small spacings between electrodes on the sample, we have succeeded in achieving sweep-out for the longer-lived carriers; the data yield a mobility of ≈ 5 cm^2/Vs at room temperature in the ns regime. These results demonstrate that the Onsager geminate recombination model, previously used extensively for the polydiacetylenes, is not applicable to PDA-TS.

A prediction of remarkably high mobility [1], of the order of 2×10^5 cm^2/Vs, has been advanced by analyzing transient transport experiments with the assumption that the quantum efficiency for electric-field dependent carrier creation could be derived from the Onsager theory of geminate recombination.

In this paper, we report fast transient photoconductivity measurements on PDA-TS, which enable us to address both the photogeneration mechanism (Is it limited by the geminate recombination processes usually described by the Onsager theory for localized states?) and the magnitude of the carrier mobility. We find that the photocurrent decay consists of a temperature-independent fast (picosecond) initial component and a longer time (nanosecond) component with magnitude that is strongly temperature dependent. Using very small spacings between electrodes on the samples, we have succeeded in achieving sweep-out of the longer-lived carriers in relatively high electric fields (> 3×10^4 V/cm); the data yield a field-independent mobility of ≈ 5 cm^2/Vs at room temperature in the ns regime, far below the previously inferred value. The temperature independence of the initial photocurrent is interpreted as evidence for "hot" carriers.

The transient photoconductivity was measured using the Auston microstrip transmission line switch technique [2,3]. Single crystal samples used in these experiments were grown at Queen Mary College, cleaved to an approximate thickness of 100 μm and mounted on the alumina substrate. The gold microstrip was evaporated directly onto the single crystal with gaps of L = 200 μm, L = 10 μm, and L = 2.5 μm. The crystals were oriented with the PDA chains parallel to the electric field within the gap.

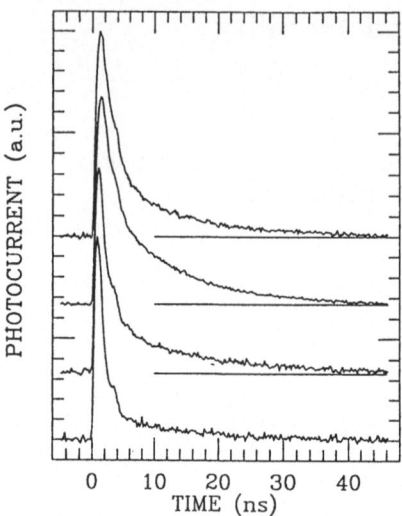

Fig. 1. Transient photocurrent waveforms (resolution 400 ps) for PDA-TS at various temperatures: from top to bottom; 300 K, 180 K, 60 K, and 15 K. The horizontal lines indicate the zero for the upper three waveforms.

In Fig. 1, we show the transient photocurrent decay following a 2.9 eV photon pulse at a series of temperatures (electric field of 2.5×10^4 V/cm across a gap of 200 μm). Experiments at a lower field, 1.5×10^3 V/cm, yield the same temperature dependence of the waveform. Each waveform is characterized by a fast initial response followed by a fast decay to a more slowly decreasing "tail". The rise time is limited by the temporal resolution of the measuring system, since the initial photogeneration process is instantaneous on this time scale. For the 2.9 eV pump, as the temperature is lowered, both the peak value and the magnitude of the tail initially increase, reaching a gentle maximum at about 180 K; at lower temperatures, the tail decreases much more rapidly going toward zero at the lowest temperatures. With excitation at 2.58 eV, both the peak value and the magnitude of the tail appear to decrease monotonically with decreasing temperature. While the peak value decreases by only a factor of two, the tail has almost completely disappeared at 15 K. With our best time resolution (50 ps without preamplifier), an initial fall time of 300 ps was measured at 15 K. This unusually fast initial decay may be in part due to the one-dimensionality (1d) of the PDA π-electrons; since the photogenerated carrier transport is principally along the PDA chains, the carriers cannot efficiently escape recombination by three-dimensional delocalization. Although the tail portion of the transient photocurrent extends out to relatively long times, the initial decay of the tail can be adequately fitted to a single exponential of the form $i = i_0 \exp(-t/\tau)$ where $\tau \sim 40$ ns. As shown in Fig. 1, the amplitude (i_0) is found to be strongly temperature dependent, while τ is insensitive to the temperature.

The data of Fig. 1 indicate two decay mechanisms with different decay rates. In order to more quantitatively separate the two, we have integrated the respective areas. The functional form of the tail was fitted at times well beyond the initial decay (> 5 ns), the form extrapolated back to t = 0, and the tail current was integrated to obtain Q_{tail}. The integrated charge associated with the fast decay was obtained by subtracting Q_{tail} from the area under the whole transient curve to obtain $Q_{initial}$. Within our experimental accuracy, $Q_{initial}$ is independent of temperature, while Q_{tail} is temperature dependent, extrapolating to zero at T = 0 K. The temperature independence of $Q_{initial}$ implies that the product ($\eta\phi$) of the photocarrier quantum yield. (η) and the probability of escaping early-time recombination (ϕ) is independent of temperature.

We were able to successfully measure the transport mobility at room temperature in the ns regime by achieving carrier sweep-out using a narrow gap ($\leq$ 10 μm). Photocurrent waveforms at various electric fields (E) are shown in Fig. 2. In these experiments, the light intensity was reduced as E was increased so as to keep the peak signal approximately constant. The tail gradually decreases in relative magnitude as E is increased until at E = 3 x 10⁴ V/cm, it completely disappears.

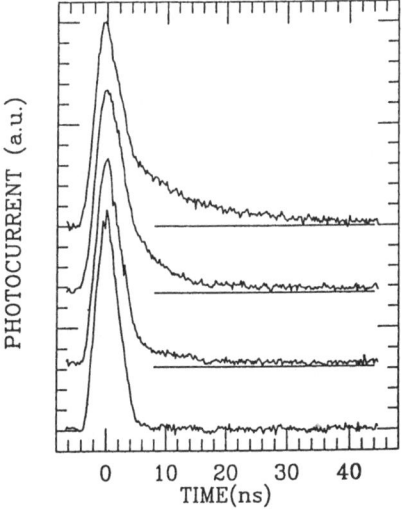

Fig. 2. Electric field dependence of the transient photocurrent waveforms (resolution 3 ns) for L=10 μm where complete carrier sweep-out is achieved. The applied electric fields are (from top to bottom) 10^3 V/cm, 3 x 10^3 V/cm, 10^4 V/cm and 3 x 10^4 V/cm.

The data imply an increase in average drift velocity with increasing E to the point where the carriers are swept out to the contacts. The tail will completely disappear when all the carriers leave the sample during a time interval equal to the integration time of the measuring system. This time is set by the particular choice of the signal amplifier that is used and is determined by the measured risetime (3 ns in Fig. 2) of the initial photocurrent. For a sample length of 10 μm and approximately uniform illumination, the typical carrier must drift 5 μm to the contact, implying a drift velocity of ~ 2 x 10^5 cm/s for E = 3 x

10^4 V/cm and a mobility of about 5 cm^2/Vs (at room temperature). The same value (within experimental error) was inferred from experiments with L = 2.5 μm where the sweepout field is correspondingly lower. This value is in agreement with that inferred in the 200 - 400 ns time scale by Reimer and Bassler in their attempt to carry out time-of-flight measurements from contact-injected carriers [4].

Because of the sweep-out, Q_{tail} saturates at fields above 3x10^4 V/cm; $Q_{initial}$, however, continues to increase even at the highest fields. Since the unresolved initial peak decays in less than 300 ps, there is not sufficient time to sweep out the charge before it thermalizes and traps, etc. into the tail.

We have ruled out the possibility that this effect is due to a field-induced increase in the decay rate [5]. In that case, the decrease in relative photocurrent arises since at high fields the carriers reach recombination centers more rapidly, an effect which is particularly effective in 1d. In PDA-TS, however, the sweep-out time (or, equivalently, the sweep-out field) depends on L as noted above. Moreover, the sweep-out phenomenon is found <u>only</u> when the length of the sample is comparable to the distance that a carrier can travel during its average lifetime. Studies at the same electric field strengths on longer samples with L = 200 μm do not indicate any change in shape of the waveforms (see Fig. 1).

Having determined the mobility in the tail, we can obtain the ($\eta\phi$) product. The photocurrent measured at the beginning of the tail, 1.3x10^{-5} amp resulting from an absorbed photon density of N=3x10^{20} cm^{-3}, leads to a value for the photoconductivity of 3x10^{-3} S/cm at ~ 4 ns. Using $\sigma = (\eta\phi)\mu eN$, we find ($\eta\phi$) ≈ 10^{-3} at the beginning of the tail. The sweep-out experiments with L = 10 μm and L = 2.5 μm indicate a field-independent mobility (within our experimental accuracy). This result and the linear dependence of the photocurrent on E for L = 200 μm (in the field range of the sweep-out measurements) imply a field independent ($\eta\phi$).

<u>The field and temperature-independent ($\eta\phi$) are in sharp disagreement with the Onsager theory of geminate recombination.</u> The theory would predict that the quantum yield, which determines the initial photoconductive response, should be limited by the probability to escape geminate recombination: ($\eta\phi$) should increase linearly in the applied electric field strength and should decrease exponentially as the temperature is lowered [1,6]. In addition, the carriers which undergo geminate recombination would produce zero net photocurrent (contributions from the geminate electron and hole would cancel). Thus, the rapid decay of the initial peak into the longer-lived tail is not due to geminate recombination. We conclude that, although early-time recombination is clearly important ($\eta\phi$ ≈ 10^{-3} at 4 ns), this is not properly described by the Onsager theory of geminate recombination.

Traditionally, the dominant mechanism for recombination (e.g., bimolecular vs. monomolecular) has been studied by measuring the dependence of the photoconductive response on the illumination intensity. We have measured the intensity dependence of the peak photocurrent as well as $Q_{initial}$ and Q_{tail}. At relatively high intensity (> · 10^5 W/cm^2), the exponents for the peak photocurrent, $Q_{initial}$ and Q_{tail}, are 1.0, 1.1 and 0.74, respectively [7]. In PDA-TS, the precise determination of the recombination mechanism(s) is made difficult by the known existence of trap levels (significant photoconductivity has been observed for photon energies well below the single-particle energy gap [8]) which can drastically affect the illumination intensity–dependence of the photocurrent [9].

We suggest that the initial fast peak in the photocurrent is due to "hot" carriers in extended band states that have acquired excess energy due either to the difference in photon excitation energy and the minimum band state energy at the bottom of the conduction band, or to ballistic acceleration of the carriers by the external field prior to the first trapping (or scattering) event. Calculations of the latter effect for traditional semiconductors demonstrate [10] overshoot of the drift velocity in the subpicosecond regime; a quick estimate shows that an overshoot by a factor of ten could be obtained at $E = 4 \times 10^4$ V/cm at times of order 0.1 ps after acceleration over a distance of order 10 Å. The larger initial photocurrent would, therefore, be due to a greater drift velocity than that inferred from μE (with $\mu \sim 5$ cm^2/Vs) at longer times. If this excess energy is much greater than $k_B T$, the typical lattice energy, then the lattice temperature is not important to the initial decay rate; rather, the initial decay comes from a decrease in drift velocity as the hot carriers thermalize. As the carriers thermalize (and recombine), a significant fraction fall into traps that govern their transport at longer times. It is this trap-dominated transport which we associate with the longer-time "tail" photoconductivity. At low temperatures, the probability of emission from traps is drastically reduced and, consequently, the tail should go to zero, in agreement with the experimental observations.

We note that since the sweep-out drift velocity in the tail is $\sim 2 \times 10^5$ cm/s, the much higher conductivity in the peak may result from a drift velocity greater than the sound velocity (consistent with hot carriers). This would imply that the time for thermalization to a polaron configuration is in the picosecond regime.

The inapplicability of Onsager geminate recombination theory to PDA-TS is perhaps not surprising, since one expects such concepts to be accurate only when the photoexcited carriers are localized. Although localization is to be expected in amorphous materials or in very narrow band molecular crystals, the broad π-bands of conjugated polymers tend toward extensive delocalization. Although a geminate pair can be self-localized by the Coulomb interaction, this effect can play no role If there are no bound excitons (as appears to be the case in polyacetylene [3]). Since bound excitons are known [8,11] to exist in PDA-TS, geminate recombination may be involved in the initial decay of photoexcitations. Our results on PDA-TS imply, however, that the role of geminate recombination, as developed within the Onsager formulation [1,6] (which has been often invoked but seldom proven) must be re-examined more generally.

Acknowledgments
We are grateful to Dr. K.J. Donovan for participating in the experiments during his stay in Santa Barbara (and for providing the PDA samples), and Professor E. G. Wilson for his encouragement and cooperation. We thank Dr. F. Yamagishi for providing samples which were used in the early phase of this study. The research was supported by the Office of Naval Research.

References
1.   K.J. Donovan, P.D. Freemen and E.G. Wilson: Mol. Cryst. Liq. Cryst 118, 395 (1985); K. J. Donovan and E.G. Wilson, Phil. Mag. B 44, 9, (1981).

2.   D.H. Auston: In Picosecond Optoelectronic Devices (Academic Press, New York, 1984) ed. by C.H. Lee.
3.   M. Sinclair, D. Moses and A.J. Heeger: Solid State Commun., 59, 343 (1986).
4.   B. Reimer and H. Bassler: Phys. Stat. Sol.(b) 85, 145 (1978).
5.   D. Haarer and H. Mohwald: Phys. Rev. Lett. 34, 1447 (1975).
6.   (a) R. Haberkorn and M.E. Michael-Beyerle: Chem. Phys. Lett. 23, 128 (1973).
      (b) F.D. Blossey: Phys. Rev.B 9,5183 (1974).
7.   This rules out the possibility that the changes in waveform shown in Fig. 3 are due to light intensity effects.
8.   H. Bassler: In Polydiacetylene: Synthesis, Structure and Electronic Properties (Martinus Nijhof, Dordrech, 1985) Ed. by D. Bloor and R. R. Chance, p. 135.
9.   Y. Marfaing: Handbook on Semiconductors, Vol. 2, Ch. 7 (North Holland, Amsterdam, 1981), Ed. by C. Hilsum.
10.   Semiconductors Probed by Ultrafast Laser Spectroscopy,Vol. I , ed. by R.R. Alfano (Academic Press, New York, 1984).
11.   L. Robins, J. Orenstein, and R. Superfine: Phys. Rev. Lett. 565,1850, (1986); and references therein.

Polarised Photoexcitation in Oriented Polyacetylene

P.D. Townsend and R.H. Friend

Cavendish Laboratory, Madingley Road,
Cambridge CB3 0HE, United Kingdom

Polarisation-dependent measurements of photoinduced absorption (PA) have been carried out on highly oriented films of polyacetylene prepared by the Durham precursor route. PA features at 0.48 eV and 1.36 eV, associated with charged and neutral excitations respectively, show differing behaviour with respect to pump and probe beam polarisation. The 0.48 eV feature is preferentially excited for perpendicular polarisation of the pump beam, but is strongly polarised along the chains. We consider that interchain charge separation is necessary to achieve the long-lived charged solitons observed. The 1.36 eV PA feature is, in contrast, isotropic with respect to the probe beam polarisation. We consider that this arises from neutral bound soliton - anti-soliton pairs which survive for the millisecond timescales probed here only if stabilised at appropriate chain defects. In view of the isotropic response with respect to probe beam polarisation we suggest that these are found where the chains are disordered and randomly oriented.

1. Introduction

Photoexcitation experiments have given some of the most direct information about relaxation of the polyacetylene chain to form 'soliton' defect states to accommodate charges added to the chain [1]. Since radiative recombination is very weak, photo-induced absorption (PA) is a more effective probe than luminescence, and since the first report by ORENSTEIN and BAKER [2] many studies of PA have been made; much of this is reviewedby ORENSTEIN [3]. It is expected that photoexcitation of an electron-hole pair will produce oppositely charged solitons [4,5]. The PA feature at 0.48 eV is indeed associated with these charged solitons. There is now considerable evidence however, that the photoexcitation mechanism in polyacetylene cannot be described fully by the one electron, single chain model of SU and SCHRIEFFER [1]. In particular, inter-chain motion of electrons [6] may strongly affect both the nature of the photoexcited states and their generation mechanism. A demonstration of the importance of correlation effects has been the observation in photoinduced absorption experiments of a long-lived (t > μs) overall-neutral state which gives rise to a PA peak at 1.4 eV [2]. This PA peak is identified with a neutral photoexcitation since, unlike the absorption feature seen at about 0.48eV which is attributed to charged solitons, it has no associated IR activity [8,9].

The nature of these neutral excitations is not well understood, although it is widely held that the defect state consists of an intra-chain pair of bound neutral solitons [5,10]. This requires a breakdown of the theorem which requires photogeneration of charged solitons [4], and some mechanism to keep at least some of these pairs metastable. There are two mechanisms put forward. One possible source of this effect is electron correlation, and it has been suggested that the bound S^o- S^o state is analogous to the lowest-lying A_g excited state in short polyenes [11]. Alternatively, KIVELSON and WU [5] have shown that the photogeneration of neutral soliton pairs can occur in polyacetylene even without strong electron-electron interactions. The presence of even weak interactions which break charge conjugation symmetry are predicted to lead to a rapid (t ~ 10^{-13} s) conversion of the charged solitons into bound neutral soliton pairs. KIVELSON and WU propose these pairs may be rendered metastable at low temperatures by interaction with existing structural defects such as remnant cis linkages.

Stretch orientation of films of Durham-route polyacetylene during thermal conversion from the precursor can give films with a very high degree of orientation [12-16]. These films are non-fibrous and are well suited for polarised optical measurements. Information about the anisotropy of intra- and inter-chain electron motion can be obtained from such measurements. We have shown elsewhere [6] that optical absorption above the π - π* gap for light polarised perpendicular to the chains is quite strong, with a peak absorption coefficient of about 10^4 cm^{-1}. This does not arise from parallel absorption from misaligned chains since the degree of alignment is too high (HWHM = 3° [17]), and also the temperature derivatives for the two polarisations have different spectral dependences. We consider that the overlap of π orbitals on adjacent chains is sufficient that the transverse-polarised absorption results in interchain excitation of electron-hole pairs. The PA measurements discussed here give further evidence for the interchain separation of charge following photoexcitation.

2. Results

Details of sample preparation and orientation are discussed elsewhere [12-17]. The excitation sources used for the PA measurements were either a Nd:YAG and dye laser combination, producing 12 nsec pulses at 2.2 eV, or a CW Ar$^+$ laser with a photon energy of 2.6 eV, chopped at 80 Hz. For both sources the change in sample transmission, ΔT, was measured by integrating the signal with a lock-in amplifier. Measurements of PA near the π-π* band edge were complicated by thermal modulation of the absorption edge by the light pulses. Separation of the thermal modulation signal from the true PA can be achieved by chopping the light faster than the thermal response of the sample, and rejecting the quadrature component in ΔT; this was possible with the Ar$^+$ laser, but limited by the 10 Hz repetition rate of the Nd:YAG laser.

PA at energies below 1 eV is shown in Fig. 1. We find the single, asymmetric absorption feature (with interference structure near the peak at 0.48 eV) that is reported for Shirakawa samples [2,9,10]. This feature is considered to arise from transitions between the band edges
and charged soliton, $S^{\pm}$, levels. This PA band is strongly polarised parallel to the chains. The variation with probe beam polarisation is shown in Fig. 2; the anisotropy of about 50 is limited by the polariser efficiency. We find that this PA is observed with both polarisations of the excitation beam, but is more strongly induced by perpendicular excitation. The anisotropy ratio is intensity dependent, as shown in Fig. 3. The PA signal saturates at high intensities, but while the dependence on pump intensity at lower power levels is sublinear for chopped CW

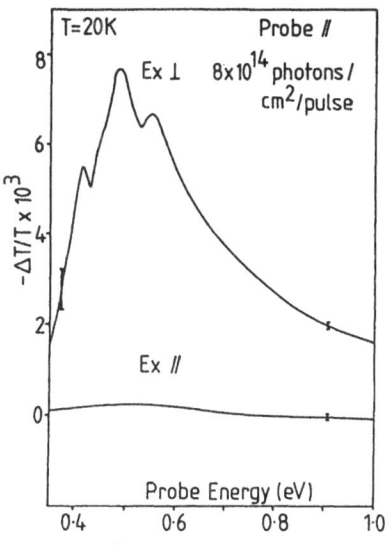
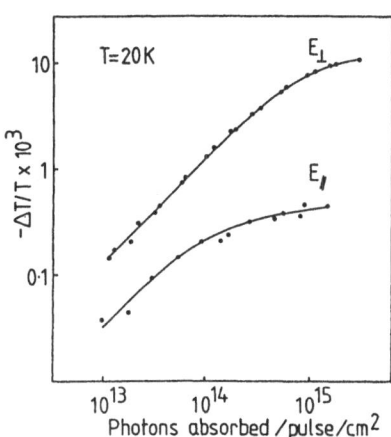

Fig. 1 PA spectra with probe parallel to chains (pump beam 12 nsec pulses at 2.2 eV)

Fig. 2 Intensity dependence of the PA near 0.5 eV. Reflectivity correction of 50% for parallel polarisation [6] is made to obtain absorbed photon flux

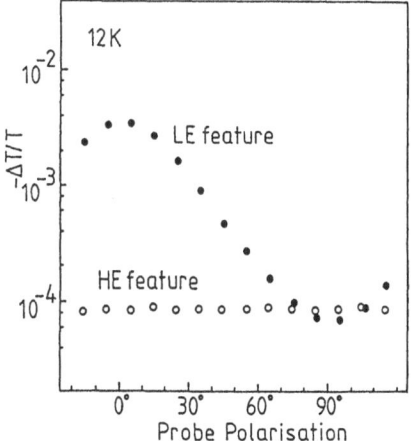

Fig. 3 Intensity of the PA peaks at 0.48 (LE) and 1.36 eV (HE) as a function of probe beam polarisation. Probe beam polarised parallel to chains = 0°. Pump at 2.6 eV polarised perpendicular to chains, intensity 66 mW/cm².

pump beam [9,10], the integrated signal following excitation with the 12 nsec pulse from the Nd:YAG/dye laser is linear at low intensities, as is shown in Fig. 3. At saturation, presuming that the density of photogenerated defects within the absorption depth, $1/\alpha$, is the same for $E_{//}$ and $E_{\perp}$, we simply probe the ratio of the optical absorption coefficients at the pump energy, $\alpha_{//}/\alpha_{\perp}$, which we find to be ≈ 25 at 2.2 eV. However, at intensities where the response is linear for both polarisations, the true anisotropy in the efficiency of photogeneration of defect states is measured, and we find a value of ≈ 4 in favour of excitation with $E_{\perp}$.

Measurements above 1 eV shown below were obtained with the chopped CW laser. The PA spectrum of oriented polyacetylene at 12 K is shown in Fig. 4(a) for the probe beam polarised perpendicular to the chains and in Fig. 4(b) for the probe beam polarised parallel to the chains. The neutral state absorption (1.36 eV) feature can be clearly seen peaking at 1.36 eV for both polarisations of the probe beam, and is preferentially induced by the perpendicular polarisation of the excitation beam. The bleaching features seen above 1.4 eV and the dip at 1.1 eV in Fig. 4(b) can be identified from their temperature, intensity and time dependences as arising mainly from thermal modulation effects . In contrast to the PA from the charged solitons at 0.48 eV, the neutral excitations give rise to an absorption peak which is independent of the probe polarisation (Fig. 2). This result is very unexpected since all other optical absorption features in oriented samples, photoinduced or otherwise, show marked anisotropies. A further important point is the relative weakness of the 1.36 eV feature in the Durham material. For example, ORENSTEIN et al.[18] find a $-\Delta T/T$ value of 3×10^{-3} for the 1.36 eV absorption in Shirakawa polyacetylene at 10K when irradiated with a photon flux of 5×10^{16} photons cm^{-2} s^{-1}. Under similar conditions our samples show a response some 200 times smaller, indicating a lower quantum yield or lifetime for the neutral states. Generation of the $S^{\pm}$ states is not affected however,

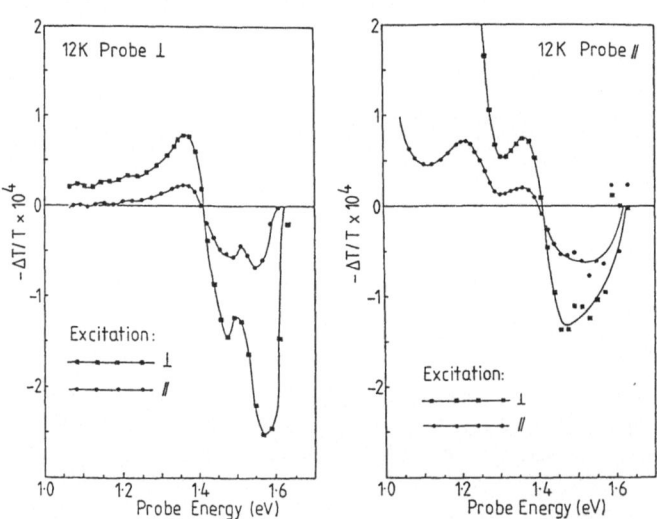

Fig. 4   PA spectrum with probe beam polarised perpendicular (a) and parallel (b) to the chains showing the peak at 1.36 eV due to neutral excitations. CW laser intensity 66 mW/cm^{-2}

since the magnitude of the 0.48 eV feature is comparable in both the Shirakawa and Durham materials.

We have also measured the dependence of the 1.36 eV absorption feature on excitation intensity, I, and temperature. At low intensities the PA exhibits a sublinear dependence of the form I^n with n=0.8 and n=0.7 for $E_\perp$ and $E_{//}$ excitation respectively. Throughout the intensity range investigated the 1.36 eV peak appears preferentially induced by $E_\perp$ excitation, although after correction in the absorbed flux for the high reflectivity (50%) for parallel polarisation, the anisotropy ratio is small at low intensities (e.g. ≈ 1.8 for 4 mW/cm^2). Dominance of the thermally induced bleaching signal at high excitation intensities explains our previous failure to observe the 1.36 eV feature [6]. In these experiments, with the high peak intensity of the 12 ns laser pulses, together with the long decay time for the thermally induced signal (~10 ms), only bleaching was observed. The temperature dependence of the 0.48 and 1.36 eV absorption features for the Durham samples investigated here is very similar to that reported for Shirakawa samples. The 1.36 eV feature is present only at low temperatures, and can no longer be resolved above about 100 K. In contrast, the 0.48 eV feature falls in intensity only by about 40% over the temperature range 15 - 100 K, and is observable up to 200 K and above.

The differences between the temperature dependence of the 0.48 eV and 1.36 eV features have been exploited by VARDENY and TAUC [7] to investigate the the onset of photoinduced bleaching (PB) associated with the 0.48 eV feature. Their measurements at 200 K, at which temperature the 1.36 eV feature is absent, show PB at 1.4 eV, below the band edge. They attribute this bleaching to the eventual trapping of charged excitations on the neutral soliton-like defects, present in as-made trans-polyacetylene in a concentration of typically 10^{19} cm^{-3}. The net process is summarised as $2S^0 \rightarrow S^+ + S^-$, and VARDENY and TAUC suggest that the effect of on-site Coulomb repulsion, U, shifts optical transitions associated with $S^\pm$ below mid-gap (PA at 0.48 eV) and transitions from band edges to S^0 to above mid-gap (PB at 1.4 eV). However, we have previously reported [6] that we can find PB associated with the 0.48 eV feature only above the band edge. We have carefully measured the PA spectra at 200 K here, and again find no evidence for PB at 1.4 eV other than that associated with thermal modulation of the absorption edge. We conclude that for samples of Durham polyacetylene the PB associated with the 0.48 eV photoinduced absorption occurs above the band edge, and does not involve the participation of the neutral soliton defects present.

3. Discussion

There is continuing controversy as to whether the charged or the neutral states probed here are the intrinsic photoexcitations of the polyacetylene chain. Our measurements indicate that neither of the states probed on the long time scales used here are intrinsic. For the charged states associated with PA at 0.48 eV and slow photoconductivity [6,13], the preferential excitation for perpendicular polarisation of the pump beam indicates that the long lived charged excitations are initially separated between chains. We consider the explanation for this is that whereas intra-chain electron hole excitations are prone to geminate recombination, charges separated between chains are able to diffuse away from each other. Inter-chain excitation of electron hole pairs must initially produce positive and negative polarons, although on the time scales probed here those remaining must have combined to form solitons, probably as like charged pairs on the same chain. The isotropic behaviour of the 1.36 eV feature and its much weaker intensity than in Shirakawa samples strongly suggest that the excitation seen is stabilised at chain defects, as modelled by KIVELSON and WU [5], and that these are found in disordered regions of the sample where the chains are randomly oriented. Recent picosecond resolved experiments on unoriented samples provide support for some of these views [19], and similar experiments on oriented samples are now required.

References

1. W. P. Su and J. R. Schrieffer: Proc. Natl. Acad. Sci USA **77**, 5626 (1980)
2. J. Orenstein and G. L. Baker: Phys. Rev. Lett. **49**, 1043 (1982)
3. J. Orenstein: In Handbook of Conducting Polymers, Vol. 2, p 1297, ed. by T. Skotheim (Marcel Dekker, New York 1986)
4. R. Ball, W. P. Su and J. R. Schrieffer: J. Phys. (Paris) **44**, C3, 429 (1983)
5. S. Kivelson and W. K. Wu: Phys. Rev. B**34**, 5423 (1986)
6. P. D. Townsend and R. H. Friend: Synthetic Metals **17**, 361 (1987)
7. Z. Vardeny and J. Tauc: Phys. Rev. Lett. **54**, 1844 (1985)
8. B. Horovitz: Solid State Commun. **41**, 729 (1982)
9. G. B. Blanchet, C. R. Fincher, T. C. Chung and A. J. Heeger: Phys. Rev. Lett. **50**, 1958 (1983)
10. Z. Vardeny, E. Ehrenfreund and O. Brafman: Solid State Sciences **63**, 91 (Springer, New York 1985)
11. B. Hudson and B. Köhler: Synthetic Metals **9**, 241 (1984)
12. G. Leising: Polymer Bull. **11**, 401 (1984)
13. P. D. Townsend, D. D. C. Bradley, M. E. Horton, C. M. Pereira, R. H. Friend, N. C. Billingham, P. D. Calvert, P. J. S. Foot, D. C. Bott, C. K. Chai, N. S. Walker and K. P. J. Williams: Solid State Sciences **63**, 50 (Springer, New York 1985)
14. G. Leising, H. Kahlert and O. Leitner: Solid State Sciences **63**, 56 (Springer, New York 1985)
15. M. Sokolowski, E. A. Marseglia and R. H. Friend: Polymer **27**, 1714 (1986)
16. H. Kahlert, O. Leitner and G. Leising: Synthetic Metals **17**, 467 (1987)
17. D. D. C. Bradley, R. H. Friend, T. Hartmann, E. A. Marseglia, M. M. Sokolowski and P.D. Townsend: Synthetic Metals **17**, 473 (1987)
18. J. Orenstein, G. L. Baker and Z. Vardeny: J. Phys. (Paris) **44**, C3, 407 (1983)
19. L. Rothberg, T. M. Jedhu, S. Etemad and G. L. Baker: Phys. Rev. Lett. **49**, 3229 (1986)

Radiative and Non-radiative Recombination Processes in Photoexcited Poly(p-phenylene vinylene)

D.D.C. Bradley[1], R.H. Friend[1], K.S. Wong[2], W. Hayes[2],
H. Lindenberger[3], and S. Roth[3]

[1]Cavendish Laboratory, Madingley Road,
Cambridge CB3 0HE, United Kingdom
[2]Clarendon Laboratory, Parks Road, Oxford OX1 3PU, United Kingdom
[3]Max-Planck-Institut für Festkörperforschung,
Heisenbergstr. 1, D-7000 Stuttgart 80, Fed. Rep. of Germany

In this paper we report studies of the luminescence from photoexcited poly(p-phenylene vinylene) [PPV]. A strong decrease in the yield and lifetime is observed as the extent of unsaturation is increased by thermal conversion from the precursor to fully converted PPV. Such behaviour can be explained by non-radiative recombination processes that compete with the luminescence and which become increasingly efficient as the conjugation length increases. The non-radiative decay is separated into monomolecular and bimolecular components both of which are seen to increase with conjugation length; with the latter of greater relative importance for the more highly converted samples. The monomolecular component appears consistent with recent theoretical treatments whilst the origin of the bimolecular component is at present unclear.

1 Introduction

The photoexcitation of conjugated polymers has attracted considerable interest following the predictions of the numerical simulations by SU and SCHRIEFFER [1] that photogenerated electron-hole pairs ought to relax within an optic phonon period (~0.1ps) to give the same self-localised defect states that are formed by charge transfer doping.

The basic processes that can follow photogeneration of an electron-hole pair in a non-degenerate ground state polymer are outlined in Fig.1. The simple models [2] based on the SU-SCHRIEFFER-HEEGER [SSH] Hamiltonian [3] do not consider the possibility of charge separation prior to lattice relaxation and take the relaxed polaron-exciton as their starting point. Subsequent charge separation into two polarons is said to be impossible on account of the confinement associated with the non-degenerate ground state. On this basis it would be expected that the polaron-exciton state will undergo geminate recombination and since non-radiative decay is forbidden by the charge conjugation symmetry [CCS] of the SSH Hamiltonian [4], the recombination should be exclusively radiative. Hence, a strong luminescence with a near 100% quantum yield and a lifetime of the order of nanoseconds should arise. Experimental observations on a variety of different materials are, however, in marked contrast to these predictions and the luminescence yields and lifetimes are respectively found to be $\leq 1\%$ and ≤ 300ps [5, 6, 7, 8].

As indicated in Fig.1 there are a number of alternative decay paths not considered in the simple models. Charge separation, either through inter-chain transfer or by intra-chain trapping processes will initially give rise to polarons which can diffuse and, if two like charged states meet, can subsequently coalesce to form bipolarons. Evidence for charge separation is provided by the observation of photoinduced absorption from electronic transitions involving the intra-gap levels of polaron or bipolaron states and in particular by the observation of photoinduced infrared active modes associated with the charge on the defects. Such absorption is indeed seen in PPV [9, 10, 11] and other materials [12, 13]. Furthermore, the inclusion of a finite second nearest neighbour hopping integral (a term which should occur in any real system) breaks the CCS of the SSH Hamiltonian and allows non-radiative processes [4]. Thus, as shown in Fig.1, the polaron-exciton would also be expected to decay non-radiatively. The presence of such competing pathways is consistent with the observation of reduced yields and lifetimes for

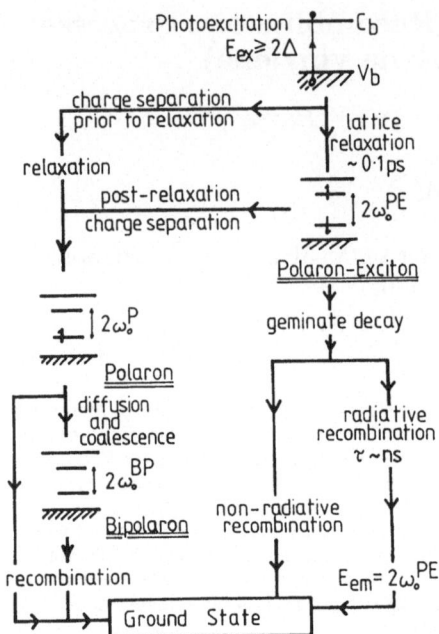

Photoexcitation $E_{ex} \geq 2\Delta$ — C_b

V_b

charge separation prior to relaxation

relaxation

lattice relaxation ~0·1ps

post-relaxation charge separation

$2\omega_0^{PE}$

Polaron–Exciton

geminate decay

$2\omega_0^P$

Polaron

diffusion and coalescence

radiative recombination $\tau \sim ns$

$2\omega_0^{BP}$

Bipolaron   non-radiative recombination

recombination $E_{em} = 2\omega_0^{PE}$

Ground State

Figure 1: Recombination processes following photoexcitation of a non-degenerate ground state conjugated polymer

radiative recombination. In this paper we concentrate on studies of the non-radiative decay processes as probed through their effect upon the luminescence emission. The results of photoinduced absorption studies on photogenerated charged bipolaron states are reported elsewhere [9,10,11].

DANIELSEN [4] has theoretically investigated the monomolecular non-radiative decay of polaron-excitons through multiphonon emission and tunneling processes and has calculated the decay rate for cis-$(CH)_x$ which is found to be consistent with the experimental lifetime. Generalisation to other non-degenerate ground state polymers indicates that the decay rate should show a strong dependence upon the electron-phonon coupling α, the band-gap 2Δ and the confinement parameter γ [4]. Of these parameters, the band-gap is the most readily accessible experimentally and the expected dependence on Δ is in qualitative agreement with the observed trend [9] amongst the different materials studied to date, where the radiative yield and lifetime are seen to decrease as the band-gap decreases.

The polymer chosen for the present study is poly(p-phenylene vinylene) which has been prepared via the sulphonium polyelectrolyte precursor route [14, 15]. The precursor polymer poly(p-xylene-α-(dimethyl sulphonium chloride)) was synthesised as detailed by KARASZ et al. [16]. The thermal conversion process from this precursor to PPV has been discussed elsewhere [9, 17, 18]. One particular advantage of the synthesis that is of importance for the present work is that the conjugation length and hence band-gap can be controlled through the use of different conversion temperatures and heating times. The extent of conjugation has been monitored by changes in the optical absorption spectra [9, 17] and in the relative intensities of certain infrared bands [9, 18]. Whilst the control is not quantitative and the conversion always results in a distribution of conjugation lengths, it is sufficient to allow qualitative examinations of the effects of band-gap on the non-radiative recombination processes. In fact, as seen later, this distribution of conjugation lengths does not cause problems since the observed emission appears to occur preferentially from the longest conjugated segments in the sample following a presumed rapid energy migration. The latter postulate is supported by the observation that the emission does not shift as a function of excitation energy.

2 Luminescence from Fully Converted PPV

Figure 2 shows the luminescence emission from a fully converted (300°C for 7 days) film of PPV. The spectrum was excited by 15 ns pulses from a frequency-trebled (355 nm) Nd:YAG laser and recorded at 25 K using an S20 response photomultiplier tube. The same spectrum is observed when the 457.9 nm line of a CW Ar$^+$ laser is used as excitation source. The lifetime has been measured with a streak camera following excitation by 4 ps pulses from a frequency doubled (322 nm) DCM dye laser and is found to be ca 300 ps [19]. The quantum yield is estimated to be < 0.1%.

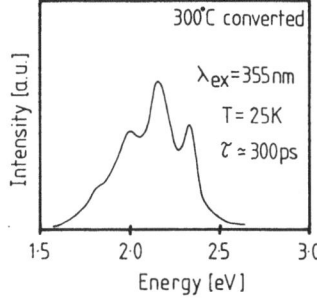

Figure 2 : Luminescence emission from fully converted PPV

The presence of a set of resolved phonon lines with energy spacing of ca 0.16 eV (consistent with a stretching vibration of the carbon backbone [9, 10]) indicates that the emission centre is in thermal equilibrium with the lattice and hence that the short lifetime does not simply arise from a hot emission process. The phonon lines show further that the polymer is highly ordered, in agreement with the all-trans· configuration deduced from infrared spectra and with the high crystallinity detected by X-ray diffraction [18, 20]. Moreover, they suggest,as mentioned above, that following photoexcitation and prior to emission there is a rapid migration of excited states to the long conjugation length segments. If this were not the case, the distribution of emission centres would be expected to smear out the vibronic structure. Similar energy transfer processes are commonly seen in mixed molecular crystals and are found there to be extremely rapid [21, 22].

The magnitude of the Stokes' shift between the maxima in the excitation and emission spectra measures the extent of relaxation in the excited state. For fully converted PPV it has been determined by TOKITO et al. [23] and at room temperature is found to be 0.18 eV. This corresponds to a value for $2\omega_0^{PE}$ (see Fig.1) of 2.32 eV (using $2\Delta = 2.5$ eV [9, 17]) and thus suggests that the polaron-exciton is relatively weakly relaxed (cf $2\omega_0^{BP} = 1.0$ eV [9, 10, 11]). One explanation for this behaviour would be that the Coulomb attraction between the two charges increases γ and thus causes a strong localisation of the defect.

Measurements of the excitation-intensity dependence provide additional information on recombination processes. For excitation with the CW Ar$^+$ laser line at 457.9 nm, the integrated emission intensity is found to increase linearly with the excitation intensity. This shows that the recombination processes are monomolecular as expected for geminate recombination. Use of the pulsed Nd:YAG laser (355 nm), for which the excitation densities are expected to be some six orders of magnitude higher, leads instead to a markedly sub-linear dependence [9, 10]. The integrated emission intensity increases initially as I to the power 0.7 going steadily over to I to the power 0.5 at the highest powers used.

This behaviour clearly shows that non-linear decay processes are also important at high excitation density and suggests modelling the recombination dynamics by a rate equation of the form

$$dn / dt = - [n(t) / \tau] - \beta n^2(t) , \tag{1}$$

where $n(t)$ is the excited state density at time t, τ is the monomolecular lifetime (including both radiative and non-radiative processes) and β is a bimolecular non-radiative rate constant. The solution of (1) is given by

$$n(t) = C / [\exp(t / \tau) - \beta \tau C] , \tag{2}$$

where

$$C = n(o) / [1 + \beta \tau n(o)] . \tag{3}$$

Equation (2) is found to give a very good fit to the non-exponential decay observed in picosecond lifetime measurements [19] and allows separation of the contributions from monomolecular and bimolecular processes.

3 Conjugation Length Dependence

We now turn our attention to investigations of the conjugation length dependence of recombination. Firstly, it is interesting to consider the emission from

$$CH_3-CH_2-CH_2-[C_6H_4-CH=CH]_2-C_6H_4-CH_2-CH_2-CH_3 \qquad [BPSB]$$

which is an effective dimer of PPV with additional saturated chain-end groups. The luminescence from trace amounts of BPSB in a solution of cyclohexane has been reported by BERLMAN [24] and consists of an emission band that is identical in shape to that shown in Fig.2 but with the whole band shifted ca 0.8 eV to higher energy in agreement with the shift in energy-gap. The equivalence of the emission from fully converted PPV films to that from BPSB in solution implies that the former must involve an intra-chain state. The appearance of the same phonon structure is in stark contrast to the large shift in electronic energy levels and tends to support the surmise that the defect state involved is a highly localised one with a small extension along the chain.

Strong differences between BPSB and fully converted PPV do however exist in that the radiative yield and lifetime of BPSB are respectively 94% and 1.09 ns. These values are as predicted by the simple models discussed earlier in which charge separation and non-radiative recombination processes are not considered. Charge separation by trapping or inter-chain transfer would clearly not be expected for BPSB in solution but the absence or negligible effect of non-radiative recombination is not such an obvious result.

Figure 3 shows the effect of increasing conjugation upon the emission intensity from PPV. All the spectra were recorded with a constant excitation intensity of 30 mW using the 457.9 nm line from a CW Ar^+ laser. Under the same conditions the emission from the fully converted film was undetectable. The relatively weak energy shifts with increasing conversion support the suggestion that the emission occurs in each case from the longer chain segments (with more than ca 5 to 6 monomer units) for which the variation in the energy gap is small [25]. Nevertheless, it

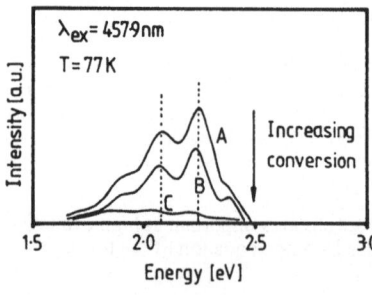

Figure 3 : Luminescence emission from samples with increasing degrees of conversion (A) weakly converted precursor, (B) 70 hours at 130°C and (C) 10 hours at 170°C. The dashed lines indicate the positions of the two strongest phonon peaks for sample (A) and serve to highlight the weak shifts towards lower energy of the emission from more highly converted samples.

is clear that even for such small changes in the electronic energy gap there are very dramatic changes in emission intensity. For the excitation conditions used here the excitation-intensity dependence was in the linear regime for all samples so that the observed effect must be related to the monomolecular component in the non-radiative decay. This very strong decrease in luminescence for a small change in energy gap is in qualitative agreement with the strong dependence predicted by DANIELSEN [4] whereby the monomolecular non-radiative decay rate should vary as $\exp(-\Delta)$.

Confirmation that the decrease in emission intensity arises as a result of an increase in non-radiative decay is obtained from picosecond lifetime measurements [19]. Experiments on three oriented samples heated respectively at 80°C for 2 hours, 133°C for 148 hours and 300°C for 7 days show a strong decrease in room-temperature lifetime on going toward the fully converted state [19]. There are however added complications in interpretation, since at the excitation densities used for the picosecond measurements, bimolecular non-radiative decay is significant and hence the monomolecular lifetime has to be determined from the fit to (2) as discussed above. The relative importance of the bimolecular non-radiative decay is found to increase strongly as the extent of conversion is increased [19]. Perpendicular excitation of the 80°C converted sample yields an exponential decay (lifetime of ca 460 ps [19]) that is indicative of purely monomolecular processes. The 300°C converted sample, however, shows a strongly non-exponential decay, implying that here the dominant contribution arises from bimolecular processes. The effective lifetime in the latter case is ca 60 ps (time for signal to fall to 1/3 of its maximum value) and the monomolecular lifetime determined from the fit to (2) is ca 235 ps [19]. The results of the fitting confirm that the monomolecular lifetime decreases with increasing conversion [19]. The origin of the bimolecular non-radiative decay is at present unknown although its increasing relative importance for more highly converted samples can be understood in that the mobility of excited states should be higher in such samples, thus allowing a greater chance of interaction between the states.

4 Summary and Discussion

The measurements reported above for luminescence from photoexcited samples of PPV with varying extents of conversion clearly show the importance of non-radiative decay in rapidly depopulating the excited state manifold. This behaviour is contrary to the proposals of simple models based upon the SSH Hamiltonian and requires the CCS of this Hamiltonian to be broken. The results show that the monomolecular non-radiative decay has a strong dependence upon the energy gap with the decay rate increasing as the gap decreases. This behaviour is qualitatively as predicted by recent theoretical studies [4].

The presence of a bimolecular non-radiative decay may also be inferred from the data. This bimolecular component is found to increase with the extent of conversion. Furthermore, it is found to be of increasing relative importance for more highly converted samples, thus indicating an even stronger dependence upon the energy gap than that seen for the monomolecular decay. There have not as yet been any theoretical investigations of bimolecular non-radiative decay in conjugated polymers.

In all of the above there has been no mention of temperature dependences, polarisation dependences or the effects of extrinsic decay processes involving defect-mediated trapping. These aspects are currently under investigation and have already shown some interesting results including a strong low–temperature fatiguing of the luminescence from weakly converted samples. This work will be published elsewhere.

5 Acknowledgements

It is a pleasure to acknowledge useful discussions with P.D. Townsend and Dr. A.D. Yoffe. We thank SERC and BP plc for financial support and the organisers of IWEPP'87 for the opportunity to present this work.

6 References

1. W.P. Su and J.R. Schrieffer : Proc. Natl. Acad. Sci. USA 77, 5625 (1980)
2. L. Lauchlan, S. Etemad, T-C Chung, A.J. Heeger and A.G. Macdiarmid : Phys.Rev. B24, 3701 (1981) and references therein
3. W.P. Su , J.R. Schrieffer and A.J. Heeger : Phys. Rev. B22, 2099 (1980)
4. P.L. Danielsen : Synthetic Metals 17, 87 (1987) and references therein
5. W. Hayes, C.N. Ironside, J.F. Ryan, R.P. Steele and R.A. Taylor : J.Phys.C16, L729 (1983)
6. S. Etemad, G.L. Baker, J. Orenstein and K.M. Lee : Mol.Cryst.Liq.Cryst. 118, 389 (1985)
7. K.S. Wong, W. Hayes, T. Hattori, R.A. Taylor, J.F. Ryan, K. Kaneto, Y. Yoshino and D.Bloor : J.Phys.C18, L843 (1985)
8. W.J. Feast, I.S. Millichamp, R.H. Friend, M.E. Horton, D. Phillips, S.D.D.V. Rughooputh and G.Rumbles : Synthetic Metals 10, 181 (1985)
9. D.D.C. Bradley : PhD Thesis, University of Cambridge (1987)
10. D.D.C. Bradley, R.H. Friend and W.J. Feast : Synthetic Metals 17, 645 (1987)
11. D.D.C. Bradley, R.H. Friend, F.L. Pratt, K.S. Wong, W. Hayes, H. Lindenberger and S. Roth : these proceedings
12. Z. Vardeny, E. Ehrenfreund, O. Brafman, A.J. Heeger and F. Wudl : Synthetic Metals 17, 183 (1987) and references therein
13. F.L. Pratt, K.S. Wong, W. Hayes and D. Bloor : J.Phys.C20, L41 (1987) and references therein
14. R.A. Wessling and R.G. Zimmermann : US patents #3,401,152 (1968) and #3,706,677 (1972)
15. M. Kanbe and M. Okawara : J.Polymer Sci. A-1 6, 1058 (1968)
16. F.E. Karasz, J.D. Capistran, D.R. Gagnon and R.W. Lenz : Mol.Cryst.Liq.Cryst. 118, 327 (1985)
17. D.D.C. Bradley, G.P. Evans and R.H. Friend : Synthetic Metals 17, 651 (1987)
18. D.D.C. Bradley, T. Hartmann, R.H. Friend, E.A. Marseglia, H. Lindenberger and S. Roth : these proceedings
19. K.S. Wong, D.D.C. Bradley, W.Hayes, J.F. Ryan, R.H. Friend, H. Lindenberger and S. Roth : J.Phys.C in press (1987)
20. D.D.C. Bradley, R.H. Friend, T.Hartmann, E.A. Marseglia, M.M. Sokolowski and P.D. Townsend : Synthetic Metals 17, 473 (1987)
21. M.J. Davies, A.C. Jones, J.O. Williams and R.W. Munn : J.Phys.Chem.87, 541 (1983)
22. M.D. Lumb : Organic Luminescence Chapter 3 in Luminescence Spectroscopy Academic Press, New York and London (1978)
23. S. Tokito, S.Saito and R. Tanaka : Makromol.Chem.Rapid Commun.7, 557 (1986)
24. I.B. Berlman : Handbook of Fluorescence Spectra of Aromatic Molecules Academic Press, New York and London (1971)
25. P.R.Surjan : private communication

Photoinduced Absorption in Poly(p-phenylene vinylene)

D.D.C. Bradley[1], *R.H. Friend*[1], *F.L. Pratt*[2], *K.S. Wong*[2], *W. Hayes*[2], *H. Lindenberger*[3], *and S. Roth*[3]

[1]Cavendish Laboratory, Madingley Road,
Cambridge CB3 0HE, United Kingdom
[2]Clarendon Laboratory, Parks Road, Oxford OX1 3PU, United Kingdom
[3]Max-Planck-Institut für Festkörperforschung,
Heisenbergstr. 1, D-7000 Stuttgart 80, Fed. Rep. of Germany

In this paper we report the results of preliminary investigations of photoinduced absorption in the non-degenerate ground state conjugated polymer poly(p-phenylene vinylene) [PPV]. We present photoinduced spectra for both electronic and vibrational transitions which closely resemble the corresponding spectra observed following charge-transfer doping. The photogenerated charged defects are identified as bipolarons.

1. Introduction

As discussed in another contribution to these proceedings [1] and elsewhere [2], consideration of simple models [3] based upon the SU-SCHRIEFFER-HEEGER [SSH] Hamiltonian [4] suggests that photogeneration of charged defect states should not be expected for non-degenerate ground state polymers. However, inter-chain electron transfer and intra-chain trapping processes should ensure that geminate recombination of the electron-hole pair may in fact be avoided and hence that charged defect states can be formed. Following the separation of photoexcited electron-hole pairs we would expect an initial formation of self-localised polarons which may move through the sample, and if they meet like charged states may subsequently coalesce to form doubly charged bipolarons. The predicted [5] thermodynamic stability of a bipolaron with respect to two polarons does not allow for the effect of the Coulomb repulsion between the two charges. It is thus of obvious interest to know whether bipolarons or polarons are actually favoured and photoexcitation measurements, unlike doping experiments, provide the opportunity to answer this question without having to consider the effects of dopant counter ions.

The presence of photoexcited polaron or bipolaron defect states is expected to lead to transient induced absorption arising from electronic transitions involving their intra-gap energy levels. In addition to this photoinduced absorption [PA], the presence of charge should also lead to photoinduced infra-red activity [6]. Indeed, the detection of such infra-red modes provides the strongest evidence for the formation of charged defects. Consideration of these photoinduced infra-red modes within the "amplitude-mode" formalism of HOROVITZ and coworkers [6, 7] can provide considerable information upon microscopic electronic properties. In particular, the frequencies of the modes provide information on the pinning of the defect charges and hence give a qualitative measure of the extent of π-electron delocalisation.

The results that are presented below detail experiments on PPV prepared via the precursor sulphonium polyelectrolyte route [8, 9]. This route has a number of advantages. Highly oriented samples may be readily prepared [2, 10, 11, 12, 13] that enable investigations of anisotropy to be undertaken. Furthermore, the crystallinity [3, 14, 15, 16, 17] and conjugation length [2, 13, 18] may be controlled through the use of suitable conversion conditions. The control of the conjugation length is especially useful since many properties such as the defect dynamic mass are expected to depend upon the extent of π-electron delocalisation.

2. Electronic Transitions of Photoexcited Defects

Figure 1 shows the PA spectrum of a fully converted (i.e. 300°C), stretch aligned ($1 / 1_0 = 5$) PPV sample for excitation perpendicular and probe beam parallel to the stretch direction. The excitation source was the pulsed (15 ns) frequency-tripled (355 nm) output from a Nd:YAG laser operating at a 10 Hz repetition rate. The probe beam was provided by a tungsten lamp and a glow bar.

The data was recorded using solid-state detectors and a lock-in amplifier which was gated to take out the short-lived luminescence [1, 2, 19, 20] that partially overlaps the spectrum. The induced absorption is strongly polarised along the chain axis with negligible absorption for the probe beam perpendicular. The appearance of only two PA peaks, at 0.6 eV and 1.6 eV suggests that bipolarons, rather than polarons, are the energetically favoured, charge carriers generated by above-gap photoexcitation [2]. This is in agreement with the results of chemical and electrochemical doping for which two peaks are seen at ca 0.9 eV and ca 2.1 eV [2, 13]. The assignment to bipolarons is further supported by the value of the ratio ω_o / Δ, where $2\omega_o$ is the inter-level spacing and 2Δ the band-gap. For 2Δ = 2.5 eV [2, 13, 21] and with $2\omega_o$ = 1.6 - 0.6 = 1 eV, we have ω_o / Δ = 0.40 which is too small to be consistent with polarons [22]. We note that recent results for polythiophene [23] and 1-hydroxy-2,4-hexadiyne polydiacetylene [PDA-1OH] [24, 25] also indicate that bipolarons are the dominant photoexcited charge carriers. This implies that the Coulomb repulsion contribution to the energy of a bipolaron is insufficient to prevent its formation. There is thus an indirect attractive interaction between two like charges in the presence of a local structural distortion.

The splitting of the levels by 1 eV (i.e. ω_o = 0.5 eV) corresponds within the FESSER-BISHOP-CAMPBELL [FBC] model [22] to a confinement parameter of γ = 0.27 which is the value that was also deduced for polythiophene [23]. For this value of γ, the FBC theory predicts an intensity ratio I(0.6) / I(1.6) = 12 which is clearly incompatible with the observed ratio. Similar behaviour is seen for polythiophene [23]. The reason for this discrepancy is not well understood. It appears that the intensity ratio can vary considerably for different dopants [2] and this tends to suggest that the model used to calculate the absorption strengths is oversimplified. The model only explicitly treats a single defect on an isolated chain. The inclusion of the effects of interactions between defects and between chains is likely to considerably alter the absorption characteristics. The discrepancy has recently been commented upon by CAMPBELL et al. [26] and SUM et al. [27] who have considered the effect of including electron-electron interactions in their model Hamiltonian. CAMPBELL et al. [26] report that for weak on-site electron-electron coupling, the predicted difference in intensity for the two bipolaron transitions is largely unaltered from the FBC result in the absence of correlations. In the strong coupling limit however, it is suggested that the two peaks may be of comparable strength. SUM et al. report similar results for calculations based upon a Hamiltonian that includes both on-site and inter-site correlation terms [27].

Another conclusion of the work by CAMPBELL et al. [26] is that on-site electron-electron effects should cause a shift in the positions of the two defect levels relative to the conduction and valence band edges. They should not however alter the inter-level spacing. As a result, the two levels would be expected to lie asymmetrically about mid-gap and have transition energies whose sum is less than the full gap by an amount equal to the shift due to correlation. The Coulomb interaction energy may thus be estimated from the displacement of the two levels and following

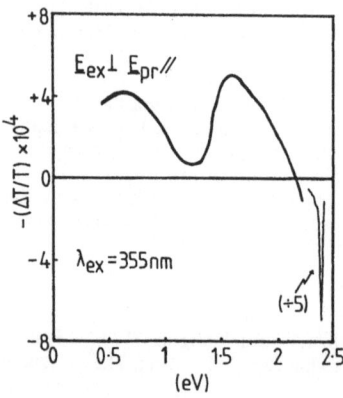

Figure 1 : Photoinduced absorption spectrum of fully (300°C) converted PPV. The spectrum was recorded at 23K with the probe beam polarised parallel to the stretch direction and the excitation perpendicular.

VARDENY et al. [23] this gives in PPV a value for U_B [23] of + 0.15 eV. The latter is somewhat smaller than found for either polythiophene [23] or PDA-1OH [24, 25].

The PA shown in Fig. 1 has a strong temperature dependence [2, 19]. The signal rapidly decreases on warming and becomes undetectable (i.e. $-\Delta T / T < 10^{-5}$) above ca 160 K. The dependence is similar to that observed for the soliton band at 0.45 eV in trans-$(CH)_x$ where the strong decrease in magnitude of the PA above 150K has been attributed [28] to a reduction in lifetime for the charged defect state. A similar dependence is also reported for the photoinduced infra-red absorption in polythiophene [29]. It is interesting to note that PA signals persist to much higher temperatures in weakly converted samples [2]. Thus if the origin of the fall-off in $-\Delta T / T$ with increasing temperature is, as proposed for trans-$(CH)_x$, related to a reduction in the excited state lifetime due to an increase in mobility, then the behaviour for the weakly converted samples suggests a greatly reduced mobility. This would be consistent with the reduced conjugation in such samples [2].

The strong sharp bleaching (i.e. negative PA) feature that is seen in Fig. 1 (reduced by a factor of five) is identified by its linear intensity dependence and isotropic excitation as being the result of thermal modulation of the absorption edge due to sample heating [2].

3. Vibrational Transitions of Photoexcited Defects

The photoinduced measurements in the infra-red were performed using a Perkin Elmer 1710 FTIR spectrometer with excitation provided by a CW He-Cd laser (422 nm). Interferograms were recorded with the laser alternately on and off the sample and co-added until a satisfactory signal-to-noise ratio was obtained. Figure 2 shows the results for a fully (i.e. 300°C) converted, oriented ($l / l_0 = 6.3$) sample of PPV. The spectrum was recorded at 77K with probe beam parallel to the stretch direction and was the result of 6000 scans with laser on and laser off. No signal was detectable for the probe beam perpendicular. The main photoinduced infra-red modes are indicated in the figure by arrows and appear at 1470 cm^{-1}, 1398 cm^{-1}, 1274 cm^{-1} and 1100 cm^{-1}. There is a close correspondence between this spectrum and those obtained following charge-transfer doping [2, 13, 16] but the photoinduced modes are shifted to lower energy. This behaviour is consistent with a reduction in the pinning of the defect charge which would be expected in the absence of electrostatic interactions with dopant counter ions.

Preliminary investigations on partially converted samples failed to detect a signal. The cause of this behaviour, which appears at odds with the strong signal seen in such samples for electronic transitions, is unclear. However, it is expected that the defect dynamic mass will be high in these partially converted samples due to their reduced π-electron delocalisation, and this would tend to

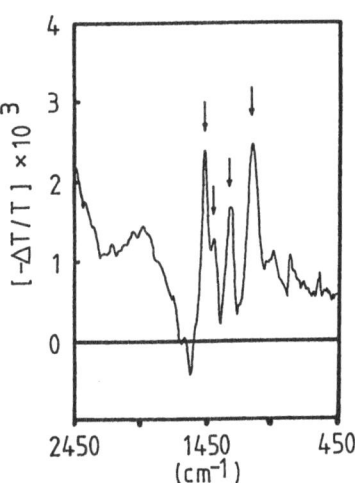

Figure 2 : Photoinduced infra-red absorption in fully (300°C) converted PPV. The spectrum was recorded at 77K with the probe beam polarised parallel to the stretch direction.

reduce the relative strength of the infra-red modes. Another possible explanation is that the lifetime of the defect states is long compared to the chop frequency used as the reference for phase-sensitive detection, and hence that the change detected per cycle represents only a small fraction of the total magnitude of the induced absorption. Evidence for very long lived photoexcited states, with τ of the order of hours, is obtained from investigations of the luminescence fatigue that is observed in weakly converted samples at low temperatures [30]. Moreover, photoinduced infra-red measurements on PDA-1OH also show indications of very long lifetimes for some of the photoexcited charge carriers [25]. Further experiments are underway to try and resolve this point.

4. Summary and Discussion

The appearance of both photoinduced intra-gap absorption which may be attributed to electronic transitions involving self-localised defect levels and more specifically the photoinduced infra-red absorption associated with the defect charge is a clear indication that contrary to the expectations of simple models [3,4], charge separation does take place following photoexcitation of a non-degenerate ground state conjugated polymer. The predominant long-lived photoexcited charge carriers are identified as bipolarons. This implies that the Coulomb energy of these defect states is insufficient to prevent their formation in preference to two polarons. The Coulomb energy is estimated to be + 0.15 eV.

The associated infra-red modes have frequencies that are lower than those seen in doping experiments, which suggests a reduced pinning, as expected in the absence of dopant counter ions. The failure to detect photoinduced infra-red modes in weakly converted samples is unexpected in the light of the strong electronic PA signal seen for such samples. This absence might be taken to indicate that the defects present in these samples are neutral, but in our opinion is more likely to be related to an increased dynamic mass and / or lifetime arising from reduced π-electron delocalisation.

Further detailed studies of PA associated with both electronic and vibrational transitions are at present underway, and should allow a greater understanding of the processes of charge generation and the formation of metastable defect states following photoexcitation.

5. Acknowledgements

It is a pleasure to acknowledge useful discussions with P.D. Townsend and Dr. A.D. Yoffe. We thank the SERC and BP plc for financial support and the organisers of IWEPP'87 for the opportunity to present this work.

6. References

1. D.D.C. Bradley, R.H. Friend, K.S. Wong, W. Hayes, H. Lindenberger and S. Roth : these proceedings
2. D.D.C. Bradley : PhD Thesis, University of Cambridge (1987)
3. L. Lauchlan, S. Etemad, T-C Chung, A.J. Heeger and A.G. Macdiarmid : Phys.Rev. B24, 3701 (1981) and references therein
4. W.P. Su , J.R. Schrieffer and A.J. Heeger : Phys. Rev. B22, 2099 (1980)
5. J.L. Bredas, R.R. Chance and R.H. Baughman : J.Chem.Phys.76, 3673 (1982) and references therein
6. B. Horovitz : Solid State Commun. 41, 729 (1982)
7. B. Horovitz, Z. Vardeny, E. Ehrenfreund and O. Brafman : Synthetic Metals 9, 215 (1984)
8. R.A. Wessling and R.G. Zimmermann : US patents #3,401,152 (1968) and #3,706,677 (1972)
9. M. Kanbe and M. Okawara : J.Polymer Sci. A-1 6, 1058 (1968)
10. I. Murase, T. Ohnishi, T. Noguchi and M. Hirooka : Polymer Commun. 25, 327 (1984)
11. J.D. Capistran, D.R. Gagnon, S. Antoun, R.W. Lenz and F.E. Karasz : ACS Polymer Preprints 25, 282 (1984)

12. D.D.C. Bradley, R.H. Friend, H. Lindenberger and S. Roth : Polymer 27, 1709 (1986)
13. D.D.C. Bradley, G.P. Evans and R.H. Friend : Synthetic Metals 17, 651 (1987)
14. D.D.C. Bradley, R.H. Friend, T.Hartmann, E.A. Marseglia, M.M. Sokolowski and P.D. Townsend : Synthetic Metals 17, 473 (1987)
15. T. Hartmann : MPhil. Thesis, University of Cambridge (1986)
16. D.D.C. Bradley, T. Hartmann, R.H. Friend, E.A. Marseglia, H. Lindenberger and S. Roth : these proceedings
17. T. Granier, E.L. Thomas, D.R. Gagnon, F.E. Karasz and R.W. Lenz : J.Polymer Sci. B24, 2793 (1986)
18. J. Fink, H. Lindenberger, B. Scheerer, A. vom Felde and S. Roth : these proceedings
19. D.D.C. Bradley, R.H. Friend and W.J. Feast : Synthetic Metals 17, 645 (1987)
20. K.S. Wong, D.D.C. Bradley, W.Hayes, J.F. Ryan, R.H. Friend, H. Lindenberger and S. Roth : J.Phys.C in press (1987)
21. P.R.Surjan : private communication
22. K. Fesser, A.R. Bishop and D.K. Campbell : Phys.Rev.B27, 4804 (1983)
23. Z. Vardeny, E. Ehrenfreund, O. Brafman, M. Nowak, H. Schaffer, A.J. Heeger and F. Wudl : Phys.Rev.Lett. 56, 671 (1986)
24. F.L. Pratt, K.S. Wong, W. Hayes and D. Bloor : J.Phys.C20, L41 (1987)
25. F.L. Pratt, K.S. Wong, W. Hayes and D. Bloor : these proceedings
26. D.K. Campbell, D. Baeriswyl and S. Mazumdar : Synthetic Metals 17, 197 (1987)
27. U. Sum, K. Fesser and H. Buttner : Solid State Commun. 61, 607 (1987)
28. Z. Vardeny, J. Orenstein and G.L. Baker : J.de Physique 44 c3, 325 (1983)
29. H.E. Schaffer and A.J. Heeger : Solid State Commun. 59, 415 (1986)
30. D.D.C. Bradley : unpublished results

Spectroscopy of Photo-Induced Solitons in *cis*-rich and *trans*-Polyacetylene

N. Colaneri, R.H. Friend*, H.E. Schaffer, and A.J. Heeger

Department of Physics and Institute for Polymers and Organic Solids, University of California, Santa Barbara, CA 93106, USA

We present detailed measurements of the vibrational and electronic transitions of photogenerated solitons in cis-rich and all-trans polyacetylene prepared by the Shirakawa method. We find that the saturation concentration of excitations on the trans sequences in cis-rich material (30% trans) can be substantially higher than in all-trans and consider that these long-lived states are pairs of like-charged solitons. For both this cis-rich material and also for all-trans we find that photoinduced bleaching occurs only above the band edge, as expected if the photogenerated soliton states are formed from band states. In addition to the 'translational' IR modes previously reported, we find weaker absorptions from the shape modes of the soliton. These are present in both cis-rich and all-trans polyacetylene, and we consider that we have observed the IR-active 'third bound mode' calculated by WADA and colleagues.

1. Introduction

The relaxation of the dimerised chain of polyacetylene to localise added charges at solitons, or domain walls, separating sections of chain with opposite senses of bond alternation has been well known for some time [1]. Solitons produced in this way have a characteristic spectrum of localised vibrational excitations and electronic states which are accessible spectroscopically [2], and very detailed information can be obtained though sub-gap optical absorption measurements on chemically doped or photoexcited samples. We have carried out such photoinduced absorption (PA) measurements on cis-rich (30% trans) and all-trans polyacetylene by comparing absorption spectra alternately with and without band-gap illumination (at 488 nm). We used an FTIR spectrometer in the energy range 400-8000 cm^{-1} [3,4] and for higher energies (2μm - 500 nm) we used a lock-in amplifier to detect the response of the probe beam, dispersed through a grating spectrometer, at the chop frequency of the band-gap excitation [5-8]. We report here on two aspects of this work, the detection of shape-modulating vibrational modes of the soliton, and the mechanism for photoproduction of the long-lived (msec) excitations probed in these experiments.

2. Shape Modes of the Soliton

The photoinduced absorption spectra for an as-prepared cis-rich film of Shirakawa polyacetylene (30% trans) and for the same sample after thermal isomerisation to all-trans polyacetylene are shown in Fig. 1. In both cases the features seen are due to

---

* Permanent address: Cavendish Laboratory, Madingley Road, Cambridge CB3 OHE, U.K.

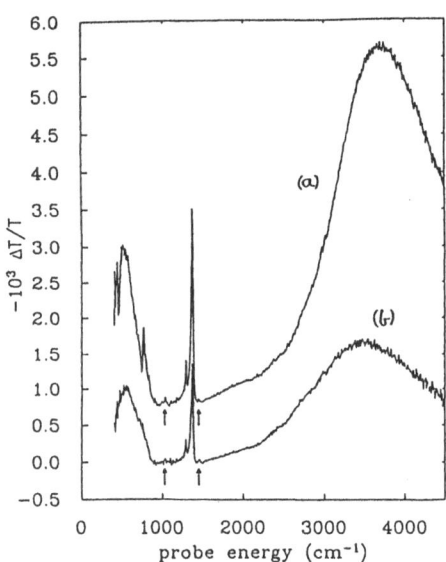

Fig. 1 PA spectra for (a) cis-rich (30% trans) and (b) all-trans polyacetylene measured at 77 K

charged solitons on the trans sequences in the samples. The feature peaking near 3600 cm^{-1} is due to electronic transitions from the soliton 'mid-gap' states to the band edges [3]. Vibrational modes associated with the solitons are seen at lower energies. Three modes with large oscillator strength, at 536, 1288 and 1365 cm^{-1}, previously observed [3,5] are clearly seen. In the amplitude mode formalism due to HOROVITZ [9] these correspond to the coupling of the three lattice degrees of freedom of the polyacetylene chain that are resonantly enhanced in Raman scattering [10] to the zero-frequency uniform translational freedom of the charge. In addition to these soliton-translation (T) modes, there have been predictions from calculations, based on the SSH Hamiltonian and its continuum version, that there are couplings of these same three lattice modes to soliton shape modulations [11-16]. NAKAHARA and MAKI [16] showed that there is a 'breathing' mode, or amplitude oscillation of the soliton, which is of opposite symmetry to the T mode and is not expected to be IR active. In a later calculation based on a one-component continuum model, ITO et al. [11] showed that there is also a third localised mode, with the same (even) symmetry of the T mode and which is thus IR active. TERAI et al. [15] extended these calculations to include coupling the the three lattice degrees of freedom. The third bound mode of the one-component model is thus found to generate three peaks in the IR absorption, of low intensity ($\approx$ few % of the T modes).

The low noise level of the spectra in Fig. 1 allows detection of two new absorption features of low intensity, at 1034 and 1438 cm^{-1} for the all-trans sample (indicated by arrows in Fig.1). The energies and relative intensities of these two peaks are in good agreement with the calculated values of TERAI et al., and we consider that these are, therefore, the third bound modes of the soliton, corresponding to the lowest and highest of the three lattice modes. The middle peak is predicted to have lower oscillator strength than the other two, and would be expected to be hidden under the stronger T modes. In a recent paper, VARDENY et al. [17] reported weak absorptions, at energies slightly lower than those of the modes we report here, that disappear upon isomerisation of the sample from cis-rich to all-trans. They consider that the solitons confined to short trans sequences in the cis-rich material show IR activity of the second (amplitude oscillation) mode of the soliton through the breaking of translational symmetry of the polyacetylene chain and an associated relaxation of the selection rule preventing IR activity in this mode. We have not found similar behaviour under our experimental conditions, and as seen in Fig. 1, there is very little change in the intensities of the 1034 and 1438 cm^{-1} absorptions upon isomerisation of the sample.

3. Stabilisation of Photogenerated Charged Excitations

The photogenerated charged solitons, $S^{\pm}$, detected under the experimental conditions used in this work are known to have lifetimes of order milliseconds [8]. It is unlikely that S^+ S^- pairs formed on one chain can avoid recombination for this length of time, and there is evidence for a recombination time of no more than a few picoseconds [18]. It was suggested by ORENSTEIN et al. [7] that interchain charge separation is necessary to allow separation of the photogenerated charged pairs, and there is direct evidence for this from polarisation dependent measurements on highly oriented Durham polyacetylene films [19]. Charges separated between chains are expected to relax to form polarons, P^+ and P^-, but the appearance of a single mid-gap electronic transition indicates that these polarons are subsequently converted to charged solitons. There are two proposed mechanisms for this.

ORENSTEIN et al. [7] suggested that the neutral, spin 1/2 soliton-like defects, S^0, present in trans polyacetylene in a concentration of typically 10^{19} cm^{-3}, act as traps for the photogenerated charges, so that the net reaction is $2S^0 \rightarrow S^+ + S^-$. This process can be tested in two ways. Firstly, the removal of S^0 states should give a negative light-induced ESR (LESR) signal. However, at present there is no direct experimental evidence for an LESR signal associated with the charged solitons [20,21]. Secondly, associated with the appearance of photo-induced absorption (PA) at 0.43 eV from charged solitons, there should be a corresponding bleaching (PB) of the optical transitions from the S^0 states removed. VARDENY and TAUC [8] show that the effect of on-site Coulomb repulsion, U, is to shift optical transitions associated with $S^{\pm}$ below mid-gap (PA at 0.43 eV) and transitions from S^0 to above mid-gap. They report that at a temperature high enough to suppress the PA at 1.35 eV due to a neutral excited state [6] they find PB at 1.4 eV, and they attribute this to the bleaching of absorption due to band-to-S^0 transitions.

The alternative route for interchain-separated charges to form solitons is the energetically favoured conversion of a like-charged pair of polarons present on the same chain to a like-charged pair of solitons on that chain, e.g. $P^+ + P^+ \rightarrow S^+ + S^+$ [20]. Motion of this soliton pair to an adjacent chain requires transfer of both charges, and as for bipolaron hopping this will be very much slower than motion of single charges. Soliton states thus stabilised are generated from band states, and PB is expected at the π - π^* band edge. No LESR signal is expected, and the saturation PA signal will correspond to a density of order one soliton pair per conjugated length of chain. For all-trans polyacetylene this is comparable to the S^0 density.

The PA spectra for the sample in Fig.1 before and after isomerisation were measured with the same pump beam intensity, and in both cases the PA signal measured was close to saturation. Contrary to earlier reports of weak PA from cis-rich samples [17,21], we find a larger saturation density of photogenerated $S^{\pm}$ than in the all-trans sample. This is in spite of an S^{0} concentration an order of magnitude lower than in the all trans material, and we consider that the photogenerated charged solitons are stabilised as like-charged pairs on the trans sequences. The higher saturation density of $S^{\pm}$ then correlates with the larger number of disconnected trans sequences in the cis-rich material than in the all-trans. The lifetimes for the photoexcitations are increased in the cis-rich sample, and at low temperatures are typically 10 ms. In Fig. 2 we show the PA spectrum for a sample similar to that in Fig.1 at 15 K with the pump beam chopped at 23 Hz and lock-in detection in phase with the 0.45 eV PA (phase lag of 55°), and in quadrature (phase lead of 35°). We see electromodulation characteristic of the trans chains (1.4-1.7 eV) and characteristic of the cis chains (1.9-2.4 eV) [6] in phase with the 0.45 eV PA; the PA feature at 1.45 eV [22] has a faster decay time and is seen clearly in the quadrature response. There is no sign of PB associated with the 0.45eV PA below the trans polyacetylene band edge. Figure 3 shows the PA spectrum for a similar sample isomerised to all-trans. The PA response is now in-phase with the pump beam at the low chop frequency used here. The spectrum is taken at a temperature (180 K) high enough to eliminate the high-energy PA feature. We find no evidence for the 1.4 eV PB observed by VARDENY and TAUC [10] and ORENSTEIN [23], but rather the PB expected from interband bleaching, with a peak near 1.7 eV. We conclude that under the conditions for sample preparation and handling used here, photogenerated charged excitations in both cis-rich and all-trans material are stabilised as pairs of like-charged solitons. This is equivalently described as a 'weakly-bound bipolaron', and we thus have a single mechanism for the stabilisation of long-lived PA for both polyacetylene and also polaron-supporting

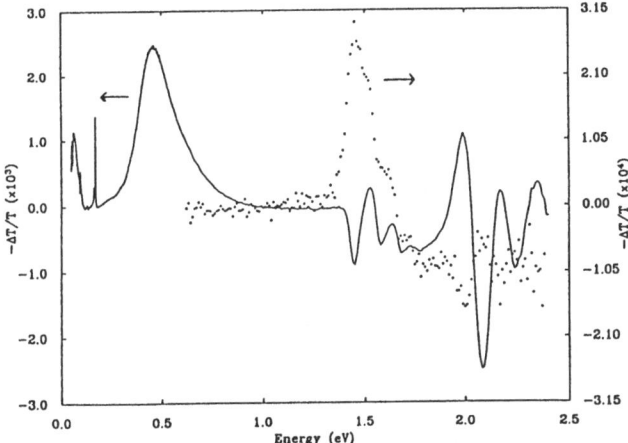

Fig. 2 PA spectrum for cis-rich (30% trans) polyacetylene. Data above 0.6 eV were measured at 15 K at a chop frequency of 23 Hz. The solid line shows the response at a phase angle 55° behind the pump beam, the data points show the response in quadrature to this, at 35 ° ahead of the pump beam. Data below 0.6 eV are as shown in Fig.1, scaled to match in the region of overlap.

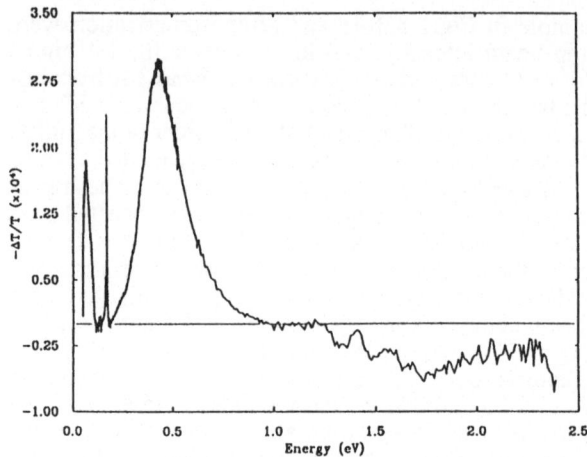

Fig. 3 PA spectrum for trans polyacetylene. Data above 0.6 eV were measured at 180 K, data below 0.6 eV are as shown in Fig.1 scaled to match in the region of overlap.

polymers such as polythiophene [24] and polyphenylenevinylene [25]. These polymers show the two absorption features in the PA spectrum expected from bipolarons, but in other respects (lifetimes, temperature and intensity dependence) the observed PA is very similar to that in polyacetylene.

References

1. W. P. Su, J. R. Schrieffer and A. J. Heeger: Phys. Rev. Lett. **42**, 1698 (1979); Phys. Rev. B22, 2099 (1980); Err. B28, 1138 (1983).
2. Handbook of Conducting Polymers: ed. by T. J. Skotheim (Marcel Dekker, New York 1986).
3. G. B. Blanchet, C. R. Fincher, T-C Chung and A. J. Heeger: Phys. Rev. Lett. **50**, 1938 (1983).
4. H. E. Schaffer, R. H. Friend and A. J. Heeger: preprint.
5. Z. Vardeny, J. Orenstein and G. L. Baker: Phys. Rev. Lett. **50**, 2032 (1983).
6. J. Orenstein, G. L. Baker and Z. Vardeny: J. Phys. (Paris) **44**, C3, 407 (1983).
7. J. Orenstein, Z. Vardeny, G. L. Baker, G. Eagle and S. Etemad: Phys. Rev. B30, 786 (1984).
8. Z. Vardeny, and J. Tauc: Phys. Rev. Lett. **54**, 1844 (1985).
9. B. Horovitz: Solid State Commun. **41**, 729 (1980).
10. Z. Vardeny, E. Ehrenfreund, O. Brafman and B. Horovitz: Phys. Rev. Lett. **51**, 2326 (1983).
11. H. Ito, A. Terai, Y. Ono and Y. Wada: J. Phys. Soc. Japan **53**, 3519 (1984).
12. H. Ito and Y. Ono: J. Phys. Soc. Japan **54**, 1194 (1985).
13. A. Terai, H. Ito, Y. Ono and Y. Wada: J. Phys. Soc. Japan **54**, 4468 (1984).
14. A. Terai and Y. Ono: J. Phys. Soc. Japan **55**, 213 (1986).
15. A. Terai, Y. Ono and Y. Wada: J. Phys. Soc. Japan **55**, 2889 (1986).
16. M. Nakahara and K. Maki: Phys. Rev. B25, 7789 (1982).
17. Z. Vardeny, E. Ehrenfreund, O. Brafman, B. Horovitz, H. Fujimoto, J. Tanaka and M. Tanaka: Phys. Rev. Lett. **57**, 2995 (1986).

18. L.Rothberg, T. M. Jedhu, S. Etemad and G. L. Baker: Phys. Rev. Lett. **49**, 3229 (1986).
19. P. D. Townsend and R. H. Friend: Synthetic Metals **17**, 361 (1987).
20. F. Moraes, Y. W. Park and A. J. Heeger: Synthetic Metals **13**, 113 (1986).
21. C. G. Levey, D. V. Lang, S. Etemad, G. L. Baker and J. Orenstein: Synthetic Metals **17**, 569 (1987).
22. Z. Vardeny, E. Ehrenfreund and O. Brafman: Synthetic metals **17**, 349 (1987).
23. J. Orenstein: ref. 2, page 1297.
24. Z. Vardeny, E. Ehrenfreund, O. Brafman, M. Nowak, H. E. Schaffer, A. J. Heeger and F. Wudl: Phys. Rev. Lett. **56**, 671 (1986).
25. D. D. C. Bradley, R. H. Friend and W. J. Feast: Synthetic Metals **17**, 645 (1987).

Photoinduced Infrared Absorption in Polydiacetylene

F.L. Pratt[1], *K.S. Wong*[1], *W. Hayes*[1], *and D. Bloor*[2]

[1]Clarendon Laboratory, University of Oxford, Parks Road, Oxford OX1 3PU, United Kingdom
[2]Department of Physics, Queen Mary College, Mile End Road, London E1 4NS, United Kingdom

1. Introduction

Photoinduced absorption (PA) is a valuable technique for studying the excitations of conjugated polymer systems. Both vibrational and electronic properties and the interaction between them may be observed for the photoinduced states. Polydiacetylene (PDA) is a particularly good material for such a study, as highly aligned crystalline films are available. Electron correlation effects are generally agreed to be significant in conducting polymers and are particularly strong for PDA. We report here infrared PA measurements on the polydiacetylene PDA–1OH.

2. Photoinduced Absorption

Infrared measurements were obtained at 77K on free-standing films using a Perkin Elmer 1710 Fourier Transform spectrometer. A He–Cd laser (442nm, $\hbar\omega = 2.8$ eV) was used as the photoexcitation source. The PA spectrum for PDA–1OH is shown in Fig.1. The PA is about an order of magnitude stronger than that of polythiophene or polyacetylene and is strongly polarised parallel to the chains.

Above 1500 cm^{-1} the PA spectrum consists principally of broad electronic absorption while the spectrum below 1500 cm^{-1} consists of a number of new vibrational peaks, the strongest appearing at 913, 1171, 1257, 1310 and 1398 cm^{-1}. Some of the weaker PA peaks such as those at 725, 967, 1008 and 1078 cm^{-1} correspond to enhancement of existing absorption peaks [1]. The dependence of the steady state PA on the excitation

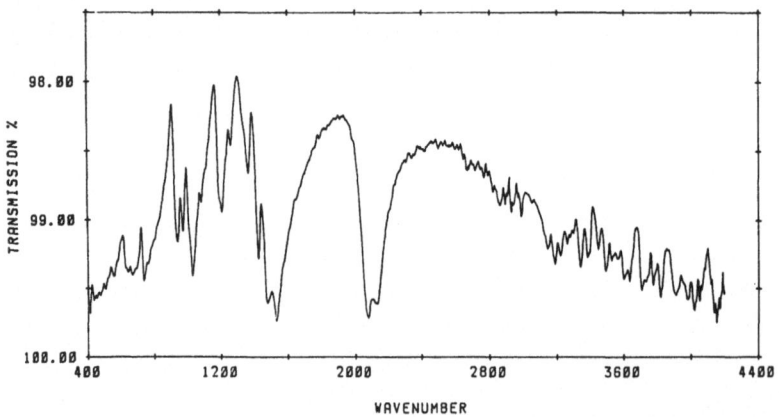

Fig. 1   Photoinduced absorption in PDA–1OH at 77K for infrared radiation polarised parallel to the polymer chain

intensity I for levels up to approximately 0.5W/cm^2 has been observed to vary as $I^{1/2}$ indicating that bimolecular recombination kinetics is dominant under steady photoexcitation conditions.

Decay of the PA at 77K after termination of the excitation has been measured over a timescale from minutes to hours and is plotted in Fig.2. There is an immediate drop by some 40% , indicating the presence of a fast initial relaxation mechanism. Subsequently the PA decays more slowly at a rate consistent with the $\exp(-Kt^{1/3})$ one-dimensional deep trapping decay law observed in the photocurrent for PDA−1OH [2]. At times greater than around 60 minutes a much slower recombination mechanism takes over, so that even after 260 minutes the PA is still around 30% of its steady initial value. This persistent absorption rapidly disappears on warming the sample to room temperature indicating the presence of a potential barrier preventing carrier recombination.

As the PA decays, the shape of the PA spectrum remains essentially constant apart from an absorption around 1700 cm^{-1} which grows in strength relative to the rest of the spectrum. This suggests that this mode may be associated with a trapping centre.

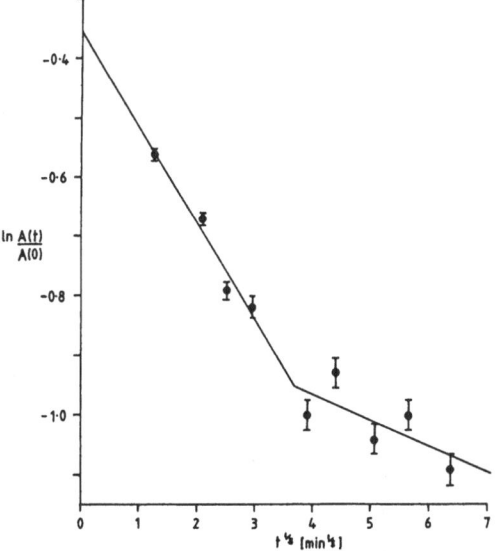

Fig. 2 Decay of the photoinduced absorption A(t) at 77K after termination of the laser excitation. A(0) is the initial steady state value for the PA. The lines indicate separate fits to the short- and long-time data.

3. Phonon Resonance Effects in the Electronic PA

The PA above 1500 cm^{-1} shows the broader features associated with electronic transitions. At first sight the spectrum appears to be made up of two peaks at 1900 and 2500 cm^{-1}. However, a much more plausible interpretation of the spectrum is that it consists of a broad electronic absorption with a high-energy tail extending beyond 4000 cm^{-1}, similar to the PA bands observed in this region for polyacetylene and polythiophene, but with photoinduced transmission bands superimposed at 1500 cm^{-1} and 2100 cm^{-1}. These windows show transmission peaks at 1483 and 1541 cm^{-1} and at 2079 and 2138 cm^{-1}. The higher peak within each window corresponds to the measured absorption for C=C and C≡C respectively (see Fig. 3).

125

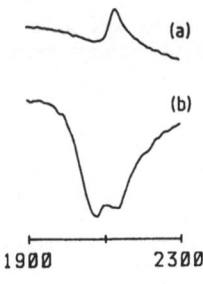

(a)

(b)

Figure 3 A comparison of (a) the ground state absorption in the region of the C≡C mode with (b) the PA in the same spectral region

|———————|————————|
1900   2300

WAVENUMBER

We suggest that these windows are primarily due to a form of Fano resonance [3] resulting from interaction between the excited electronic and vibrational states of the defect, but with bleaching of the C=C and C≡C ground state phonons superimposed. The separation of the peaks within each window is around 60 cm^{-1}, giving a measure of the softening of these modes in the excited electronic state. This softening is consistent with the results of resonant Raman scattering experiments [4]. The width of the resonance gives a measure of the strength of the electron–phonon interaction and it corresponds to a lifetime for the excited electronic state of order 25 fs before it relaxes into its original state by phonon emission.

4. The Nature of the Photogenerated Defect

Current models for nondegenerate conjugated polymers suggest that the stable electronic excitations are in the form of polarons and bipolarons each of which introduce two localised energy levels in the bandgap. A transition between these levels is expected for the polaron but not for the bipolaron. A PA band has previously been observed in PDA–1OH at 1.32 eV [5]. Both this band and the band at 0.25 eV have an asymmetric lineshape with a long high–energy tail, indicating transitions from localised states to a continuum. These two transitions also display a similar variation in intensity with temperature. Thus if we assume that they have a common origin we conclude that they are likely to be bipolaron transitions. This conclusion is further supported by the value for the level spacing $2\omega_0$ in relation to the bandgap $2\Delta_0$; taking $2\omega_0 = 1.32 - 0.25 = 1.07$ eV and $2\Delta_0 \approx 2.4$ eV we get $\omega_0/\Delta_0 = 0.45$. This value of ω_0/Δ_0 is too small for a polaron for which ω_0/Δ_0 is expected to be greater than $1/\sqrt{2}$ [6]. The relative intensity of the two bands is predicted to be 15:1 for this value of ω_0/Δ_0 from the theory of Fesser et al [6]. This is a difficult comparison to make reliably as it involves two separate measurement systems, however, we estimate that the intensity ratio for the 0.25 eV to the 1.32 eV band is of order 10:1; this is in reasonable agreement with the calculated value for the bipolaron.

5. Correlation Effects

For a bipolaron the sum of the two transition energies ω_1 and ω_3 should be $2\Delta_0$ in the absence of correlation effects. These correlation effects are expected to shift the transitions to lower energy [7,8] so that we can write [7]

$$\omega_1 + \omega_3 = 2\Delta_0 - 2U_{bp}$$

where U_{bp} is the difference in Coulomb correlation energy between initial and final states. Since we have $\omega_1 + \omega_3 = 1.57$ eV, we find $U_{bp} \sim 0.4$ eV and $U_{bp}/\Delta_0 \sim 0.35$. The correlation energy is related to the on–site Hubbard energy U for weak U by [8]

$$U = 3 U_{bp} l_{bp}$$

126

where l_{bp} is the width of the bipolaron. For a calculated width of order 7 units [9] this implies a value of around 8eV for U which is of the same order as the π electron bandwidth. This suggests that the electron correlation is too large to be treated as a small perturbation in this system.

The stability of a bipolaron compared to a polaron will depend on the difference between the electron–electron correlation energy and the electron–phonon energy associated with the distortion of the chain at the defect. The electron–phonon term has been calculated to be of order 0.36 eV for a bipolaron in polydiacetylene compared with 0.05 eV for a polaron [9]. Given our estimated value of $U_{BP} \sim 0.4$ eV for the bipolaron, it would appear that the calculated value for the electron phonon term is not large enough to account for the stability of the bipolaron.

References

1. F.L. Pratt, K.S. Wong, W. Hayes and D. Bloor, J. Phys. C 20, L41 (1987)
2. I.G. Hunt, D. Bloor and B. Movaghar, J. Phys. C 16, L623 (1983)
3. U. Fano and J.W. Cooper, Phys. Rev. 137, A1364 (1965)
4. D.N. Batchelder and D. Bloor, J. Phys. C 15, 3005 (1982)
5. T. Hattori, W. Hayes and D. Bloor, J. Phys. C 17, L881 (1984)
6. K. Fesser, A.R. Bishop and D.K. Campbell, Phys. Rev. B 27, 4804 (1983)
7. Z. Vardeny, E. Ehrenfreund, O. Brafman, M. Nowak, H. Schaffer, A.J. Heeger and F. Wudl, Phys. Rev. Lett. 56, 671 (1986)
8. D.K. Campbell, D. Baeriswyl and S. Mazumdar, Synth. Metals 17, 197 (1987)
9. N.A. Cade and B. Movaghar, J. Phys. C 16, 539 (1983)

Towards Solid State Investigations on Polyacetylene

G. Leising and M. Filzmoser

Institut für Festkörperphysik, Technische Universität Graz,
Petersgasse 16, A-8010 Graz, Austria

Introduction

Even when single crystals of conventional semiconductors and
metals became available, quantitative, reproducible measurements of their optical properties posed serious problems to the
investigator. For a long time, reflectivity data have been
dominated by the surface properties (roughness, adsorbed
impurities etc.) until improved preparation techniques were
developed. Many of the optical measurements on polyacetylene
have been performed under very ill-defined conditions (surface
as prepared, in air etc.) and and also neglecting the specific
morphology of the Shirakawa-type material. Our improved
synthesis allows the preparation of highly oriented
polyacetylene (cis and trans) with well-defined bulk properties
as well as a sample surface with a sufficient quality for
optical investigations in a wide spectral range. Careful
experiments allowing a quantitative evaluation of optical
properties are decribed, which are indispensible for the
determination of other physical properties.

Experimental

Highly oriented trans-polyacetylene films were prepared by the
well-known stretch-orientation process of a precursor polymer
during the thermal conversion to polyacetylene /1/.
Polyacetylene films with stretching ratios of 10 - 20 have been
used in this investigation. The thickness of the polymer films
ranged from one to about five micrometer. Highly oriented cis-
polyacetylene is achieved by a distinct choice of time and
temperature for the conversion reaction which will be described
in detail in a forthcoming publication. The attainable maximum
value of cis content was about 90 %, as evaluated by the ratio
of the integrated absorptions of the cis and trans out-of-plane
deformation vibrations. Special care was invested in the
preparation of the precursor polymer film, because the surface
quality of the precursor obviously determines the surface
quality of the oriented polyacetylene film. An optimum quality
could be achieved by carefully filtering the precursor polymer
solution and slowly drying the cast films under dust-free
pure argon atmosphere. After the stretching and conversion
reaction is completed, the oriented polyacetylene film is
clamped between the two jaws of a stainless steel ring where it
remains plane and flat during all subsequent measurements
keeping full orientation and a smooth surface for the optical

investigations. The polarized reflectivity measurements in the range between 200 cm^{-1}(0.025 eV) to 4000 cm^{-1}(0.5 eV) have been carried out on a Perkin Elmer 684 grating instrument equipped with AgBr wire-grid polarizer. A Perkin Elmer Lambda 9 spectrophotometer covered the range from 3800 cm^{-1}(0.47 eV) to 33000 cm^{-1} (4.1 eV) utilizing depolarizers and prism polarizers in the sample and reference beam. Special specular reflectance accessories have been designed in our laboratory for both instruments to meet the need of carrying out the measurements in inert atmosphere without any contact of the samples with air. The angle of incidence of the reflectance accessories is about 10° in both instruments. The reflectivity is measured relative to aluminum mirrors, which are calibrated according to the literature values of nickel, chromium and platinum.

Results and Discussion

In Fig. 1 the reflectance spectra of trans-polyacetylene for light polarized parallel or perpendicular to the chain direction are shown. The parallel spectrum is dominated by the strong reflectivity maximum around 2 eV arising from the direct interband ($\pi - \pi^*$) transition, and the value of the reflectivity is 0.63 in the maximum. The oscillations which occur for energies below 1 eV are caused by multiple beam interference effects in the planeparallel polymer film. The perpendicular reflectance spectrum is flat in region of the $\pi - \pi^*$ transition with a nearly constant value of 0.04. Again, pronounced interference fringes develop for energies below 1.5 eV.

There exist different experimental methods for the determination of the optical constants of a solid. A

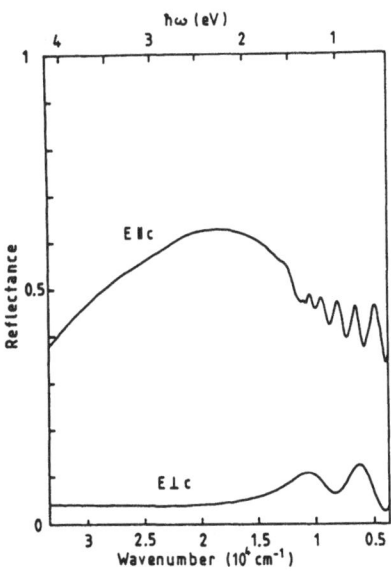

Fig.1 The polarized optical reflectance spectrum of trans-polyacetylene

transmission (T) and a reflection (R) measurement on a sample with known thickness (d) would allow one to deduce the absorption coefficient (K) via the formula

$$T = (1 - R)^2 \exp (-Kd). \tag{1}$$

However, one has to keep in mind that equ. 1 is only applicable in cases where multiple reflection and multiple beam interference effects can be neglected. For highly oriented polyacetylene films, these conditions are not satisfied, as one can see from Fig. 1 where we have to deal with both effects. In the present case one has to use the more complex formula system for an ideally planeparallel plate

$$T_M = T (1 + (k/n)^2) / D, \tag{2}$$

$$R_M = R (1 + \exp(-2Kd) - 2 \cdot \exp(-Kd) \cdot \cos(2nk_0d)) / D, \tag{3}$$

where $D = (1 + R^2\exp(-2Kd) - 2 \cdot R \cdot \exp(-Kd) \cdot \cos(2nk_0d+2\delta))$, $k_0 = 2\pi\bar{v}$, $K = 2k_0k$ and the phase angle $\delta = \arctan(2k/(n^2+k^2-1))$. Here n is the refractive index and k the absorption index and $\bar{v}$ is the wavenumber. R_M and T_M are the measured reflectivity and transmission, respectively. The evaluation of the optical constants of polyacetylene by equ. 2 and 3 is faced with serious problems, especially in the energy range of the interband transition, where even for very thin free-standing films (minimum thickness of about 2300 Å) the absorption is too high to be measured accuratly (pinholes etc). An attempt to get rid of these problems was to prepare extremely thin trans-polyacetylene films on top of a stretchable transparent substrate, and to perform the stretching, the elimination-reaction and the isomerization of this composite sample /2/. On samples of this type reliable transmission measurements could be performed, but the surface quality did not allow accurate reflectivity measurements. Another method for evaluating the optical constants of a solid is to measure the reflectivity over a wide energy range (theoretically from zero to infinity) and apply the Kramers-Kronig relation /3/. The phase angle can be calculated from the measured reflectivity by

$$\delta (\bar{v}) = -1/2 \cdot \pi \int_0^\infty \log((\bar{v}+\bar{v}_0) / (\bar{v}-\bar{v}_0)) \cdot (d\log R(\bar{v})/d\bar{v}) /d\bar{v}. \tag{4}$$

Since the actual measurement covers a limited spectral region only, one has to use an extrapolation for the high wavenumber part of the spectrum. The integral can be evaluated by means of a computer . Although it is not obvious from equ. 4, a significant contribution to the integral comes from the reflectance at wavenumbers far away from $\bar{v}_0$, when $dR/d\bar{v}$ is large. Thus the extrapolation of the reflectance to regions outside the range of experimental measurements must be done with great care, and will be described in detail in a forthcoming publication. Another serious difficulty arises from the fact that equ. 4 is not valid in regions where interference and multiple reflection effects appear. To overcome this difficulty, the method we applied for the caculation of the optical constants from the reflectance spectrum was the following: In the interference-free region of the spectrum the reflectivity data were taken as measured. In the region

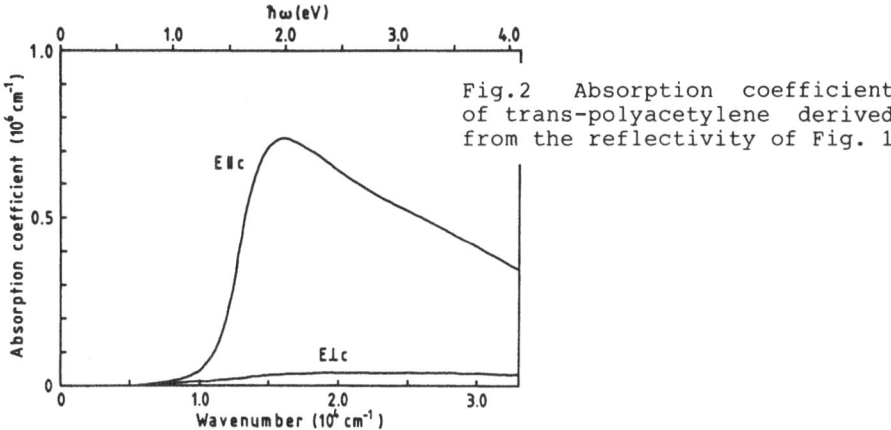

Fig.2 Absorption coefficient of trans-polyacetylene derived from the reflectivity of Fig. 1

of multiple-beam interference we start with an assumption about the average reflectivity without interference, and the calculation is carried out according to equ. 4. From the phase angle one can deduce the optical constants n and k. Introducing these values in equ. 3 should give the measured reflectivity spectrum if the assumption about the spectral dependence of the reflectivity in the interference-region was chosen right. If not, one has to repeat the above procedure until the measured and the calculated reflectivity spectra coincide. This has been done for both polarization directions of Fig.1 and in Fig. 2 we present the calculated spectral dependence of the absorption coefficient of trans-polyacetylene. The expected qualitative behaviour of the polarized absorption and reflectance of oriented trans-polyacetylene demonstrating the strong optical anisotropy of the material is already known from the literature but this work claims to present the first quantitative results from optical investigations /4/.

The parallel absorption coefficient for the π - π^* transition is $7.4 \cdot 10^5 \mathrm{cm}^{-1}$ in the maximum, which is somewhat lower than the value of $8.5 \cdot 10^5 \mathrm{cm}^{-1}$ we found by electron-energy loss-spectroscopy on very thin samples /5/. The small difference between the two values may be explained by the fact that the energy loss spectrum was taken up to 48 eV and there might be some contribution of the σ - σ^* transition (appearing around 12 eV) to the Kramers-Kronig calculation in the 2 eV region, which would increase the optically derived value. However, since the results for the absorption coefficient have been obtained by two very different methods on different samples, the agreement is quite satisfactory for a polymer system. The energy position of the absorption maximum in the parallel polarization is 1.9 eV, which is expected also from the electron-loss results for zero momentum transfer /5/. In addition, the 1.9 eV position coincides with the value from optical investigations on Shirakawa-type polyacetylene /6/, proving that the conjugation length of our oriented polyacetylene is comparable to that of Shirakawa-type polyacetylene, a fact which has been questioned for quite a time. The origin of the small feature which shows up in the

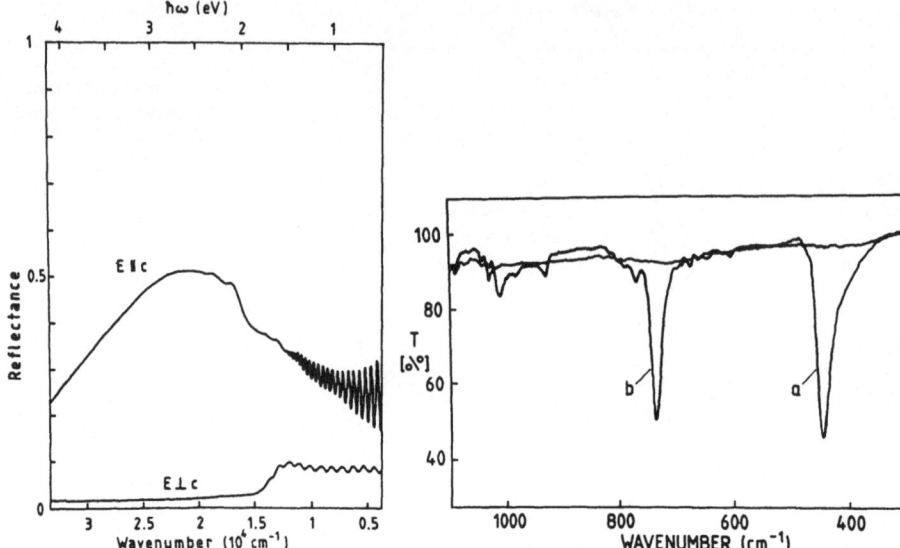

Fig.3 The polarized optical reflectance spectrum of cis-polyacetylene

Fig.4 The polarized infrared transmission spectrum of cis-polyacetylene (a:E‖c, b:E⊥c)

reflectance spectrum for parallel polarization around 1.5 eV is not yet known, but one could suppose that it is excitonic or vibronic in nature or even a result of the bandstructure /7/. In Fig. 4 we present the optical reflectance spectrum of cis-polyacetylene. As expected, the interband transition is shifted to higher energies compared to trans-polyacetylene and distinct vibrational fine structure appears. The anisotropy seems to be somewhat higher in the cis-case, and the small features belonging to already present trans-sequences are visible around 1.7 eV. Pronounced interference patterns appear in the low energy region, proving the good surface quality of the cis-samples. The reason for the lower value of the maximum reflectivity of the cis-polyacetylene compared to the trans-polyacetylene is not known at present. In Fig. 4 we show the polarized infrared transmission spectra of highly oriented cis-polyacetylene, where the perfect orientation of the polymer chains is evidenced by the very high dichroitic ratio for the C-C-C deformation vibration (a) as well as for the C-H deformation vibration (b). A cis-content of about 90 % has been calculated for this sample from the ratio of the integrated absorption for the 740 cm^{-1}-band (cis) to the 1015 cm^{-1}- band (trans). The calculation of the optical constants for cis-polyacetylene is in progress.

Acknowledgement

This work was supported by the Austrian Science Research Fund under the project Nr. 6198.

References

1. G. Leising: Polymer Bulletin $\underline{11}$, 401 (1984)
2. G. Leising, H. Kahlert and O. Leitner:
   Springer Series in Solid-State Sciences $\underline{63}$, 56 (1985)
3. F. Wooten: Optical Properties of Solids, Academic Press
   (New York, London, Toronto, Sydney, San Francisco 1972)
4. G. Leising, R. Uitz, B. Ankele, W. Ottinger and F. Stelzer:
   Mol. Cryst. Liqu. Cryst. $\underline{117}$, 327 (1985)
   P.D. Townsend and R.H. Friend: Synth. Metals $\underline{17}$, 361 (1987)
5. J. Fink and G. Leising: Phys.Rev. $\underline{B34}$, 5320 (1986)
   J. Fink, H. Fark, N. Nücker, B. Scheerer, G. Leising and
   R. Weizenhöfer: Synth. Metals $\underline{17}$, 377 (1987)
6. C.R. Fincher, M. Ozaki, M. Tanaka, D. Peebles, L. Lauchlan,
   and A.J. Heeger: Phys.Rev. $\underline{B20}$, 1589 (1980)
7. P. Vogl: private communication

Raman Scattering of Highly Oriented Polyacetylene

P. Knoll[1], H. Kuzmany[2], and G. Leising[3]

[1]Institut für Experimentalphysik, Karl-Franzens-Universität Graz,
Universitätsplatz 5, A-8010 Graz, Austria
[2]Institut für Festkörperphysik der Universität Wien and
Ludwig Boltzmann Institut für Festkörperphysik,
Strudlhofgasse 4, A-1090 Wien, Austria
[3]Institut für Festkörperphysik, TU Graz,
Petersgasse 16, A-8010 Graz, Austria

1. Introduction

A considerable amount of Raman work has been done on polyacetylene. In particular, the C–C and C=C modes of trans-$(CH)_x$ have attracted experimental and theoretical Raman investigations to this material because of their unusual resonance behaviour (Fig. 1). However, most of the work has been done on the unoriented material and only little attention has been paid to the Raman intensities themselves and their dependence on the excitation energy /1-3/, although much information can be obtained from such measurements. With the success in preparing fully oriented polyacetylene /4/ a great improvement in the knowledge of the Raman properties is expected. First Raman spectra of such highly oriented samples are shown in /5/, and first quantitative measurements are reported in /3/. In this contribution we present a detailed quantitative Raman analysis performed on highly oriented trans-polyacetylene in several polarization directions and with several laser excitations.

2. The gap-phonon relation of trans-polyacetylene

The Raman spectrum of trans-polyacetylene (Fig. 2) is dominated by the two A_g modes, which are denoted as C–C (around 1100 cm^{-1}) and C=C (around 1500 cm^{-1}) /6/. The remaining two A_g modes are at 1285 cm^{-1} (weak) and at 2990 cm^{-1}, and the mode around 1000 cm^{-1} is assigned to the out-of-plane B_g mode.

The dominating C–C and C=C modes show the same unusual behaviour on tuning the laser from the red to the blue emission. Figure 1 shows the exciting frequency dependence for the C=C mode. This behaviour is described in detail by e.g. KUZMANY and KNOLL /5/. From experiments with artificially introduced defects /7/ it is clear that disorder will shift both the vibrational frequencies and the electronic transitions to higher values, which together with the resonance Raman process explains why the satellite line appears when spectra are excited in the blue spectral region.

The distribution in disorder is the reason why each polyacetylene film shows its individual Raman features. A more intrinsic property, which excludes the distribution of disorder, is the relation between change of vibrational frequencies and change of electronic transition energies. There are two ways to obtain such a gap-phonon relation from the experimental data:

1) The maximum of the satellite line indicates the place where the incident laser frequency matches the electronic transition energy.

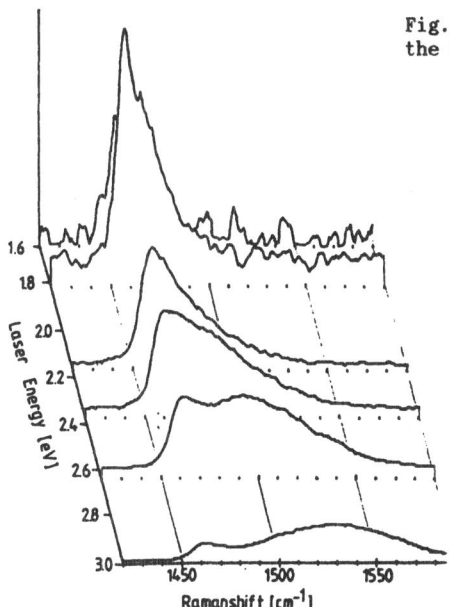

Fig.1 The resonant Raman behaviour of the C=C mode in trans-polyacetylene

Monitoring the shift of the satellite line by exciting with different laser lines will correlate a vibrational frequency with an electronic energy.

2) Observing the intensity of a particular position within the Raman profile while changing the laser energy (sliced excitation profile) will give a maximum just at the energy where the electronic transition lies for this particular vibration frequency. If this is done for several positions within the Raman profile one will also get a gap-phonon relation.

In this simple description the gap-phonon relations obtained by both methods should be the same.

So far, two models have been applied to describe Raman scattering of trans-polyacetylene. The first quantitative description of the Raman behaviour was given by the conjugation length model /8-11/. The disorder is caused by a distribution of effective conjugation lengths N. The vibrational frequencies ω are taken from measurements of finite polyenes and scale with $1/N$. The electronic transitions ε are evaluated in the Hückel approach and also follow a $1/N$ law:

$$\omega = \omega_0 + \frac{a}{N} \qquad \varepsilon = \varepsilon_0 + \frac{b}{N} \qquad =====> \qquad \omega = \omega_0 + \frac{a}{b}(\varepsilon - \varepsilon_0) \ . \qquad (1)$$

This leads to a linear gap-phonon relation $\omega(\varepsilon)$ for each Raman band.

The second model starts from a one-dimensional system with a charge density wave causing the dimerization gap ε /12,13/. Disorder is described by a distribution of the electron-phonon coupling λ which renormalizes the square of the phonon frequencies. The gap-phonon ($\tilde{\lambda}$) relation is given by

$$\varepsilon = 4 \ E_c \ \exp \ (-1/2\tilde{\lambda}) \qquad \text{with} \qquad 2\tilde{\lambda} = \Pi \ \frac{\omega^2}{\bar{\omega}^2} \ . \qquad (2)$$

The influence of the electron-electron interaction was also investigated within this model /14,15/, but there is no dramatic change in (2).

3. Experimental results and correction of the intensities

The Raman spectra were taken with a Spex double monochromator with holographic gratings and photomultiplier-photoncounting detection. A very careful correction of the response function of the whole equipment was performed by comparing the Raman intensities with the emission of black-body radiation for each polarization direction. For the correction of absorption, reflection etc., optical reflectivity measurements were performed on the same sample from the IR up to the UV. Using a Kramers - Kronig analysis, absorption, refractive index, amount of multiple reflection and interference are obtained as a function of frequency. For the backscattering geometry the Raman intensity has to be corrected by /16/

$$\frac{\alpha_i + \alpha_s}{(1-R_i)(1-R_s)[1-\exp(-\alpha_i d - \alpha_s d)]} \quad . \tag{3}$$

Only in the case of strong absorbance ($\alpha d \gg 1$), does (3) reduce to a simpler form given in /17/, which is not applicable for the (--) polarization because of the small thickness d of the sample.

Spectra for the blue and red laser excitation for the ($||$) and (--) polarizations are shown in Fig. 2. (The first and second positions in a symbol like ($|$-) describe the directions of polarization, either parallel ($|$) or perpendicular (-) to the chain, for the incident and scattered light, respectively). From Raman line profiles of the C-C and C=C modes of the blue ($||$) excitation the material can be classified as good quality (a small amount of disorder). The (--) polarizations show Raman intensities for the C-C and C=C modes although they should be very weak. A careful correction of the intensities shows that the (--) spectra are three orders of magnitude weaker than the ($||$) spectra. However, the (--) intensities show interesting line profiles (e.g. a double peak for the red excitation) and are

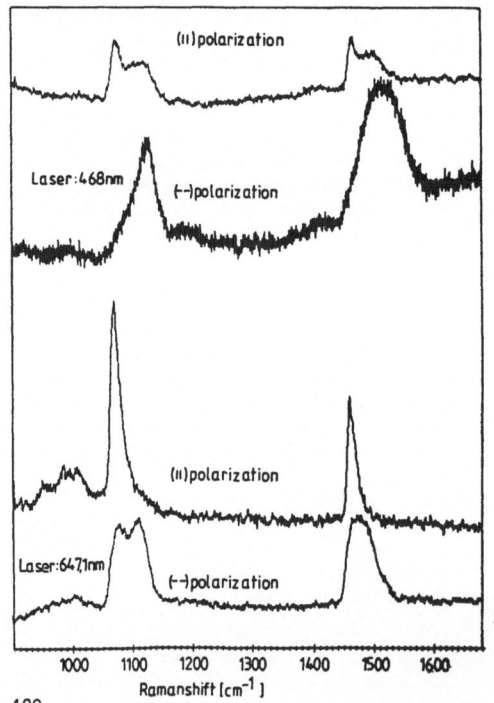

Fig.2 Raman spectra for the oriented trans-polyacetylene

still not really understood. A short-chain effect /17/ may explain the (--) line profiles of the blue excitation, but not of the red excited spectrum. To clarify this situation it is necessary to look at the excitation profiles which are shown for 1460 cm^{-1} in Figs. 3 and 4. If imperfect chain orientation is the reason for the (--) polarized intensity, then identical excitation profiles for (II) and (--) polarization are expected, whereas if short chains (or any other kind of disorder) are the reason, then the (--) excitation profile should have the maximum at higher energies. As Fig. 3 shows, the Raman intensity in the (--) polarization cannot arise from imperfect orientation or any short-chain effect. Following the (--) excitation profile to lower laser energies the intensity rises, indicating a resonance lower in energy than the π-π^* gap. This is a very surprising result as electronic transitions of such an energy with dipole moments perpendicular to the chain have not been considered or predicted by any theory so far. An increase of the excitation profile was also found in the unoriented material /18/, but at lower energies. A possible explanation may be a 2^1A_g state as known from the oligomers /19/, which can give an allowed transition by symmetry breaking. However, further investigations are necessary to clarify this problem.

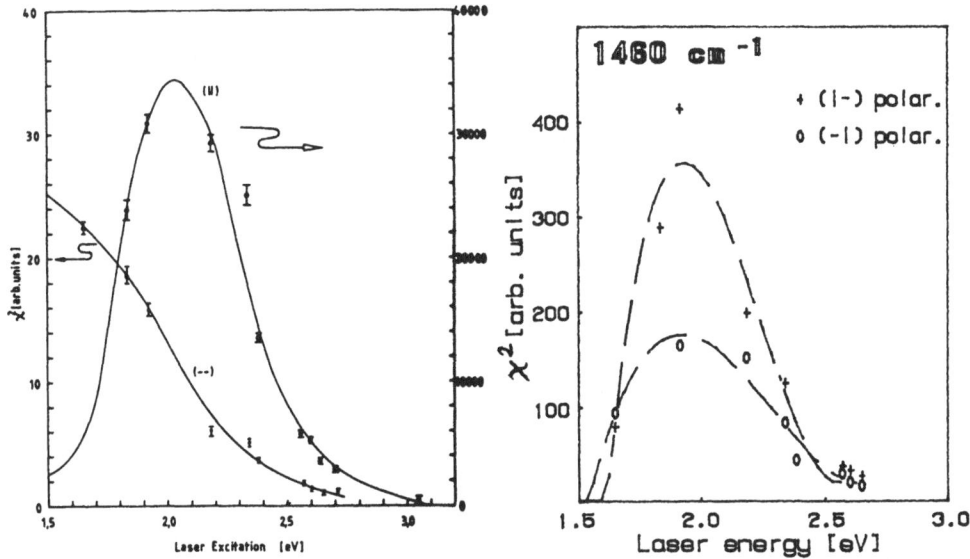

Fig.3 Excitation profiles of 1460cm^{-1} for the (II) and (--) polarization

Fig.4 Excitation profiles for the off-diagonal elements

In Fig. 4 the two off-diagonal elements of the scattering tensor are compared. In the off resonance region both curves fit together, whereas in the case of resonance they split, indicating an antisymmetric part of the scattering tensor, as is expected from resonance Raman theory. The symmetry between (I-) and (-I) for all vibrational frequencies (not only 1460 cm^{-1}) regardless of the frequency range of the scattered light is proof of the proper correction of the intensities.

Table 1 shows the integrated intensities (in arbitrary units) and depolarization ratios as obtained for several laser excitations. Although not too much information lies in these values (the individual distribution of

TABLE 1: Integrated intensities and depolarizations

Laser wavelength [nm]	C - C			C = C		
	$\frac{I_{\|\|}}{\omega_s^4}$	$\frac{I_{=}}{I_{\|\|}}$	$\frac{I_{\|-}}{I_{\|\|}}$	$\frac{I_{\|\|}}{\omega_s^4}$	$\frac{I_{=}}{I_{\|\|}}$	$\frac{I_{\|-}}{I_{\|\|}}$
468.0	204 135	0.0010	0.0097	166 841	0.0026	0.017
476.2	281 603	0.0011	0.013	254 984	0.0021	0.017
482.5	319 196	0.0011	0.015	264 635	0.0023	0.017
520.8	571 766	0.0010	0.0097	448 663	0.0015	0.013
530.9	1102 249	0.0007	0.0084	791 793	0.0011	0.012
568.2	1144 425	0.0006	0.013	809 849	0.0010	0.013
647.1	1073 698	0.0010	0.015	728 376	0.0017	0.019
676.4	822 947	0.0015	0.017	531 026	0.0023	0.020

disorder enters), they are shown for comparison with measurements done on an uncalibrated spectrometer /17/. In general, our values of the depolarizations are about a factor of ten lower. This may be due to our better sample quality and the more precise correction done with (3). Also, the claimed decrease of the depolarizations /17,20/ with decreasing laser excitation is only true for laser energies down to 2.2 eV; for lower energies gap-states become dominant.

4. Interpretation of the Raman data

Following the two ways of evaluating the gap-phonon relation suggested in Sect. 2 we obtain the data points of Fig. 5. In contrast to the given simple explanation, the values obtained from the excitation profile will always give higher vibrational frequencies than values obtained from the shift of the satellite line. This is due to damping of phonons and

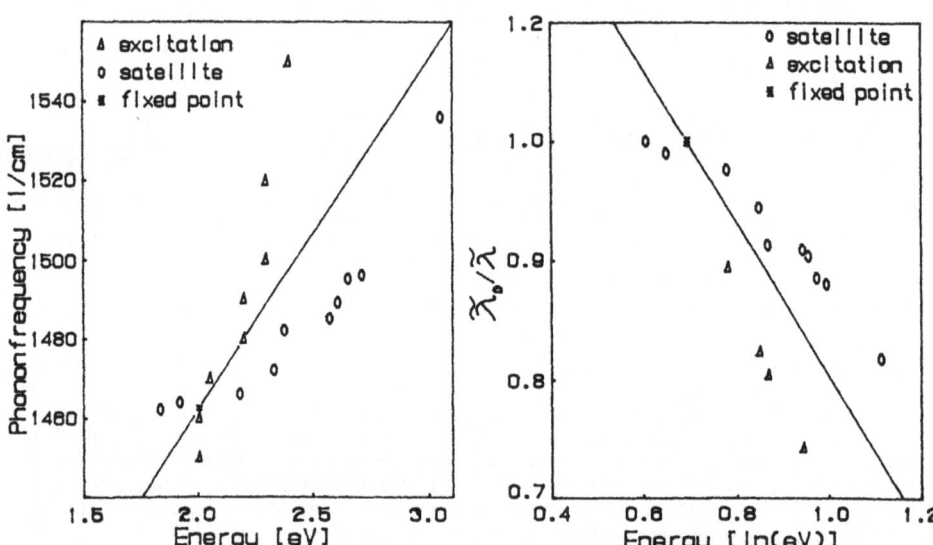

Fig.5 Gap-Phonon relation for the C=C mode

Fig.6 Logarithmic gap-phonon plot normalized by the fixed points

electronic states, which cause the distribution of disorder to influence the shift of the satellite line and the change of maximum of the excitation profile in a different way. Going to low laser energies the shift of the satellite line will saturate at the lowest possible phonon frequency; in the same way the maximum of the excitation profile will saturate at the lowest possible electronic transition. (This saturation causes the two relations to cross, giving the diagram a fishtail-like shape.) The two resulting saturation values are the phonon frequency and the gap of the least disturbed material available in the sample and are a "fixed point" of the gap-phonon relation. The most probable gap-phonon relation is obtained by a least squares fit to all values through the fixed point. The straight line in Fig. 5 is the result of the fit. Table 2 summarizes the fit parameters corresponding to (1) and (2) where $\varepsilon_0/\omega_0^1, \omega_0^2$ are the fixed points for the C-C and C=C modes, respectively. As the same gap-phonon relations are obtained from the ($|$-) and (-$|$) polarizations they are not listed separately.

TABLE 2: Parameters of the gap-phonon relations

Polari- zation	ε_0 [eV]	ω_0^1 [cm^{-1}]	ω_0^2 [cm^{-1}]	$\frac{a^1}{b}$ [cm^{-1}/eV]	$\frac{a^2}{b}$ [cm^{-1}/eV]	E_C [eV]
($||$)	2.0	1070	1462	125	105	1.9
(--)	<1.6	1074	1465	85	72	3.8
off-diagonal	1.9	1073	1464	102	92	1.4

In the interpretation of the charge density wave model the gap-phonon relation is given by (2) where all phonons contribute to λ. This relation is shown in Fig. 6. Of course, again different relations are obtained whether the values are taken from the satellite line or from the excitation profile. This fact is very important as, so far, only the shift of the satellite line has been taken into account /21,22/, which will give wrong cutoff energies E_C. We normalized to the frequencies of the fixed points, as they are the best defined values, and fitted to the data points of Fig. 6, resulting in the full line. The fitting parameter E_C is also listed in Table 2 and is much smaller than fitting only to the values of the satellite line (6.2eV for ($||$)). As in a tight binding approximation this value should be around 10 eV, we claim that within this model extrinsic disorder plays a dominant role also in pure trans-polyacetylene.

In the interpretation of both models the parameters of Table 2 show that ($||$) and cross-polarization have nearly the same gap-phonon relation, whereas the (--) polarization is quite different because of the strange behaviour observed with red laser excitation. From this fact we can conclude that ($||$), ($|$-) and (-$|$) polarized Raman spectra are produced by the same physical mechanism, whereas the (--) component has to be treated separately.

Acknowledgment

One of us (P. K.) would like to acknowledge Prof. W. Kiefer for stimulating discussions. This work was supported by the "Stiftung Volkswagenwerk".

References

1. L. Lauchlan, S.P. Chen, S. Etemad, M. Kletter, A.J. Heeger and MacDiarmid: Phys. Rev. B 27, 2301 (1983)
2. H. Kuzmany and P. Knoll: J. Raman Spectrosc. 17, 89 (1986)

3. P. Knoll, G. Leising and H. Kuzmany:
    Proc. X^{th} intern. conf. on Raman spectrosc. 12-33 (1986)
4. G. Leising: Polymer Bulletin 11, 401(1984)
5. P.D. Townsend et al., p.50
    G. Leising, H.Kahlert and O. Leitner, p.56
    H. Kuzmany and P. Knoll, p.114
    all in: Springer Series in Solid-State Sciences 63 (1985)
6. For details of the normal coordinate see e.g.:
    F.B. Schügerl and H. Kuzmany: J. Chem. Phys. 74, 953 (1981)
7. P. Knoll and H. Kuzmany: Mol. Cryst. Liqu. Cryst. 106, 317 (1984)
8. H. Kuzmany: phys. stat. sol. (b) 97, 512 (1980)
9. H. Kuzmany, E.A. Imhoff, D.B. Fitchen and A. Sarhangi:
    Phys. Rev. B26, 7109 (1982)
10. G.P. Brivio and E. Mulazzi: Chem. Phys. Lett. 95, 555 (1983)
11. G.P. Brivio and E. Mulazzi: Phys. Rev. B30, 876 (1984)
12. Z. Vardeny, E. Ehrenfreund, O. Brafman and B. Horovitz:
    Phys. Rev. Lett. 51, 2326 (1983)
13. B. Horovitz, Z. Vardeny, E. Ehrenfreund and O. Brafman:
    Synth. Metals 9, 215 (1984)
14. B. Horovitz and J. Sólyom: Phys. Rev. B32, 2681 (1985)
15. D. Baeriswyl and K. Maki: Phys. Rev. B31, 6633 (1985)
16. P. Knoll: to be published in Appl. Spec.
17. E. Faulques, E. Rzepka, S. Lefrant et. al.: Phys. Rev. B33, 8622 (1986)
18. J. Berrehar, J.L. Fave, C. Lapersonne, M. Schott and H. Eckhard:
    Mol. Cryst. Liqu. Cryst. 117, 393 (1985)
19. B. Hudson, B.E. Kohler and K. Schulten: Excited States 6, 1 (1982)
20. E. Mulazzi: Solid State Commun. 55, 807 (1985)
21. R.H. Friend, D.D.C. Bradley, C.M. Pereiro, P.D. Townsend,
    D.C. Bott and K.P.J. Williams: Synth. Metals 13, 101 (1986)
22. Z. Vardeny, E.Ehrenfreund, O. Brafman and B. Horowitz:
    Phys. Rev. Lett. 54, 75 (1985)

Experimental Raman Investigation of Oriented Undoped and Iodine Doped Polyacetylene

S. Lefrant[1] and E. Mulazzi[2]

[1]Laboratoire de Physique Cristalline, Université de Nantes,
 F-44072 Nantes Cedex 03, France
[2]Dipartimento di Fisica dell'Università di Milano,
 I-20133 Milan, Italy

1. Introduction

For several years, Resonance Raman Scattering (RRS) spectra have been studied on $(CH)_x$ samples issued from the synthesis initiated by Ito et al. /1/. The as-prepared polymer films consist in randomly oriented fibers with a diameter of $\simeq 200-500$ Å. The attention was focused on the band shape of the main Raman bands enhanced by resonance effects and observed at frequencies around 1100 and 1500 cm^{-1}. Their unusual behaviour when the exciting wavelength is changed from the red to the violet range has been described by many groups /2,3,4/. In particular, satellite components were observed whose frequencies are shifted upwards as the excitation energy becomes higher. This peculiar behaviour has been interpreted differently. One model considers a bimodal distribution of conjugated segments /5/ whereas another one uses a charge density wave approach /6/ to describe the experimental data. All these theoretical considerations have been applied to Raman data obtained on unoriented samples.

More recently, highly oriented samples were prepared by using procedures for the polymerization different from that usually used. One of them initiated by Feast et al. /7/ and then slightly modified /8/ is the so-called "Durham route" which uses a precursor polymer. Another way to obtain oriented trans-$(CH)_x$ is to use a new catalyst for the synthesis as proposed in Ref. /9/. Both methods lead to samples which exhibit strong anisotropic properties. Therefore, the oriented films can be studied in RRS recorded in polarized light.

In this paper, we describe experimental data on samples prepared by the two techniques. Results are theoretically analyzed in the frame of the bimodal distribution model for which details are given elsewhere /5/ and we show the good agreement obtained between experimental data and theoretical predictions. As a test, we also describe and interpret data obtained in the case of iodine-doped oriented $(CH)_x$ films for which the distributions of conjugated segments are drastically modified.

2. Experimental Results

In Figure 1, we show polarized Raman spectra of highly oriented $(CH)_x$ samples (stretching ratio $\Delta \ell/\ell = 14$) for the two characteristic wavelengths $\lambda_L = 676.4$ nm and 457.9 nm. Both // // polarized spectra (the first sign refers to the polarization of the incident light with respect to the chain axis, whereas the second one refers to the polarization of the scattered light), exhibit features which indicate a rather low concentration of long conjugated segments. This is evidenced by the relatively symmetric band

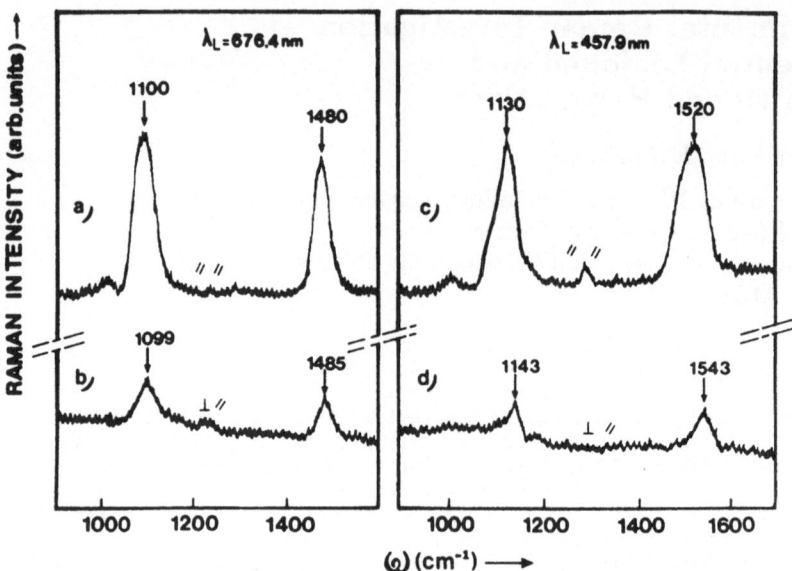

Fig. 1 : Experimental RRS polarized spectra of oriented trans-(CH)$_x$;
Durham route synthesis /8/.
λ_L = 676.4 nm : a) // //, b)$\perp$// ; λ_L = 457.9 nm : c) // // ; d)$\perp$//.

shapes observed with λ_L = 676.4 nm and also by the lack of low frequency
components in the spectrum recorded with λ_L = 457.9 nm. The main point
which has to be emphasized is the shift of the band peaks observed in the
polarized spectra with crossed polarization ($\perp$//,$\perp \perp$ and //$\perp$). This shift
is much more important with the violet excitation. Also, the depolarization
ratios (ρ_1, ρ_2, ρ_3) have been measured by considering the integrated intensities of each main Raman bands. In Table 1, we have collected data obtained for the two excitation wavelengths. The different values are compared
to the theoretical predictions made from the bimodal distribution model on
which we have tentatively performed a correction by using the existing
preliminary optical parameters /10/. Despite the roughness of the corrections, we can see on Table 1 that the agreement between experimental and
theoretical results is rather satisfactory.

Only the main points concerning the theoretical model, which has been
presented elsewhere, are recalled here. It considers two distributions of
conjugated segments, one peaked on short segments (N : number of double
bonds, N < 30) and the other peaked on long conjugated segments (N > 30).
The electronic and vibrational properties and the electron-phonon interaction in the excited electronic states of the conjugated segments are studied as a function of N. In particular, the electric dipole moments for the
transitions are evaluated and it turns out that for conjugated segments
with N > 10, they are oriented along the chain axis. This is not the case
when 3 < N < 10 for which the electric dipole moments for the transitions
are much more localized on the double bonds, this effect being enhanced for
N decreasing from 9 to 3. This property is very important in order to
understand the behaviour of the RRS spectra of oriented trans-(CH)$_x$, since
the polymer is composed of segments of different conjugation lengths.

142

Table 1 : Depolarization ratios - a) Experimental values are from the data shown on Figure 1 ; b) Corrected values with the use of optical parameters determined in Ref. /10/ ; c) Theoretical evaluations.

Laser excitation λ_L		$\rho_1 = I_\perp{}_{//}/I_{//}{}_{//}$	$\rho_2 = I_\perp{}_\perp/I_{//}{}_{//}$	$\rho_3 = I_{//}{}_\perp/I_{//}{}_{//}$
$\lambda_L = 676.4$ nm	a) Exp. :	0.033		
	b) Corr. :	0.012		
	c) Theo. :	-		
$\lambda_L = 457.9$ nm	a) Exp. :	0.052	0.12	0.048
	b) Corr. :	0.02	0.013	0.019
	c) Theo. :	0.03	0.008	0.03

We have evaluated the polarized Raman spectra in the // // and $\perp$// configurations for the two excitation energies $\Omega_L = 1.83$ eV and $\Omega_L = 2.7$ eV which are close to the experimented conditions. The best fit in the case corresponding to the experimental data shown in Fig. 1 led to the following parameters :
$N_1 = 20$; $\sigma_1 = 10$; $N_2 = 8$; $\sigma_2 = 4$ and $G = 0.40$.

Another set of experiments was carried out on oriented samples with a stretching ratio $\Delta\ell/\ell = 7$, prepared by using a new catalyst as described in Ref. /9/. In figures 2a and b, we show these experimental spectra for both

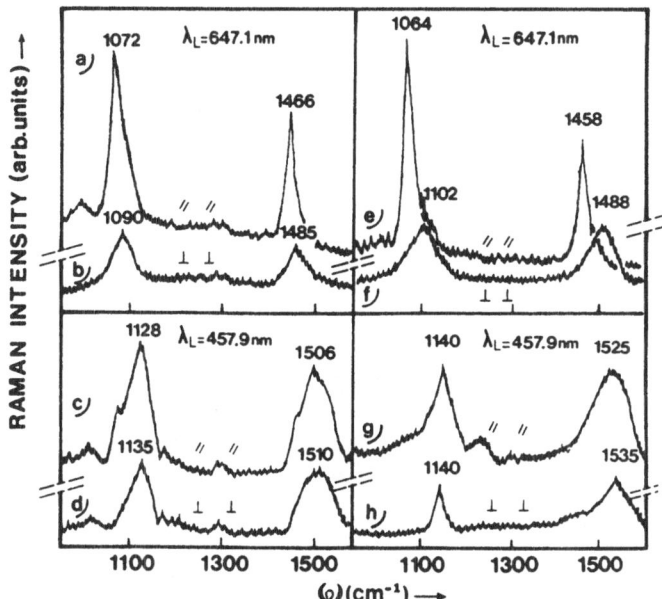

Fig. 2 : Experimental RRS polarized spectra of stretched trans-$(CH)_x$; Synthesis as in Ref. /9/.
Undoped : $\lambda_L = 647.1$ nm : a) // // ; b)$\perp\perp$; $\lambda_L = 457.9$ nm : c) // // ; d)$\perp\perp$
doped : $\lambda_L = 457.9$ nm : e) // // ; f)$\perp\perp$; $\lambda_L = 457.9$ nm : g) // // ; h)$\perp\perp$

143

λ_L = 647.1 nm and 457.9 nm in // // and $\perp\perp$ polarizations. A behaviour similar to the one described in the first part of this paper is observed and the parameters used to fit the polarized spectra of the undoped samples are the following :
N_1 = 100 ; σ_1 = 50 ; N_2 = 14 ; σ_2 = 7 and G = 0.5.

The calculated spectra are shown in Figures 3a,b,c and we can see that the main features are reproduced in very good agreement with the experiments, especially for the $\perp\perp$ polarized spectrum for λ_L = 457.9 nm. It is seen in particular that the low-frequency components of the main Raman bands due to the scattering of long conjugated segments do not contribute in the $\perp\perp$ polarized spectrum whereas the satellite components, due to the scattering of shorter conjugated segments, do. This property is even confirmed if we analyze in details the polarized spectra recorded on iodine doped sample (CHI$_{0.023}$)$_x$ (Figs 2e,f,g,h, Figs 3d,e,f). The main effect of the slight doping is to modify the two distributions of conjugated segments which are described by other parameters :
N_1 = 100 ; σ_1 = 50 ; N_2 = 6 ; σ_2 = 3 and G = 0.4.

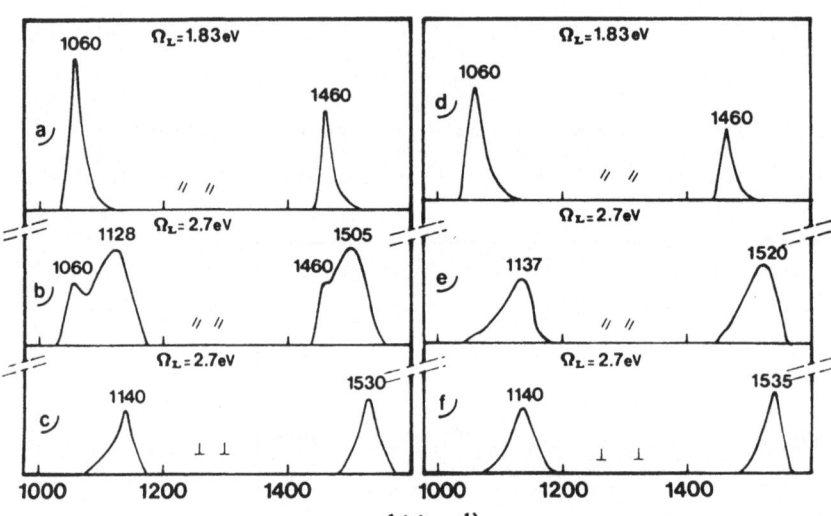

Fig. 3 : Calculated RRS polarized spectra (parameters in the text).
Undoped : a) Ω_L = 1.83 eV, // //; b) Ω_L = 2.7 eV, // //; c) Ω_L = 2.7 eV, $\perp\perp$
Doped : d) Ω_L = 1.83 eV, // //; e) Ω_L = 2.7 eV, // //; f) Ω_L = 2.7 eV, $\perp\perp$

In this case, we note a complete lack of conjugated segments of intermediate length and a lower concentration of long conjugated segments. As a consequence, for λ_L = 457.9 nm, the Raman bands in the $\perp\perp$ polarized spectrum do not exhibit any shift as observed before. Since mainly very short conjugated segments are in resonance for this excitation energy, they have components for the electric dipole moments for the transitions parallel and perpendicular to the chain axis and, therefore, they contribute to the scattering of light in both // // and $\perp\perp$ spectra. The theoretical spectra are shown in Figures 3e,f.

144

3. Discussion and Conclusion

In this paper, we have reported the properties of the polarized RRS spectra carried out on fully oriented $(CH)_x$ samples prepared by two different procedures. Both samples exhibit a similar behaviour as we change the excitation wavelength λ_L. The main experimental features can be described as follows : i) experimental spectra are highly polarized but show different depolarization ratios for $\lambda_L = 647.1$ nm and 457.9 nm (see Table 1) ; ii) a shift is observed in the Raman peaks in the $\perp\perp$, $//\perp$ and $\perp//$ polarized spectra with respect to those in the $//$ $//$ polarized spectrum. Both shifts and depolarization ratios of integrated intensities are sample dependent.

All these spectra have been theoretically analyzed in the frame of the bimodal distribution model and it is shown that the above results depend strongly on the different parameters which describe the distributions of the short and long conjugated segments respectively. Concerning the depolarization ratios, we have experimentally shown that $\rho_1 > \rho_2$ whatever λ_L and that ρ_1 and ρ_2 are always stronger for $\lambda_L = 457.9$ nm than for $\lambda_L = 647.1$ nm. This was predicted by the theory /11/ and these results are qualitatively similar to those reported recently by Knoll et al. /12/.

We have shown that the contribution to the $\perp\perp$, $//\perp$ and $\perp//$ polarized spectra when $\lambda_L = 457.9$ nm comes mainly from conjugated segments shorter than 10 double bonds which have components for the electric dipole moments for the transitions perpendicular to the chain axis. In the red range, since this property does not exist for the long conjugated segments, the contribution to the scattering of light in the $\perp\perp$, $//\perp$ and $\perp//$ polarized spectra can be interpreted by using the following arguments : i) the short segments induce some scattering of light in the tail of each Raman band ; ii) a small disorder of long conjugated segment in preresonance conditions may exist ; iii) another contribution may come from other electronic states located in the gap with electrical dipole moments perpendicular to the chain axis. It is worth noting that these states must be intrinsic states of the conjugated segments because they must interact with the Raman active stretching vibrational modes with the same frequency as those of the long conjugated segments.

In conclusion, we have shown that the bimodal distribution model provides a good and consistent analysis of the RRS spectra in highly oriented trans-$(CH)_x$.

Acknowledgments : We thank Dr. G.P. Brivio for helping in the computer programme and Dr. E. Faulques and E. Perrin for their contribution to the experimental work.

1. T. Ito, H. Shirakawa and S. Ikeda: J. Polym. Sci., Polym. Chem. Ed. **12**, 11 (1974)
2. H. Kuzmany: Phys. Status Sol. (b) **97**, 521 (1980)
3. H. Kuzmany, E.A. Imhoff, D.B. Fitchen and A. Sarhangi: Phys. Rev. B**26**, 7109 (1982)
4. S. Lefrant: J. Phys. (Paris) Colloq. **44**, C3-247 (1983)
5. G.P. Brivio and E. Mulazzi: Phys. Rev. B**30**, 676 (1984)
6. B. Horovitz, Z. Vardeny, E. Ehrenfreund and O. Brafman: Synth. Met. **9**, 215 (1984)
7. D.C. Bott, C.K. Chai, J.H. Edwards, W.J. Feast, R.H. Friend and M.E. Horton: J. Phys. (Paris) Colloq. **44**, C3-143 (1983)
8. G. Leising: Polym. Bull. **11**, 401 (1984)
9. G. Lugli, V. Pedretti and G. Perego: Mol. Cryst. and Liq. Cryst. **117**, 43 (1985)
10. G. Leising, R. Uitz, B. Ankele, W. Ottinger and F. Stelzer: Mol. Cryst. and Liq. Cryst. **117**, 327 (1985)
11. E. Mulazzi: Sol. St. Commun. **55**, 807 (1985)
12. P. Knoll, H. Kuzmany and G. Leising: Proceedings of this conference.

Polarization Properties of the Raman Spectra in *trans*- and *cis*-polyacetylene

S. Fuso[1], *C. Cuniberti*[1], *G. Dellepiane*[1], *S. Luzzati*[2], and *R. Tubino*[2]

[1]Istituto di Chimica Industriale, Università,
 Corso Europa, I-16132, Genova, Italy
[2]Istituto di Chimica delle Macromolecole, C.N.R.,
 Via Bassini 15, I-20133 Milano, Italy

1. INTRODUCTION

The interpretation of resonant Raman scattering data from conventional isotropic films of polyacetylene is considerably complicated by the insolubility of the film and by the lack of any preferred orientation.

We have studied the polarized Raman scattering of a highly oriented film of trans polyacetylene which exhibits an almost perfect alignment of the polymer chains along the stretching direction [1]. Moreover we have measured the resonance cross section and depolarization ratio of the resonantly enhanced Raman bands of a soluble form of trans and cis polyacetylene. The soluble samples have macroscopic random orientation while the oriented films offer the possibility to disentangle the contributions of the different polarizability tensor components to the Raman scattering.

2. HIGHLY ORIENTED POLYACETYLENE

Resonance Raman measurements at 386, 488 and 609.6 nm have been performed on the resonantly enhanced Raman bands [2,3] of the highly oriented film of trans polyacetylene (draw ratio 7). Data on the polarization properties of the Raman bands have been obtained in the four possible scattering configurations, namely X(ZZ)Y, X(ZX)Y, X(YZ)Y and X(YX)Y (Z = stretching direction; for the definition of the scattering configurations see ref.[2]).

The change of the shapes of the Raman bands with the exciting laser frequency is reported for the X(ZZ)Y scattering configuration in fig.1. The polarization properties of the Raman bands measured using the λ = 609.6 nm exciting line, are illustrated in fig.2.

Because of the high anisotropy of the optical properties of the sample, a simple model to roughly account for the large correcting factors due to reflectivity and absorption has been worked out, yielding for λ = 609.6 nm the following values of the corrected intensity ratios [3] :

$$I_{X(ZX)Y}/I_{X(ZZ)Y} = I_{X(YZ)Y}/I_{X(ZZ)Y} = 0.03; \quad I_{X(YX)Y}/I_{X(ZZ)Y} = 0.01 . \quad (1)$$

The changes of the observed bandshapes and of the corrected intensities with the scattering configuration are consistent with a distribu-

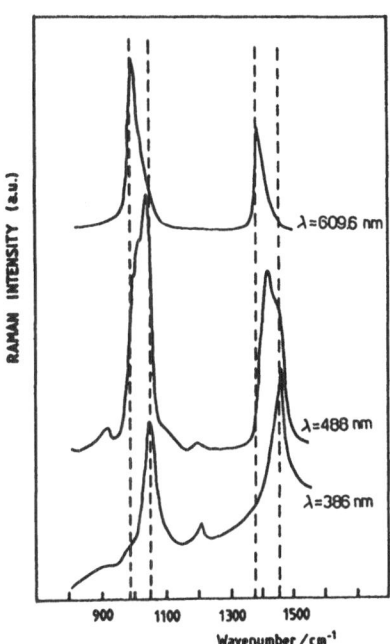

Fig.1 - Raman intensities of
highly oriented trans polyacetylene in the X(ZZ)Y configuration for various exciting
laser lines.

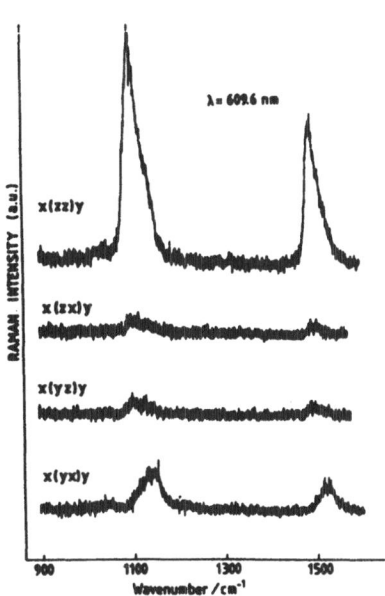

Fig.2 - Raman intensities of
highly oriented trans polyacetylene for the four different scattering configurations (λ_L = 609.6 nm).

tion of conjugation length [4-8] and seem to indicate that the direction of the electric transition moment varies with the conjugation length
of the trans-segment. Indeed, while for long conjugated segments the
transition moment seems to be strictly aligned along the chain axis,
for short conjugated segments some misalignment is likely to occur
[2,3,9].

The changes of the corrected Raman intensities of trans polyacetylene with the scattering configuration are in a pleasing agreement with
the results of a preliminary study [10] performed on a sample of trans
β-carotene stretch-oriented in a polyethylene film (Fig.3). In this case, due to the very small concentration of β-carotene, it is likely that
only very small correcting factors due to the anisotropy of the absorption should be applied to the observed Raman intensities. Further work
is in progress to clarify this point.

3. SOLUBLE POLYACETYLENE

A form of polyacetylene soluble in the common organic solvents has been
obtained by grafting polyenic chains onto activated sites of polybuta-
diene which acts as the soluble carrier [11]. The solvent lines can be

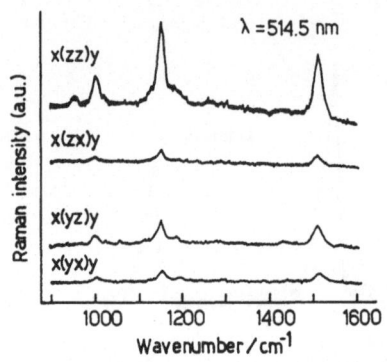

λ =514.5 nm

x(zz)y

x(zx)y

x(yz)y

x(yx)y

Raman intensity (a.u.)

Wavenumber / cm⁻¹

Fig.3 - Raman intensities of trans β-carotene, stretch-oriented in a polyethylene film, for the four different scattering configurations (λ_L = 514.5 nm).

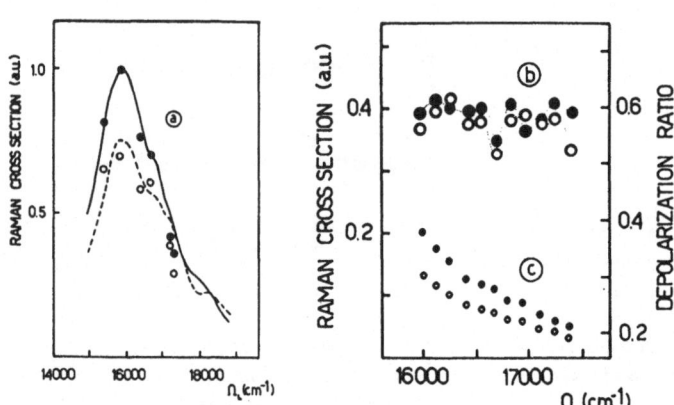

Fig.4 - Experimental and calculated Raman intensities for one sample of soluble trans $(CH)_x$ (a) [12] and intensities (c) and depolarization ratio (b) for another sample of soluble trans $(CH)_x$ [14]. (● 1080 cm⁻¹; ○ 1460 cm⁻¹).

used as an internal standard for the Raman intensities, thus allowing a straightforward determination of the excitation profiles for the totally symmetric modes of trans [12] (fig.4a) and cis polyacetylenes [13] (fig.5a).

Very recently additional Raman experiments have been carried out, namely, the depolarization ratio has been measured as a function of the exciting frequency for the two most intense Raman active vibrations of trans polyacetylene (fig.4b) and for the three totally symmetric modes of cis polyacetylene (fig.5b).

The observed dispersions of the depolarization ratios and their large values are inconsistent with a scattering in resonance with a single electronic transition of an infinitely long polyenic chain, whose transition moment coincides with the molecular axis. Indeed in this case only one component of the scattering tensor would contribute to the observed intensity, yielding the value 1/3 for the depolarization ratio.

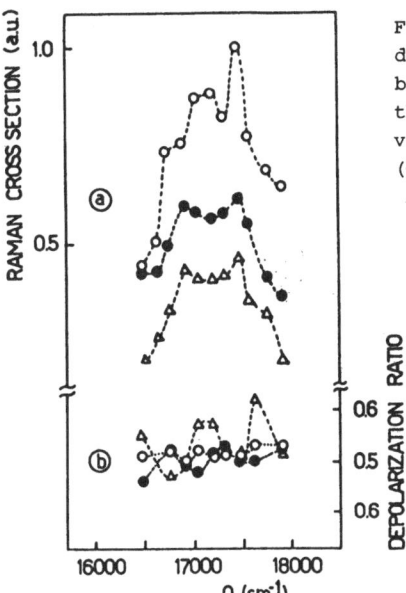

Fig.5 - Raman intensity (a) and depolarization ratio (b) of soluble cis polyacetylene for the three most intense Raman active vibrations.
(O 1250 cm^{-1}; ● 1520 cm^{-1}; Δ 900 cm^{-1}).

At present we believe that the depolarization data can be interpreted in terms of a model which assumes two overlapping electronic states in the resonance region, which are vibronically coupled via the totally symmetric vibrations exhibiting Raman activity. Calculations are under way for detailed comparisons with experiment. The results will be reported elsewhere.

1. U. Pedretti, G. Perego, G. Lugli: Italian Patent Appl. n.22722 A/82, n.21349 A/83 to Assoreni
2. G. Masetti, E. Campani, G. Gorini, L. Piseri, R. Tubino, P. Piaggio, G. Dellepiane: Solid State Commun. 55, 737 (1985)
3. G. Masetti, E. Campani, G. Gorini, R. Tubino, P. Piaggio, G. Dellepiane: Chem.Phys. 108, 141 (1986)
4. H. Kuzmany: Phys.Stat.Sol. 97b, 521 (1980)
5. S. Lefrant: J.Phys. 44, C3-247 (1983)
6. L.S. Lichtmann, A. Sarhangi, D.B. Fitchen: Solid State Commun. 36, 869 (1980)
7. L. Piseri, R. Tubino, E. Mulazzi, G. Dellepiane: in Proceedings of the 8th ICORS, J.Lascombe and P.V.Huong eds. (Wiley, New York 1982)
8. G.P. Brivio, E. Mulazzi: Chem.Phys.Letters 95, 555 (1983)
9. E. Mulazzi: Solid State Commun. 55, 807 (1985)
10. R. Cavalli, C. Cuniberti, P. Piaggio, G. Dellepiane, G. Masetti: to be published
11. S. Destri, M. Catellani, A. Bolognesi: Makromol.Chem., Rapid Commun. 5, 353 (1984)
12. R.S. Cataliotti, G. Paliani, G. Dellepiane, S. Fuso, S. Destri, L. Piseri, R. Tubino: J.Chem.Phys. 82, 2223 (1985)
13. R.S. Cataliotti, G. Paliani: to be published
14. C. Cuniberti, P. Piaggio, G. Dellepiane, R. Tubino: to be published.

Interpretation of the Raman Spectra of n-doped $trans$-$(CH)_x$ Films

E. Mulazzi[1] *and S. Lefrant*[2]

[1]Dipartimento di Fisica dell'Università di Milano,
 via Celoria 16, I-20133 Milano, Italy
[2]Institut de Physique, Université de Nantes,
 F-44072 Nantes Cedex, France

1. Introduction

Trans polyacetylene has been extensively studied these last years by a great number of techniques including Resonant Raman Scattering (RRS) /1-4/. It has been shown in particular that the Raman band shapes exhibit drastic modifications when the laser excitation wavelength is changed from the red to the violet range. Also the history of the sample has been proved to be an important factor since the peak positions and the band shapes are strongly related to the average conjugation lengths of the polymer chains, via the vibrational and electronic properties /5/.

In doped samples, RRS spectra show modified features which have been also studied in detail experimentally for both p and n dopants /6-11/. It appears that the first stage of doping leads to a material in which two phases coexist: a doped phase and an undoped one and, as pointed out in /9/, because of resonance effects of the incident light frequency, only the undoped segments contribute to the scattering of light. As a consequence, the bimodal distribution model /5/ proposed and applied to the undoped $trans(CH)_x$ as a good approach to interpret the RRS spectra can be used to determine the average conjugation length of the undoped segments in the doped films. This was done in /9/ where the RRS spectra of $(CHBr_y)_x$ are analyzed .

For the n doped system $(CHn_y)_x$, when $y < 3.5\%$ Raman spectra recorded in many different samples exhibit similar results, i.e. a general shortening of the undoped conjugated segments /7,8,11/. When the dopant concentration level becomes higher, new intense features appear in the Raman spectra /7,8,11/. The purpose of this paper is to describe the experimental Raman spectra of the n-doped $trans(CH)_x$ and to show by using the results of the perturbed Green function approach applied to this system that these new bands are due to doped induced vibrational modes of the polymer chain perturbed by the dopant.

2. Experimental Results

As reported in the Introduction, the RRS spectra induced by n-doped $(CH)_x$ samples have been studied by many groups /6-11/. The new features which appear in the RRS spectra , when the dopant concentration level becomes sufficiently high, consist in two bands at ~ 1270 cm^{-1} and 1580-1600 cm^{-1}, respectively. Most of the previous studies were done in chemically doped samples for which very high reactivity at ambient atmosphere prevented precise measurements of the dopant concentration level. Later on, the electrochemical procedure was developed allowing direct determination of the number of charges passing through the polymer electrodes. It was then reported that the new features were observable in $(CHn_y)_x$ for y of the order of 5 or 6% /7-8/ for excitation wavelengths λ_L = 604 nm and λ_L = 676 nm.

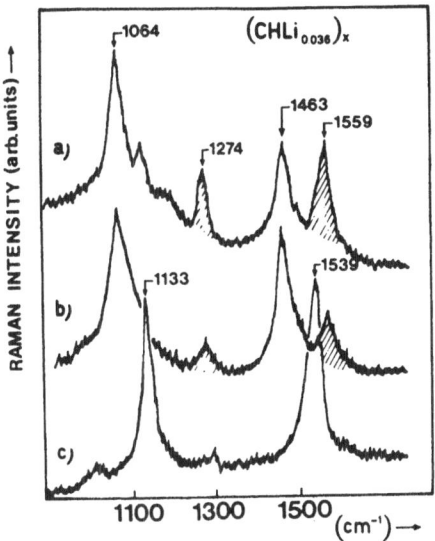

Fig.1 Raman spectra of (CH)$_x$ electrochemically doped with Li at 3.6% level; T=20°C. a) λ_L=676.4 nm; b) λ_L^x=600 nm; c) λ_L=457.9 nm.

More recently, more accurate experiments were reported in which the new Raman bands are observed in (CHLi$_y$)$_x$ for y=3.6% with the laser excitation wavelength λ_L=676 nm /11/. See Fig.1a. In the Fig.1 there are also reported the spectra obtained for λ_L=600 nm and 457.9 nm (Fig.1 b,c). From these figures it should be noticed that the doping-induced bands decrease in intensity by increasing the excitation light frequency and they become unobserved for λ_L=457.9 nm. For this excitation wavelength, the spectrum is similar to that of the previously described case of (CHBr$_y$)$_x$ /9/ and to the spectrum of (CHI$_y$)$_x$ /12/.

As reported by other authors, a higher concentration level of the dopant y(y>>10%) leads to RRS spectra in which the new features at $\sim$1270 cm^{-1} and 1600 cm^{-1} are observable for all the excitation wavelengths in the visible frequency region /7,8/. One of the main characteristics of these spectra is the very weak intensity of the Raman bands due to the remaining undoped segments. This is illustrated in Fig.2 which shows the Raman spectrum of (CHLi$_y$)$_x$ for y>>10% for λ_L=676 nm and λ_L=547.9 nm. The prominent peaks are observed at $\sim$1280 cm^{-1} and 1600 cm^{-1} for λ_L=457.9 nm. As in the case of all n-type dopants these two peaks behave differently when the laser excitation wavelengths change from the red to the violet region (see Fig.2).

3. Theoretical Model

We use the model proposed in /13,14/ which has been developed in order to explain the features of the photoinduced /15/ and doping induced /16/ bands in the infrared spectra of trans(CH)$_x$ and (CD)$_x$. Following that model the photoinduced and doping induced infrared active vibrational modes are calculated by using the perturbed Green Function formalism. In this scheme, the perturbation induced by the trapped charges on the lattice dynamics of the conjugated segments is taken into account through Λ, the change in the force constants with respect to the unperturbed lattice system. In order to evaluate the new vibrational modes, we have considered that the perturbation

151

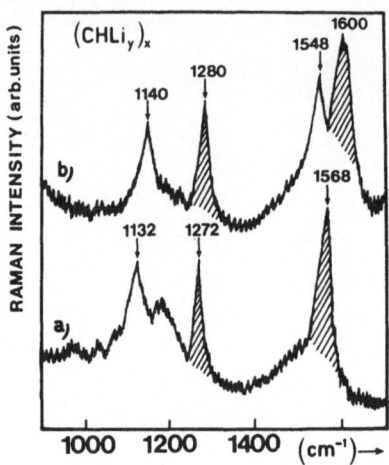

Fig.2 Raman spectra of $(CH)_x$ highly doped with Li(chemical doping); T=20°; a) λ_L=676 nm; b) λ_L=457.9 nm.

determined by the trapped charges gives a positive change in the force constants. Then the Λ values considered are always positive and we have found /13,14/ that $\Lambda_{phot.} < \Lambda_{dop.}$. This is to be expected since the Λ values are related to the change of electron-vibration and electron-electron interactions determined by the presence of the trapped charges on the conjugated segments.

By following the perturbed Green function formalism and the approximations used in /13/, we write the density of perturbed vibrational states $\rho(\omega^2)$ in terms of Λ, $\tilde{\rho}^o(\omega^2)$ and $\rho^o(\omega^2)$ which are the real part and imaginary parts of the unperturbed Green functions $G^o(\omega^2+io^+)$ respectively (see /13/):

$$\rho(\omega^2) = \frac{\rho^o(\omega^2)}{|1-\Lambda\tilde{\rho}^o(\omega^2)|^2 + |\pi\Lambda\rho^o(\omega^2)|^2} . \tag{1}$$

In eq.(1) $\rho^o(\omega^2)$ the imaginary part of the unperturbed Green function and the $\tilde{\rho}^o(\omega^2)$ the real part of the unperturbed Green function, are given in /13/. Both the functions are evaluated by considering the q-dispersion of the vibrational frquencies of trans$(CH)_x$ in the approximations described in /13/. Following the arguments given in the same reference the new vibrational modes, which are either infrared active or Raman active, are determined by the zeros of the following equation:

$$\frac{1}{\Lambda} = \tilde{\rho}^o(\omega^2) . \tag{2}$$

We show in Fig.3 the behaviour of $\tilde{\rho}^o(\omega^2)$ as function of ω for trans$(CH)_x$ system. We indicate in the figure four different regions where $\tilde{\rho}^o(\omega^2)$ is different from zero. For $\Lambda>0$ (positive values on the vertical axis), which is the case we consider here, the regions (I, II, III) and (II, IV) in Fig.3, are those where the new infrared active modes and the new Raman active modes, respectively, have to be found following eq.(2). This result comes from symmetry arguments. In /13/ and /14/ we have calculated only the new vibrational modes which appear in the infrared spectra. In this paper we want to evaluate the doped induced Raman active modes which have therefore

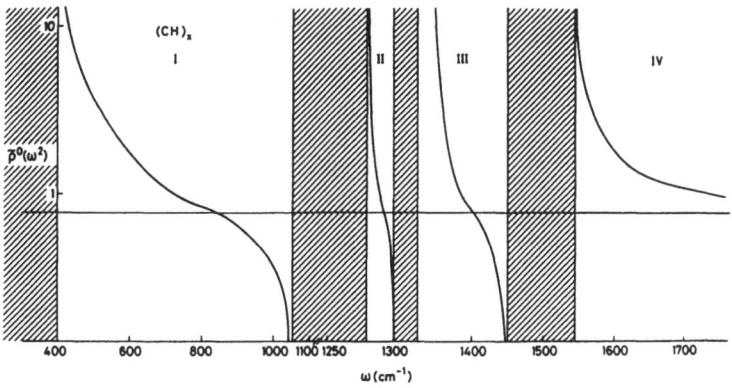

Fig.3 Calculated $\tilde{\rho}^0(\omega^2)$ as function of ω for trans(CH)$_x$. Note the change of scale in the intervals 1100-1200 cm^{-1} and 1200 - 1700 cm^{-1}. The values of $\tilde{\rho}^0(\omega^2)$ are given in 10^{-6} cm^{+2}.

to be found in the regions II and IV. From the calculations it is possible to obtain that the new Raman active frequencies ω_{1R} and ω_{2R} have the following values: $\omega_{1R} \sim 1275$-1280 cm^{-1}; $\omega_{2R} \sim 1560$-1610 cm^{-1} by considering a perturbation $\Lambda \sim 3$-$5 \, 10^6$ cm^{-2}.

One has to notice that while in II the variation of Λ induces only a very small change in frequency of ω_{1R}, in IV ω_{2R} is much more dependent on the value of Λ.

We would like to point out that the appearance of the Raman active doping induced vibrational modes in the spectra is related only to the pre-resonance or resonance condition of the incident laser excitation frequency with the doping-induced electronic transitions, which lie in the gap of the pristine trans(CH)$_x$ system. Moreover it should be noticed that the intensities of these two doping-induced modes in the RRS spectra depend mainly on the linear electron vibration couplings of the two modes in the doping induced electronic states excited by the incident laser frequency.

4. Discussion

As already mentioned in the previous section, the appearance in the RRS spectra of the doping-induced modes with frequencies ω_{1R} and ω_{2R} is related to the resonance condition of the incident laser frequency with the doping induced electronic transitions. Then the investigation of the electronic properties of the doped system is very important in order to understand the features of the RRS spectra. Recent experimental results have shown fundamental differences in the absorption curves recorded for some p-doped /17/ and the n-doped /10/ trans(CH)$_x$. All the n-doped systems exhibit similar absorption changes upon doping as described in /10/. The main modification consists in a new absorption band whose intensity increases in the near infrared by increasing the doping concentration level, while the main absorption band of the pristine polymer decreases.

In the case of p-doped system with iodine /17/ and bromine, the main features of the absorption spectra are: i) the band in the near i.r. whose peak shifts toward lower frequencies by increasing the dopant concentration level, ii) the main band in the visible region which decreases a little in intensity and changes its shape. This absorption is probably due to both the electronic transitions ($\pi \rightarrow \pi*$) of the remaining undoped segments of the pristine polymer and to the absorption of I_3^- and I_5^- species /6/.

In view of these results, the new doping-induced vibrational features begin to appear in the RRS spectra of the n doped polymer, when the excitation wavelength is in (pre)-resonance condition with the doping induced electronic transitions. This explains why in the spectra of $(CHLi_{0.036})_x$ shown in Fig.1, the new Raman bands appear first for $\lambda_L = 647$ nm and are unobserved for $\lambda_L = 457.9$ nm. By increasing the dopant concentration level, the resonance condition is achieved for all the incident wavelengths in the visible region, yielding the observation of the doping induced vibrational modes, even for an excitation wavelength in the violet (Fig.2). By using similar arguments, one can explain why the doping-induced modes are not observable so clearly in the I_2 and Br_2 doped systems, aside from the peak at 1600 cm^{-1} which is still very weak at high doping level.

References

1. D.B. Fitchen: Mol. Cryst. Liq. Cryst. 83, 95 (1982).
2. H. Kuzmany: Phys. Stat. Sol.(b) 97, 521 (1980) and H. Kuzmany, E.A. Imhoff, D.B. Fitchen and A. Sarhangi, Phys. Rev. B 26, 7109 (1982).
3. S. Lefrant: J. Phys. (Paris) Colloq. 44, C3-247 (1983).
4. E. Mulazzi, G.P. Brivio, E. Faulques and S. Lefrant: Sol. St. Comm. 46, 851 (1983).
5. G.P. Brivio and E. Mulazzi: Phys. Rev. B30, 676 (1984) and R. Tiziani, G.P. Brivio and E. Mulazzi: Phys. Rev. B 31, 4015 (1985).
6. I. Harada, V. Furukawa, M. Tasumi, H. Shirikawa and S. Ikeda: J. Chem. Phys. 73, 4746 (1980).
7. E. Faulques, S. Lefrant, F. Rachdi and P. Bernier: Synt. Metals 9, 53 (1984).
8. H. Eckhardt, L.W. Shacklette, J.S. Szobota and R.H. Baughman: Mol. Cryst. Liq. Cryst. 117, 401 (1985).
9. S. Lefrant, E. Faulques, G.P. Brivio and E. Mulazzi: Sol. St. Comm. 53, 583 (1985); E. Mulazzi, G.P. Brivio, S. Lefrant and E. Faulques: Mol. Cryst. Liq. Cryst. 117, 351 (1985).
10. J. Tanaka, Y. Saito, H. Shimizu and M. Tanaka: Synt. Metals 17, 307 (1987).
11. S. Lefrant, E. Faulques, A. Chentli, F. Rachdi and P. Bernier, Synt. Metals 17, 325 (1987).
12. E. Mulazzi, G.P. Brivio, S. Lefrant, E. Faulques and E. Perrin: Synt. Metals 17, 325 (1987).
13. G.P. Brivio and E. Mulazzi: Sol. St. Comm. 60, 203 (1986).
14. P. Piaggio, G. Dellepiane, E. Mulazzi and R. Tubino, to be published in Polymer (1987).
15. G. B. Blanchet, C.R. Fincher, T.C. Chung and A.J. Heeger: Phys. Rev. Lett. 50, 1938 (1983).
16. B. Francois, M. Bernard and J.J. Andre: J. Chem. Phys. 75, 4142 (1981).
17. H. Fujimoto, J. Tanaka, M. Tanaka and T. Kishi: Synt. Met. 16, 133 (1986).

Part IV

Magnetic Resonance

Pulsed ENDOR and TRIPLE Resonance on *trans*-Polyacetylene à la Durham Route

A. Grupp[1], P. Höfer[1], H. Käss[1], M. Mehring[1], R. Weizenhöfer[2], and G. Wegner[2]

[1]2. Physikalisches Institut, Universität Stuttgart,
Pfaffenwaldring 57, D-7000 Stuttgart 80, Fed. Rep. of Germany
[2]Max-Planck-Institut für Polymerforschung, Postfach 3148,
D-6500 Mainz, Fed. Rep. of Germany

1. Introduction

Electron Nuclear Double Resonance (ENDOR) has proven to be very important in the investigation of neutral soliton and polaron structures [1-3]. Dalton and co-workers [4-6] have demonstrated in a series of publications that ENDOR provides much more decisive information on soliton structure in polyacetylene than ESR or any other type of spectroscopy. Others followed this route and have presented similar ENDOR results with partially different interpretations [7-10]. However, the interpretation of ENDOR spectra of powders or amorphous materials is rather complicated, although much simpler than ESR spectra. At the IWEPP 85 [11] and in a later publication [12] it was shown that virtually any distribution function of spin-density ρ_j at position j, i.e. soliton structure, fit the observed ESR spectra, whereas the ENDOR spectra vary significantly. Although ENDOR spectra represent the direct distribution of hyperfine interactions, thus spin-density distributions, their analysis is hampered by several complications: (a) The hyperfine interaction of the soliton spin with the surrounding nuclei is not a scalar quantity but a second-rank tensor. Powder spectra are therefore a superposition of different spectra for different orientations. (b) The connection between the hyperfine interaction and the spin-density is only approximately known within the limits of the McConnell relation. (c) Spin dynamical effects due to different relaxation mechanisms can severely alter the ENDOR spectra. (d) Hyperfine enhancement up to second order influences the transition probabilities.

Let us now turn to the question how we have solved those complications: (a) The hyperfine tensor for sp^2 radicals has been established in a number of aromatic molecules and one can be fairly confident about the numbers. We have used: A_{11}=-32.2 MHz, A_{22}=-64.4 MHz and A_{33}=-96 MHz with a_{iso}= $(A_{11}+A_{22}+A_{33})/3$=-64.2 MHz, where the 1-axis is pointing along the CH-bond, the 2-axis points perpendicular to the C=C-C plane and the 3-axis is perpendicular to the former. (b) We have used a_{iso}=$Q \cdot \rho_j$ for the hyperfine interaction a_{iso} at position j of the soliton according to McConnell. The McConnell factor of Q = -64 MHz used here is typical for protons in aromatic compounds. (c) We have avoided to a large extend most relaxation mechanisms by applying pulsed-ENDOR techniques [13] which have a number of advantages with respect to conventional cw-ENDOR techniques used so far. (d) The correct transition probabilities have been calculated up to second order and were incorporated in the simulated spectra.

2. Experimental

Pulsed ENDOR spectra and TRIPLE resonance spectra with 9.7 GHz electron spin Larmor frequency and 2 - 50 MHz ENDOR frequency were obtained with different pulse schemes, to be discussed elsewhere. Samples of trans-polyacetylene were prepared according to the Feast-Durham route from a precursor polymer which was partially stretch oriented.

3. Results and Discussions

Typical pulsed ENDOR spectra are shown in Fig.1, for two different pulse schemes. Fig.1 left shows the typical double peak pulsed ENDOR spectrum with a "distant-ENDOR" central peak. This double peak structure has also been observed in cw-ENDOR experiments on Shirakawa type $(CH)_x$ and was interpreted as due to a spectral feature caused by hyperfine values from negative spin-density [5,8]. In fact, together with some other structural features in the spectrum, it has been taken as evidence for just two different hyperfine tensors, i.e. a rectangularly shaped soliton extended over 50 CH-units. Fig.1 left, however, shows a series of spectra for one sample with different microwave pulse strengths for a Davies-ENDOR sequence [14]. It is evident that experimental parameters determine the peak separation, rather than spectral features. Moreover, Fig.1 right presents an alternative pulsed ENDOR scheme of the SEDOR-type (Spin Echo Double Resonance [13]). This experiment does not show the double peak structure, which is clear evidence for a wide distribution of different hyperfine values. The spin-dynamical effect of the different pulsed ENDOR experiments have been properly taken into account in all simulations of the spectra. Details of this will be published elsewhere.

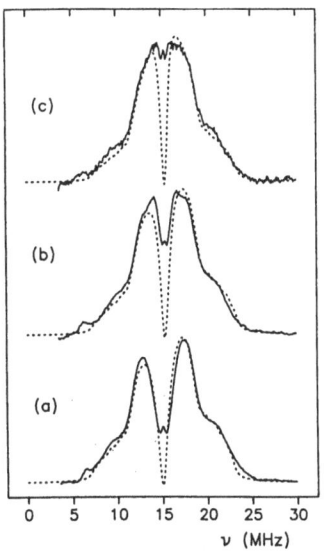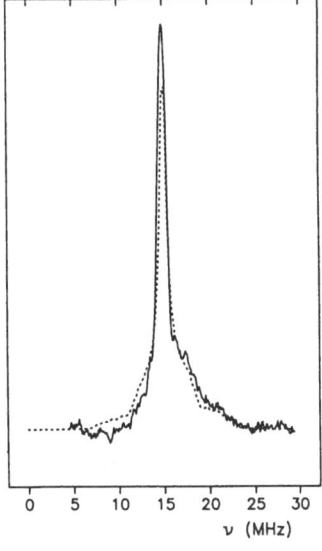

Fig.1 Experimental and calculated (---) ENDOR spectra at 10 K of a sample which was stretched by a factor of 3 and oriented with the stretch direction parallel to the external magnetic field. On the left a Davies sequence with microwave pulse strengths of 0.9 G (a), 0.45 G (b), and 0.22 G (c) was applied. The simulations are calculated including the spin alignment effect according to different microwave pulse strengths. As is evident, the spin alignment effect causes a spectral hole around the free proton frequency. The simulation of the SEDOR spectra (right) is calculated with the same parameters, but without spin alignment which is not expected to be present in SEDOR.

157

Once having established that there is a wide distribution of hyperfine interactions, one could try to simulate the spectra by some spin density distribution function. By obeying the rules stated in (a)-(d) rigorous calculations of powder spectra can be performed for any given spin density distribution with the normalization condition $\Sigma_j \rho_j = 1$. We have chosen for convenience the spin density distribution function

$$\rho_j = \frac{1}{\ell} \operatorname{sech}^2 \{\frac{j}{\ell}\} \left[g \cos^2\{\frac{j\pi}{2}\} - u \sin^2\{\frac{j\pi}{2}\} \right] \qquad (1)$$

where 2ℓ is the full width at half maximum of the soliton. In order to differentiate between even (gerade g) and odd (ungerade u) sites of the soliton the parameters g and u were introduced [12]. We note, that the observation of negative spin density at odd sites is direct evidence for Coulomb correlation. An important parameter is therefore ug=u/g, which is the ratio of negative to positive spin density. Although ug might be inferred from spectral fitting, we have instead performed pulsed TRIPLE-resonance experiments (not shown here) which clearly show the appearance of positive and negative hyperfine interactions. In a number of different experiments we have obtained always the same value, ug=0.43. In keeping this value fixed, we have simulated the ENDOR spectra with the spin density distribution shown in Fig.2, by varying the soliton width ℓ. The best fit was obtained with ℓ=11. This has to be compared with the SSH values ℓ=7 and ug=0. It is evident that large electron-electron correlations are observed in trans-polyacetylene which are even larger

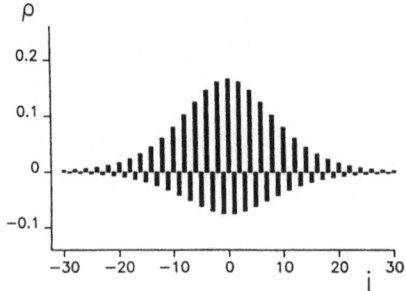

Fig.2 Spin density distribution with ℓ=11 and ug=0.43 used for simulations of the spectra in Fig.1

than considered before. The width of the soliton seems to be somewhat increased compared with the SSH model. We propose that this is due to electron-electron correlation. Comparing our ug value with Quantum Monte Carlo calculations of Hirsch and Grabowski [15] we arrive at $U/4t_o$=0.6 (for V=0) or $U/4t_o \approx 1$ (for V=U/2).

Further theoretical investigations seem to be very important in understanding these values.

Acknowledgement: This project was partially supported by the Sonderforschungsbereich 329 der Deutschen Forschungsgemeinschaft and Stiftung Volkswagenwerk.

1. W.P. Su, J.R. Schrieffer and A.J. Heeger, Phys.Rev.Lett. 42, 1698 (1979)
2. D.K. Campbell and A.R. Bishop, Phys.Rev. B 24, 4859 (1981)

3. J.L. Brédas, R.R. Chance and R. Silbey, Phys.Rev. B 26, 5843 (1982)
4. H. Thomann, L.R. Dalton, Y. Tomkiewicz, N.S. Shiren and T.C. Clarke, Mol.Cryst.Liq.Cryst. 83, 1065 (1982)
5. J.F. Cline, H. Thomann, H. Kim, A. Morrobel-Sosa, L.R. Dalton and B.M. Hoffman, Phys.Rev. B 31, 1605 (1985)
6. H. Thomann, L.R. Dalton, M. Grabowski and T.C. Clarke, Phys.Rev. B 31 3141 (1985)
7. S. Kuroda and H. Shirakawa, Solid State Commun. 43, 591 (1982)
8. S. Kuroda, H. Bando and H. Shirakawa, Solid State Commun. 52, 893 (1984)
9. G.J. Baker, J.B. Raynor and A. Pron, J.Chem.Phys. 80, 5250 (1984)
10. A. Bartl and A. Stasko, phys.stat.sol. (a) 95, 659 (1986)
11. M. Mehring and P.K. Kahol: In Electronic Properties of Polymers and Related Compounds, ed. by H. Kuzmany, M. Mehring and S. Roth, Springer Ser. Solid-State Sci., Vol. 63 (Springer, Berlin, Heidelberg, 1985)
12. P.K. Kahol and M. Mehring, J.Phys. C: Solid State Phys. 19, 1045 (1986)
13. P. Höfer, A. Grupp and M. Mehring: In Electronic Resonance of the Solid State, ed. by J.A. Weil, Canadian Chemical Society, Symposium Series, 1987
14. E.R. Davies, Phys. Lett. 47A, 1 (1974)
15. J.E. Hirsch and M. Grabowski, Phys.Rev.Lett. 52, 1713 (1984)

ESR Study of Metallic Complexes of Alkali-Doped Polyacetylene

F. Rachdi and P. Bernier

Groupe de dynamique des phases condenseés (CNRS), U.S.T.L., F-34060 Montpellier, France

We have followed the dependence of ESR linewidth (ΔH_{pp}) of highly alkali-metal doped polyacetylene on the nature of the dopant and on the temperature in the range from 4 K to 300 K. The room-temperature linewidth increases with increasing atomic number (z) of the dopant and seems to follow a z^{α}-law with $\alpha \simeq 2.3 \pm 0.7$ which is close to the z^4 law expected for metals. This linewidth behaviour suggests a significant contribution from the spin-orbit coupling of the unpaired electrons on the dopant site. Thus, according to Elliott's model, we have been able, in the case of Li-, K- and CS-doped films, to estimate the expected electronic g-factor deviations from the free-electron g-value. The linewidth decreases quasi-linearly with decreasing temperature down to ∼80 K where it starts to increase. This temperature behaviour can be analyzed in terms of gradual transition from delocalized states at high temperature, where interchain electron hopping plays a dominant role, to localized states at low temperature where electrons are mainly confined along the chains.

1. Introduction

Highly doped polyacetylene with n- or p-type dopants exhibits a high electrical conductivity and a very low thermoelectric power, which are associated with a temperature-independent Pauli-type magnetic susceptibility (1,2). The system is then commonly assimilated to a metal. Nevertheless, the temperature dependence of its electrical conductivity is not metal-like. The conductivity decreases with decreasing temperature for both p- and n-type doped polyacetylene (3,4).

Using Electron-Spin-Resonance (ESR), we have studied the evolution of the magnetic properties of the highly alkali-metal doped polyacetylene, as a function of the nature of the dopant and the temperature. The observed dependence of the room-temperature ESR linewidth ΔH_{pp} and the g-factor with the atomic number z of the alkali dopant is analyzed in terms of spin-orbit coupling of the unpaired electrons on the dopant site, according to the Elliott model (5). The ΔH_{pp} temperature dependence can be attributed to a transition from a metallic state with delocalized spin carriers at high temperature to a state with localized spins at low temperature.

2. Experimental

cis-rich polyacetylene $(CH)_x$ films prepared according to the Shirakawa technique were chemically doped to saturation (dopant concentration y > 15%) by immersing them in a 0.3 M solution of alkali-metal naphthalene complexes in tetrahydrofuran (THF) (6). ESR spectra were recorded in the temperature range from 4 to 300 K, using an X-band (9.5 GHz) ER 200 D Bruker, provided with an ESR 900 Oxford Instruments with a continuous flow helium system.

3. Results and discussion

We present in Fig. 1 experimental data of ΔH_{pp} versus the dopant atomic number z at room temperature for highly alkali-metal doped polyacetylene. These data include our values as well as values reported by various authors (7)(8). For Li-doped $(CH)_x$, a large range of ΔH_{pp} values is obtained related to the cation solvation effect which influences significantly the width (9). In the case of highly Na-doped films and K-doped ones, the observed room-temperature ΔH_{pp} values vary in the range 6 to 10 G and 5 to 35 G, respectively. The large range of ΔH_{pp} values for K-doped films is mainly due to the influence of the annealing on the line broadening (10)(11). Highly Rb- and Cs-doped $CH)_x$ exhibit a large asymmetric line characterized by $\Delta H_{pp} \simeq 50 \pm 10$ G for Rb-doped films and $\Delta H_{pp} \simeq 300 \pm 50$ G for Cs-doped ones

As shown in Fig. 1, increasing atomic number z of the dopant results in an increase of the room-temperature linewidth ΔH_{pp}.

A similar behaviour has been observed in the cases of alkali-

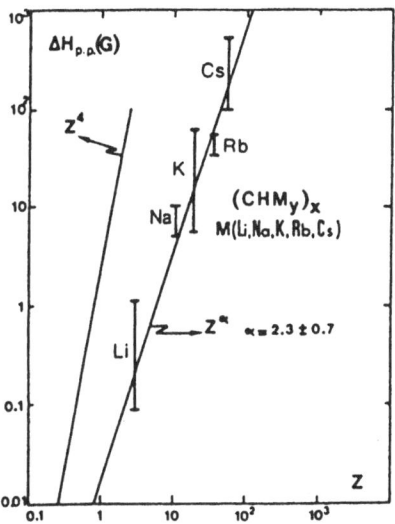

Fig. 1 Room-temperature dependence of the ESR linewidth of highly alkali-metal doped $(CH)_x$ vs the atomic number z of the dopant (dopant concentration y > 15%)

161

metal doped polyphenylene (12) and graphite (13). By analogy
with these two cases, we expect that one important contribution to the linewidth comes from the spin-orbit interaction
on the alkali-metal dopant atoms. According to the theory
developed by Elliott (5) and Yafet (14) indicating that the
dominant process inducing electronic spin reversal is a
spin-lattice relaxation, which comes from the modulation of
the spin-orbit interaction by the phonons, the ESR linewidth
dependence on the dopant atomic number z should follow a z^4-
law. The observed room-temperature ΔH_{pp} variation versus z
follows a z^α-law with $\alpha \simeq 2.3 \pm 0.7$ (see Fig. 1). This linewidth behavior suggests that spin-orbit coupling on the alka-
li dopant plays the dominant role in defining the width. However, the discrepancy with the z^4-law mentioned above could
be interpreted as a consequence of the contribution to the
width of other relaxation mechanisms depending on the nature
of the dopant.

On the other hand, Elliott's theory (5) predicts a
deviation of the electronic g-factor from the free-electron g-
value, resulting from the efficiency of the spin-orbit coupling to induce electronic spin reversal for simple metals.
Computer simulation of the observed ESR spectra of metallic
complexes of $(CH)_x$ with alkali dopants enables us to estimate
the expected deviation from the free-electron g-value. For
Li-doped $(CH)_x$, the estimated g-value is very close to the
free-electron one. For K- and Cs-doped films, the obtained g-
values are 2.0039 ± 0.0003 and 2.0041 ± 0.0003, respectively.
Thus , we can conclude that the observed ΔH_{pp} versus z behavior and the g-shifts suggest a significant contribution to
the ESR linewidth from the spin-orbit coupling of the unpaired electron on the alkali-dopant site.
As shown in Fig. 2, increasing the temperature from 100 to
300 K for Li-doped films and from 50 to 300 K for K-doped

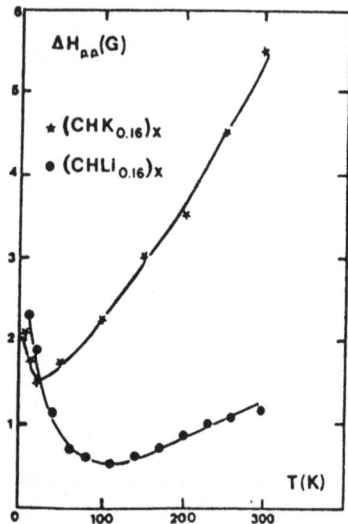

Fig. 2
ESR linewidth of highly Li- and
K-doped cis-$(CH)_x$ as a function
of temperature.
Solid lines are guides to the
eye.

ones results in an increase of the linewidth which is nearly proportional to temperature. A similar behavior has been
observed in the case of highly Na- and Rb-doped $(CH)_x$ (15)
(11). Such a temperature dependence will be interpreted
using the structural model developed by Baughman et al.(16),
concerning metallic complexes of polyacetylene with alkali-
metal dopant atoms, which predicts that the $(CH)_x$ chains
form a host lattice in which the alkali metal ions are present in channels. We expect such a structure to result in a
spin-orbit coupling on the alkali-ions, which is more efficient for electronic motion normal to the chains than along
the chain. With increasing temperature above 50 K, the electron activated transverse hopping becomes more and more efficient. Then spin-lattice relaxation via the spin-orbit
coupling becomes the dominant relaxation process in defining
the linewidth, yielding the observed ΔH_{pp} increase with temperature.

On the other hand, for metallic systems in which the
electronic longitudinal and transverse relaxation time, respectively T_1 and T_2, are similar, the theory of Yafet (14)
predicts that $T_1^{-1} \simeq T_2^{-1} \alpha T$ for T above the Debye temperature.
Thus according to the latter relation, the observed line-
width behavior as a function of temperature seems to follow
a metallic T-law, in agreement with the fact that the origin
of the obtained ESR signal can be attributed to the delocalized electrons.

Contrary to the preceding case, for low temperatures
between 4 and 100 K for Li-doped films and between 4 and 50 K
for K-doped ones, the linewidth increases when the temperature decreases (see Fig. 2). Such a behavior could be interpreted as a consequence of electron localization on the
chains. Therefore, when the temperature decreases, the electron spends more and more time on a given chain and the hopping time between chains ($\tau_\perp$) becomes very large compared to
the longitudinal scattering time ($\tau_\parallel$). consequently, spin-
lattice relaxation via the spin-orbit coupling is overcome
by other relaxation mechanisms. The exchange between quasi-
localized spins and the hyperfine interaction with the nuclear spins could explain the observed broadening of the line
with decreasing temperature.

References

1.   S. Ikehata, J. Kaufer, T. Woerner, A. Pron, M.A. Druy,
     A. Sivak, A.J. Heeger and A.G. MacDiarmid, Phys. Rev.
     Lett. 45, 1123 (1980)

2.   Y.W. Park, A. Denenstein, C.K. Chiang, A.J. Heeger,
     A.C. MacDiarmid, Solid St. Comm., 29, 747 (1979)

3.   A.J. Epstein, H. Rommelmann, R. Bigelow, H.W. Gibson,
     D.M. Hoffmann and D.B. Turner, J. Physique Paris, C3,
     44, 61 (1983)

4.   M. Audenaert, F. Rachdi, R. Bernier, Synth. Met. 15, 91
     (1986)

5.   R.J. Elliott, Phys. Rev., 96, 266 (1954)

6. F. Rachdi, R. Bernier, E. Faulques, S. Lefrant, F. Schué, J. Chem. Phys., 80, 6285 (1984)

7. B. Francois, M. Bernard, J.J. André, J. Chem. Phys., 75, 4142 (1981)

8. A. Elkhodary, R. Bernier, J. Phys. (Paris) Lett., 45, 551 (1984)

9. F. Rachdi, R. Bernier, F. Schué, Mol. Cryst. Lip. Cryst. 117, 121 (1985)

10. R.L. Elsenbaumer, R. Delannoy, G.G. Miller, C.E. Forbes, N.S. Murphy, H. Eckhardt, R.H. Baughman, Synth. Met. 11, 251 (1985)

11. F. Rachdi, R. Bernier, Phys. Rev. B. 33, 7817 (1986)

12. L.D. Kispert, J. Joseph, G.G. Miller, R.H. Baughman, J. Chem. Phys. 81, 2119 (1984)

13. R. Lauginie, H.Estrade, J. Conard, D. Guerard, R. Lagrange, M.El. Makrini, Physica B 99, 514 (1980)

14. Y. Yafet, Solid St. Phys. 14, 1 (1963)

15. F. Moraes, J. Chen, T.C. Chung, A.J. Heeger, Synth. Met. 11, 271 (1985)

16. R.H. Baughman, N.S. Murphy, G.G. Miller, J. Chem. Phys. 79, 515 (1983)

In situ ESR Study
During the Electrochemical Intercalation
of Potassium in Polyacetylene

C. Fite and P. Bernier

GDPC-USTL Pl. E. Bataillon, F-34060 Montpellier Cedex, France

We describe and analyze in situ ESR measurements of the
electrochemical doping of cis-rich $(CH)_x$ with K. From the
evolution of the peak-to-peak linewidth (ΔH_{pp}) versus the doping
level y we suggest the existence of successive ordered phases
during the doping and undoping processes. A simplified model of
a pseudo-biphasic system (one phase is created while the other
disappears) can explain qualitatively the variations of the
linewidth during the doping and the undoping. We also discuss
the hysteresis appearing at each cycle and the difference
between the first cycle and the following cycles.

1 INTRODUCTION:

The existence of well—organized structures in doped
polyacetylene $(CH)_x$ with alkali metals has been previously
demonstrated. The ion intercalation is suggested to occur via a
sequence of crystalline phases /1/2/. The ESR linewidth of such
a system is very sensitive to the nature of the dopant, varying
from 0.1G for Li to more than 300G for Cs /3/. The origin of
such widely spread values has been demonstrated to be the
spin-orbit coupling of the conduction electrons on the dopant
site /3/4/. So the ESR study, through the intermediary of the
spin-orbit coupling effect, can give information on the local
electronic properties on the dopant site and on the likely
structural transitions or modifications occurring during the
intercalation process. In this paper we describe the results of
such a study performed in the case of potassium doping, using an
electrochemical cell designed for in situ observation.

2 EXPERIMENTAL TECHNIQUE:

The electrochemical cell consisted of a cis-rich (85% cis-15%
trans) polyacetylene $((CH)_x)$ film (380 μm thick) as one

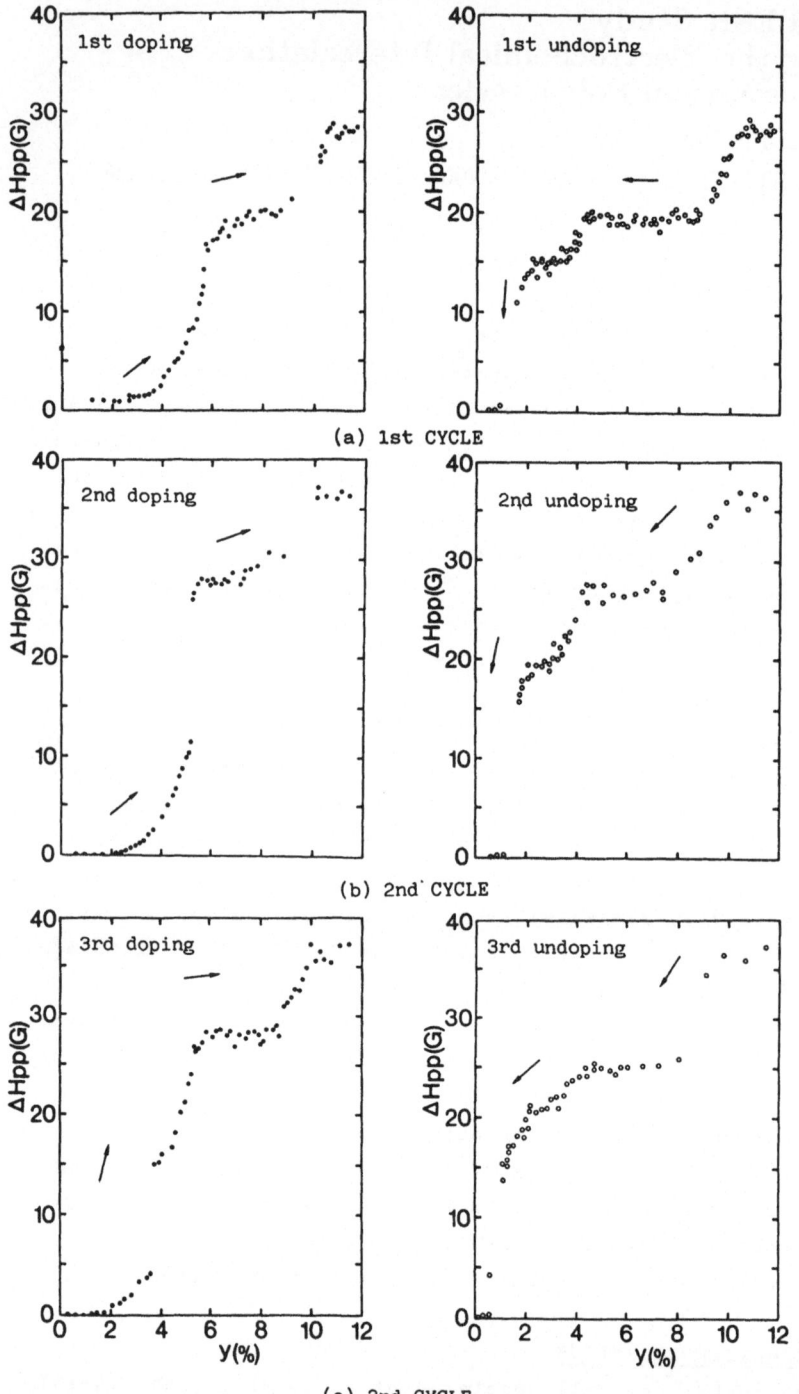

(a) 1st CYCLE

(b) 2nd CYCLE

(c) 3rd CYCLE

Caption see opposite page

electrode, of metallic potassium as the counter electrode, and
of a 1M solution of potassium cyanide in tetrahydrofuran (THF)
with triethyl boron $((C_2H_5)_3B)$ as the electrolytic solution. A
detailed description of the cell construction has been presented
elsewhere /5/.

Prior to doping, the potential difference between the $(CH)_x$
electrode and the alkali metal was in the range 1.6 to 1.8V. The
electrochemical doping occured during a galvanic process,
involving a spontaneous current flowing between the two
electrodes via an external circuit, leading to the system
$(CHK_y)_x$. The doping level y was determined from the charge
injected into the $(CH)_x$ and the weight of the polymer film. The
electrochemical cell was placed in the ESR cavity and ESR
spectra were recorded continuously during doping or undoping.

3 EXPERIMENTAL RESULTS:

The neutral cis-rich $(CH)_x$ had an ESR signal with a peak-to-peak
linewidth $\Delta H_{pp} \sim 6G$ characteristic of the cis-isomer. For very
light doping levels a narrow signal of linewidth $\Delta H_{pp} \sim 1G$ was
superimposed on the broad signal. At a doping level y of 0.3%
the broad signal had almost totally disappeared, indicating a
state of almost complete isomerization /6/7/. The behaviour of
ΔH_{pp} for y>0.3% presented successive plateaus related by sharp
variations of ΔH_{pp} : a $\sim 1G$ plateau for y between 0.3% and 3%, a
$\sim 19G$ plateau for y between 6% and 8% and a $\sim 28G$ plateau for
y>10%. During the undoping process the variations of ΔH_{pp} also
presented successive plateaus: a $\sim 28G$ plateau for y>10%, a $\sim 19G$
plateau for y between 8% and 4.5% and a plateau of 15G for
doping level from 3% to 2%. At doping level less than 2% the
linewidth decreased to 0.2G for y=0.5% (Fig.1).

In the case of subsequent cycles, we note that during the
doping and the undoping the behaviour of ΔH_{pp} showed plateaus
respectively at the same doping levels observed during the first
doping and undoping.But the values of these plateaus,for doping
levels larger than 3%, are $\sim 8G$ higher in comparison with the
values measured during the first cycle (Fig.1).

Fig. 1 - Evolution of the peak to peak linewidth ΔH_{pp} versus the doping
level y. (a): 1st cycle, (b): 2nd cycle, (c): 3rd cycle.

4 DISCUSSION:

We observe that during doping and undoping of each cycle the evolution of ΔH_{pp} has remarkable features around 6% and 10% during doping and around 10%, 4.5% and 2% during undoping. We suggest that the behaviour of ΔH_{pp} can be associated with the presence of intercalation stages /8/ which appear successively during the doping and undoping processes at roughly the same values (the case of 2% is not yet understood).

The general behaviour of ΔH_{pp} can be explained with a two—phases model where the ESR spectrum is the sum of two ESR lines of fixed but different widths issued from each phase. In this case, it is easy to show that the measured linewidth cannot have a linear dependence versus the composition but presents marked plateaus with sharp variations between them /9/. This model leads us to consider the doping with K as being intrinsically inhomogeneous, in good agreement with the model of Conwell /10/ on the structural and electronic properties of the doped $(CH)_x$.

The fact that ΔH_{pp} is systematically larger during the second and third cycles compared with the first one suggests a more closely packed structure in these cases, yielding a stronger spin—orbit interaction. Remaining short cis sequences embedded in the trans isomer, present during the first cycle but disappearing after, can explain this difference. Such an imperfect isomerization even after strong doping has been already observed /11/.

5 REFERENCES:

1. R.H Baughman, N.S. Murthy and G.G. Miller, J. Chem. Phys. 79, 515 (1983).
2. L.W. Shacklette and J.E. Toth, Phys. Rev. B32, 5892 (1985).
3. F. Rachdi and P. Bernier, Phys. Rev. B33, 7817 (1986).
4. R.J. Elliot, Phys. Rev. 96, 266 (1954).
5. A. El Khodary and P. Bernier, J. Chem. Phys. 85(4), 2243 (1986).
6. A. ElKhodary and P. Bernier, J. Phys. Lett. (Paris) 45, 551 (1984).
7. F. Rachdi, P. Bernier, E. Faulques, S. Lefrant, F. Schue, Polymer 23, 173 (1982).
8. C. Fite and P. Bernier, to be published.

9. This is due to the fact that $\Delta H_{pp}(A+B) \neq \Delta H_{pp}(A) + \Delta H_{pp}(B)$.

10. E.M. Conwell, Phys. Rev. B <u>33</u>(4), 2465 (1986).

11. R.L. Elsenbaumer, P. Delanoy, G.G. Miller, C.E. Forbes, N.S. Murthy, H. Eckhardt and H. Baughman, Synth. Met. <u>11</u>, 251 (1985).

^{7}Li NMR Study of the Electrochemical Doping of Poly(acetylene)

A.K. Whittaker, C. Fite, K. Zniber, and P. Bernier

GDPC-USTL Pl. E. Bataillon, F-34060 Montpellier Cedex, France

We have made the first NMR measurements during the
electrochemical doping of poly(acetylene). Changes in the ^{7}Li NMR
linewidth and chemical shift were observed during doping and
undoping. A hysteresis in the values of linewidth was due to the
presence of two lines on undoping. The presence of these two
resonances suggests two types of lithium and could be evidence
for two distinct intercalation stages during undoping.

1 Introduction:

The electrochemical doping of poly(acetylene) is of interest
since the doping level can be accurately determined from the mass
of polymer and the amount of current passed during doping. The
changes in conductivity /1/, open-circuit potential /1/, ESR /2/
and Raman /3/ spectra, and X-ray diffraction patterns /4/ have
been studied extensively, and several models for the insertion of
dopant into the polymer have been postulated. However, although
much progress has been made, the precise mechanism of doping is
still unclear.

There have appeared several reports of the NMR
characterization of poly(acetylene) in the undoped state /5,6/,
after cis - trans isomerization /7/, and after doping by chemical
methods /8,9/. The ^{7}Li NMR spectra of poly(acetylene) doped
chemically with lithium has been reported by Rachdi et al. /10/.
Changes in the ^{7}Li T_1 were explained in terms of a diffusion
model.

The aims of this study were to observe the ^{7}Li NMR parameters,
i.e chemical shift, linewidth, and T_1, on the electrochemical
doping of poly(acetylene) with Li. Changes in these parameters
on doping can be related to the arrangement of dopant species
within the polymer matrix, and provide information on the
mechanism of doping.

170

2 EXPERIMENTAL:

The specially designed electrochemical cell, containing poly(acetylene) (film thickness 120 μm) and lithium metal electrodes in a $LiClO_4$ (0.5M) in THF solution /11/, could be inserted into the horizontal coil of a standard broad band probe. Before NMR measurements the electrolyte solution was transfered to a reservoir situated outside the NMR coil, and thus the contribution of the solution to the NMR signal of the doped poly(acetylene) was minimal. Inversion of the cell after NMR measurement allowed immersion of the two electrodes in the electrolyte solution and further increments of doping or undoping. NMR measurements were taken three hours after doping was halted.

^{7}Li NMR spectra were recorded with a Bruker CXP 200 spectrometer operating at 77.7MHz for [7]Li. Typical $\pi/2$ pulse times were 24 μs. The chemical shift reference was Li_2SO_4 in water.

3 RESULTS AND DISCUSSION:

^{7}Li NMR spectra obtained during doping and undoping (Fig. 1) had approximately Lorentzian lineshapes and showed evidence of a second component from the assymetry during doping, and the resolved second peak during undoping. We suggest that the presence of two peaks is due to two types of lithium, and could result from two lithium environments in intercalation stages. Shacklette and Toth /1/ have previously observed maxima in plots of $-dVoc/dy$ for poly(acetylene) doped with potassium and sodium

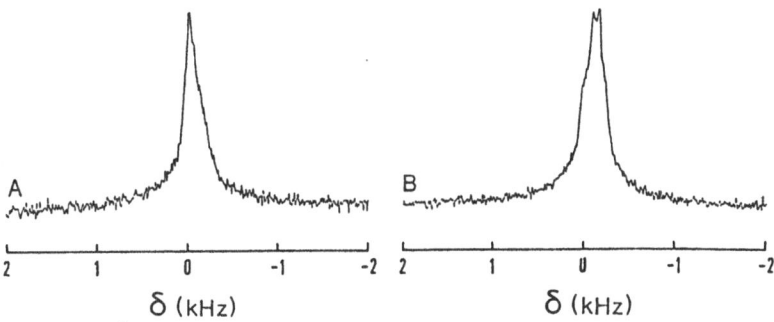

Fig. 1 ^{7}Li NMR spectra of doped $(CH)_x$: A) 7.6% Li during doping, B) 5.5% Li during undoping.

(Voc is the open-circuit cell potential, y is the doping level) which they claim correspond to the compositions of stages. X-ray diffraction studies /5/ of doped poly(acetylene) lend some support to this model. They claim however, that these structures are unlikely for lithium doping since the large size of the heavily solvated Li^+ ion would be unsuitable for their model.

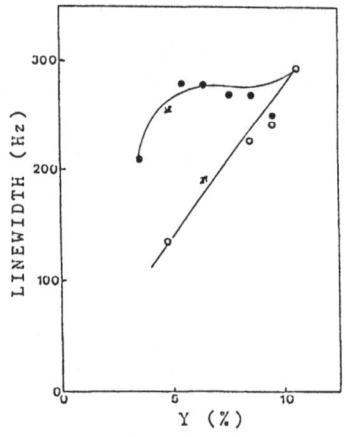

Fig. 2 Variation in NMR line-width during doping and undoping.

The NMR linewidth, measured at half peak height, increased linearly with y during doping (Fig. 2). This is consistent with an increase in the $^7Li-^7Li$ dipole-dipole interactions as the average separation of the dopant species decreased during doping. During undoping a plateau with linewidth of 270Hz was observed (Fig. 2). Computer simulation of the spectra revealed that the plateau was due to the presence of two peaks, with decreasing linewidth but increasing separation.

Spin-lattice relaxation times (T_1) of 0.2 ± 0.05s were obtained at three concentrations of lithium, during doping and undoping. T_1 values of this magnitude have been reported /12/ for ^7Li in solid solutions of Li_5NI_2-LiOH, in which diffusive motion dominates the relaxation. Furthermore these values are an order of magnitude less than T_1 reported for poly(acetylene) doped chemically with lithium /10/, and probably reflect the differences between the two doping methods. Electrochemical doping is generally assumed to produce a more homogeneous distribution of dopant, and better defined dopant-polymer structures, than chemical doping.

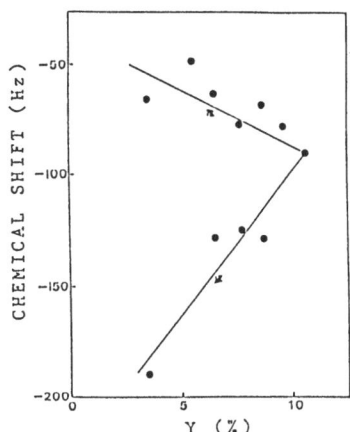

Fig. 3. Variation in chemical shift during undoping.

The chemical shifts of the two peaks observed during undoping (Fig. 3) diverged with decreasing lithium concentration, suggesting a progressive, but small, change in the conduction electron densities in the environments of the two types of lithium nuclei. Rachdi et al. /13/ have shown that the ESR linewidths of $(CH)_x$ doped with alkali metals are due to spin-orbital coupling between the dopant and the conduction electrons. The magnitude of the changes in chemical shift is small compared with the Knight shift observed for lithium metal (260ppm) /14/, however the lithium nuclei could be largely shielded from the conduction electrons by the solvent sheath /14/. Experiments are continuing to determine the precise origin of the two peaks.

4 ACKNOWLEDGEMENTS:

The NMR computer simulations were performed using the program LINESIM written by Dr. P.F. Barron, Griffith University, Nathan 4111, Australia.

5 REFERENCES:

1. L.W. Shacklette, J.E. Toth, Phys. Rev. B, 32, 5892 (1985).
2. C. Fite, P. Bernier, Solid State Commun., accepted.
3. E. Faulques, S. Lefrant, F. Rachdi, P. Bernier, Synth. Met., 9, 53 (1984).
4. R. H. Baughman, L. W. Shacklette, N. S. Murthy, G. G. Miller, R. L. Elsenbaumer, Mol. Cryst. Liq. Cryst., 118, 253 (1985).

5. M. Nechtstein, P. Devreux, R. L. Greene, T. C. Clarke, G. B. Street, Phys. Rev. Lett., $\underline{44}$, 356 (1980).

6. M. M. Maricq, J. S. Waugh, A. G. MacDiarmid, H. Shirakawa, A. J. Heeger, J. Am. Chem. Soc., $\underline{100}$, 7729 (1978).

7. M. Audenaert, P. Bernier, Mol. Cryst. Liq. Cryst., $\underline{117}$, 83 (1985).

8. F. Masin, G. Gusman, R. Deltour, Solid State Commun., $\underline{40}$, 513 (1981).

9. M. Peo, H. Foster, K. Menke, J. Hocker, J. A. Gardner, S. Roth, K. Dransfield, Solid State Commun., $\underline{38}$, 467 (1981).

10. F. Rachdi, M. Audenaert, P. Bernier, R. J. Schweizer, In Electronic Properties of Polymers and Related Compounds, ed. by H. Kuzmany, M. Mehring, S. Roth, Springer Ser. Solid-State Sciences, vol. 63, (Springer, Berlin, Heidelberg 1985) p. 278.

11. A. El Khodary, P. Bernier, J. Chem. Phys. $\underline{85}$(4), 2243 (1986).

12. T. Asai, S. Kawai, R. Nagai, S. Mochizuki, J. Phys. Chem. Solids, $\underline{45}$, 173 (1984).

13. F. Rachdi, P. Bernier, these proceedings.

14. P. Bernier, F. Rachdi, A. El Khodary, M. Audenaert, R. J. Schweizer, In Electronic Properties of Polymers and Related Compounds, ed. by H. Kuzmany, M. Mehring, S. Roth, Springer Ser. Solid-State Sciences, vol. 63, (Springer, Berlin, Heidelberg 1985) p. 281.

p_z-Radical Electron Structure in Polydiacetylene (PDA) Molecules

H. Sixl* and C. Kollmar

3. Physikalisches Institut der Universität Stuttgart,
Pfaffenwaldring 57, D-7000 Stuttgart 80, Fed. Rep. of Germany

By ESR spectroscopy it is shown that the transient absorption and the delayed luminescence observed in PDA single crystals is due to a metastable
triplet state of the individual conjugated chains. This triplet state is
described by a configuration model in analogy to the ground-state carbene
reaction intermediates of the solid-state polymerization reaction. ESR fine
structure and hyperfine structure data are consistent with a transfer integral of $t = -1.1$ eV and an energy difference of $2\epsilon = 0.4$ eV between butatriene and acetylene structure per unit cell.

1. Introduction

Upon doping or optical excitation of conjugated chains p_z-radical electrons
(kinks/antikinks) representing bond alternation defects and/or charges are
generated. For the electrical conductivity only the charged polarons or bipolarons [1] are of interest. In order to obtain more insight into the
shape of the polaron wavefunction it is important to have a triplet state
configuration due to the high sensitivity of the fine structure interactions. Fine structure interactions are absent in doublet one-electron systems. Triplet radical electron pairs are present in ground-state carbene
reaction intermediates, as has been shown in previous presentations [2], as
well as in the optically excited triplet state of the polydiacetylene
chains, which has been reported on very recently [3,4].

The high attractivity of the PDA system is given by its single crystal
structure. The resulting highly anisotropic spectroscopic data reveal a manifold of fine details of the spin systems in their ground and excited states. In the experimental part the transient triplet-triplet absorption and
the delayed phosphorescence emission as well as the pulsed spin-echo ESR
spectra corresponding to the metastable triplet state of the PDA-chains are
reviewed. The experimental data will be explained by means of a one-electron configuration model. The optically excited triplet state is described
consistently as a triplet polaron. The close relationship of the carbene
p_z-electron configuration to that of the p_z-triplet state electron pair
configuration will be demonstrated.

2. Summary of the Experiments

ESR and optical experiments concerning the carbene reaction intermediates
as well as the metastable triplet state have been performed in our group as
well as in the groups of D. Bloor (QMC London), M. Schwoerer (Univ. Bay-

---

*) present address: Hoechst AG, Angewandte Physik,
6230 Frankfurt am Main 80, West Germany

reuth) and J. Orenstein (Bell Labs.). The results have already been published [4-7]. It has been shown by ESR and optical absorption experiments
that the short chain reaction intermediates in the solid-state polymerization reaction are given by diradicals with butatriene structure and the
long chains by dicarbenes with acetylene structure. For a better unterstan-
ding of the theory and the conclusions discussed below we want to focus our
interest on the dicarbene states observed at chain-lengths n > 5.

Using the data of the ESR-fine structure and hyperfine structure a simple
configuration model by Sixl et al. [8] gives us the p_z-radical electron
distribution shown in Fig. 1. There are two different species of radical
electrons on the left and right side of the oligomer. The odd radical electron is distributed on odd carbon atom positions (1, 3, 5 ...). The even
radical electron is distributed on even carbon atom positions (24, 22,
20,...). The partners of the p_z-radical electrons are σ-radical electrons
on each side of the chain.

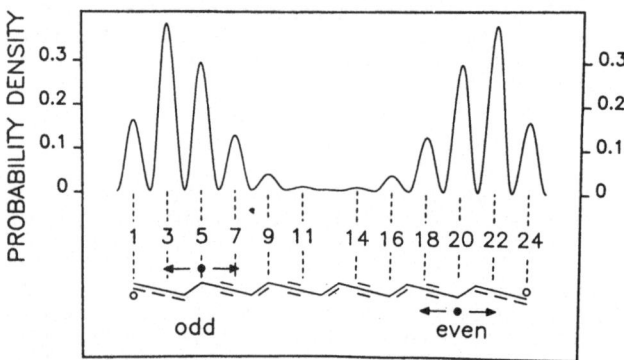

Fig. 1: p_z-radical electron distribution in a dicarbene hexamer molecule.
The p_z-radical electron is symbolized by a full circle, the σ-radical electron by an open circle.

Let us take a distinct radical electron distribution on the left and right
side of the chain as shown in Fig. 2 and add it in a Gedanken experiment to
a closed ring (n → ∞). We thus end up with a neutral polaron-like configuration of the chain. This clearly is an excitation of the chain, which
might have singlet or triplet character. If the two radical electrons have
antiparallel spin they will attract each other, due to the unfavourable butatriene structure, and thus will recombine immediately to form a perfect
diacetylene chain. However, if the two radical electrons have parallel spin
they are unable to recombine due to spin selection rules.

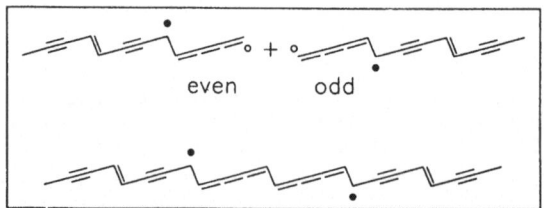

Fig. 2: Combination of two carbene-like chain ends to give a neutral
polaron-like configuration.

The existence of the above metastable triplet on a PDA chain has been proven by Robins et al. [4] by ODMR experiments. They showed that the transient absorption observed in PDA crystals is dependent on the magnetic field strength and orientation. In addition they were able to detect the ESR transitions via the transient optical absorption. Some more details of this effect are shown in this conference proceedings [9].

A direct proof of the triplet character of this PDA chain excitation has been given by Winter et al. [3] using pulsed ESR spin-echo techniques. There it was demonstrated that the triplet state sublevels are equally populated with $p_x = p_y = p_z = 1/3$. A strong selectivity of the decay of the triplet sublevels and an absent spin lattice relaxation causes a very strong spin polarization of the triplet state. The average lifetime of $\tau \approx 80$ μs of the ESR is consistent with that of the transient absorption [10] and with the observation of the delayed phosphorescence emission [11].

3. Theoretical Models

The theoretical methods used for the explanation of the fine-structure and hyperfine-structure of the carbene states as well as of the metastable triplet state are based on the simple assumptions of the configuration model which has been described in a previous work [8,12] and is elucidated in Fig. 3. Moving a p_z-radical electron located at the carbene chain end of a PDA-chain towards the centre of the chain the system requires energy due to the creation of the energetically unfavourable butatriene structure. The energy difference between acetylene and butatriene structure per unit cell amounts to $2\epsilon = 0.4$ eV. In order to allow the radical electron to move along the chain we introduce a transfer integral t. The numerical value of t = -1.1 eV is obtained by fitting the experimentally observed fine structure of an asymmetric carbene chain [12]. Diagonalization of the energy matrix finally leads to the energy eigenvalues and wavefunctions of the ground and excited state configurations as shown in Fig. 4. The energy step function and the transfer integral define the extension of the radical electron wavefunction in the ground and excited states. The wavefunctions are only defined on the odd carbon atom positions.

The configuration model can also be applied to the triplet polaron state obtained by connecting two carbene chain ends as shown in Fig. 2. The resulting configurations are shown in Fig. 5. It can be proven that each electron of the triplet pair state can be treated in the same way as a sin-

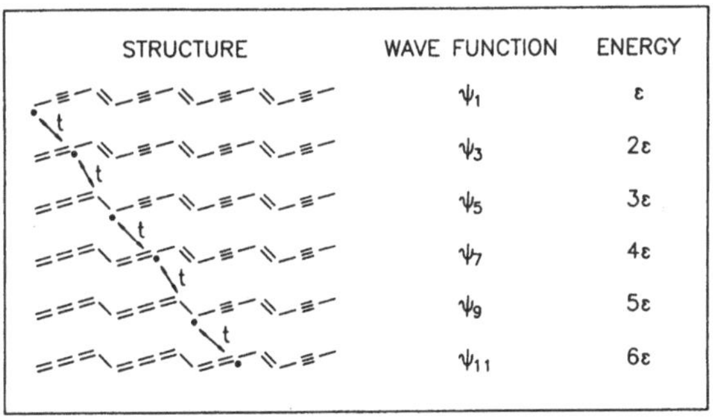

STRUCTURE	WAVE FUNCTION	ENERGY
	ψ_1	ϵ
	ψ_3	2ϵ
	ψ_5	3ϵ
	ψ_7	4ϵ
	ψ_9	5ϵ
	ψ_{11}	6ϵ

Fig. 3: Radical electron configurations of a carbene-like chain end on a PDA-chain with the corresponding symbols of the wavefunctions and the energies. The mobile p_z-electron is symbolized by a full circle.

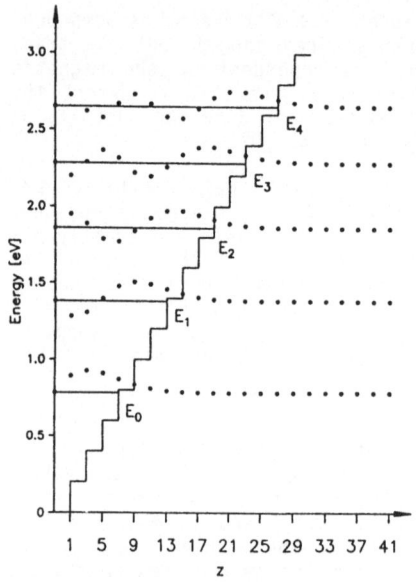

Fig. 4: Energy eigenvalues and corresponding wavefunctions of the p_z-radical electron of a carbene chain end obtained by the configuration model. The points symbolize the coefficients of the different configurations for the ground state and four excited states. z denotes the position of the radical electron. The height of one step of the energy ladder is given by ϵ.

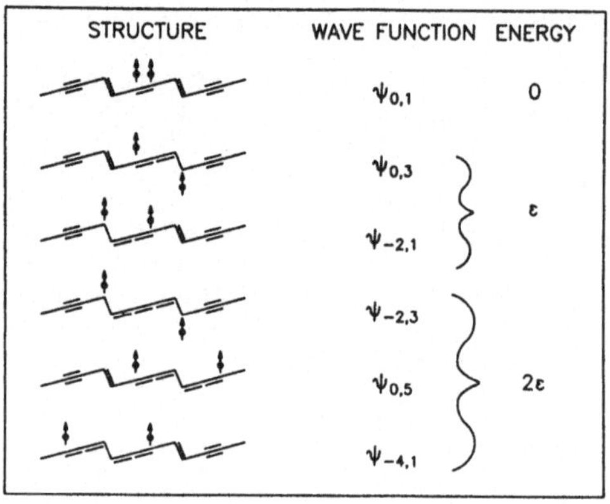

Fig. 5: Radical electron configurations of the polaron-like triplet pair state on a PDA-chain with the corresponding symbols of the wavefunctions and the energies.

gle p_z-radical electron of a carbene chain end. The spin density of the polaron is shown in Fig. 6. The fine structure parameters D and E obtained by this spin distribution are also included in Fig. 6 and can be compared with the experimentally observed values of Winter [3]:

$$|D| = 0.0654 \ cm^{-1}$$

$$|E| = 0.0096 \ cm^{-1}$$

Especially the D-values agree well. The very small E-values which are burdened with great uncertainty differ by one order of magnitude. The calculation of the fine structure of the polaron also yields the angle between the axis of the polymer molecule and the axis of largest fine structure splitting which amounts to 6^0. This is in good agreement with the experimental observations of Orenstein [13].

Summarizing, we conclude that the configuration model which has been developed for the radical electrons of a carbene chain end also proves to be a suitable method for describing the spin distribution of a triplet polaron on a PDA-chain.

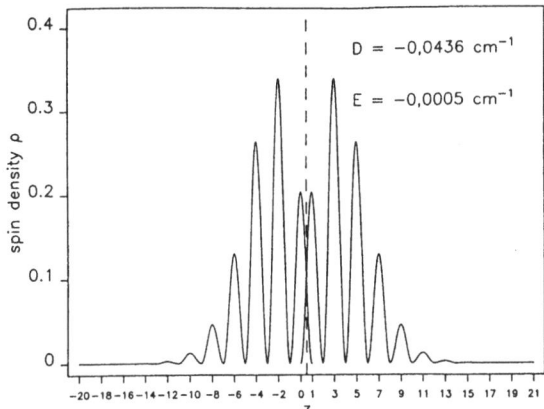

Fig. 6: Spin density of a triplet polaron on a PDA-chain. z denotes the positions of the odd and even radical electrons.

Acknowledgements

This work has been supported by the Deutsche Forschungsgemeinschaft (SFB 329) and by the Stiftung Volkswagenwerk.

References

1. J.L. Brédas, R.R. Chance and R. Silbey, Phys. Rev. B26, 5843 (1982)
2. H. Sixl, Adv. Polym. Sci., Vol. 63, 49, Springer Verlag (1984)
3. M. Winter, A. Grupp, M. Mehring and H. Sixl, Chem. Phys. Lett., 133, 482 (1987)
4. L. Robins, J. Orenstein and R. Superfine, Phys. Rev. Lett., 56, 1850 (1986)
5. D. Bloor in "Developments in Crystalline Polymers", ed. by D.C. Bassett, Appl. Sci. Publ., London, 151 (1982)
6. H. Niederwald and M. Schwoerer, Z. Naturforsch., 38a, 749 (1983)
7. R.A. Huber and M. Schwoerer, Chem. Phys. Lett., 72, 10 (1980)
8. H. Sixl , W. Neumann, R. Huber, V. Denner and E. Sigmund, Phys. Rev. B31, 142 (1985)
9. W. Rühle and H. Sixl, in these proceedings
10. H. Sixl, R. Jost and R.Warta, J. Chem. Phys., in press
11. R. Warta and H. Sixl, Chem. Phys. Lett., 116, 307 (1985)
12. C. Kollmar and H. Sixl, in press
13. J. Orenstein, private communication

Triplet State ODMR
of Polydiacetylene Crystals

W. Rühle and H. Sixl*

3. Physikalisches Institut der Universität Stuttgart,
Pfaffenwaldring 57, D-7000 Stuttgart 80, Fed. Rep. of Germany
*Present address: Hoechst AG, Angewandte Physik,
 D-6000 Frankfurt a.M. 80, Fed. Rep. of Germany

The triplet state of the polymer-backbone in the polydiacetylene TS-6
(toluene-sulphonate) was investigated by ODMR (optically detected magnetic
resonance) detected on the triplet-triplet-absorption. From the angular
dependence the fine-structure constants D and E, as well as the orientation
of the triplet state in the plane of the polymer-backbone, are obtained. The
signals are detected in the temperature range between 5 K and 100 K showing
changes in lineshape and a decrease of intensity with increasing temperature.

1. Introduction

The polydiacetylene TS-6 (toluene-sulphonate) forms nearly perfect polymer
single crystals by a solid state polymerization reaction. The orientation
of two neighbouring polymer chains is shown in Fig. 1. In partially and
fully polymerized TS-6 crystals a transient absorption is observed in the
near infrared, which is only present during UV-excitation [1,2]. For fully
polymerized TS-crystals the peak of this absorption is located at 909 nm
(see Fig.2) [3]. Previously reported ODMR experiments by ROBINS et al. [4]
revealed the triplet nature of the observed state. A direct observation of
this state by pulsed ESR gave additional information about the formation
and selective decay of the triplet sublevels [5].

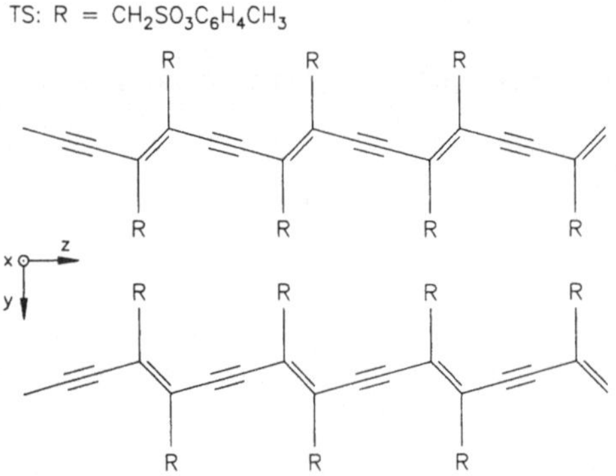

TS: R $= CH_2SO_3C_6H_4CH_3$

Fig. 1: Structure of the polymerized PDA TS-6. The two types of neighbouring
polymer chains are converted into each other by a symmetry operation
(crystallographic space group $P2_1/c$)

*
 present address: Hoechst AG, Angewandte Physik,
                  6230 Frankfurt/Main 80, Fed. Rep. Germany

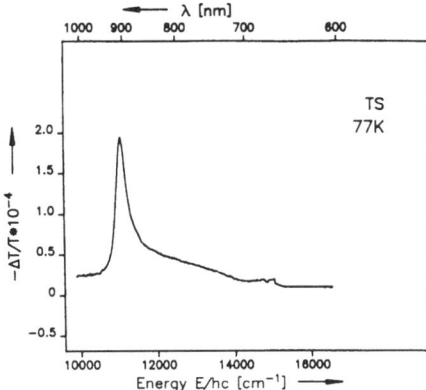

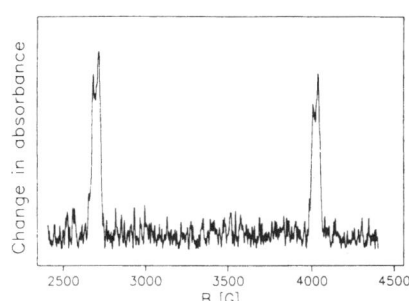

Fig. 2: Transient absorption of a fully polymerized TS-crystal during UV-excitation (260-380 nm)

Fig. 3: ODMR spectrum detected on the transient absorption of TS at 909 nm. Magnetic field parallel to the polymer chain (z-axis), T=10 K, microwave power 300 mW

2. Experimental

In all experiments thermally fully polymerized crystals of TS-6 were used. The UV-excitation was performed by a high pressure mercury arc (HBO 500) and a filter UG 11 (Schott) with transmission between 260 and 390 nm. The transient absorption was detected by a silicon photodiode OSI-5K, using a tungsten lamp as light source and a 909 nm interference filter in the detection pathway.

During the transient absorption measurement of Fig. 2 the UV-excitation was modulated by a chopper at 75 Hz and the signals were detected using lock-in technique. The ODMR measurements (Fig. 3 ff.) were performed in a X-band ESR spectrometer at a microwave frequency of 9.4 GHz. During these experiments a cw-UV-excitation was used and the signal was detected by lock-in technique modulating the microwaves with a PIN-modulator at 1 kHz. The microwave power was enhanced by use of a travelling-wave tube amplifier. The measurements were performed in the temperature range between 5 K and 100 K using a helium gas continuous flow cryostat of the type Oxford ESR-900.

3. Results

Figure 3 shows the ODMR spectrum detected at 10 K on the triplet-triplet-absorption (909 nm) with magnetic field parallel to the polymer chain (z-axis). The mirror symmetric high- and low-field signals show at least a double-line structure. The angular dependence of the signals was measured for magnetic fields perpendicular to the polymer chain axis (sample rotated around z, left side of Fig. 4) as well as for B within the polymer-backbone plane (sample rotated around x, right hand side of Fig. 4). Computer fits yielded the fine-structure constants $|D| = 0.0608$ cm^{-1} and $|E| = 0.0096$ cm^{-1} (curves in Fig.4).

Note that the notation of the axes is different from Refs. [4] and [5], because in triplet-ESR usually the axis of widest fine-structure splitting is defined as z-axis (see also Fig. 1).

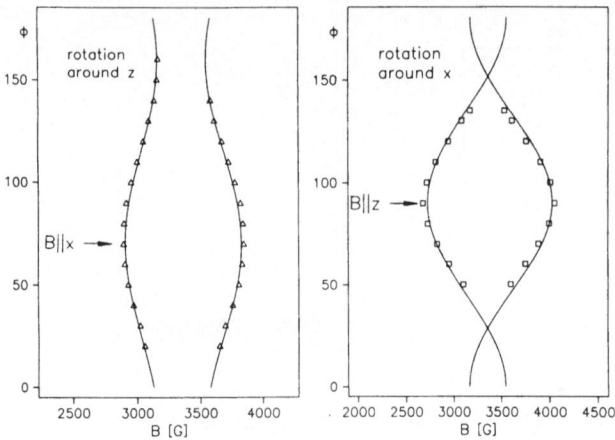

Fig. 4: Angular dependence of the ODMR spectrum: Experimental points and fitted curve. The fine structure constants obtained by the fit are $|D| = 0.0608$ cm^{-1} and $|E| = 0.0096$ cm^{-1}

With increasing temperature the signals show a decrease of intensity. ODMR signals are observed in the temperature range from 5 K up to 100 K. The signal shape shown in Fig. 5 changes from a double-line structure at 10 K to a single-line structure at 60 K.

The ODMR signals are strongly enhanced by additional magnetic field modulation at 100 kHz. In this way more spins of the inhomogeneous line are saturated simultaneously (see Fig.6). Without any field modulation the signal is very weak at the low microwave power of 5 mW (bottom curve in Fig. 6). With increasing amplitude of the 100 kHz magnetic field modulation

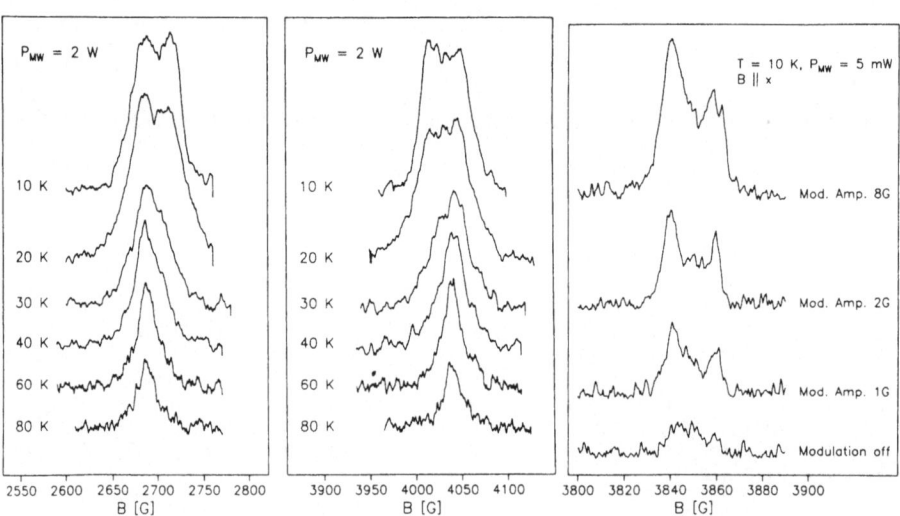

Fig. 5: Temperature dependence of the ODMR signals for orientation $B \| z$

Fig. 6: Enhancement of the ODMR signal by magnetic field modulation at 100 kHz

the signal-to-noise ratio of the ODMR signals is improved and at least two lines with a linewidth of about 8 Gauss can be well resolved for magnetic field parallel x. At the highest modulation amplitude of 8 Gauss (top curve in Fig. 6) an additional line broadening because of the magnetic field modulation occurs, so that the optimum value was a field modulation of 2 Gauss. With this technique future measurements can be performed at lower microwave power so that no power broadening occurs and the spectra are better resolved.

4. Discussion

The angular dependence of the ODMR signals demonstrates that the triplet fine-structure tensor is oriented nearly parallel to the polymer-backbone. Therefore the triplet state is an attribute of the polymer chain but not of the sidegroup. The values of the fine-structure constants obtained by the fit in Fig. 4 are in agreement with those determined by ROBINS et al. [4] and WINTER et al. [5]. The ODMR signals show the same angular dependency and intensity as the signals observed by pulsed ESR in Ref. [5] so that there is no doubt that the same state is observed. At low microwave power the lineshapes of the observed ODMR signals are also similar to those observed by pulsed ESR [5].

On increasing the temperature to 60 K, a part of the signals seems to disappear. Future measurements at low microwave power, and therefore higher resolution, will show if there is an additional line narrowing with increasing temperature. There are also still open questions concerning the formation of the observed triplet states, the extension of their wavefunctions and their mobility.

Acknowledgements

This work has been supported by the Deutsche Forschungsgemeinschaft (SFB 329) and by the Stiftung Volkswagenwerk. Helpful discussions with R. Warta, C. Kollmar and A. Grupp are gratefully acknowledged.

References

1. J. Orenstein, S. Etemad and G.L. Baker, J. Phys. C 17, L 297, (1984)
2. T. Hattori, W. Hayes and D. Bloor, J. Phys. C 17, L 881, (1984)
3. H. Sixl, R. Jost and R. Warta, J. Chem. Phys., to be published
4. L. Robins, J. Orenstein and R. Superfine, Phys. Rev. Letters 56, 1850, (1986)
5. M. Winter, A. Grupp, M. Mehring and H. Sixl, Chem. Phys. Letters 133, 482, (1987)

Part V

Theory

Electronic and Nonlinear Optical Properties of Conjugated Polymers: A Quantum Chemistry Approach

J.L. Brédas[a]

Laboratoire de Chimie Théorique Appliqueé, Centre de Recherches sur les Matériaux Avancés, Facultés Universitaires Notre-Dame de la Paix, B-5000 Namur, Belgium

In the first part of this paper, we give a broad overview of the quantum chemistry techniques that have been mostly used to describe the electronic structure of conjugated polymers. Without going into a detailed presentation, our main purpose is to provide some insight into the approximations which are involved and the ways the different methods are related to one another. We then briefly show some applications of these techniques to two polymers of high current interest, polyaniline and polythiophene. Finally, we focus on the nonlinear optical properties of conjugated systems. We briefly state the main principles of nonlinear optics and express the reasons why organic compounds, in particular conjugated molecules and polymers, are extensively investigated as very promising nonlinear optical materials.

I. Quantum chemistry approach to the electronic structure of conjugated systems [1]

The goal of most quantum chemistry calculations performed on an N-electron system is to find the solution of the time-independent Schrödinger equation in the so-called Born-Oppenheimer (adiabatic) approximation (in which the electrons are considered as moving around fixed nuclei). The Schrödinger equation then reads

$$\mathbf{H} \ \Psi(1,...N) = \mathbf{E} \ \Psi(1,...N)$$

and corresponds to an eigenvalue problem. $\mathbf{H}$ is the Hamiltonian describing the system and $\mathbf{E}$ (the eigenvalue) and Ψ (the eigenfunction) are the total energy and N-electron wavefunction of the system, respectively. The square of the value of the wavefunction at a given point in space represents the electron probability density at this point.

Unfortunately, as soon as the number of electrons is larger than one, the Schrödinger equation cannot be solved exactly and one has therefore to rely on various approximations to try and approach at best the exact wavefunction. Within quantum chemistry, the most popular approximations are based on the Hartree-Fock (HF) method. In this context, the N-electron Schrödinger equation is replaced by a set of N one-electron Hartree-Fock equations:

---

a. Chercheur Qualifié of the Belgian National Fund for Scientific Research (FNRS).

$$h^{HF}(1) \quad \varphi_1(1) \; = \; \varepsilon_1 \; \varphi_1(1)$$

$$\cdots$$

$$\cdots$$

$$h^{HF}(N) \quad \varphi_N(N) \; = \; \varepsilon_N \; \varphi_N(N) \quad .$$

$\varphi_i(i)$ denotes a molecular (spin-)orbital (MO) and represents a one-electron wavefunction; ε_i is the one-electron energy corresponding to that MO. The set of HF equations has to be solved iteratively (to produce a self-consistent field solution) since the solution φ_i depends on the other solutions $\varphi_{j \neq i}$ and vice versa. The N-electron wavefunction $\Psi^{HF}(1,...N)$ is then constructed as an antisymmetrized product of the occupied molecular orbitals:

$$\Psi^{HF}(1,...N) \; = \; (1/\sqrt{N!}) \; \hat{A} \; |\varphi_1(1)......\varphi_N(N)| \; .$$

Such a wavefunction is called a single-determinant wavefunction since it is based on a single electron configuration represented by one Slater determinant.

The one-electron HF Hamiltonian usually contains four terms:

$$h^{HF}(1) \; = \; \frac{-1}{2} \nabla^2(1) \; - \; \sum_A^{\text{atoms}} \frac{Z_A}{r_{A1}} \; + \; \sum_j^{\text{occ. MO's}} J_j(1) \; - \; \sum_j^{\text{occ. MO's}} K_j(1) \; ,$$

which represent respectively (in atomic units): (i) the kinetic energy of electron 1; (ii) the nuclear attraction between electron 1 and the nuclei present in the system (the summation runs over all nuclei, Z_A is the charge of nucleus A, and r_{A1} is the distance between electron 1 and nucleus A); (iii) the Coulomb repulsion between electron 1 and the other electrons (J_j is the Coulomb operator and the summation runs over all occupied MO's); and (iv) the nonlocal exchange between electron 1 and the other electrons with parallel spin (K_j is the Pauli operator).

As is apparent, a lot of electron-electron interactions are explicitly considered. For instance, the exchange term takes care of all interactions between electrons with parallel spin. There is, however, a problem related to the Coulomb repulsion term because it only contains the average repulsion between electron 1 and the mean field due to the other electrons. Therefore, the highly energetic situation where a second electron with antiparallel spin comes nearby electron 1 is not forbidden to occur. In other words, the Coulomb term does not include the instant correlation of the relative motions of the two electrons when they approach each other. This feature constitutes the reason why the HF total energy is always larger than the exact total energy of the N-electron system. By definition, the difference between these total energies is called <u>correlation energy</u>. (At this stage, it is worth stressing that the expressions "electron-electron interaction" or "electron correlation" have different meanings, as should be clear from the above discussion, and should not be confused with one another).

187

Although the correlation energy amounts in absolute numbers to a small portion of the total energy ($\approx 1\%$), it can sometimes play a major role when looking at the energy difference between two similar systems. (Note that in most methods which go beyond Hartree-Fock and try to introduce some of the correlation energy, for instance the Configuration Interaction method, the wavefunction is described by as large a number of Slater determinants as possible.)

The HF equations can in principle be solved numerically but this quickly represents a very tedious task even if the system contains only a few atoms. Therefore, one usually takes profit of a further approximation by expressing the MO's as Linear Combinations of Atomic Orbitals (LCAO approximation):

$$\varphi_i = \sum_p c_{ip} \chi_p \, ,$$

where χ_p represents an atomic orbital (AO) or its appropriate simulation. In an actual calculation, each AO can be described by one or more so-called basis functions. It is obvious that the larger the number of basis functions (i.e. the larger the summation over p), the closer the result of the calculation comes to the Hartree-Fock numerical limit. The set of basis functions taken into account in a calculation is referred to as the basis set and it is possible to distinguish between different basis set types:

— a minimal basis set: a single basis function describes each of those AO's that belong to electronic shells occupied in the free atom (1s for H; 1s, 2s, and $2p_{x,y,z}$ for C, N, or O);

— a split valence basis set: each of the valence AO's is described by two basis functions, a feature which adds much flexibility to the basis set;

— an extended basis set: a basis function of higher quantum number is added on each atom (e.g. a 2p-like function on H; a 3d-like function on C, N, or O).

When the molecular orbitals are written in terms of the basis functions χ_p, a matrix element of the HF Hamiltonian is then expressed as (when use is made of atomic units):

$$< \chi_p(1) \,|\, h^{HF} \,|\, \chi_q(1) > \; = \; < \chi_p(1) \,|\, \frac{-1}{2} \nabla^2(1) \,|\, \chi_q(1) >$$

$$- \sum_A Z_A \; < \chi_p(1) \,|\, \frac{1}{r_{A1}} \,|\, \chi_q(1) >$$

$$+ \sum_r^{AO} \sum_s^{AO} C_{jr} C_{js} \; < \chi_p(1) \, \chi_r(2) \,|\, \frac{1}{r_{12}} \,|\, \chi_s(2) \, \chi_q(1) >$$

$$- \sum_r^{AO} \sum_s^{AO} \frac{1}{2} C_{jr} C_{js} \; < \chi_p(1) \, \chi_q(2) \,|\, \frac{1}{r_{12}} \,|\, \chi_s(2) \, \chi_r(1) > \, ,$$

where the first two terms (kinetic and nuclear attraction) correspond to one-electron integrals because they depend on the coordinates of a single electron and the last two terms (Coulomb repulsion and exchange) correspond to two-electron integrals. In the expression of a two-electron integral (more simply denoted as (pq|rs)), four basis functions do appear. As a result, the number of those integrals to be calculated roughly goes as the fourth power of the number of basis functions and quickly reaches high values on the order of millions or billions or more (e.g. $\approx 10^7$ for a split-valence basis set calculation on a molecule of the size of benzene; $\approx 10^9$ for a minimal basis set calculation on a molecule such as phthalocyanine).

The calculation of the two-electron integrals represents the major computational effort in what we have been presenting so far, i.e. the Hartree-Fock *ab initio* all-electron method. Therefore, more simple techniques based on the HF Hamiltonian have been introduced, whose aim is to reduce significantly the computational task related to two-electron integrals. Among these techniques, we can mention those which rely on the Zero Differential Overlap (ZDO) approximation. Here, only valence electrons are explicitly taken into account and the major simplification resides in the fact that, in the ZDO approach, a two-electron integral (pq|rs) is retained only if χ_p is equal to χ_q and χ_r to χ_s:

$$(pq|rs) = \delta_{pq} \, \delta_{rs} \; .$$

In this case, the two-electron integrals to be evaluated are only two-center integrals and their number only goes as the square of the number of atoms. Among the methods which use the ZDO approximation, we can cite the following and their multiple adaptations:
→ CNDO (Complete Neglect of Differential Overlap);
→ INDO (Intermediate Neglect of Differential Overlap);
→ MNDO (Modified Neglect of Differential Overlap).
These techniques are usually referred to as semiempirical because they involve some parameterization performed on the basis of experimental data. For instance, in MNDO, most of the two-electron integrals are discarded as described above and those which are retained are parameterized in order to reproduce good geometries and heats of formation in a large set of model organic molecules. (Note that, in doing so, some of the correlation energy is, at least partly, implicitly introduced.)

Other popular methods only treat π electrons and are devoted exclusively to the investigation of conjugated molecules and polymers. This is the case of the Pariser-Parr-Pople (PPP) and Hückel techniques, where only one π atomic orbital per site is considered. In the PPP approach, the ZDO approximation is used for the two-electron integrals and the one-electron integrals are also parameterized. It is worth pointing out that in the PPP method, a fair amount of electron-electron interactions is retained since all on-site interactions (which correspond to the U terms of the Hubbard model) and all two-center interactions (which correspond to the V terms of the Extended Hubbard model) are explicitly calculated [2].

In the Hückel approximation, the Hamiltonian matrix elements are expressed in the most simple way:

$$\rightarrow \; <\chi_p \,|\, h \,|\, \chi_q> \; = \; \alpha \quad \text{if p is equal to q .}$$
$$\rightarrow \; <\chi_p \,|\, h \,|\, \chi_q> \; = \; \beta \;(\text{or t}) \; \text{if p is connected to q ,}$$
$$\rightarrow \; <\chi_p \,|\, h \,|\, \chi_q> \; = \; 0 \quad \text{otherwise .}$$

The on-site matrix element is set equal to α, the Coulomb integral; when the two atoms carrying the p and q π-orbitals are directly bonded to one another, the matrix element is set equal to β, the resonance integral (in chemistry terminology) or t, the transfer integral (in physics terminology); in all other instances, the matrix element is neglected. In more refined versions of the Hückel method, the resonance (transfer) integral can be made to vary as a function of the bond length between the atoms carrying the p and q π-orbitals (a feature which allows one to take account of the electron-phonon coupling) and the energy of the σ-framework can be included in a parameterized way. It is interesting to point out that the so-called Su-Schrieffer-Heeger Hamiltonian [3] just corresponds to such a refined Hückel model.

We conclude this section by briefly presenting the nonempirical pseudopotential Valence Effective Hamiltonian (VEH) technique which we have heavily used in the context of conducting polymers [4]. In this method, we build an effective HF Hamiltonian containing the kinetic term and a sum of atomic potentials which depend on each atom type and simulate the nuclear attraction, Coulomb repulsion, and nonlocal exchange terms:

$$F_{eff} = \frac{-1}{2}\; \nabla^2 \; + \; \sum_{A}^{atoms} V_A \quad .$$

The atomic potentials are expressed as linear combinations of Gaussian projectors:

$$V_A = \sum_{l} \sum_{i,j} C_{l,ij}^A \; |G_{il}^A> \; <G_{jl}^A | \quad ,$$

where l denotes the projector angular dependence (l corresponding to the second quantum number); the sums over i and j define the complexity of the atomic potential and are usually carried up to two. G_{il} is a normalized Gaussian with exponent α_j. The parameters of the atomic potentials are thus the linear coefficients $C_{l,ij}$ and the nonlinear exponents α_j. These parameters are optimized for each atomic potential type on model molecules (so that the influence of the chemical environment is explicitly included) in such a way as to reproduce the one-electron energy levels obtained on these molecules from high-quality Hartree-Fock *ab initio* split-valence basis set calculations. The VEH calculations are very fast (the required computer time is roughly on the same order as that for a semiempirical MNDO-like calculation) because only products of one-electron integrals have to be evaluated.

II. Electronic-structure calculations on polyaniline and polythiophene

We now briefly describe the application of some of the theoretical techniques described in the previous section, to the electronic-structure investigation of two polymers currently extensively studied, polyaniline and polythiophene. Polyaniline [5] is a family of polymers of increasing interest because the electronic properties can be modified through variation of either the number of protons, the number of electrons, or both. The emeraldine base form of the polymer, which corresponds to $[-C_6H_4-NH-C_6H_4-NH-C_6H_4-N=C_6H_4=N-]_x$, is semiconducting and presents a first optical transition maximum at about 2.0 eV. Upon protonation of the formerly unprotonated $-N=$ sites, the conductivity of polyemeraldine increases by about ten orders of magnitude to reach 1 S/cm despite the unchanged electron concentration [6]. A study of the temperature-dependent static and dynamic magnetic susceptibility of polyemeraldine shows that the Pauli susceptibility is essentially linearly proportional to the percent protonation in agreement with the proposal that protonation leads to phase segregation of un- and full-protonated domains, i.e. a first order transition to a metallic phase [7]. A model based upon a transition from isolated doubly charged spinless bipolarons to a polaronic metal (corresponding to a polaron lattice of the form $[-C_6H_4-NH-C_6H_4-NH\cdot+-]_x$) has therefore been suggested [7].

We have theoretically investigated the nature of the metallic phase by performing a number of calculations aimed at describing the geometric and electronic structures of the polaron lattice. The geometry optimizations have been carried out using the MNDO technique, as detailed elsewhere [8]. The polaron lattice electronic structure has then been calculated at the VEH level on the basis of the optimized MNDO geometry. The polaron lattice band structure, which is displayed in Figure 1, yields electronic transitions which are in very good agreement with the optical absorption data for 50% proton-doped polyemeraldine [8]:

(i) we obtain a polaron band which is only half-occupied, allowing for intraband absorption and thus for far-infrared absorption, as is experimentally observed [8, 9]. The calculated polaron bandwidth is 1.1 eV, in excellent agreement with the 1.0 eV estimate from the intraband optical absorption at low frequency;

(ii) direct optical transitions from band $\underline{b}$ to the polaron band are calculated to be at 1.8 eV, which compares favorably with the first optical absorption at about 1.5 eV;

(iii) direct optical transitions from band $\underline{c}$ to the polaron band are calculated to be at 2.6 eV, which is in very good agreement with the second absorption observed at 2.8 eV;

(iv) the first electronic transitions involving the upper defect band $(\underline{x}')$ and the conduction band $(\underline{x})$ are predicted to occur at 4.1 eV, in agreement with the location of the third absorption. As a result of the excellent accord between the calculated band structure and the optical absorption data for 50% proton-doped polyemeraldine, the calculations offer a strong support of the polaronic metal model which has been proposed on the basis of the magnetic studies.

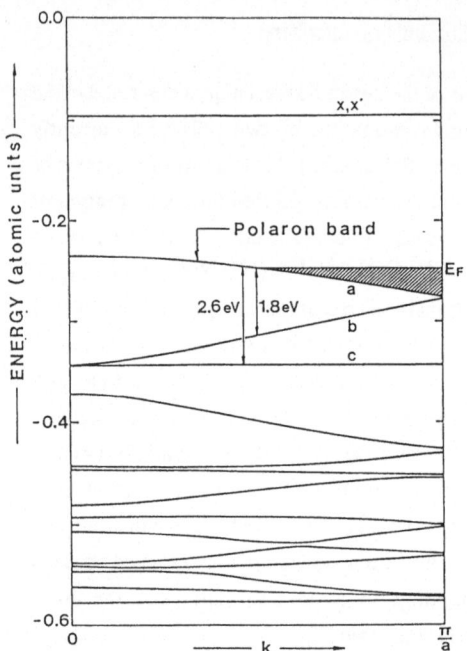

Figure 1: VEH band structure for the polaron lattice in 50% proton-doped polyemeraldine.

We now turn to a discussion of recent calculations we have devoted to polythiophene in order to determine the polaron and bipolaron wavefunctions [10] on the basis of the SSH Hamiltonian. In the case of a single oxidation process, polaron formation is energetically favorable as soon as the increase in $\pi + \sigma$ (total) energy due to lattice deformation around the polaron is more than compensated by a lowering in ionization energy. This difference in energy corresponds to the polaron binding energy. As one electron is extracted from a polythiophene chain, we obtain the formation of a polaron whose binding energy is calculated to be 0.20 eV in our model. As in the case of poly(p-phenylene) [11] and polypyrrole [12], we find that the polaron lattice deformation extends mostly over four rings. The presence of a polaron results in the appearance of two localized electronic states within the gap. The wavefunction (spin/charge density) associated with the occupied polaron level is schematically depicted in Figure 2. The localized character of the polaron electronic state is illustrated by the fact that the spin/charge has a density probability of 80.4% within the four rings of the geometric defect. The maximum spin density is calculated to be on the order of 0.08 and occurs on carbon atoms which are located four sites away from the middle of the defect, a behavior similar to that found for the polaron in trans-polyacetylene [13].

The calculated maximum spin density amounts, however, to only about 40% of the 0.2 maximum spin density estimated from the ESR hyperfine splitting in lightly doped

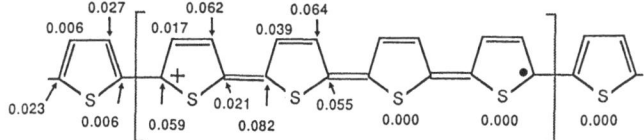

Figure 2: Spin (charge) densities for the polaron in polythiophene.

poly(3-hexylthiophene) solutions [14]. Within our simple theoretical approach, agreement with the maximum spin densities experimentally observed in poly(3-hexylthiophene) solutions is obtained only if the polaron is forced to extend over a chain segment which contains just two rings and has no electronic overlap with the rest of the chain due to strong conformational twists. The polaron in solution might thus be related with a small rigid section of an otherwise flexible chain and possibly correspond to a "conformon" defect proposed by Pincus *et al.* [15]. More experimental data are, however, needed in order to explore the validity of this possibility.

We have also studied the most stable electronic configuration when two electrons are removed from a polythiophene chain. The results indicate the formation of a bipolaron (i.e. a localized dication associated with a very strong lattice distortion), which is thermodynamically favored with respect to the formation of two polarons by some 0.26 eV. It is important to stress that the stability of bipolarons along polythiophene chains has been demonstrated by photo-induced absorption experiments on neutral polythiophene [16]. These measurements performed upon laser illumination of polythiophene under resonant conditions show the appearance of two subgap electronic absorptions related to bipolarons and the photobleaching of the π-π^* interband transition. Thus, upon illumination, there occur major shifts in the oscillator strengths of the optical absorptions; this constitutes an essential feature with regard to nonlinear optical properties, as we will develop in the next section.

III. Nonlinear optical properties of conjugated systems

The purpose of this section is to provide a very general introduction to nonlinear optics and mostly to try and show how important a role conjugated molecules and polymers can potentially play in this field. Interested readers can find more thorough treatments of the subject in references [17-19].

In bulk media, the macroscopic polarization $\mathbf{P}$ induced by the electric field $\mathbf{E}$ of incident light can be expanded as a power series of $\mathbf{E}$:

$$\mathbf{P} = \chi^{(1)}\,\mathbf{E} + \chi^{(2)}\,\mathbf{E}\,\mathbf{E} + \chi^{(3)}\,\mathbf{E}\,\mathbf{E}\,\mathbf{E} + \dots \,,$$

where $\chi^{(1)}$ is the linear optical susceptibility (and represents the only term taken into account in classical linear optics) and $\chi^{(2)}$ and $\chi^{(3)}$ are the nonlinear susceptibilities of order 2 and 3, respectively. If the nonlinear susceptibilities are large, nonlinear optical effects can be observed when the medium is illuminated by a very intense source, such as that produced by a laser. The odd-parity susceptibilities $\chi^{(1)}$ and $\chi^{(3)}$ are always present in any material but the $\chi^{(2)}$ term exists only in noncentrosymmetric systems. This is the reason why a major part of the efforts aiming at the production of large $\chi^{(2)}$ effects has also to deal with the development of a strategy to ensure the noncentrosymmetric character of the medium.

It is well known that in linear media, when two light waves of frequency ω interact:

$$y_1 = A \sin(\omega t) + \varphi_1 ,$$
$$y_2 = B \sin(\omega t) + \varphi_2 ,$$

the superposition principle is in effect and leads to the production of a wave of same frequency ω:

$$y = y_1 + y_2 = C \sin(\omega t) + \varphi .$$

However, in nonlinear media, this superposition principle is no longer valid and the light waves can interact in such a way that new waves appear, other waves disappear, intensities can be strongly modified,... A simple illustration of these effects can be given by looking at the second order term and expressing the electric field E as equal to $[E_0 \sin(\omega t)]$:

$$\chi^{(2)} E \; E = \chi^{(2)} E_0^2 \sin^2(\omega t) = (1/2) \chi^{(2)} E_0^2 [1 - \cos(2\omega t)] .$$

Thus, we observe that, from a wave of frequency $\underline{\omega}$, a wave of frequency $\underline{2\omega}$ can be obtained. In other words, by shining a laser light in the near-infrared on a medium with a large second-order susceptibility, one could observe the emission of a green light from the medium.

There is a very large number of second-order and third-order processes that have been discovered. To cite only a few, we can mention among second-order processes:

$\rightarrow$ the second harmonic generation (SHG) $[\omega + \omega \rightarrow 2\omega]$, which leads to frequency doubling, as has been depicted above;

$\rightarrow$ the frequency transposition $[\omega_1 \pm \omega_2 \rightarrow \omega_3]$, used for optical mixing;

$\rightarrow$ the parametric amplification $[\omega \rightarrow \omega_1 + (\omega - \omega_1)]$, which is the basis of optical parametric oscillators;

$\rightarrow$ the linear electrooptic effect, also called Pockels effect $[\omega + 0 \rightarrow \omega]$, used for electrooptic modulators;

and among third-order processes:

$\rightarrow$ the third harmonic generation (THG) $[\omega + \omega + \omega \rightarrow 3\omega]$;

$\rightarrow$ the four-wave mixing $[\omega_1 + \omega_2 - \omega_1 \rightarrow \omega_2]$, which is for instance at the origin of the multiple applications based on optical phase conjugation (a process also named time reversal of light), such as dynamic holography, optical microlithography or optical pattern recognition; in the case of degenerate four-wave mixing, all four waves have the same frequency ω;

$\rightarrow$ the quadratic electrooptic effect (Kerr effect) [$\omega + 0 - 0 \rightarrow \omega$], used for optical bistability and optical computing.

It is interesting to notice that the refractive index of a nonlinear medium varies as a function of light intensity I:

$$n(\omega) = n_0(\omega) + n_2(\omega) I(\omega) ,$$

$$n_2(\omega) = \frac{3 n_0(\omega) \pi \chi^{(3)}}{2 + 8 \pi \alpha(\omega)} ,$$

which means that the more intense the incident light, the more optically dense the medium. It can be shown that the $\chi^{(3)}$ and thus n_2 terms are large whenever there is a large shift in oscillator strengths upon illumination in resonance conditions. As we have seen in the previous section, this is precisely the case in a conjugated polymer such as polythiophene [16]. Upon intense illumination, the formation of a large number of bipolarons provoke an important redistribution of oscillator strengths. The same behavior has also been observed in polydiacetylene [20] and polyparaphenylene vinylene [21] as well as in polyacetylene [22] where laser illumination leads to midgap soliton absorption. As a result, because of the strong electron-phonon coupling, the conjugated polymers are inherently highly nonlinear optical materials [22]. This conclusion, driven from experiments conducted in the resonant mode, remains valid for off-resonance processes (which are technologically the most sought after) as will be clear from the discussion below.

Currently, inorganic materials are mostly used in nonlinear optical applications. Systems of interest are for instance lithium niobate (LiNbO$_3$), potassium dihydrogeno-phosphate KDP (H$_2$KPO$_4$), and the III-V semiconductors such as GaAs or InSb. Organic materials are however very promising, and are given a serious try in many academic and industrial laboratories worldwide. This is the case for instance of molecular crystals based on polar molecules such as 2-methyl-4-nitroaniline (MNA) or 3-methyl-4-nitropyridine-1-oxide (POM) for $\chi^{(2)}$ effects or conjugated polymers such as polydiacetylene, polyacetylene, or polythiophene for $\chi^{(3)}$ effects.

The organic materials present a number of advantages:

(i) they possess extremely large nonlinear optical responses. For instance, the SHG coefficient of MNA and POM are significantly larger than those of the inorganic compounds we have mentioned;

(ii) they possess very large optical damage thresholds. The POM molecular crystal can sustain laser powers from some 50 MW/cm^2 up to about 2GW/cm^2. Polyacetylene has been reported as resisting to powers as high as 10 GW/cm^2 [22]. For comparison, the damage threshold of lithium niobate is about 15 MW/cm^2;

(iii) another advantage is the relative ease of synthetic variations from organic chemistry. A major aspect is the possibility of performing a real molecular engineering to design the material which is best suited for a given nonlinear optical effect (this is especially interesting when it is realized that there exist more than 20 third-order processes);

(iv) the ease of processing of organic materials in many ways can turn out to be also important: organic systems can be prepared as crystals, liquid crystals, polymeric liquid crystals, or thin films (which can be synthesized namely through Langmuir-Blodgett or organic molecular-beam epitaxy technologies).

A common feature of all organic materials which are investigated for nonlinear optics is that they possess a high degreee of π conjugation. In nonresonant processes, conjugated materials are expected to give rise to:

on the one hand, a very high nonlinear optical response, because of the ease of polarization of the π-electron cloud upon appearance of the light electric field;

and, on the other hand, a very fast nonlinear response because the origin of the response is purely electronic. Responses as fast as the femtosecond range can be expected. Even in resonant processes, the response remains fast, on the order of a tenth of a picosecond. In inorganic materials, the fastest responses are in the picosecond range.

In summary, we can say that the materials which are most interesting in the field of conducting polymers are also potentially those which are most promising in the context of nonlinear optics. Finally, we would like to stress that quantum chemistry has a role to play in the development of new materials for nonlinear optics. Appropriate calculations can indeed help in guiding the synthetic chemist and the physicist towards the best materials. Much in the same way as in the conducting polymers area, an interdisciplinary approach is required in order to achieve significant developments.

Acknowledgements

I wish to thank all my colleagues that have at some point or another contributed to the work that have been presented here, in particular J.M. André, J. Delhalle, S. Stafström, V.P. Bodart, and C. Barbier in Namur, A.J. Epstein at OSU, A.J. Heeger and F. Wudl at UCSB. This work has been supported by the Belgian National Fund for Scientific Research (FNRS). The collaborative work with UCSB is supported by NATO Scientific Affairs Division through Research Grant No. 407/84. I am indebted to FNRS, IBM Belgium, and the Facultés Universitaires Notre-Dame de la Paix for the use of the Namur Scientific Computing Facility (SCF).

1. Excellent presentations of the quantum chemistry techniques alluded to in this section can be found in the following books:
– A. Szabo, N.S. Ostlund: Modern Quantum Chemistry (MacMillan, New York, 1982);
– R. Daudel, G. Leroy, D. Peeters, M. Sana: Quantum Chemistry (Wiley, New York, 1983);
– W.J. Hehre, L. Radom, P. v.R. Schleyer, J.A. Pople: Ab Initio Molecular Orbital Theory (Wiley, New York, 1986);

- J. Sadlej: Semiempirical Methods of Quantum Chemistry (Ellis Horwood, Chichester, 1985);

- L. Salem: Molecular Orbital Theory of Conjugated Systems (Benjamin, New York, 1966).

2. D. Baeriswyl, these proceedings.

3. W.P. Su, J.R. Schrieffer, A.J. Heeger: Phys. Rev. Lett. 42, 1698 (1979); Phys. Rev. B 22, 2209 (1980).

4. G. Nicolas, Ph. Durand: J. Chem. Phys. 70, 2020 (1979) ; J.M. André, L.A. Burke, J. Delhalle, G. Nicolas, Ph. Durand: Int. J. Quantum Chem. S13, 283 (1979); J.L. Brédas, R. Silbey, D.S. Boudreaux, R.R. Chance: J. Am. Chem. Soc. 105, 6555 (1983); J.L. Brédas: in Handbook of Conducting Polymers, edited by T.A. Skotheim (Marcel Dekker, New York, 1986), vol. 2, pp. 859-913.

5. J.C. Chiang, A.G. MacDiarmid: Synth. Met. 13, 193 (1986) and references therein; J.P. Travers, J. Chroboczek, F. Devreux, F. Genoud, M. Nechtschein, A. Syed, E.M. Genies, C. Tsintsavis: Mol. Cryst. Liq. Cryst. 121, 195 (1985); A.G. MacDiarmid et al. and E.M. Genies et al.: these proceedings.

6. A.G. MacDiarmid, J.C. Chiang, A.F. Richter, A.J. Epstein: Synth. Met. 18, 285 (1987).

7. J.M. Ginder, A.F. Richter, A.G. MacDiarmid, A.J. Epstein: to be published; A.J. Epstein, J.M. Ginder, F. Zuo, R.W. Bigelow, H.S. Woo, D.B. Tanner, A.F. Richter, W.S. Huang, A.G. MacDiarmid: Synth. Met. 18, 303 (1987).

8. S. Stafström, J.L. Brédas, A.J. Epstein, H.S. Woo, D.B. Tanner, W.S. Huang, A.G. MacDiarmid: to be published; S. Stafström: these proceedings.

9. N.S. Sariciftci, H. Neugebauer, H. Kuzmany: these proceedings; A.P. Monkman, D. Bloor: these proceedings.

10. J.L. Brédas, F.Wudl, A.J. Heeger: Solid State Commun., to be published.

11. J.L. Brédas, R.R. Chance, R. Silbey: Phys. Rev. B 26, 5843 (1982).

12. J.L. Brédas, J.C. Scott, K. Yakushi, G.B. Street: Phys. Rev. B 30, 1023 (1984).

13. D.S. Boudreaux, R.R. Chance, J.L. Brédas, and R. Silbey, Phys. Rev. B 28, 6927 (1983).

14. M. Nowak, S.D.D.V. Rughooputh, S. Hotta, A.J. Heeger: Macromolecules, in press.

15. P.A. Pincus, G. Rossi, M.E. Cates: to be published.

16. Z. Vardeny, E. Ehrenfreund, O. Brafman, M. Nowak, H. Schaffer, A.J. Heeger, F. Wudl: Phys. Rev. Lett. 56, 671 (1986).

17. Nonlinear Optical Properties of Organic and Polymeric Materials, ed. by D.J. Williams (American Chemical Society, Washington D.C., 1983).

18. D.J. Williams, Angew. Chem. Int. Ed. Engl. 23, 690 (1984).

19. Nonlinear Optical Properties of Organic Molecules and Crystals, ed. by D.S. Chemla and J. Zyss (Academic, Orlando, 1987).

20. F.L. Pratt, K.S. Wong, W. Hayes, D. Bloor: these proceedings.

21. D.D.C. Bradley, R.H. Friend, H. Linderberger, S. Roth: these proceedings.

22. A.J. Heeger: in Conducting Polymers (part 2), ed. by H. Sasabe (CMC, Tokyo, 1987), pp. 5-54.

On the Role of the Coulomb Interaction in Conjugated Polymers

D. Baeriswyl*

Max-Planck-Institut für Festkörperforschung,
Heisenbergstr. 1, D-7000 Stuttgart 80, Fed. Rep. of Germany

There is currently a lively discussion about the relative importance of electron-electron and electron-phonon interactions in conjugated polymers. It is the purpose of this paper to summarize the interplay between these two types of interaction on the basis of a simple Hamiltonian and to relate the properties of this model to experiments in polyacetylene and polydiacetylene. Three regimes can be discussed according to the strength of the (on-site) Coulomb interaction (U): small, intermediate and large U. It is concluded that conjugated polymers belong to the intermediate regime. More details can be found in a forthcoming publication [1].

1. The Hubbard-Peierls model

The most simple model for describing the combined effects of electron-electron and electron-phonon interactions for a single polymer chain is defined in terms of the electronic Hamiltonian

$$H = - \sum_{n\sigma} t_{n,n+1} (c_{n\sigma}^+ c_{n+1\sigma} + c_{n+1\sigma}^+ c_{n\sigma}) + U \sum_n n_{n\uparrow} n_{n\downarrow} , \quad (1)$$

where n denotes the site and σ the spin projection of π-electrons, $n_{n\sigma} = c_{n\sigma}^+ c_{n\sigma}$ and the resonance integrals $t_{n,n+1}$ vary with varying bond length. The total energy for a given configuration of bond lengths includes both the expectation value of H and the elastic energy of the σ-bonds. Two limiting cases are well understood. For $t_{n,n+1} = t_0$, i.e. in the absence of any coupling to the lattice, the Hamiltonian (1) corresponds to the one-dimensional Hubbard model for which several properties are known exactly [2,3]. For an average density of one electron per site (half-filled band) the (many-particle) excitations are gapless spinwaves and current-carrying states with a gap (for U > 0). Thus the system is insulating however small U. For all other band fillings it is conducting however large U. The other limit where Coulomb interactions are neglected altogether but the coupling to the lattice is taken into account corresponds to the Hückel or SSH model and is also well understood [4]. In the half-filled band case the lattice is dimerized and produces a (Peierls) gap at the Fermi energy.

---
* Permanent address: Institut für theoretische Physik,
ETH-Hönggerberg, CH-8093 Zürich, Switzerland

Adding charge results first in a periodic array of charged solitons which is continuously transformed into a sinusoidal modulation of the lattice as doping proceeds. The accompanying incommensurate charge-density wave can move and thus produce a d.c. current although the single-particle spectrum still has a gap at the Fermi energy.

2. Theoretical procedures

The important numbers characterizing the interaction strengths are the dimensionless electron-phonon parameter λ and U/W for the electron-electron interaction where W is the π-bandwidth ($\lambda \simeq 0.1$ and $W \simeq 10$ eV for polyacetylene (4)). Varying U and keeping λ small we can distinguish three regimes. For $U < \frac{1}{2}\pi\lambda W$ ("weak coupling") the single-particle picture is a good starting point and the Coulomb effects can be calculated perturbatively (5). For $U \gg W$ ("strong coupling") spin- and charge-degrees of freedom are decoupled (6). In the special case of one electron per site the Hamiltonian (1) is transformed to a spin $\frac{1}{2}$ Heisenberg model with antiferromagnetic exchange $I_{n,n+1} = 4t^2_{n,n+1}/U$. The spin-phonon coupling leads again to a dimerization of the lattice (spin-Peierls transition) which has often been treated by mapping the spin Hamiltonian onto that of a system of interacting spinless fermions (7). In the intermediate regime where $\frac{1}{2}\pi\lambda W < U \lesssim W$ neither the Hückel nor the Heisenberg limits are good starting points for perturbation theory and more powerful methods have to be used. Various routes have been followed. Variational calculations (8,9,10), numerical diagonalization of the full many-body Hamiltonian for short chains (11,12), quantum Monte Carlo (13,14), a real-space renormalization group method (15) and a field-theoretic approach (16) have provided at least qualitatively consistent results, especially also in this interesting intermediate regime.

3. Electronic correlation

The word "electronic correlation" is often used but unfortunately not always for the same thing. Within many-body theory it has a specific meaning which is worth to be remembered.

In Hartree-Fock theory electron-electron interactions are replaced by an effective single-particle potential (or mean field). The Hartree-Fock ground state is therefore a single Slater determinant of (quasi-)particle wavefunctions. Minimizing the expectation value of the original Hamiltonian with respect to this trial state one finds the best single-particle approximation.

Besides the effects imposed by the Pauli principle no correlations between quasi-particles are taken into account in mean-field theory, they move independent of each other in the effective potential. Correlation therefore refers to the

effects which are not taken into account by the mean field
and, correspondingly, the energy gained by going beyond Hartree-Fock is called correlation energy. More precisely, a correlated state is a superposition of Slater determinants which
cannot be reduced to a single determinant by any linear
transformation (of the single-particle basis functions).

If we accept this definition we have to conclude that the Unrestricted Hartree-Fock approximation, where the up- and down
-spins experience two different mean fields, yields also an
uncorrelated state. Indeed the energy gain with respect to
the simple Hartree-Fock theory is not a correlation energy
but rather a condensation energy due to the appearance of a
spin-density wave (broken symmetry). In fact for the simple
Hubbard model in the small U limit this condensation energy
is much smaller than the exact correlation energy [17]. For
large U the spin-density wave state goes over into the Néel
state (a single Slater determinant in configuration space)
whereas the true correlated ground state is a superposition
of many spin configurations. In the correlated state the probability that a particle occupies a particular site with a
particular spin depends on the specific configuration of
other particles and not simply on their average density or
other global quantities.

4. Dimerization and optical gap

The amount of lattice dimerization for given parameters λ
and U represents one of the fundamental issues of the Peierls
-Hubbard model. Unrestricted Hartree-Fock theory predicts
that for $\lambda > 0$ and small enough U the dimerization amplitude
is independent of U and simply given by the familiar BCS relation for the Hückel model. Above a critical value of U the
dimerization is predicted to vanish due to the appearance of
a spin-density wave [18,19]. This picture which has dominated
the discussion for a long time is simply wrong. For finite λ
the dimerization is non-zero both for the weak-coupling regime where it increases with U [5,9] and for the strong-coupling or spin-Peierls regime where it decreases with U but
tends to zero only for $U \rightarrow \infty$ [7,20]. Numerical calculations
[12,13], variational [8,10] and scaling methods [15] confirm
these results and show in addition that the dimerization amplitude has a maximum in the intermediate regime. There is
also a qualitative change in the functional dependence on λ
which shows a cross-over from the BCS-type behaviour for
small U to a superlinear power law for intermediate values
of U [16,21].

An optical gap exists both in the Peierls (U=0) and Hubbard
limit (λ=0). In general one expects that both the lattice
distortion and the Coulomb interactions are involved. However, since one has to deal with excited "ionic" many-particle states it is much harder to obtain reliable results for
the gap than for the dimerization, which is a ground-state
property, except for the strong-coupling regime where the gap
is simply equal to U. Explicit calculations indicate that a
Peierls gap (enhanced by Coulomb interaction) exists in the
weak-coupling region,whereas in the intermediate regime the

effect of the Hubbard term dominates (16,21), giving rise to a "correlation gap".

5. Solitons

The enhancement of bond alternation by electronic correlation implies that the topological solitons, the domain walls separating the two possible dimerization patterns, are also stabilized. However, their internal structure is expected to be strongly modified as compared to their simple appearance in the Hückel limit. Since these inhomogeneous states can be considered as ground states of odd-numbered rings, their structure and energetics can be studied by numerical methods (14,22). The local spin densities for neutral solitons are particularly important quantities since they can be deduced from ENDOR experiments (23). Coulomb effects produce alternating positive and negative spin densities. The Hartree-Fock approximation is found to yield good results in the weak coupling regime but to break down in the intermediate regime, grossly overestimating the spin polarization (22).

The additional optical transitions induced by inhomogeneous structures (solitons, polarons, bipolarons) provide further important informations about electron-electron interactions. Unfortunately, similarly as for the ground state, the theory for optical absorption from defect states is still somewhat rudimentary. Simple perturbation calculations give at most a semi-quantitative picture for the splittings produced by Coulomb interactions. For the Hubbard-Peierls model it is found that the soliton-induced transitions, which are predicted to occur exactly at midgap in the Hückel limit, both for neutral and charged solitons, are shifted upwards and downwards, respectively, by $\pm U/(3\ell)$ where ℓ is the coherence length (the ratio between bandwidth and bandgap) (24).

6. Conjugated polymers: intermediate U materials

The Peierls-Hubbard Hamiltonian represents an extremely simplified model for a conjugated polymer. Even admitting that π-electron models are adequate for describing the electronic structure close to the Fermi energy, one would have to include long-range interactions as for instance in the Pariser-Parr-Pople (PPP) Hamiltonian. Fitting experiments with a single Coulomb parameter U one has to keep in mind that the result will yield an effective U, somewhat reduced from its bare value by the long-range Coulomb terms.

Not all properties are equally sensitive to interaction effects. Thus the magnetic susceptibility for q=0 depends only weakly on U in the case of the simple Hubbard model at least for $U \lesssim W$, whereas $q=2k_F$ response functions (involving spin, charge or bond order operators) can be strongly enhanced with increasing U. The electron-phonon parameter λ has been consistently determined on the basis of electronic and vibrational properties of organic molecules and there is no apparent reason for keeping it as freely adjustable parameter. Unfortunately many calculations, especially the numerical simulations, use too large values of λ and should not be compared with experiment without reservation.

A value U/W $\simeq$ 0.7 has been deduced on the basis of variational calculations for the lattice dimerization, the amplitude mode frequency (showing up in Raman spectra) and the electronic gap in comparison with experiments on polyacetylene (10). A similar value has been found from the oscillator strength for $\pi-\pi^*$ transitions (25). Recent ENDOR experiments on oriented polyacetylene give a ratio of negative and positive spin densities for a neutral soliton of 0.43 corresponding to U/W $\simeq$ 0.6 (26). It is interesting to note that calculations by Ramasesha and Soos for standard PPP parameters yield 0.42 for the ratio of spin densities (27). The perturbative analysis of the soliton-induced optical absorption in polyacetylene gives U $\sim$ W (24). The triplet exciton has now been observed in polydiacetylene (28,29). Its binding energy is substantially larger than that of the singlet exciton. This can be understood as an effect of the on-site Coulomb interaction (30). All these experimental results confirm the importance of Coulomb effects and put the conjugated polymers into the intermediate U category. In this regime the optical gap is mainly a correlation gap and to a minor degree caused by electron-phonon interactions.

Acknowledgement

I have greatly benefitted from close collaboration with D.K. Campbell, J. Carmelo, K. Maki and S. Mazumdar.

References

1.   D. Baeriswyl, D.K. Campbell and S. Mazumdar: manuscript in preparation.

2.   E.H. Lieb and F.Y. Wu: Phys. Rev. Lett. 20, 1445 (1968).

3.   A.A. Ovchinnikov: Zh. Eksp. Teor. Fiz. 57, 2137 (1969) (Sov. Phys. JETP 30, 1160 (1970)).

4.   For a review see, e.g., D. Baeriswyl, in H. Kamimura (ed.): Theoretical Aspects of Band Structures and Electronic Properties of Pseudo-One-Dimensional Solids (D. Reidel, Dordrecht, 1985), p. 1.

5.   S. Kivelson and D.E. Heim: Phys. Rev. B 26, 4278 (1982).

6.   J. Bernasconi, M.J. Rice, W. Schneider and S. Strässler: Phys. Rev. B 12, 1090 (1975).

7.   M.C. Cross and D.S. Fisher: Phys. Rev. B 19, 402 (1979).

8.   I.I. Ukrainskii: Zh. Eksp. Teor. Fiz. 76, 760 (1979) (Sov. Phys. JTEP 49, 381 (1979)).

9.   P. Horsch: Phys. Rev. B 24, 7351 (1981).

10.   D. Baeriswyl and K. Maki: Phys. Rev. B 31, 6633 (1985).

11.   Z.G. Soos and S. Ramasesha: Phys. Rev. B 29, 5410 (1984).

12.   S.N. Dixit and S. Mazumdar: Phys. Rev. B 29, 1824 (1984).

13. J.E. Hirsch: Phys. Rev. Lett. $\underline{51}$, 296 (1983).

14. D.K. Campbell, T.A. De Grand and S. Mazumdar: Phys. Rev. Lett. $\underline{52}$, 1717 (1984).

15. G.W. Hayden and E.J. Mele: Phys. Rev. B $\underline{34}$, 5484 (1986).

16. V. Ya. Krivnov and A.A. Ovchinnikov: Zh. Eksp. Teor. Fiz. $\underline{90}$, 709 (1986) [Sov. Phys. JETP $\underline{63}$, 414 (1986)].

17. J. Carmelo: Ph. D. thesis, Copenhagen 1986, unpublished

18. R.A. Harris and L.M. Falicov: J. Chem. Phys. $\underline{51}$, 5034 (1969).

19. K.R. Subbaswamy and M. Grabowski: Phys. Rev. B $\underline{24}$, 2168 (1981).

20. T. Nakano and H. Fukuyama: J. Phys. Soc. Japan $\underline{49}$, 1679 (1980).

21. D. Baeriswyl and K. Maki: Synth. Met. $\underline{17}$, 13 (1987).

22. J.E. Hirsch and M. Grabowski: Phys. Rev. Lett. $\underline{52}$, 1713 (1984).

23. H. Thomann, L.R. Dalton, M. Grabowski and T.C. Clarke: Phys. Rev. B $\underline{31}$, 3141 (1985).

24. D. Baeriswyl, D.K. Campbell and S. Mazumdar: Phys. Rev. Lett. $\underline{56}$, 1509 (1986).

25. J. Fink and G. Leising: Phys. Rev. B $\underline{34}$, 5320 (1986); see also the contribution of J. Fink, N. Nücker, B. Scheerer, W. Czerwinski, A. Litzelmann and A. vom Felde in this volume.

26. A. Grupp, P. Höfer, H. Käss, M. Mehring, R. Weizenhöfer and G. Wegner: contribution in this volume.

27. S. Ramasesha and Z.G. Soos: Synth. Met. $\underline{9}$, 283 (1984).

28. L. Robins, J. Orenstein and R. Superfine: Phys. Rev. Lett. $\underline{56}$, 1850 (1986).

29. M. Winter, A. Grupp, M. Mehring and H. Sixl: Chem. Phys. Lett. $\underline{133}$, 482 (1987).

30. I.I. Ukrainskii: phys. stat. sol. (b)$\underline{106}$, 55 (1981).

Lattice Relaxation Approach to Soliton and Polaron Dynamics in Conducting Polymers

Su Zhao-bin [1] *and Yu Lu* [2]

[1]Institute of Theoretical Physics, Academia Sinica,
 Beijing, China
[2]International Centre for Theoretical Physics, Trieste, Italy
 (currently) and
 Institute of Theoretical Physics, Academia Sinica,
 Beijing, China (permanently)

The lattice relaxation theory generalized by us to include the self-consistency of multi-electron states with lattice symmetry-breaking is summarized. The discrete symmetries and corresponding selection rules for both radiative and nonradiative processes are discussed. The theoretically calculated probability of nonradiative decay of an electron-hole pair into a soliton pair and that of an electron (hole) into a polaron as well as the probability of soliton pair photogeneration is compared with results of numerical and laboratory experiments. The resonance Raman scattering data of *cis*-polyacetylene are interpreted in terms of a bipolaron model. The parameters involved are determined directly from experimental data.

1. Introduction

The soliton model proposed by SU, SCHRIEFFER, and HEEGER(SSH) /1/ and, independently, by RICE /2/ has been very successful in interpreting the peculiar electric, magnetic and optical properties of *trans*-polyacetylene, $(CH)_x$ /3/. A closely related concept – the polaron /4/ – has found even broader applications in conducting polymers, since it does not require strict degeneracy of the ground state configurations.

Pristine polyacetylene (PA for short) is a semiconductor with dimerized ground state. The energy per site as a function of the dimerization parameter u can be written as /1/

$$E_0(u) = A + B u^2 \ln\left(\frac{u}{a}\right) + 2K u^2,\qquad(1.1)$$

where a is the lattice spacing, K the spring constant and A, B are numerical coefficients. Here the symmetry breaking is "dynamical" in the sense that it is not due to the sign change of K as in the theory of phase transitions or in many field-theoretical models, but because of a logarithmic term contributed by electrons, or, in other words, the lowering of electronic energy always overwhelms the increase of potential energy caused by lattice distortion. In a pictorial way one can imagine that a chain of springs equally spaced in equilibrium, is immersed into a "magic" liquid and becomes alternately stretched and compressed. This "magic" liquid is nothing but π- electrons obeying Fermi statistics.

Trans-PA has two degenerate ground state configurations and a soliton is a defect state interpolating these configurations. The soliton formation energy consists of three parts, namely the bound state energy, the energy difference of quasi-continuum states and the change of elastic energy, and is equal to $2\Delta/\pi$. Since the creation energy of a soliton pair is smaller than the energy of an electron-hole pair, it is a stable excitation. Also, this "smallness" is due to the gain in elastic energy. Otherwise, it would be very costly. In our analogy with the chain of springs, the soliton excitation corresponds to a local relaxation of springs. The electron condensation energy is partly lost, but the elastic energy is partly recovered. On the one hand, the soliton is a local entity. On the other hand, its creation modifies the whole energy spectrum. These two complementary aspects are important for the understanding of dynamical processes. Similarly, the formation energy of a polaron/4/ $E_p = 2\sqrt{2}\Delta/\pi$ is lower than the electron (hole) energy for the same reason.

The nonradiative decay of an e-h pair into soliton pair and that of an electron into a polaron was studied numerically on a discrete chain in a pioneering paper of SU and SCHRIEFFER /7/. These authors found that such a decay process is rather fast and takes place during a few cycles of optical vibrations. The direct process of soliton pair photogeneration was considered by SETHNA and KIVELSON /8/ using an instanton approach. The lattice relaxation approach extended and applied by us to conducting polymers /6,9/ treats radiative and nonradiative processes in a unified manner.

2. Lattice Relaxation Theory

The lattice relaxation theory of multiphonon processes was developed in the early fifties by HUANG and RHYS/10/ and others. The basic idea of this approach is very simple. Consider, for example, a defect (F-centre) in an ionic crystal. Let one negative ion be missing and an electron be trapped at this site. It has since long been known that both absorption and luminescence spectra connected with such defects are very broad, exhibiting profound multiphonon structures. Huang and others realized at that time that these multiphonon processes were due to a difference in the lattice configurations for initial and final states. In configurational coordinates, a typical energy diagram might appear like that shown in Fig. 1, where Q is some characteristic parameter for the lattice configuration (relaxation), say, the average distance of the neighbouring sites to the defect centre. As shown in the diagram, the equilibrium lattice positions (or the phonon "vacua") are at 0 and Δ_{if} for initial and final states, respectively. Huang and others proposed to perform perturbation calculations between these phonon sets with different vacua, instead of using a conventional perturbation technique, to study both radiative and nonradiative processes. These two phonon vacua are connected by an origin-shifting operator given by

$$|\Delta_{if}> = \exp(\frac{i}{\hbar}P\,\Delta_{if})\,|0>, \qquad (2.1)$$

where P is the momentum operator. Expanding the exponent in a power series and expressing P in terms of phonon creation and annihilation operators, we see immediately that $|\Delta_{if}>$ contains states with an arbitrary number of phonons on the vacuum $|0>$.

In our problem the initial lattice configuration is a dimerized pattern, while the final state has a kink-antikink pair. The similarity with the F-centre issue is obvious. Moreover, in one-dimensional systems each site has only two nearest neighbours, so that the lattice relaxation effect is even more profound than in the three-dimensional case. We have generalized the lattice relaxation theory to include the self-consistency of electronic states with lattice configurations and the many-body background effects for electrons and applied this scheme to study both radiative and nonradiative processes of soliton and polaron generation in PA /6,9/. The big advantage of the present case is the availability of self-consistent analytic solutions.

The continuum version of the SSH Hamiltonian can be written as /5/

$$H = \int dx \, \psi^{\dagger}(x) \, [\, \frac{\hbar v_F}{i} \frac{d}{dx} \tau_3 + 4\alpha u(x)\tau_1 \,]\psi(x)$$

$$+\frac{\rho}{2} \int dx \, [\dot{u}(x)\dot{u}(x) + \omega_0^2 \, u(x)u(x)] \,, \tag{2.2}$$

where $\psi^{\dagger}$, ψ are pseudo-spinor electron creation and annihilation operators, u, $\dot{u}$ the displacement and velocity operators for the staggered field, v_F the Fermi velocity, $\rho = M/a$ the linear density, $\omega_0^2 = 4K/M$ the squared optical phonon frequency and τ_1, τ_3 Pauli matrices.

According to the basic idea of the lattice relaxation theory, we identically rearrange (2.2) as

$$H = H_0^{el} + H_0^{ph} + H_{int} \,, \tag{2.3}$$

$$H_0^{el} = \int dx \, [\psi^{\dagger}(x) \, (-i\hbar v_F \frac{d}{dx} \tau_3 + 4\alpha u_c(x)\tau_1)\psi(x) + \frac{\rho}{2}\omega_0^2 u_c^2(x) \,)],$$
$$\tag{2.4}$$

$$H_0^{ph} = \frac{\rho}{2} \int dx \, [\dot{u}(x)\dot{u}(x) + \omega_0^2(u(x) - u_c(x)) \, (u(x) - u_c(x))], \tag{2.5}$$

$$H_{int} = \int dx \, [4\alpha\psi^{\dagger}(x)\tau_1 \psi(x) + \rho\omega_0^2 u_c(x)] \, (u(x) - u_c(x) \,), \tag{2.6}$$

where $u_c(x)$ is the classical lattice configuration appearing in the free parts of the electron and phonon Hamiltonian.

Generally speaking, the initial and the final configurations, $u_c(x)$ and $v_c(x)$, are different (say, the dimerized pattern and S-$\bar{\text{S}}$ pair, respectively). This means that the total Hamiltonian is split into free and interacting parts in different ways for the initial and for the final states. Thus we are dealing with perturbation between different complete sets. However, as proved in the general theory of scattering /11/, the transition probability does not depend on the overlap integral provided the total energy is conserved. Also, the final result does not depend on whether the interaction Hamiltonian for the initial or final state is used as perturbation.

The general formalism for the lattice relaxation approach applied to the PA model was developed in the first paper of Ref. 6. In the low temperature $(\hbar \omega_0 \le kT)$ and strong coupling $(|S \hbar \omega_0 - W_{if}| \le S \hbar \omega_0)$ limit, the transition probability is given by

$$W = \frac{\pi \lambda v_F}{4} \left(\frac{2\pi}{S}\right)^{1/2} \exp\left\{-\frac{(W_{if} - S \hbar \omega_0)^2}{2S \hbar^2 \omega_0^2}\right\} U^2, \qquad (2.7)$$

where λ is the dimensionless electron-phonon coupling constant, U the electronic transition amplitude and W_{if} the difference between the corresponding sums of electronic and lattice relaxation energies (see Fig.1). The Huang- Rhys factor S appearing here is the average number of phonons needed to bring the initial lattice configuration into the final one and is given by

$$S = \frac{\rho \omega_0^2}{2\hbar} \int dx \, [u_c(x) - v_c(x)]^2. \qquad (2.8)$$

It is interesting to note that (2.7) can be derived in the quasi-classical approximation as a zero-point motion activated process. In a single frequency model we can replace the displacement field $u(x)$ by a single variable Q. The energy diagram is shown schematically in Fig. 1. Using the Landau formula for molecule predissociation /12/ and averaging it over the ground state oscillator wave function, one recovers (2.7) /6/.

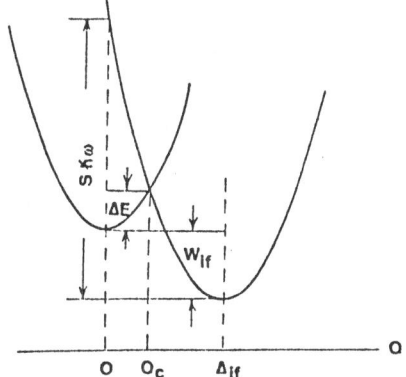

Fig.1 Energy diagram for nonradiative decay.

3. Discrete Symmetries and Selection Rules

The charge conjugation (CC) and the parity symmetries of the SSH Hamiltonian and its continuum version (2.2) have been extensively discussed /6,9,13-16/ in connection with selection rules for both radiative and nonradiative processes.

We first consider the CC symmetry. In the earlier discussions /6,9,13/ a single particle formulation was used. The CC transformation is defined as

$$c_n \to (-1)^n c_n \qquad\qquad \psi \to \tau_2 \psi \qquad\qquad (3.1)$$

for the discrete and continuum models, respectively. The electronic Hamiltonian (first line in (2.2)) changes sign under this transformation. Also, the wave functions become complex-conjugate when the energy changes sign. It was argued in Ref. 13 that the nonradiative decay of a charged soliton pair corresponding to single occupation of bonding and antibonding soliton levels, to a neutral soliton pair corresponding to double occupancy of the bonding level, is forbidden. The question of nonradiative decay of an e-h pair into a charged soliton was not addressed in that paper. However, if one sticks to this philosophy, such a process would also not be allowed. On the other hand, this is the fundamental process simulated in Ref. 7. To resolve this controversy, we pointed out that the total Hamiltonian as well as its remnant part (2.6) contains terms with mixed CC symmetry, and therefore such process should be allowed. The actual calculation in Ref. 6 was done for nonradiative decay of an e-h pair into a charged soliton pair, but the statement made there about the possibility of nonradiative decay of an e-h pair into a neutral soliton pair was incorrect. As we see, the single particle formulation of CC symmetry is not complete and may give rise to misleading conclusions.

The many-particle formulation of CC symmetry was considered in Ref. 16 for the continuum model and in Refs. 14-15 for the discrete version. The CC transformation is defined as

$$C\psi(x)\, C^\dagger = [\psi^\dagger(x)\tau_3]^T, \qquad \text{or} \quad \psi_1 \to \psi_1^\dagger,\ \psi_2 \to -\psi_2^\dagger \qquad (3.2)$$

in pseudo-spin components for the continuum version and as

$$c_n \to (-1)^n c_n^\dagger \qquad\qquad (3.3)$$

for the discrete model. Using expansion into a complete set, one can show that this definition is consistent with the standard formulation of the particle-hole symmetry in many-body theory as well as in the quantum field theory. It is straightforward to show /16/ that the SSH Hamiltonian, even generalized to include Coulomb interactions, is invariant under this CC transformation. The charge and current operators as well as the e-h pair and soliton pair states are CC odd, while the ground state and singlet neutral soliton pair are CC even.

As a consequence of this CC symmetry, the photoexcitation of an e-h pair and a charged soliton pair is allowed, while the photogeneration of a neutral soliton pair is forbidden. The same is also true for the reverse process via photoluminescence. On the other hand, the nonradiative decay is caused by the remnant electron-phonon interaction (2.6) which is CC even. Thus nonradiative decay of an e-h pair into a charged soliton pair as well as that of a neutral soliton pair into a dimerized state is allowed, whereas other processes like nonradiative decay of a charged soliton pair into a neutral one is not permitted. Hence the many-body formulation shows unambiguously that both direct and indirect (via nonradiative decay of an e-h pair) processes of neutral soliton pair photogeneration is forbidden by CC symmetry.

The parity symmetry of the SSH Hamiltonian was considered in the third paper of Ref.6. The parity transformation is defined as

$$P\psi(x)P^{\dagger} = \tau_1 \psi(-x), \qquad Pu(x)P^{\dagger} = u(-x). \qquad (3.4)$$

It is obvious that the SSH Hamiltonian is invariant under this transformation. The symmetry-broken free Hamiltonian (2.4) has the same symmetry if the topology of the system is even. Since the current operator is parity odd, whereas the neutral soliton pair and the dimerized state are parity even, the photoproduction of neutral soliton pair is also forbidden by the parity symmetry.

4. Soliton and Polaron Generation in *trans*-Polyacetylene

A detailed calculation for the nonradiative decay rate of an electron (hole) into a polaron based on the general formalism described in Sect.2 was carried out in the second paper of Ref. 6. Towards this end, the self-consistent solutions of the BdeG equation and the gap equation for the polaron case were used /4/. To simplify the calculation, a product wave function (Hartree approximation) was assumed for the many-electron state. One can show, however, that a better approximation (determinant wave function) does not change the essential features of the obtained results. Using the same set of parameters as of Ref. 7 in numerical simulations, the nonradiative decay rate for such a process was estimated as 10^{13} sec^{-1}, in fairly good agreement with simulation results. It is important to emphasize that not only the polaron bound state itself but also the whole quasi-continuum of the background states do contribute to the transition probability as mentioned in Sect.1. A similar calculation was also done there for nonradiative decay of an e-h pair into a charged soliton pair (although the fact that only a charged soliton pair is involved was not stated there explicitly). For this purpose the polaron solution with the distance between two kinks $2x_0$ as a running parameter was used as a transient state. The numerical estimate for the nonradiative decay rate thus obtained, also of the order 10^{13} sec^{-1}, agrees well with numerical simulations.

As for radiative processes, there are two ways of soliton photoproduction, one being the direct process and the other being the indirect process of first photoexciting an e-h pair with its subsequent nonradiative decay into a soliton pair. The second process is allowed only if the energy of exciting photons exceeds the optical gap 2Δ, whereas the first process is possible below this value but above the soliton pair energy $4\Delta/\pi$. Within the lattice relaxation approach the multiphonon effects for radiative processes are included as a Frank-Condon overlap integral. In the low temperature, strong coupling limit (in the sense similar to that for nonradiative processes), the cross section for the direct process is given by /6/

$$\sigma_d = \frac{16\pi e^2 v_F^2}{cn\omega\hbar\omega_0} \left(\frac{2\pi}{S}\right)^{1/2} |<e_i|e_f>|^2 \exp\left\{-\frac{(\hbar\omega - 4\Delta/\pi - S\hbar\omega_0)^2}{2S\hbar^2\omega_0^2}\right\},$$

(4.1)

where c is the speed of light and n the refractive index. Apart from a pre-exponential factor, this is a Gaussian distribution in frequency peaked at the soliton pair energy plus lattice relaxation energy difference.

On the other hand, the nonradiative decay of an e-h pair into a soliton pair gives rise to life-time effects which modify the cross section of absorption due to band-to-band transitions as /6/

$$\sigma_i^b = \frac{4\pi^2 e^2 v_F^2}{cn\omega} \sum_k \frac{\Delta^2}{\varepsilon_k^2} \frac{1}{\pi} \frac{\Gamma_k/2}{(2\varepsilon_k - \hbar\omega)^2 + \Gamma_k^2/4},$$

(4.2)

where Γ_k is the line-width due to nonradiative decay. We see that the multiphonon effects remove the singularity at the absorption edge due to density of states in the one-dimensional case. Also, we can estimate the ratio of directly and indirectly (via e-h pair) photoproduced charged soliton pairs as

$$\sigma_d/\sigma_i^b \approx 10^{-3} \quad \text{at} \quad \hbar\omega \approx 2\Delta.$$

(4.3)

This means that the direct process of soliton pair photoproduction dominates. We would like to mention that this result was misinterpreted as branching ratio of neutral versus charged soliton pair photoproduction in Refs. 6 and 9. As shown in Sect. 3, only charged solitons are allowed in both direct and indirect processes.

The evidence for photoproduction of charged soliton pairs first came from photoconductivity measurements /17/. The onset of photocurrent appeared at about 1 eV, well below the optical gap. The frequency dependence of the photocurrent from Ref. 21 is compared with theoretical results given by (4.1) in Ref. 9. There are no adjustable parameters involved in this comparison apart from the arbitrariness of photocurrent units. We should also mention that because of (4.3) it is legitimate to compare the experimental results with the cross section of charged soliton pair photogeneration due only to direct processes. Of course, one may argue that photoconductivity is a very complicated process depending on many factors such as mobility, recombination rate and so on. However, the photoinduced IR absorption experiments showed /18/ that the excitation profile of the IR absorption due to charged defects coincides with the frequency dependence of photocurrent, which makes our comparison more justified. Furthermore, very recent measurements on fast photoconductivity (of the order of 50 ps) show /19/ that the frequency dependence of such a transient process is similar to the stationary photoconductivity, although it is much larger than the latter in magnitude.

Other more direct evidence of charged soliton photogeneration comes from a very recent study of picosecond dynamics of photoinduced absorption at 0.45 eV /20/. These authors clearly identify two types of charged solitons. First, there are charged solitons which are directly photogenerated from intrachain absorption and geminately recombine on a subpicosecond time scale. At a later time (≥ 50 ps), there are charged solitons formed by an interchain mechanism where polarons are formed and diffuse to convert residual neutral solitons to charged ones.

An important question is whether only charged solitons are photogenerated. The early photoinduced ESR measurements /21/ showed that the ratio of photoinduced spins to photoinduced charges was very small, less than 10^{-2}. In fact, this served as a stimulus to formulate the selection rules. However, later measurements of photoinduced ESR with controversial results, sometimes no change, sometimes negative change of spin signal /22/, have cast doubts on this conclusion. Very recently, a more careful experiment excluding heating effects has shown a positive change of ESR signal upon laser illumination /23/. These authors interpret their results in terms of photogenerated and trapped neutral solitons. From the theoretical point of view the photogeneration of neutral solitons would be allowed if one included terms violating the CC symmetry, like the next nearest neighbour hopping /14,15/ or considered the triplet state /24/.

Furthermore, we would like to mention that the absolute measurements of optical absorption /25/ showing frequency dependence identical to that of photoconductivity are consistent with our interpretation of the absorption edge as due to soliton pair photogeneration. Also, the multiphonon effects giving rise to life-time effects improve the agreement of theory with optical absorption experiments showing no singularity.

Finally, we briefly compare our results with other theoretical approaches. The equation of motion method used by MELE /26/ to study the optical absorption for a discrete model is close in spirit to our approach and yields similar results. The instanton approach used by SETHNA and KIVELSON /8/ is elegant but more complicated. These authors emphasize the role of quantum fluctuations in subgap soliton pair photogeneration. In our quasi-classical description, the quantum nature appears as a zero point motion activated process. The instanton approach incorporates the virtual nonlinear process in a straightforward way, so the frequency dependence near threshold is quite precise, but it is very difficult to include the real phonon process, so the frequency dependence away from threshold is not correct. The lattice relaxation approach treats the virtual and real phonon processes on an equal footing, hence the overall frequency dependence is correct. We have used the steepest descent method to obtain compact formulas like (2.7) or (4.1) which do not show sharp onset at the threshold. However, if one is interested in the precise shape of the onset, it can be easily calculated from the starting formula without using the steepest descent approximation.

5. Bipolaron Interpretation of Resonance Raman Scattering in *cis*-Polyacetylene

Up to now we have been discussing mainly systems with a degenerate ground state. It is obvious that the lattice relaxation approach can be applied to systems with nondegenerate ground state configurations as well. The generalization of calculations for both radiative and nonradiative transitions to this case is straightforward and will not be elaborated here. As an illustrative example, we discuss resonance Raman scattering in *cis*-PA.

The resonance Raman spectra in *cis*-PA exhibit a series of overtones on the background of a broad emission /27/. It is obvious that multiphonon processes are involved. Unlike optical absorption or luminescence, the initial and final lattice configurations in Raman case are the same. Therefore, the lattice relaxation effects are important only if the intermediate state is in resonance or near resonance with some eigenstate of the system with a distinct lattice configuration. Put another way, the difference in lattice symmetry breaking can show up only if the system is spending a "long enough" time in this state.

Previously, there have been several attempts to explain these features of the Raman spectra /28-30/. GUO and YU /31/ proposed to interpret the resonance Raman spectra of *cis*-PA in terms of a microscopic bipolaron model. The amplitude mode theory of HOROVITZ /32/ is combined with the lattice relaxation approach to calculate the Raman spectra. A reasonably good agreement with experiments has been achieved without adjustable parameters. The Huang-Rhys factors are calculated from the analytic solutions, while other parameters are determined directly from the observed IR and Raman frequencies.

6. Concluding Remarks

To summarize, the lattice relaxation approach is a natural framework to study the soliton and polaron dynamics in conducting polymers. The results obtained so far for both radiative and nonradiative processes as well as for Raman spectra are in good agreement with numerical and laboratory experiments. This method has also been used by other research groups to discuss various problems in conducting polymers, like luminescence /14/, interchain polaron tunneling /33/ and so on. Other problems currently under consideration include nonlinear optical effects, two-photon excitation of breathers and interplay of impurity states with soliton and polaron excitations.

7. References

1. W.P. Su, J.R. Schrieffer, and A.J. Heeger, Phys. Rev. Lett. **42**, 1698 (1979); Phys. Rev. **B22**, 2099 (1980).
2. M.J. Rice, Phys. Lett. **71A**, 152 (1979).
3. See, e.g.,Yu Lu, *Solitons and Polarons in Conducting Polymers*, to be published by World Scientific Publ. Co., Singapore.
4. S.A. Brazovskii & N. Kirova, Sov. Phys. JETP Lett. **33**, 6 (1981); D. K. Campbell & A.R. Bishop, Phys. Rev. **B24**, 4859 (1981).
5. H.Takayama, Y.R. Lin-Liu, and K. Maki, Phys. Rev. **B21**, 2388 (1980); S.A. Brazovskii, Sov. Phys. JETP **51**, 342 (1980).
6. Su Zhao-bin & Yu Lu, Commun. Theor. Phys. **2**, 1203, 1323, 1341 (1983).
7. W.P. Su & J.R. Schrieffer, Proc. Natl. Acad. Sci. (USA), **77**, 5526 (1980).

8. J.P. Sethna & S. Kivelson, Phys. Rev. **B26**, 3513 (1982); See also, A. Auerbach & S. Kivelson, Phys. Rev. **B33**, 8171 (1986).
9. Z.B. Su & L. Yu, Phys. Rev. **B27**, 5199 (1983), **B29** (E), 2309 (1984); Mol. Cryst. & Liq. Cryst. **117**, 219 (1985).
10. K. Huang & A. Rhys, Proc. Roy. Soc. **A204**, 406 (1950).
11. See, e.g., M.L. Goldberg & K.M. Watson, *Collision Theory*, John Wiley & Sons, N.Y., 1964.
12. L.D. Landau & E.M. Lifshitz, *Quantum Mechanics*, 3rd. ed. Pergamon, 1977, § 90.
13. R.C. Ball, W.P. Su, and J.R. Schrieffer, J. de Phys. (Paris) **44**, C3-429 (1983).
14. P.R. Danielsen & R.C. Ball, J. de Phys. (Paris) **46**, 1611 (1985).
15. S. Kivelson & W.K. Wu, Phys. Rev. **B34**, 5423 (1986).
16. Z.B. Su & L.Yu, Phys. Rev. **B35**, 3020 (1987).
17. S. Etemad, T. Mitani, M. Ozaki, T.C. Chung, A.J. Heeger, and A.G. MacDiarmid, Solid State Commun. **40**, 75 (1981).
18. G.B. Blanchet, C.R. Fincher, and A.J. Heeger, Phys. Rev. Lett. **51**, 2132 (1983).
19. M. Sinclair, D. Moses, and A.J. Heeger, Solid State Commun. **59**, 343 (1986).
20. L.R. Rothberg, T.M. Jedju, S. Etemad, and G.L. Baker, Phys. Rev. Lett. **57**, 3229 (1986) and to be published in Phys. Rev. **B**.
21. J.D. Flood & A.J. Heeger, Phys. Rev. **B28**, 2356 (1983).
22. F. Moraes, Y.W. Park, and A.J. Heeger, Synth. Met. **13**, 113 (1986); J. Orenstein, Invited Talk at ICSM'84, Abano Terme, Italy, 1984.
23. C.G. Levey, D.V. Lang, S. Etemad, G.L. Baker, and J. Orenstein, Synth. Met. **17**, 569 (1987).
24. W.P. Su, Phys. Rev. **B34**, 2988 (1986).
25. B.R. Weinberger, C.B. Roxlo, S. Etemad, G.L. Baker, and J. Orenstein, Phys. Rev. Lett. **53**, 86 (1984).
26. E.J. Mele, Synth. Met. **9**, 207 (1984).
27. L.S. Lichtmann, E.A. Imhoff, A. Sarhangi, and D.B. Fitchen, J. Chem. Phys. **81**, 168 (1984).
28. L. Piseri, R. Tubino, and E. Mulazzi, Mol. Cryst. & Liq. Cryst. **83**, 135 (1982); R. Tubino, L. Piseri, G. Dellepiane, J.L. Birman, and U. Pedretti, Solid State Commun. **49**, 161 (1984).
29. W. Siebrand & M.Z. Zgierski, J. Chem. Phys. **81**, 185 (1984).
30. E. Mulazzi, in *Polydiacetylene: Synthesis, Structure and Electronic Properties,* ed. by D. Bloor & R.R. Chance, Martinus Nijhoff, Dordrecht, 1985, P.233.
31. Guo Youjiang & Yu Lu, Solid State Commun. **58**, 411 (1986).
32. B. Horovitz, Solid State Commun. **41**, 729 (1982).
33. S. Jeyadev & J.R. Schrieffer, Phys. Rev. **B30**, 3620 (1984).

Part VI

Polyaniline

Photoelectrochemistry of Polyaniline

Mei Xiang Wan[1], *A.G. MacDiarmid*[1], *and A.J. Epstein*[2]

[1]Department of Chemistry, University of Pennsylvania,
  Philadelphia, PA 19104, USA
[2]Department of Physics and Department of Chemistry,
  Ohio State University, Columbus, OH 43210, USA

1. INTRODUCTION

We have proposed (1) that the intense dark blue base form of the emeraldine
oxidation state of polyaniline has the constitution represented by

(y = 0.5) being composed of equal

numbers of oxidized, —◯—N=◯=N— and reduced, —◯—N—◯—N— groups.

This structure has recently been confirmed by a step-wise synthesis of the
polymer (2). Reduction of the polymer ultimately results in the formation
of the pale yellow leucoemeraldine oxidation state (y = 1); oxidation
results ultimately in the formation of the unstable intense violet pernigraniline oxidation state (y = 0). Protonation of the emeraldine base in
~ 1M aqueous HCl (pH = 0) results in an increase in conductivity by a
factor of ~ 10^{10} to a conductivity of ~ 5 S/cm^2 with the formation of
the emeraldine hydrochloride salt,

, which we believe is actually the

polysemiquinone radical cation,

As the pH is increased from 0 to ~ 6, the extent of protonation decreases,
resulting in the presence of an increasingly greater number of base repeat
units in the polymer with final reformation of emeraldine base (y = 0.5).

It is well established (3) that NN' diphenyl-p-phenylene-diamine,

, the model monomer corresponding to a reduced group

in the polyanilines, undergoes photo-oxidation in an ethanol/acetic acid
glass at liquid nitrogen temperatures with the formation of a blue color.
The spectrum of the new species is that of the semiquinone of the amine-
imine, viz.,

$$+ e ----- (1)$$

the ejected electron being trapped in the solvent (3).

The present study was carried out in order to ascertain if forms of
polyaniline of known constitution, particularly the emeraldine and leucoemeraldine oxidation states, both of which contain reduced repeat units
would undergo analogous photo-oxidation in the solid state. In order to
study this effect, dissolved oxygen in the electrolyte was used in this
study as the "trapping" agent for the ejected electron in an electrochemical

cell configuration. Previous studies using unknown forms of polyaniline have exhibited pronounced photoelectrochemical effects. The very rapid photo response and decay time (300 μsecs) in one such study has been attributed to thermal effects (4). In a second study, forms of polyaniline which were not clearly defined exhibited photo responses in aqueous electrolytes under a variety of conditions and behaved as p-type semiconductors (5).

2. RESULTS AND DISCUSSION
2.1 Synthesis of Polyaniline
All potentials given in this report are vs. an SCE reference electrode. The polyaniline film was electrochemically polymerized on a platinum foil electrode from a solution (pH ~ 2.5) consisting of 2 ml. of aniline and 50 ml of 1M HCl by cycling ~ 50 times between −0.20V and + 0.65V at a scan rate of 50 mv/sec. The film, which was removed at a potential of +0.43V was dark green, characteristic of emeraldine salts. It was washed with HCl (pH ~ 2.5) and its cyclic voltammogram (Fig. 1) was recorded at pH = 2.5 at a scan rate of 50 mv/sec. It was identical to that of polyaniline of the type described above at this pH (6). Moreover, the color-changes of the film on cycling from −0.2V to +0.8V (pale yellow – green – blue) were also characteristic of polyaniline. Sometimes the polyaniline film was synthesized at a constant potential of 0.65V. The cyclic voltammograms of such films were identical to those obtained by the recycling procedure above. They differed greatly from polyaniline synthesized electrochemically according to the procedure used for the previous photoelectrochemical study (5) in an aqueous solution (pH = 6) consisting of 0.1M aniline and 0.2M $LiClO_4$ at a constant potential of 1.0V (vs Ag/AgCl i.e. 0.96V vs SCE) which gave a black film. Its cyclic voltammagram recorded in HCl (pH ~ 2.5) at a scan rate of 50 mv/sec is also given in Fig. 1. It is clearly apparent that the two forms of polyaniline are different. This difference was further supported by the observation that the film did not change color on cycling between −0.2V and +1.0V.

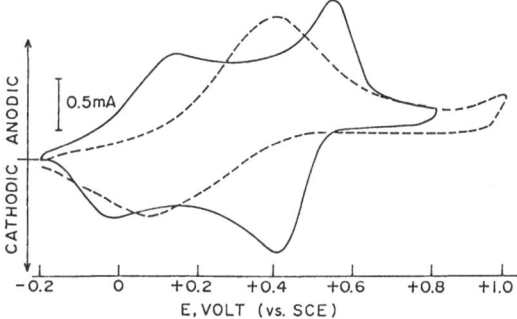

Figure 1. Cyclic voltammograms of different forms of polyaniline. This work ——— ; ---- ref. (5).

2.2 Effect of Oxygen on Dark Current
The polyaniline film was held at a potential of +0.43V in 1.0M HCl (pH = 0) for 2 hours to bring it to the approximate emeraldine oxidation state (6). It was then washed several times with a 0.1M $ZnSO_4$ solution (pH = 5.1) and was placed in a stoppered cell containing 0.1M $ZnSO_4$ and a zinc electrode. The zinc potential of the Zn^{+2}/Zn couple is −1.0 V. Electrons were found to pass from the zinc to the polyaniline in an external circuit (incorporating an ammeter) joining the two electrodes. The current was

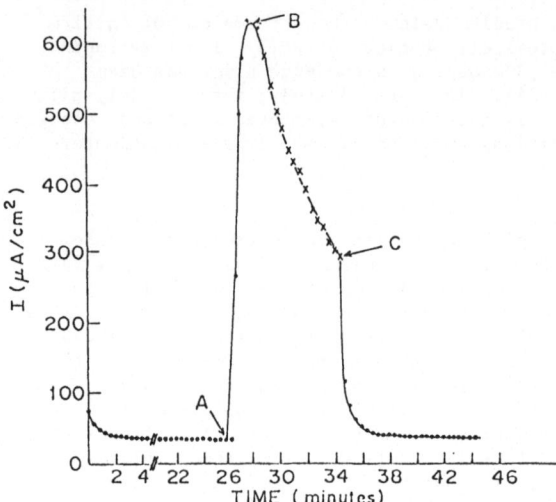

Figure 2. Effect of O_2 and argon on the dark current of polyaniline in an electrochemical cell configuration, (see text).

measured while argon was bubbled through the stirred solution in the dark. At point A (Fig. 2), O_2 was bubbled through the electrolyte. As can be seen an immediate increase in current occurred. The passage of O_2 was discontinued at point B. The current decreased to essentially the original value when argon was again passed through the electrolyte (point C).

At this pH (5.1), the emeraldine base is only very slightly protonated and hence is a poor conductor (6). Even so, some electrochemical reduction of the oxidized repeat units by the zinc occurs viz.,

$$Zn \rightarrow Zn^{+2} + 2e \quad - \quad (2)$$

resulting in the net reaction

Since the reduced repeat groups in forms of polyaniline more reduced than emeraldine (i.e. $y > 0.5$) are oxidized rapidly by oxygen

it is believed that the small background current is caused by the continuous electrochemical reduction of the polyaniline (Eqs.3, 4) followed by its chemical re-oxidation (Eq. 5) by traces of residual oxygen dissolved in the electrolyte. This is consistent with the large changes in current on passing oxygen, then argon through the electolyte.

2.3 Effect of Electrode Temperature on Dark Current

In order to ascertain the effect of electrode temperature on the dark current, a similar polyaniline film to that employed in the previous experiment was pretreated at +0.43V as before and a steady-state dark current was obtained as described above at an electolyte temperature of 24°C. The platinum/ polyaniline electrode/zinc electrode assembly was then transferred to a cell containing the electrolyte at 0°C for 5 minutes in order to cool the electrode. It was then rapidly transferred back to the 24°C electrolyte. Since all electrical connections were already in place, a very rapid current reading could readily be made. The immediate current reading was ~ 80 μA greater than the previous steady-state background current. It is believed that this increase was caused by a small amount of air oxidation of the partly reduced emeraldine polymer during exposure to air during the ~ 3 second transfer time between the two electrolytes. The current decreased to the original steady-state dark current within ~ 52 seconds. It was therefore concluded that any possible <u>decrease</u> in current caused by a decrease in the temperature of the polyaniline electrode occurred in <~ 1 minute i.e. thermal equilibration of the platinum foil/polyaniline electrode with the 24°C electrolyte took <~ 1 minute.

2.4 Photoactivity of Polyaniline
A. In a constantly decreasing oxidation state
A different piece of polyaniline film was synthesized by the recycling process and was then held at +0.43V in one 1M HCl for 2 hours to convert it to the approximate emeraldine oxidation state. After washing in 0.1M $ZnSO_4$ solution it was placed in a 0.1M $ZnSO_4$ aqueous electrolyte (whose pH was adjusted to ~ 3 by dilute H_2SO_4) together with a zinc counter electrode. The dark current and effect of light (300W tungsten unfiltered projector light source) is shown in Fig. 3. The behavior can be decomposed into the sum of two components: (i) a decay of the background current over 25 minutes during which time the oxidation state of the polyaniline changes from that of emeraldine (y = 0.5) to leucoemeraldine (y = 1; pale yellow with a characteristic open circuit potential of - 0.4 V) <u>and</u> (ii) a photo-response term associated with the reduced repeat units (see below). The reduction of the oxidized groups in the emeraldine initially placed in the cell to ultimately give leucoemeraldine is proposed to follow (Eqs. 2-4). Note that in the more acidic solutions the protons in Eqs. 3 and 4 will actually be attached to the oxidized repeat units prior to reduction by the electrons. The approximate intermediate oxidation states ("y" values)

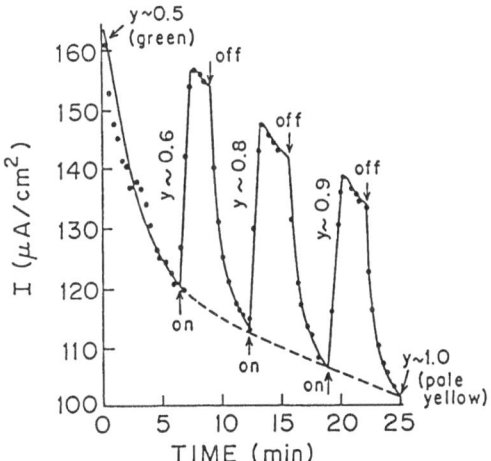

Figure 3. Photocurrent of polyaniline as a function of decreasing oxidation state at pH = ~ 3 in an electrochemical cell configuration.

can be estimated from the coulombs passed in the changing background current as a function of time. The photocurrent decreases slightly as the polyaniline becomes more reduced, consistent with the known decreasing conductivity of the polymer in its more reduced state (7). The small photo-voltages (~ 50 μV) observed under all conditions suggests that band bending at the polymer-electrolyte interface is small.

It seems unlikely that the increase in current on exposure to light is caused by heating of the thin polyaniline film on the thin platinum foil electrode in view of the long time (~ 3 minutes) required for the photo-current to decrease to the dark current value. This conclusion is supported by the observation described in the previous section in which it was found that thermal equilibrium of the electrode assembly with the electrolyte took $< \sim 1$ minute.

The above study was repeated at different pH values in 0.1M ZnSO$_4$ by the addition of appropriate amounts of dilute H$_2$SO$_4$. The photocurrents increased from a minimum of ~ 1 μA/cm^2 at pH ~ 1 to a maximum of ~ 32 μA/cm^2 at pH ~ 3 and then decreased to ~ 12 μA/cm^2 at pH ~ 5, the highest pH attainable without incipient precipitation of Zn(OH)$_2$. The maximum at pH ~ 3 is ascribed as being due to an optimization of opposing effects, viz., a simultaneous increase in conductivity (reduction of internal resistance of the cell) and the concommitant reduction in the volume fraction of the semiconductor (nonprotonated) form as an increasing fraction of the polymer is converted to the metallic state with decreasing pH (8).

B. In the completely reduced oxidation state

In order to study the photoactivity of the polyaniline in a known oxidation state, a (thicker) film of the polymer was deposited at constant potential (pH ~ 2.5) and was then reduced electrochemically in a 0.1M ZnSO$_4$ electrolyte (pH = 3.2) by zinc as before (Eqs. 2-4). When a steady-state dark current was attained, the polymer was pale yellow and had an open circuit potential of -0.4 V, both characteristic of the leucoemeraldine oxidation state (y = 1). Exposure to the light source resulted in photocurrents > $\sim$100 μA/cm^2, (Fig. 4). It seems unlikely, as discussed earlier, that the long decay time (~ 4 minutes) of the current in the dark is caused by a cooling of the electrode. We

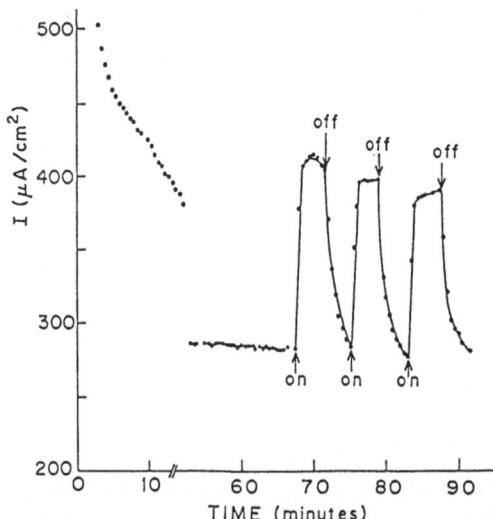

Figure 4. Photocurrent of the leucoemeraldine (y = 1) oxidation state of polyaniline at pH = 3.2 in an electrochemical cell configuration.

220

believe it is a diffusion-controlled effect caused by the reduction of the photo-oxidized polymer by the zinc as summarized below. When similar studies were performed at a pH = 5.1, much smaller photocurrents (12 μA/cm^2) were observed consistent with a reduction in the conductivity of the polymer due to smaller protonation. The proposed processes occurring are summarized below:

At Polyaniline Electrode
(i) Photo-oxidation

$$2e(\text{ejected}) + 0.5O_2 + H_2O \rightarrow 2(OH)^- \quad - - - - - - - - - - - \quad (7)$$

The rate of oxidation of reduced groups by O_2 occurs slowly in the dark but more rapidly in the presence of light.

(ii) Electrochemical reduction by electrons from zinc.

At Zinc Electrode

$$Zn \rightarrow Zn^{+2} + 2e^- \quad . - \quad (9)$$

The Zn^{+2} ions (Eq. 9) will react with the $(OH)^-$ ions (Eq.7), viz.,

$$Zn^{+2} + 2(OH)^- \rightarrow Zn(OH)_2 \quad . - - - - - - - - - - - - - - - - - - \quad (10)$$

In the more acidic electrolytes, the $Zn(OH)_2$ will be converted to $ZnSO_4$ by the free H_2SO_4.

3. CONCLUSIONS
Polyaniline of the type synthesized in the present investigation undergoes photo-oxidation under the experimental conditions employed and exhibits a pronounced photocurrent in an electrochemical cell configuration. It is concluded that the reduced repeat units of which, for example, the leuco-emeraldine (y = 1) oxidation state is solely comprised are the photoactive groups in all the polyanilines, consistent with the photo studies on the model-reduced monomeric compound, NN' diphenyl-p-phenylene-diamine (3).

4. ACKNOWLEDGEMENT
This work was supported by NSF Grant No. DMR-86-15475.

5. REFERENCES

1. A. G. MacDiarmid, J. C. Chiang, A. F. Richter and A. J. Epstein, Synth. Met., 18, 285 (1987).
2. F. Wudl, R. O. Angus Jr., F. L. Lu, P. M. Allemand, D. J. Vachon, M. Nowak, Z. X. Liu and A. J. Heeger, J. Amer. Chem. Soc., (1978) in press.
3. H. Linschitz, J. Rennert and T. M. Korn J. Am. Chem. Soc., 76, 5839 (1954); H. Linschitz, M. Ottolenghi and R. Bensasson J. Am. Chem. Soc., 89, 4592 (1967).

4.   E. M. Genies and M. Lapkowski, private communication, (1987).
5.   M. Kaneko and H. Nakamura, J. Chem. Soc. Chem. Commun., 346 (1985);
      H. Nakamura and M. Kaneko, Kobunshi Ronbunshu 43, 677 (1986); Chem.
      Abstr., 105, 236748n (1987). Translation not available.
6.   W.-S. Huang, B. D. Humphrey and A. G. MacDiarmid, J. Chem. Soc., Faraday
      Trans. 1, 82, 2385 (1986).
7.   E. W. Paul, A. J. Ricco and M. S. Wrighton, J. Phys. Chem., 89, 1441 (1985).
8.   J. M. Ginder, A. F. Richter, A. G. MacDiarmid and A. J. Epstein, Solid
      State Commun. 63, 97 (1987).

Redox Mechanisms in Polyaniline Films

E.M. Geniès* and M. Lapkowski**

Electrochimie Moléculaire, Laboratoires de Chimie,
Département de Recherche Fondamentale, Centre d'Etudes
Nucléaire, 85 x, F-38041 Grenoble, France

In situ spectroelectrochemical and EPR studies of polyaniline
(PANi) synthetized in $NH_4F,2.3HF$ were carried out in
association with cyclic voltammetry. The results are discussed
in terms of the redox reaction mechanisms. We propose that the
complete redox mechanism involves four one-electron steps and
two polaron – bipolaron states in PANi films.

I Introduction

Polyaniline (PANi) is one of the most stable conducting
polymers. Due to this fact several applications of this polymer
can be envisioned, the most important ones being associated
with its electrochemical activity. Each electrochemical
application requires however a deep understanding of the redox
reaction mechanisms. The aim of this paper is, therefore, to
discuss several aspects of electron and proton transfer
reactions which have been proposed to explain the
electrochemical behaviour of PANi films.

II Results and discussion

II-1 Overall redox reactions

Cyclic voltammetry curves of a PANi film, obtained in aqueous,
acidic /1/ or in nonaqueous, aprotic media /2/ exhibit two
couples of current peaks (I and II). Figure 1 shows the cyclic
voltammogram obtained in $NH_4F,2.3HF$ (BN) medium, in which the
redox reactions can clearly be observed. The main question to
be answered involves the correlation between the peaks observed
in the cyclic voltammograms and the redox reactions occurring
in the PANi films. In the literature all authors agree that
several forms of polyaniline may exist or co-exist in the
polymer chains /3/, i.e. aromatic rings (.), quinonic rings
(=), free amine (NH), imine (N) and protonated (NH_2^+ and NH^+)
structures. Using the above notation the fully reduced PANi
(polyleucoemeraldine) can be expressed as follows:
.NH.NH.NH.NH.. WOLF et al./4/ and CHANCE et al./5/, on the
basis of electrochemical behaviour of aniline oligomers (dimer
to decamer) and VEH calculations propose that the couple I
corresponds to the oxidation of PANi to polyradicalcation,

$$.NH.NH.NH.NH. \longrightarrow .NH^{\circ+}.NH.NH.NH^{\circ+}. + 2e-,$$

---

* Member of Grenoble University. ** Permament adress: Institut of Inorganic
Chemistry and Technology, Silesian Technical University, Gliwice, Poland.

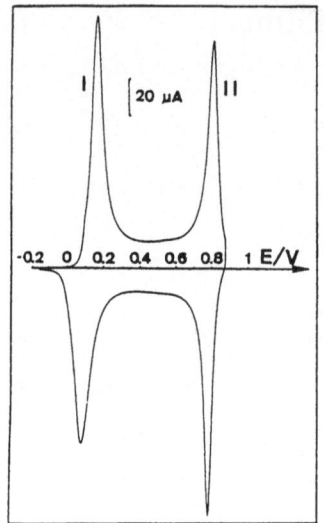

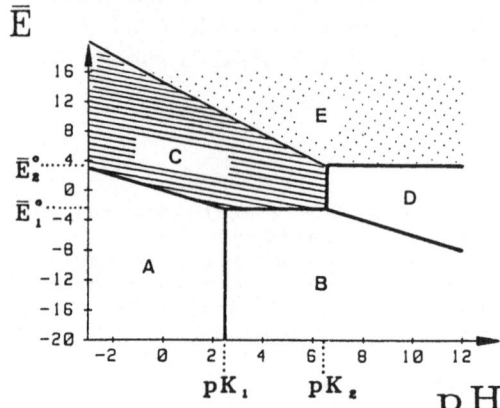

Fig.1.Cyclic voltammetry curve
of the PANi in BN medium.

Fig.2.Potential-pH diagram of
PANi. $E = E(V)/0.06$.

and the couple II is associated with the oxidation to polyimine
(polypernigraniline) with the loss of two electrons and four
protons :

$$.NH^{\bullet+}.NH.NH.NH^{\bullet+}. \longrightarrow .N=N.N=N. +2 \ e- \ + \ 4 \ H^+.$$

This is consistent with the observations of MACDIARMID /6/ who
reported that the potential of couple II changes by 120 mV per
a pH unit. Based on the results obtained, MACDIARMID et al. /6/
propose a different redox mechanism. According to these
authors, in the first step (couple I) the reaction leads to
protonated polyemeralidine whereas in the second step (couple
II), polypernigraniline is formed. Additionally MACDIARMID /7/
presented an experimental determination of the potential-pH
relationship which is consistent with our data shown in Fig. 2.
In particular he obtained the line which corresponds to
electron B to C transfer between pK1 and pK2 and the line
corresponding to C to E transfer with a slope of 60 mV/pH. The
discussed mechanism must be formulated with the following
square scheme:

$$.NH_2^+.NH.NH.NH_2^+. \quad (A)$$

$$pK1 \ \Big\downarrow \ -2H^+$$

$$.NH.NH.NH.NH. \xrightarrow[E1]{-2e-} .NH.NH^+=NH^+.NH. \quad (C)$$

$$(B)$$

$$pK2 \ \Big\downarrow \ -2H^+$$

$$(D) \quad .NH.N=N.NH. \xrightarrow[E2]{-2e-} .NH^+=N.N=NH^+.$$

$$pK2 \ \Big\downarrow \ -2H^+$$

$$.N=N.N=N. \quad (E)$$

Its starting point involves a strong acid medium with partially protonated polyleucoemeraldine.

II-2 **Hyperfine redox reactions**

A more detailed mechanism is needed to explain the results of optical and EPR spectroscopic studies carried out **in situ** in an electrochemical cell. First of all, quantitative chronocoulometric experiments on PANi film indicate that 0.4 electrons per ring are involved for the step I and 0.2 for the step II /8/. It has also been demonstrated by in situ optical and EPR studies /9-11/ that the oxidation of some conducting polymers, such as for example polypyrrole, generates initially a polaron state, which for higher charges is then transformed into a bipolaron state. In the case of polypyrrole these two steps overlap in the cyclic voltammetric peak. In the present case we observe them twice, separately for each current peak, I and II. Consequently, it is necessary to propose a new redox mechanism for PANi. In a strong acid medium, for the couple of current peaks I, it can be written as

$.NH_2^+.NH.NH.NH_2^+.$ (A)

$pK1 \downarrow -2H^+$

                           1st polaron   1st bipolaron

$.NH.NH.NH.NH. \xrightarrow[E1]{-e-} .NH.NH^{\circ+}.NH.NH. \xrightarrow[E'1]{-e-} .NH.NH^+=NH^+.NH.$

    (B)   (C)

For the couple corresponding to the current peak II it can be expressed as

$.NH.NH^+=NH^+.NH.$ (C)

$Pk2 \downarrow -2H^+$

                           2nd polaron   2nd bipolaron

$.NH.N=N.NH. \xrightarrow[E2]{-e-} .NH.N^{\circ+}=N.NH. \xrightarrow[E'2]{-e-} .NH.N^+=N^+.NH.$

    (D)   (E')

This is in agreement with the potential evolution of 120 mV per pH unit for the transfer II, because the transformation D into E' as observed by chronocoulometry involves half as many electrons as the B to C transfer.
Figure 3 shows the cyclic voltammetry curve of a PANi film at 10mV/s together with the **in situ** optical absorbance at 406, 543 and 821 nm. For each redox couple (I and II), it is observed that the absorption at 821 nm initially grows and then disappears. By analogy with polypyrrole this behaviour can be interpreted in terms of the formation and annihilation of the 1st and 2nd polarons. The shape of the voltabsorptiometric curve indicates that polarons and bipolarons have different molecular extinction coefficients. In addition such experiments indicate that the redox reactions corresponding to the 1st and the 2nd peaks have different kinetic behaviour. For instance the latter is much faster than the former.

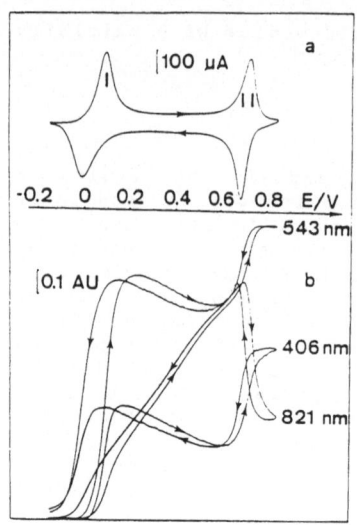

Fig.3.Cyclic voltammetry and voltabsorptiometric curves of a PANi film at 10 mV/sec.

Fig.4.Cyclic voltammetry and number of spins of a PANi film at 1 mV/sec.

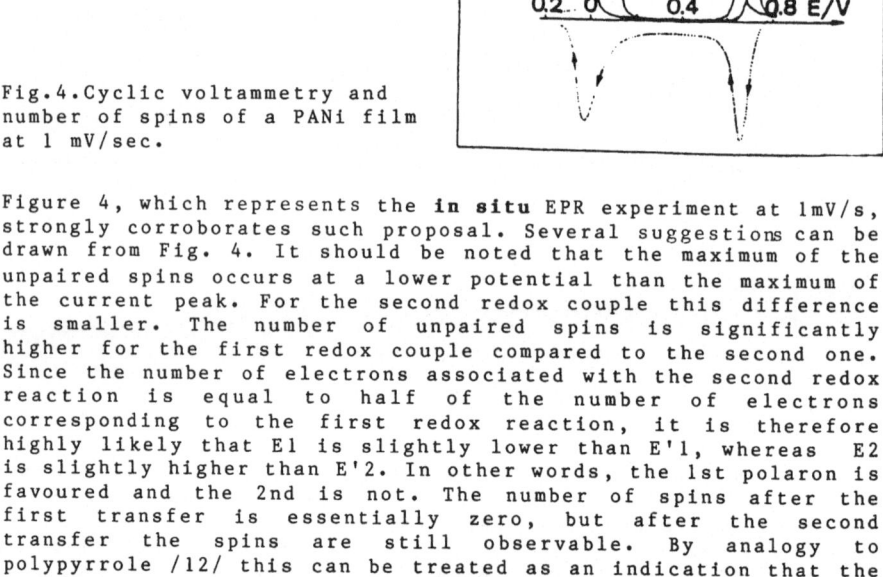

Figure 4, which represents the **in situ** EPR experiment at 1mV/s, strongly corroborates such proposal. Several suggestions can be drawn from Fig. 4. It should be noted that the maximum of the unpaired spins occurs at a lower potential than the maximum of the current peak. For the second redox couple this difference is smaller. The number of unpaired spins is significantly higher for the first redox couple compared to the second one. Since the number of electrons associated with the second redox reaction is equal to half of the number of electrons corresponding to the first redox reaction, it is therefore highly likely that E1 is slightly lower than E'1, whereas E2 is slightly higher than E'2. In other words, the 1st polaron is favoured and the 2nd is not. The number of spins after the first transfer is essentially zero, but after the second transfer the spins are still observable. By analogy to polypyrrole /12/ this can be treated as an indication that the electrostatic interaction is stronger for the second transfer than for the first one.

In optical and EPR experiments we observe a hysteresis after the inversion of the potential. This may be associated, in part, with a noncompensated ohmic drop in the electrochemical cell, but more likely with the limitation imposed by the rate of electron and proton transfers. This rate is determined by the diffusion properties of the ions within the polymer matrix and by interaction between the charges of the polymer chain and the anions.

III   Conclusion

In situ experiments involving optical and EPR spectroscopies were found to be powerful tools for the elucidation of the complex electrochemical behaviour of PANi. Very recently an interesting approach to the spectroelectrochemical determination of the potential-pH relationship on PANi was presented by CUSHMAN et al./13/. In the experiments presented here, it seems clear that the redox process of the two oxidation steps which occur in PANi are rather different. In addition, KUZMANY et al./14/ observe, in organic medium, by **in situ** FTIR, that in the first step the anions (ClO_4^-) remain in the polymer during both reduction and oxidation, but for the second step the anions (BF_4^-) are inserted during oxidation and diffuse outside during the reduction as for a classical conducting polymer. KITANI et al./15/ have also shown by chemical analysis that the number of anions inserted during the second redox transfer is roughly half the number of anions inserted in the first electron transfer. This observation is in agreement with our chronocoulometric data /2/.

References

1. E.W. Paul, A.J. Ricco, M.S. Wrighton: J. Phys. Chem., <u>89</u>, 1441 (1985)
2. E.M. Genies, C. Tsintavis: J. Electroanal. Chem., <u>195</u>, 109 (1985)
3. J-C. Chiang, A.G. MacDiarmid: Synt. Met. <u>13</u>, 193 (1986)
4. J.F. Wolf, S. Gould, L.W. Shacklette: Abstract INOR 310, 192nd A.C.S. Meeting, Anaheim, CA, Sept. 7-12 (1986)
5. R.R. Chance, D.S. Boudreaux, J.F. Wolf, L.W. Shacklette, R. Silbey, B. Themans, J.M. Andre, J.L. Bredas: Synt. Met., 15, 105 (1986)
6. W-S. Huang, B.D. Humphrey, A.G. MacDiarmid: J. Chem. Soc. Faraday Trans.1, <u>82</u>, 2385 (1986)
7. A.G. MacDiarmid: this conference.
8. E.M. Genies, C. Tsintavis: J. Electroanal. Chem., <u>200</u>, 127 (1986)
9. J.H. Kaufman, N. Colaneri, J.C. Scott, G.B. Street: Phys. Rev. <u>B30</u>, 1023 (1984)
10. E.M. Genies, J.M. Pernaut: J. Electroanal. Chem., <u>191</u>, 111 (1985)
11. F. Genoud, M. Guglielmi, M. Nechtschein, E. Genies, M. Salmon: Phys. Rev. Lett., <u>55</u>, 118 (1985)
12. M. Nechtschein, F. Devreux, F. Genoud, E. Vieil, J.M. Pernaut, E. Genies: Synt. Met., <u>15</u>, 59 (1986)
13. R.J. Cushman, P.M. McManus, S.C. Yang: J. Electroanal. Chem., <u>219</u>, 335 (1986)
14. N.S. Sariciftci, H. Neugebauer, H. Kuzmany: this conference.
15. A. Kitani, J. Izumi, J. Yano, Y. Hiromoto, K. Sasaki: Bull. Chem. Soc. Jpn., <u>57</u>, 2254 (1984)

In situ FTIR Spectroscopy of Polyaniline

N.S. Sariciftci[1], H. Neugebauer[2], H. Kuzmany[1], and A. Neckel[2]

[1]Institut für Festkörperphysik, Universität Wien,
Strudlhofgasse 4, A-1090 Wien, Austria and
Ludwig Boltzmann Institut für Festkörperphysik, A-1090 Wien, Austria
[2]Institut für Physikalische Chemie, Universität Wien,
Währingerstr. 42, A-1090 Wien, Austria

1 Abstract

The infrared absorption of electrochemically synthesized polyaniline
is measured in situ during the electrochemical redox processes in
non aqeous electrolytes with Fourier Transform (FTIR) technique. The
first oxidation process is associated with a strong increase in the
background absorption in addition to vibrational changes due to the
benzenoid-semiquinoid transition in the ring structure and loss of
protons bound to the nitrogen atoms. In the second oxidation process
intercalation of anions occurs, the background absorption decreases
and quinoid ring bands appear. From these results and neutron activation analysis of anions in the polymer we derived a structural
model of polyaniline, which explains the changes in the two redox
processes.

2 Introduction

Polyaniline (PANI) is an interesting conducting polymer because its
physical properties can be changed both by electrochemical redox
processes and by changing the pH-value of the medium. These properties and the unique proton mechanism of doping were recently investigated by many authors /1-3/. In this work we present results of
in situ FTIR measurements during the electrochemical oxidation and
reduction processes of PANI. The IR spectra were recorded with FTIR
technique simultaneously during sweeping the electrochemical potential of the PANI working electrode. The results are interpreted in a
structural model which explains the changes during the electrochemical redox reactions.

3 Experimental

The samples used in the experiments were synthesized on platinum
disc electrodes in an acidic solution of aniline (1 M $HClO_4$ with 0.1
M $NaClO_4$ and aniline) by sweeping the potential between -200 mV and
+800 mV versus SCE. The samples were then very well washed in distilled water and dried under vacuum. The electrolyte for the IR
measurements was acetonitrile with 0.5 M $LiBF_4$. In situ IR observations were performed with external reflection method using a thin
layer IR cell /4/ and a NICOLET 60 SX FTIR spectrometer. The study
of the changes of the spectra relative to a particular state of the
polymer proved to be very useful for in situ FTIR spectroscopy /5/.
The electrochemical equipment consisted of a programmable PRODIS
function generator and a JAISSLE potentiostat for the conventional
three-electrode system.

4 Results

Figure 1 displays the spectral changes during the first oxidation process relative to the fully reduced state of the polymer. A strong increase in the background absorption centered at 4000 cm^{-1} (0.5 eV) is the dominating feature, whereas vibrational changes around 3300 cm^{-1} and below 1800 cm^{-1} are also observable (Fig 1a). In addition to these bands sharp features from electrolyte incompensation appear. The absorption at 3380 cm^{-1} lags behind the background, which indicates a decrease of oscillator strength for this vibration during the first oxidation process. A decrease of absorption at 1510 cm^{-1} and an increase of absorption at 1566 cm^{-1} is observed in Fig. 1b.

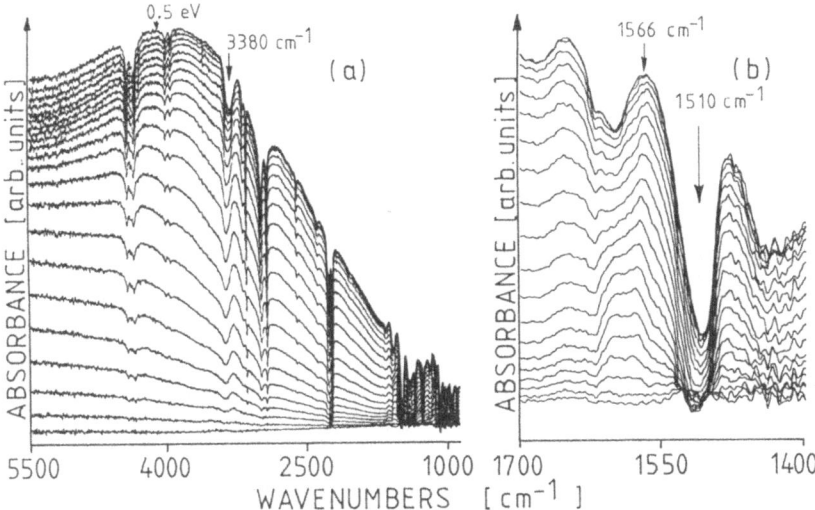

Fig. 1: Spectral changes during the first oxidation process of PANI relative to its fully reduced state (a) and a section of the spectrum in an extended scale (b)

Figure 2 shows the changes in the IR spectra during the second oxidation process relative to the first oxidized state of PANI. The strong background absorption around 4000 cm^{-1} is bleached and a strong increase of absorption intensities in the 1050 cm^{-1} region is observed (Fig. 2a). Among others Fig. 2b resolves characteristic changes of intensities at 1629 cm^{-1} and 1566 cm^{-1}.

5 Discussion

From its position and shape the broad band around 0.5 eV is considered to originate from an electronic transition. According to very recent band structure calculations /6,7/ it can be correlated to the intraband transition in the half-filled polaron band of the polaron lattice structure. This interpretation is consistent with the Pauli susceptibility observed in this material /8/. The band at 3380 cm^{-1} corresponds to the N-H vibrations of PANI /9/. The loss of absorp-

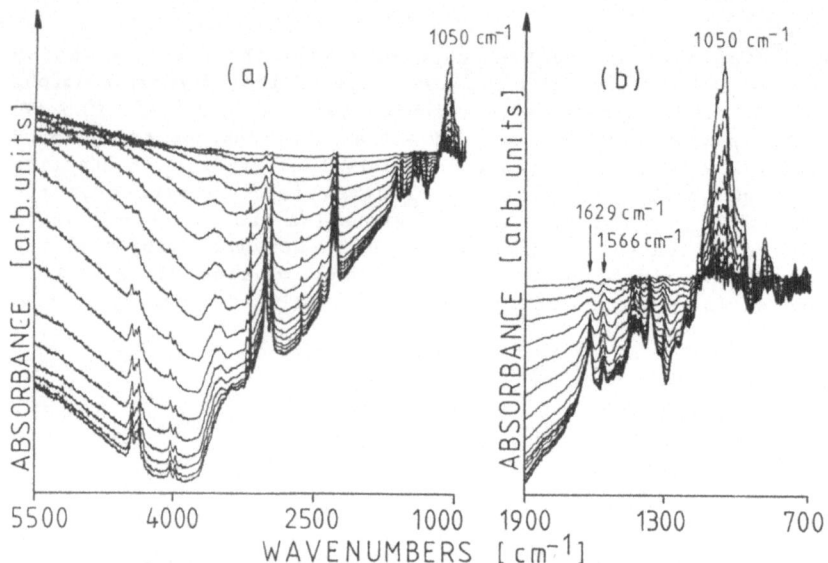

Fig. 2: Spectral changes during the second oxidation process of PANI relative to its first oxidized state (a) and a section of the spectra in an extended scale (b)

tion at this energy indicates the activation of a protonic mechanism in the first oxidation step. From the sign of the change a loss of protons bound to nitrogen atoms is assumed. In Fig. 1b the strong loss of the vibrational absorption intensity at 1510 cm^{-1} indicates the loss of benzenoid character of the aromatic ring structure in this first oxidation step. Since the formation of N=ring vibrations does not appear, the results support the assumption of a semiquinoid polaron lattice structure in the first oxidized state of PANI /8/. In the second oxidation step we observe the decrease of the free carrier absorption, which indicates the loss of the delocalization of electrons. The strong increase in the intensity of vibrational absorption at 1050 cm^{-1} corresponds to the intercalation of anions from the electrolyte due to a change in the charge of the polymer chain. This mechanism is in contrast to the first oxidation process, where no intercalation is observed. The formation of N=ring vibrations at 1629 cm^{-1} indicates the transformation of the semiquinoid polaron lattice into a real quinoid structure in this second step of oxidation. All the observed features are reversible in the corresponding reduction processes, except for the anion de-intercalation, which lags behind.

6 Summary

From FTIR measurements we found that in the first step of oxidation of PANI a protonic mechanism, whereas in the second step of oxidation an anion intercalation process, is activated. These individual processes are accompanied by a benzenoid-semiquinoid and a semiquinoid-quinoid transition, respectively. This model is in agreement with results of Raman and optical absorption spectroscopy /10,11/ and neutron activation analysis.

7 Acknowledgement

This work is supported by the "Fonds zur Förderung der wissenschaftlichen Forschung" in Austria (Proj. No. 5735).
H. N. and A. N. wish to acknowledge the financial support by the "Jubiläumsfonds der Österreichischen Nationalbank" (Proj. No. 2505). We want to thank G. Grass for the neutron activation analysis and A. G. MacDiarmid for valuable discussions during the conference.

8 Literature

1. E. W. Paul, A. J. Ricco, M. S. Wrighton: J. Phys. Chem. $\underline{89}$, 1441 (1985)
2. W. S. Huang, B. D. Humprey, A. G. MacDiarmid: J. Chem. Soc. Faraday Trans. $\underline{1}$, 82 (1986)
3. W. R. Salaneck, I. Lundström, T. Hjertberg, C. B. Duke,, E. Conwell, A. Paton, A. G. MacDiarmid, N. L. D. Somasiri, W. S. Huang, A. F. Richter: Synth. Metals $\underline{18}$, 291 (1987)
4. A. Bewick: In Trends in Interfacial Electrochemistry, ed. by A. F. Silva (D. Reidel Publ., 1986) p. 331
5. H. Neugebauer, A. Neckel, N. Brinda-Konopik: In Springer Series in Solid State Sci., ed. by H. Kuzmany, M. Mehring, S. Roth, Vol. 63 (Springer, Berlin, Heidelberg 1985) p. 227
6. S. Stafström: this volume
7. J. L. Bredas: this volume
8. A. J. Epstein, J. M. Ginder, F. Zuo, H. S. Woo, D. D. Tanner, A. F. Richter, M. Angelopoulos, W. S. Huang, A. G. MacDiarmid: Synth. Metals, in press
9. Yong Cao, Suzhen Li, Zhijiann Xue, Ding Guo: Synth. Metals $\underline{16}$, 305 (1986)
10. H. Kuzmany, N. S. Sariciftci: Synth. Metals $\underline{18}$, 353 (1987)
11. N. S. Sariciftci, H. Kuzmany: Synth. Metals, in press

Spectroscopic Investigation of Polyaniline

A.P. Monkman and D. Bloor

Department of Physics, Queen Mary College, Mile End Road,
London E14NS, United Kingdom

1. INTRODUCTION

Polyaniline (PANi) is being intensively studied and in consequence is now
a much better characterised system. With this in mind, we have been studying spectra of PANi in the UV/visible/NIR region, both in situ and as dried
films in air. These results are compared with the latest calculations of
the band structure of PANi.

Previously published spectra of PANi /1-3/ have been restricted to the
visible region. We have observed interesting new features in the UV and
NIR regions which we report in this communication.

2. EXPERIMENTAL

As reported previously /1/ the PANi samples were prepared by a single electrochemical route, since it is crucial to synthesise films which are as identical as possible so that results can be compared unambiguously. All
films were grown in an electrolyte consisting of 200ml H_2O (triply distilled) 50ml HCl (10M) containing 5g of Aniline at 0.72V vs S.C.E. ref electrode on both Pt substrates and semi transparent Pt/Quartz electrodes.
Films were then electrochemically switched at -0.2V vs S.C.E. to produce
reduced films and at 0.8V vs S.C.E. to produce oxidised films. The characteristic "polyemeraldine" form of PANi was obtained at 0.4V vs S.C.E.
The films which were grown on Pt substrates were then left to dry in air.
This resulted in some partial oxidation of the reduced sample.

To measure the absorption spectra in situ, a special electrochemical
cell was made, and for reflection measurements, the films on Pt substrates
were used. All measurements were made using a Perkin Elmer Lamda 9 spectrophotometer, and data were smoothed using a Savipzcy/Golay smoothing function routine.

3. RESULTS

The in situ results for PANi, Fig.1, show the formation of a new absorption, at about 11,750 cm^{-1} and as the oxidation potential is increased
this band grows in intensity and shifts to higher energy. There is also
evidence for the formation of a small shoulder near 24,100 cm^{-1} which becomes more pronounced as the film is oxidized but then declines again at
high oxidation levels. The large absorption at 29,400 cm^{-1} shows only a
very small blue shift as the film is oxidized.

The range of observation was much greater for the reflection spectra and
significant changes occur as the oxidation level is increased see Fig. 2.

232

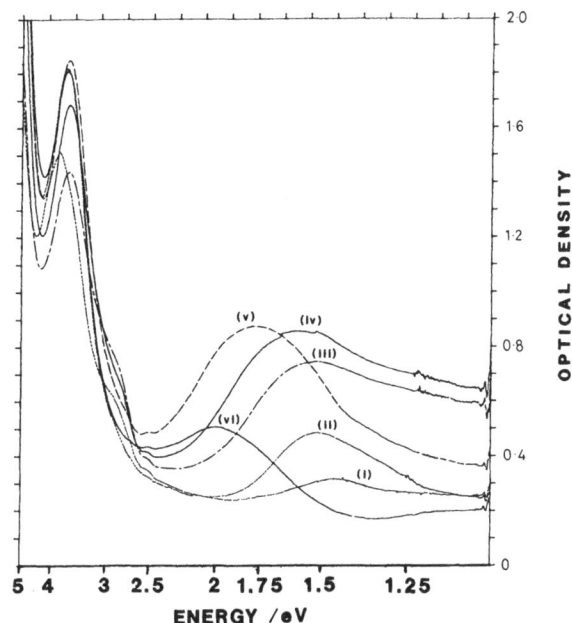

Fig. 1. *In situ* optical spectra of PANi.   Voltages are with respect to a
silver wire reference electrode.   Applied voltages; (i) = -0.2V,
(ii) = 0.4V, (iii) = 0.8V, (iv) = 1.0V, (v) = 1.2V, (vi) = 1.4V

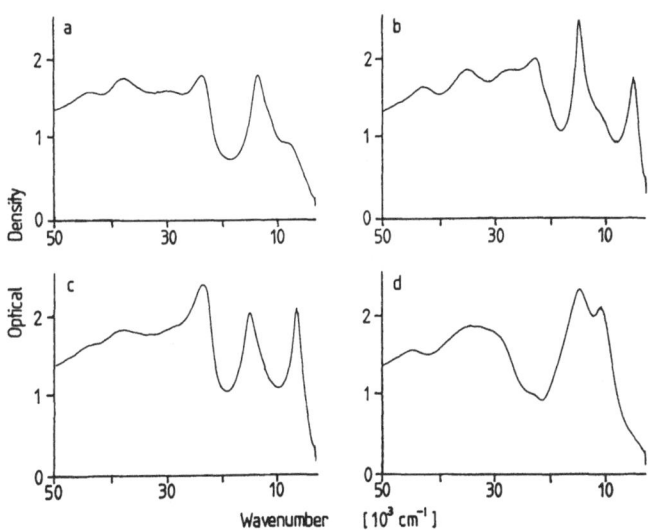

Fig. 2. Reflection spectra of PANi in various oxidation states
a) partial air oxidized "reduced" film, b) and c) green "polyemeraldine",
d) oxidized PANi

In the optical range, one can see that the spectra are very similar to those observed in the $\underline{in}$ $\underline{situ}$ measurements. However, one observes that at both extremes of the spectral range there are absorptions which are strongly dependent on the oxidation state of the PANi. In the near infra-red a band appears weakly at 8,000 cm^{-1} in the partially oxidized film. As oxidation is increased it gains strength and shifts to lower energy. In the "poly-emeraldine" film it occurs at 5,800 cm^{-1} but then as the film is oxidized further this peak shifts back to 11,000 cm^{-1} (Fig. 2(d)). It is then vir-tually coincident with the main visible band which is red shifted in the highly oxidized state. In the ultra-violet many peaks can be seen especi-ally in the polyemeraldine state. These absorptions also display sensiti-vity to the oxidation state.

4. DISCUSSION

The VEH calculations /4/ predict that the lowest unoccupied level is a flat band lying about 4eV above the filled band in the reduced state. The model also predicts a further low lying flat band. The existence of this band is consistent with ultra-violet photo-electron spectra /5/ indicating a strong peak at 6.2eV and a band at 48,200 cm^{-1} in all states of oxidation. The peak at 12,000 cm^{-1} in Fig. 2(a) is attributed to partial oxidation since it is absent from the previous $\underline{in}$ $\underline{situ}$ spectra. In the emeraldine form peaks are observed in the visible at energies close to those predicted for a polaron band, energy gap 0.5eV (4,000 cm^{-1}) and transitions to the Fermi level at 1.8eV (14,500 cm^{-1}) and 2.7eV (22,000 cm^{-1}). The higher lying transitions can be either transitions between flat bands as suggested above or to localised excitonic states. These excitations could well be self trapped excitons /6/ which are localized on 1 to 3 adjacent rings.

5. CONCLUSIONS

Excellent agreement is found between our measured data and VEH band calcula-tions for PANi. Thus we concluded that there is substantial evidence for the formation of a Polaron conduction band in PANi, and also the existence of localized excitons on the chain.

6. ACKNOWLEDGEMENTS

The SERC is thanked for providing a studentship to APM. G.E.C. Research Ltd. are thanked for the use of the Perkin Elmer Lamda 9 spectrophotometer. Dr. G.C. Stevens of Central Electricity Generating Board, Central Research Laboratory is thanked for continuing support and helpful discussions.

7. LITERATURE REFERENCES

1. D. Bloor and A.P. Monkman: Synthetic Metals in press
2. P.M. McManus, S.C. Young and R.J. Cushmai: J.CHEM.SOC.CHEM.COMM. (1985) 1556
3. M. Kuzmay and N.S. Sariciftci: Synthetic Metals in press
4. S. Stafström: in these proceedings
5. A.P. Monkman, D. Bloor, G.C. Stevens and J. Stevens: to be published
6. C.E. Duke $\underline{et}$ $\underline{al}$: CHEM.PHYS.LETT. 131 (1986) 82

In situ Ellipsometry of Early Steps of Polyaniline Electrosynthesis

B. Grodzicka, K. Brudzewski, R. Mińkowski, J. Płocharski, and J. Przyłuski

Institute of Inorganic Technology, Warsaw University of Technology, PL-00-664 Warsaw, Noakowskiego 3, Poland

Polyaniline films have been intensively studied recently. The results obtained, however, are not very clear and more work should be done to elucidate the mechanism of PANI formation, redox processes, molecular structure and film morphology. In our opinion, these problems can be at least partially solved by increasing the number of physical chemical methods adopted. Particular difficulties are connected with investigations of very early steps of polymer film growth.

In situ ellipsometry is a very promising method whose applications to PANI films are alas rather scarce to date, and only the paper by CARLIN et al. /1/ can be mentioned. This method is based on measurements of the state of polarization of light reflected from a sample and allows calculations of film thickness and optical constants. Such results can serve as a basis for the determination of polymer growth kinetics.

In our paper we present the results of investigations of electropolymerization of aniline for polymer films thinner than 1500Å.

1. Experimental

PANI films were potentiostatically formed on polished platinum disc electrodes (5mm diam.) in a 1M HCl solution of 0.1M aniline at various potentials in the range 0.65-0.80V vs SCE. The charge passed was measured. The electrodes were mounted in a cell made of a hollow equilateral glass prism (at 60° angle of incidence). An automatic ellipsometer of self-compensating type, Model EL-10D, equipped with a He-Ne laser was used. Interpretation of the data obtained was accomplished by finding values of characteristic parameters of applied models which reproduced the experimental values of Δ and ψ in the best way.

Two models of PANI films were adopted: a one-layer model that assumed homogeneity of the optical constants n and k (n and k being the real and imaginary parts of the complex refractive index $N = n - ik$, respectively), and a two-layer model that approximated the real film by two discrete homogeneous layers with different thicknesses and optical constants.

An unequivocal determination of the above model parameters was achieved by a minimalization computer procedure assuming invariable values of n and k for every series of four experimental points. The films were measured for thicknesses not exceeding 1500Å because of increasing scattering and absorption of the laser beam with increasing film thickness.

2. Results and Discussion

A reasonable fit by the one-layer model could be obtained only for films thinner than 200Å. Therefore the two-layer model was arbitrarily chosen and successfully applied up to a total film thickness of 1500Å. The in-depth inhomogeneity of the PANI films is so considerable, however, that for more precise calculations and thicker films a multilayer model should be applied.

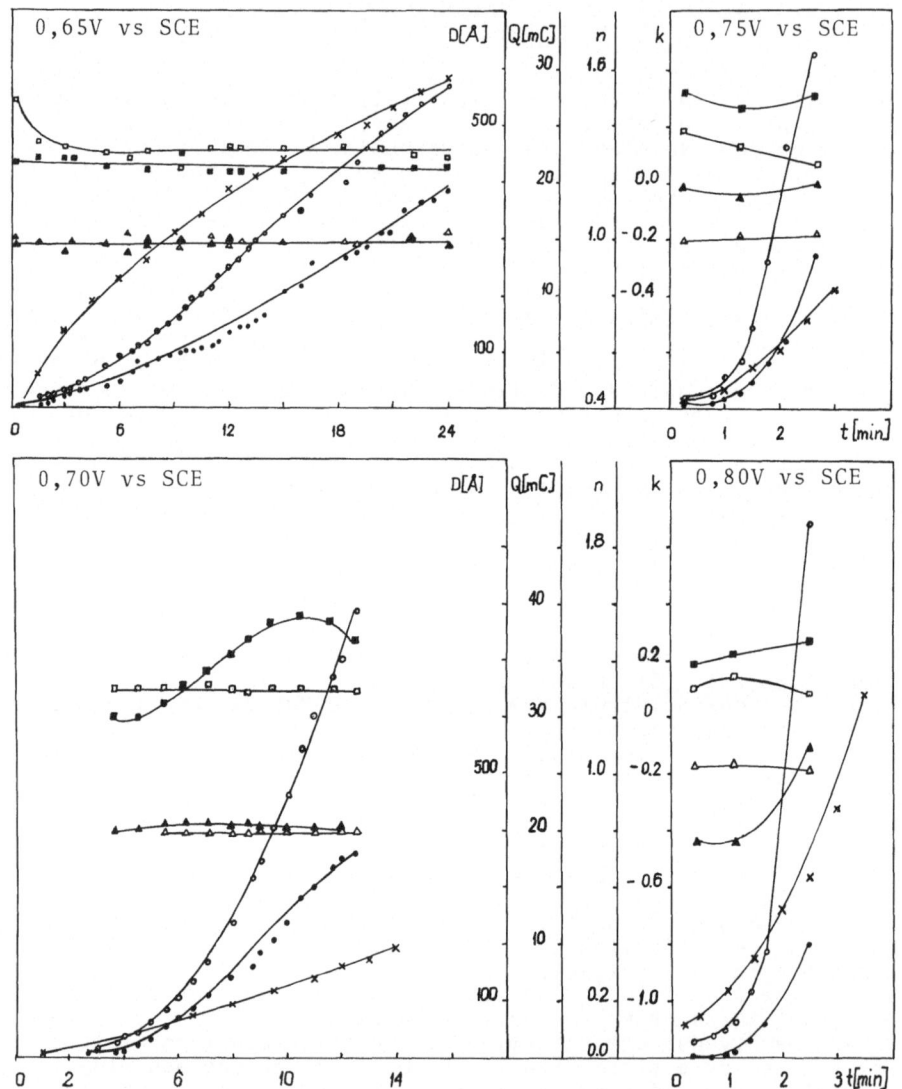

Fig.1   Charge Q (x x x), thicknesses D_1+D_2 (o o o), D_2 (• • •), refraction index n_1 (▫ ▫ ▫), n_2 (■ ■ ■), extinction coeff. k_1 (△ △ △), k_2 (▲ ▲ ▲) for PANI formation at various potentials

The calculated values of two-layer model parameters for four chosen potentials are plotted versus time of polymerization in Fig.1. The film growth rate depends obviously on the potential applied, however, nonlinear behaviour can be noticed. The measurable growth rate of the film appears after passing charge of about $10mC/cm^2$. This behaviour distinguishes PANI from, for instance, polypyrrole /2/, for which linear growth is typical. The appearance of an induction time might be connected with possible formation of soluble aniline oligomers and bypass products.

The values of both n and k characteristic of the first and second layers are close to each other for the films synthesized at potentials below 0.67V vs SCE. This suggests that these films are relatively homogeneous, although low values of n indicate their high porosity. The materials obtained at higher potentials (>0.7V vs SCE) exhibit considerable large differences between optical constants of layers, especially for the case where the total film thickness exceeds 300Å. The layers near the Pt substrate seem to be much more dense, according to their n values (1.4-1.6), than the layers near the electrolyte. The porosity of the external layers calculated on the basis of Burgemann's equation is approximately 65%. The inhomogeneous morphology of PANI films evidenced by the above observations is also confirmed by the non-linear dependence of the charge on polymerization time and film thickness.

The results presented here enable us to propose a morphological scheme for PANI films immersed in an electrolyte and subjected to an electric field (Fig.2). An external region consisting of polymer fibers bristled up by the electric field is combined with the less porous internal area. The growth of the film probably takes place throughout its whole volume, and consecutive covering of pores occurs, leading to an increase in thickness of the inner layer.

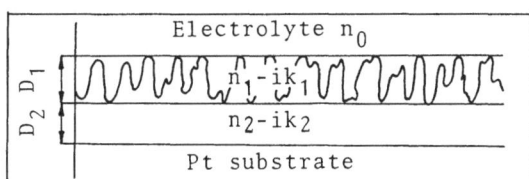

Fig.2
Morphological scheme
of PANI films

3. References

1. C.M. Carlin, I.J. Kepley, A.J. Bard: J. Electrochem. Soc. 132, 353 (1985)
2. M. Zagórska, K. Brudzewski, R. Mińkowski, J. Przyłuski: In Proc. of Int. Conf. on Electrical and Related Properties of Organic Solids, Szklarska Poręba, Poland (1987), to be published

Electronic Properties of Polyaniline

S. Stafström[a]

Laboratoire de Chimie Théorique Appliquée, Centre de Recherches
sur les Matériaux Avancés, Facultés Universitaires Notre-Dame
de la Paix, B-5000 Namur, Belgium

Geometry, band structure, and electronic excitations are studied in tetramers and polymers built up from the emeraldine structural unit. Special attention is focused on the bipolaron and polaron conformations of the conducting emeraldine salt form. It is found that, in contrast to all other conducting polymers, only one defect band appears in the gap. The calculated electronic excita-tion spectrum for the emeraldine salt tetramer having a bipolaron conformation is shown to compare favourably with the optical absorption spectrum for the same system. For poly-emeraldine salt, however, we are able to present strong evidence that the polaron conformation is stable.

1. Introduction

Neutral polyaniline consists of phenyl-amine and phenyl-imine structural units. The relative ratio between these two groups can be varied more or less continuously. Acid treatment of the polymer results in protonation of the imine nitrogens and a highly conducting salt is obtained if the amine to imine ratio in the perstine polymer is roughly one [1]. This form of polyaniline is usually termed polyemeraldine. Thus, polyemeraldine base (PEB) can be doped to become a good electric conductor ($\sigma \sim 5$ S/cm) without changing the number of electrons in the polymer chain. This is a new property for conducting polymers, which also leads to a picture different from the usual one for the creation of defect states. Instead of removing electrons from the valence band there is a mixing between the energy bands of the polymer with the discrete, unoccupied, 1s orbital of the protons. For reasons of symmetry, discussed in more detail below, the effect of this mixing on the eigenvalue distribution close to the Fermi level will appear as a single unoccupied bipolaron band, or a single half-occupied polaron band, close to the highest occupied band. Thus, we have a picture different from the usual one of two bands associated with the bipolaron/polaron lattice.

It should be noted that highly conducting polyemeraldine salt (PES) can also be obtained through electrochemical oxidation of acid-treated reduced polyaniline. Spin density studies during this kind of doping process have been reported, and on the one hand MacDiarmid et al. [2] found an initial creation of Curie spins, indicating a polaron formation, followed by a conversion into Pauli spins, which shows the formation of the polaron lattice. On the other hand, Geniès and Lapkowski [3] observed an initial increase of Curie spins followed by a vanishing spin density, i.e. the polarons recombined into bipolarons. These two studies were performed in different acidic media, which leads to different types of interactions between the anions and the polymer chain. Thus, the question of the stability of polarons versus bipolarons, in the doping context, involves not only the intrinsic properties of the polymer chain but also the properties of the anion (or cation). In this theoretical study we will present results on geometry, band structure, and electronic excitations of the base and salt forms of (poly)emeraldine. For the salt form, both the bipolaron and the polaron conformations will be considered. The geometry and the electronic excitation spectra are studied in oligomer systems but their extension to polymer equivalents are also discussed.

2. Method

In the absence of detailed experimental data related to the geometrical structure of polyaniline as well as aniline oligomers, we performed geometry optimizations using the MNDO (Modified Neglect of Differential Overlap) method [4] on systems of four up to seven aniline units. It was found that within the central four ring unit of oligomers from five up to seven rings no considerable change in the geometrical structure occurs. The geometry of this four-ring unit is therefore believed to be a good representative of the geometry of the polymer unit cell.

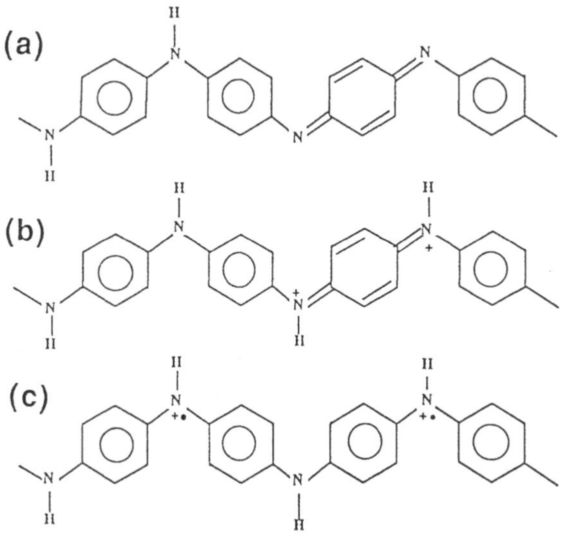

Fig. 1. The structure of (a) emeraldine base, (b) the bipolaron conformation, (c) the polaron conformation.

Oscillator strengths and excitation energies are calculated for tetra-aniline oligomers using the INDO/S-CI method [5]. In the CI, 300 singly and doubly excited configurations with lowest energy are chosen from 5000 generated [6]. The band structure is calculated using the VEH (Valence Effective Hamiltonian) method [7]. Both the electronic excitation spectra and the band structure are calculated using MNDO geometries as described above.The systems included in this study are shown in Fig. 1a-c. These structures should be regarded either as the polymer unit cells of the VEH band structure calculations or as the tetramer (with an additional hydrogen attached to the nitrogen at the end of the molecule) studied by INDO/S-CI. The tetramers (polymers) are named (poly)emeraldine base (Fig. 1a) and (poly)emeraldine salt with bipolaron (Fig. 1b) and polaron (Fig.1c) conformations respectively.

3. Results and Discussion

The geometry of the four phenyl ring unit cells of PEB and PES is, except for some end effects, very similar to the geometry of the tetra phenyl oligomers. In the emeraldine base (Fig. 1a) the C=N, C-N and the quinoid ring C=C and C-C bond lengths are 1.30 Å, 1.42 Å, 1.35 Å and 1.49 Å respectively. In the bipolaron conformation (Fig. 1b) this set of bond lengths take the following values: 1.34 Å, 1.41 Å, 1.36 Å, and 1.47 Å respectively. For the polaron conformation (Fig. 1c) we have chosen a structure symmetric with respect to the nitrogens instead. The C-N bonds are 1.41(1) Å and 1.41(4) Å for the charged and uncharged nitrogens respectively (see Fig. 1c) and as one goes away from the charged nitrogen the C-C bond lengths are 1.44 Å, 1.40 Å, and 1.42 Å. The C-C bond lengths in the benzenoid rings all take values of 1.40-1.42 Å.

In Fig. 2 we display a part of the VEH band structure for PEB (Fig. 2a) and for the bipolaron (Fig. 2b) and polaron (Fig. 2c) conformations of PES. Note that for the polaron lattice, the unit cell is only half that for the two other cases. It is seen that the overall change in the band structure in going from PEB to the bipolaron conformation is rather small. In particular, the band x is unchanged. The eigenstates in this band have zero contribution on the nitrogen atoms and on the carbon atoms linked to the nitrogens, resulting in almost no overlap between these states and the 1s orbital of the attached proton. Apparently there is only one bipolaron band present, the unoccupied band a' shown in Fig. 2b. A similar picture is observed in the polaron case, i.e. only one polaron band appears in the gap, the half-filled band a' shown in Fig. 2c. It should be stressed that this property of polyaniline is unique since in all other conducting polymers investigated so far there always appear two defect bands in the gap. Clearly this finding must result in a modified interpretation of the optical spectrum of PES, since in other conducting polymers both defect bands are involved in the subgap optical absorptions.

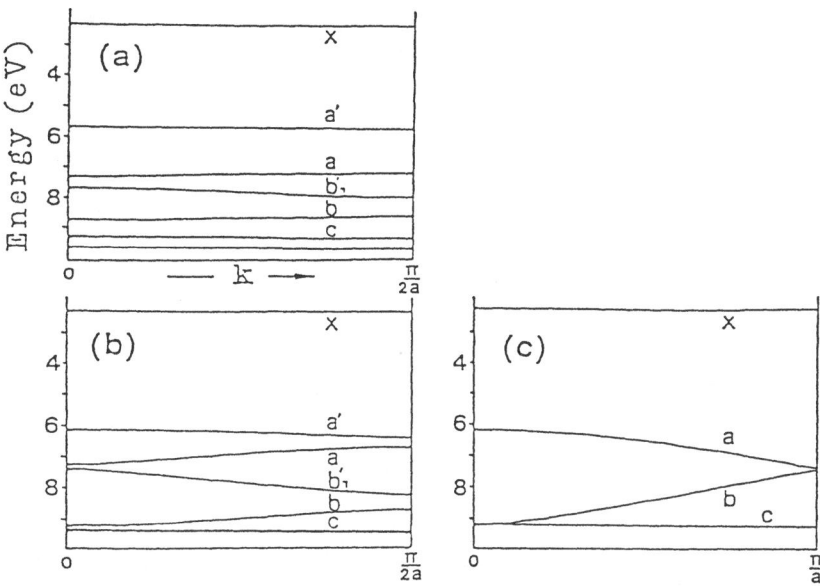

Fig. 2. The VEH band structure of (a) polyemeraldine base, (b) the bipolaron lattice, (c) the polaron lattice.

It is fruitful to make a comparison between the energy of the possible electronic excitations derived from the band structure and the experimentally obtained absorption energies. In PEB the first four interband transitions appear between bands a→a', b'→a' b→a', and c→a' with energies 1.5 eV, 2.1 eV, 3.0 eV, and 3.7 eV respectively. Experimentally, however, only a peak at 2.1 eV followed by a very broad absorption starting at ~3.5 eV is observed [8]. The 2.1 eV absorption is shown below to involve a mixture of eigenstates (see also ref. [9]) which might explain the inadequate interband transitions description. In the bipolaron case the bands involved in the low energy transitions are the same as in PEB now at energies 0.7 eV, 1.7 eV, 2.4 eV, and 2.9 eV respectively. These energies should be compared with the data of Geniès and Lapkowski [3] on the spinless (bipolaron) samples showing absorption peaks at 1.6 eV, 1.9 eV (weak) and ~3 eV. No data were reported for energies lower than 1.4 eV. Thus, a fairly good agreement is found between the interband transitions and the observed absorption peaks. In the polaron lattice the unit cell extends over two aniline units only, resulting in a degeneracy in the band structure at π/2a. Since the number of electrons is unchanged compared to the previous cases, the band a (see Fig. 2c) is now half-filled and one expects metallic-like properties of the material. Indeed high Pauli susceptibility [2] and far IR absorption [10-12] are observed in PES (treated in HCl). From experiment the width of the polaron band is estimated to be ~1.0 eV in good agreement with the width of 1.1 eV obtained from the VEH band

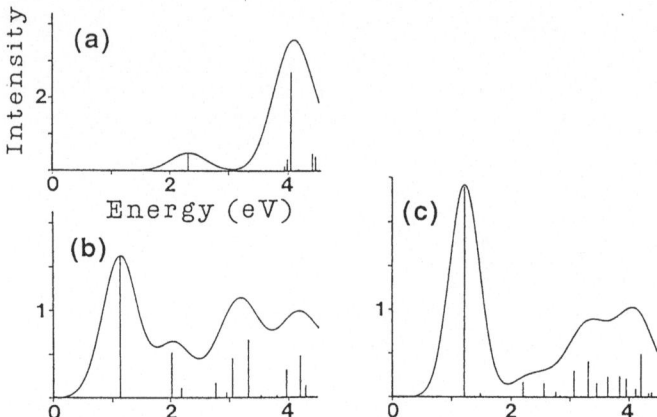

Fig. 3. The electronic excitation spectra of the tetramers (a) emeraldine base, (b) the bipolaron conformation, (c) the polaron conformation.

structure (Fig. 2c). Furthermore, optical absorption measurements on these samples show two low energy peaks at ~1.5 eV and 2.8 eV respectively. Again, these values are in excellent agreement with the corresponding interband transitions calculated to be at 1.8 eV (b→a) and 2.7 eV (c→a). Thus, the existence of a polaron lattice in PES is verified beyond doubt by this comparison between experimental and theoretical results (see also ref. [10].

The electronic excitation energies and oscillator strengths in emeraldine base and the bipolaron and polaron conformations of emeraldine salt are presented in Fig. 3a-c. The vertical bars show the oscillator strength of a particular transition and the solid line is the result when each bar has been replaced by a Gaussian with a full width at half maximum of 0.3 eV. In emeraldine base the transitions at 2.3 eV and 4.0 eV are in good agreement with the experimentally observed absorption peaks at 2.2 eV and 4.1 eV respectively for the identical system [13] and also in fairly good agreement with the polymer equivalent, PEB, where an absorption peak appears at 2.1 eV followed by a broad absorption starting at ~3.5 eV. The low energy peak corresponds to a transition from states residing on the benzeniod moieties of the tetramer to the LUMO, which is localized at the quinoid moiety. Indeed the transition is from a linear combination of a few of the highest occupied molecular orbitals, which makes the interpretation of the absorptions in PEB in terms of interband transitions less meaningful (see above).

The overall features of the electron excitations shown in Fig. 3b and c are very similar, except for a strong absorption at 2.0 eV for the bipolaron conformation (Fig. 3b) which is missing for the polaron conformation (Fig. 3c). This is crucial in the comparison with the experimental result [13] on emeraldine salt, which shows two distinct low energy peaks at 1.2 eV and 1.8 eV respectively, i.e. in reasonable agreement with our findings displayed in Fig. 3b. This suggests a bipolaron conformation for emeraldine salt. The first transition in both Fig. 3b and 3c is dominated by the HOMO to LUMO transition and the second transition in the bipolaron case is mostly HOMO-1 to LUMO. The extension to the polymer equivalents is in this case less clear than for emeraldine base. One would expect, however, that the low energy peak in Fig 3b-c should move down considerably when the system length increases and in the polymer limit it should agree with the 0.7 eV transition in Fig. 2b (bipolaron lattice) and with the intraband absorption of the polaron lattice. By inspection of the wave functions involved in the transitions one finds that in none of the excitations below ~4 eV is there any significant contribution to the final states from molecular orbitals above the LUMO. This is true for both the bipolaron and the polaron cases. Thus, the INDO/S-CI study has confirmed the interpretation of the optical absorption spectra aided by the band structure, namely, that there is only one bipolaron/polaron band involved in the excitations associated with the defect.

Acknowledgments

The author sincerely thanks Dr. J.L. Brédas and Mr. B. Sjögren for their contributions to this work. Continuous support from NFR (Swedish Natural Science Research Council) is also gratefully acknowledged.

References

a. Permanent address: Dept. of Physics, Linköping University, S-58183 Linköping (Sweden).
1. J.C. Chiang and A.G. MacDiarmid, Synth. Met. 13, 193 (1986).
2. A.G. MacDiarmid, J.C. Chiang, A.F. Richter, and A.J. Epstein, Synth. Met. 18, 285 (1987).
3. E.M. Geniès and M. Lapkowski, to be published.
4. M.S.J. Dewar and W. Thiel, J. Am. Chem. Soc. 99, 4899 & 4907 (1977).
5. J.A. Pople and G.A. Segal, J. Chem. Phys. 43, 5136 (1968).
6. B. Dick and G. Hohlneicher, Theoret. Chim. Acta 53, 221 (1979).
7. J.M. André, L.A. Burke, J. Delhalle, G. Nicolas, and Ph. Durand, Int. J. Quantum Chem. Symp. 13, 283 (1979); J.L. Brédas, R.R. Chance, R. Silbey, G. Nicolas, and Ph. Durand, J. Chem. Phys. 75, 255 (1981).
8. W.R. Salaneck, unpublished data.
9. B. Sjögren and S. Stafström, to be published.
10. S. Stafström, J.L. Brédas, A.J. Epstein, H.S. Woo, D.B. Tanner, W.S. Huang, and A.G. MacDiarmid, to be published.
11. N.S. Sariciftci, H. Neugebauer, and H. Kuzmany, these proceedings.
12. A.P. Monkman and D. Bloor, these proceedings.
13. Y. Cao, S. Li, Z. Xue, and D. Guo, Synth. Met. 16, 305 (1986).

On the Acidic Functions of Polyaniline

C. Menardo[1], F. Genoud[1;], M. Nechtschein[1;**], J.P. Travers[1;**], and P. Hani[2]*

[1]DRF-G/SPh/DSPE, Centre d'Etudes Nucléaires de Grenoble, 85 X,
  F-38041 Grenoble Cedex, France
[2]DRF/LCM/EM, Centre d'Etudes Nucléaires de Grenoble, 85 X,
  F-38041 Grenoble Cedex, France

We present investigations on the acidic sites of Polyaniline, with titration curves which point out the existence of 2 acidic functions in the protonated oxidized form. We report results showing a correlation between the
dependence of spin susceptibility and conductivity of Polyaniline versus
the equilibrium pH. We observe an hysteresis on the insulator to conductor
transition depending on the experimental conditions.

1. INTRODUCTION

It has been reported by different authors [1-4] that polyaniline undergoes
an insulator to conductor transition if treated with appropriate protonic
acids. DORIOMEDOFF et al. [1], TRAVERS et al. [2] have studied the polyaniline prepared by chemical oxidation of aniline with $(NH_4)_2S_2O_8$ in sulphuric acid solutions. They found that the conductivity remains essentially
constant at the level 10^0 S.cm^{-1} till pH4 and then decreases to 10^{-10}S.cm^{-1}
over a 4-7 pH range. On the other hand, MACDIARMID et al. [4] who had carried out their experiments on polyaniline synthesized by the same oxidant
but in HCl solutions, reported a much steeper conductivity decrease occurring over two pH units. It should be however stressed that the previous
authors equilibrated the salt form of polyaniline at a given pH whereas the
latter one started from the base form.

   These discrepancies in the position of insulator to conductor transition
for differently equilibrated samples strongly indicates the possibility of
an hysteresis effect in the properties of polyaniline. We have therefore
performed a detailed study of the equilibration of the base and the salt
form of polyaniline at different pH, i.e. we have carried out the equilibration experiment with increasing and decreasing pH.

   We also present titration curves on the protonated polyaniline in its
oxidized and reduced forms in order to define the acidic functions of
polyaniline.

2. SYNTHESIS

Experimental conditions for the synthesis of polyaniline [3,5] are briefly
described below :

---

*Also member of Université Scientifique et Médicale de Grenoble
**Also members of Centre National de Recherche Scientifique, ER 216

- Ammonium peroxydisulfate $(NH_4)_2S_2O_8$ (0.05 mole) in 1M HCl.
- Aniline (0.2 mole) in 1M HCl.

(a) The reagents were cooled to 1°C and stirred in an ice bath for 1.5 hour.
(b) The precipitate was filtered and washed with 1M HCl and then dried.
(c) The obtained powder was extracted with CH_3CN and dried.
(d) The powder was divided in two fractions. One was equilibrated (15 hours) in 1M HCl (SA), the other in 0.1M NH_4OH(SB).
(e) Samples of increasing (SA) or decreasing (SB) pH were obtained by equilibration (55 hours) of these powders in solutions of appropriate pH. During the equilibration no evolution of pH value was observed after 5 hours of equilibration.

Some samples were prepared differently than those described above, in particular :
- S1 and S2 were not dried at the end of (b) and were not extracted with CH_3CN.
- For S4, all the powder was equilibrated in 1M HCl between (c) and (d).

3. EVIDENCE FOR TWO TYPES OF ACIDIC FUNCTIONS OF POLYANILINE

3.1 Experimental

Polyaniline powder was suspended in H_2O (50 ml) and titrated by NaOH in a closed cell under argon flow. The titrations were performed very slowly with incremential additions using an automatic microprocessor controlled TACUSSEL system (TT.Processeur + Electroburex EBX2). The increment volume was adjusted between 0.01 and 0.5 ml depending on the slope of the titration curve. The delay between two consecutive additions depended on the time required for the pH to stabilize, with a minimum of 5 mn.

The values of pKa, determined from two independent experiments lasting 3 and 10 hours, were the same, whereas the concentrations of the acidic sites in polyaniline differed by 10 %. It can be therefore concluded that the system under investigation was close to equilibrium. Several titration experiments have been carried out on two types of samples :
- Polyaniline in its oxidized form prepared with $(NH_4)_2S_2O_8$ as oxidizing agent and then protonated in 1M HCl (S3A).
- Polyaniline in its reduced form obtained by the reduction of S3A with pure hydrazine and then protonated in 1M HCl.

3.2 Results and discussion

Typical titration curves and their derivatives are given in Fig.1a and 1b, for the oxidized sample and the reduced one, whereas in Table 1 the acidic site characteristics are collected.

▷ One may notice two pH steps in the reduced form of polyaniline ; the 5.5 pKa acidic function has a negligible concentration and is probably associated with protonation of residual oxidized rings. Since the proposed [3] reduced base form of polyaniline is composed of only one type of nitrogen site :

245

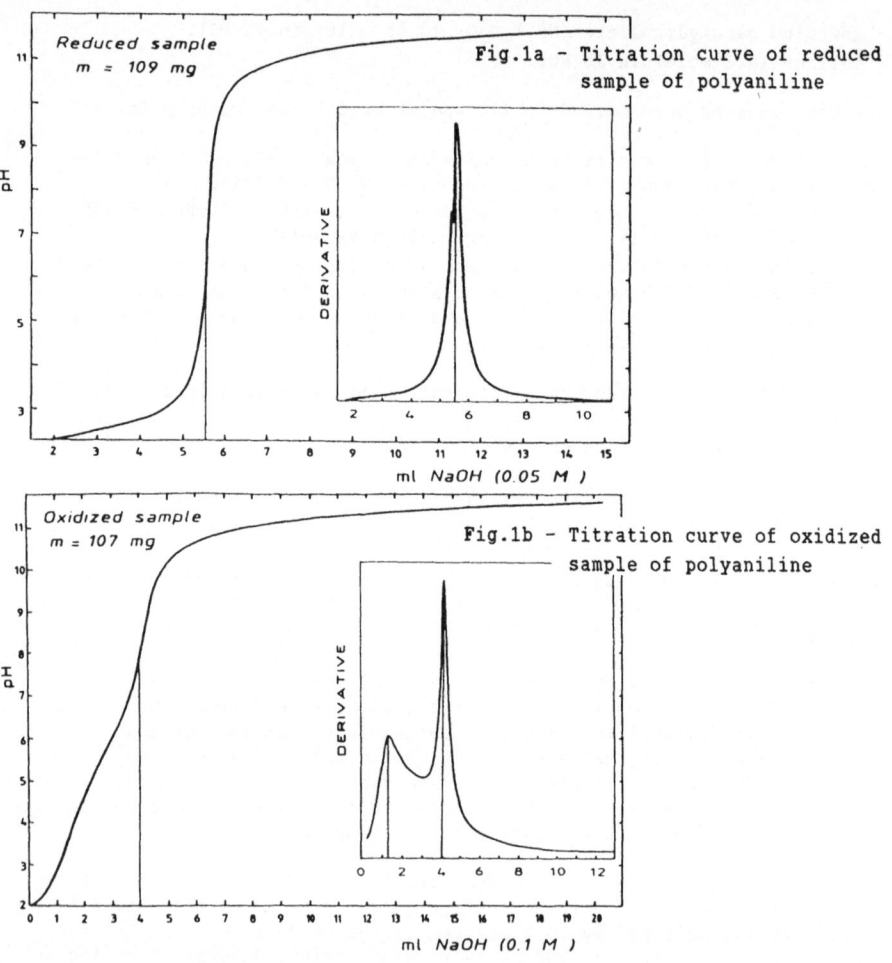

Fig.1a – Titration curve of reduced sample of polyaniline

Fig.1b – Titration curve of oxidized sample of polyaniline

Table 1 – Conductivity and acidic site characteristics obtained for the oxidized and reduced forms of polyaniline

Polyaniline form	pKa : 2.5	pKa : 5.5	conductivity S.cm^{-1}
Protonated reduced	0.25 eq/mole (1 ring)	~ 0.01	8 x 10^{-8}
Protonated oxidized	0.14	0.32	3.2

it is clear that in the protonated (pH0) reduced form of polyaniline,
1(-NH-) site over 4 is protonated into a (-NH$_2^+$-) site.

▷ Two inflexion points can be defined in the titration curve of the
oxidized form as mentioned in Ref.8. We observed two distributions of pKa
values around 2.5 and 5.5. The proposed form of the oxidized polyaniline
can be schematically depicted as follows :

Reduced base unit   Oxidized base unit

with y ≃ 0.5 [3].

By analogy to the reduced form of polyaniline we can attribute the
2.5 pKa acidic function to the protonated site (-NH$_2^+$-) of the reduced base
unit. Consequently, the acidity distribution around pKa 5.5 is related to
the protonated sites (-HN$^+$=) of the oxidized base unit of the chain.

So, in the oxidized protonated (pH0) polyaniline the (-N=) sites are
not the only ones which are protonated, therefore it is no longer possible
to derive the oxidation state of the polymer solely on the basis of the
anion concentration in the protonated salt form of polyaniline.

In the case of electrochemically synthesized polyaniline,
chronocoulometric studies [6] have shown that the product corresponding to
y = 0.6 is formed between the two oxidation peaks. In addition, the
voltammograms of polyaniline prepared chemically are very similar to those
obtained on electrochemically synthesized samples [7]. Moreover the
potential which is measured on chemically synthesized polyaniline is about
0.35 V/SCE which is an intermediate value between the 2 oxidation peaks.
If we assume y = 0.6 then the proportion of the reduced sites (amine
nitrogen) which are protonated is 0.14/0.6 i.e. about 1 site over 4, that
is the same ratio we have measured in the totally reduced material. For
the oxidized sites (imine nitrogen) the ratio is 0.32/0.4 i.e. 4 sites
over 5 are protonated on the oxidized units.

4. EVIDENCE FOR HYSTERESIS EFFECT IN THE pH DEPENDENCE OF CONDUCTIVITY

Conductivity and spin susceptibility have been measured on samples
described in part 2.

We notice (see Fig.2a) that the insulator to conductor transition
occurs between pH 4-2 for samples of decreasing pH(SB) and pH 4-7 for
samples of increasing pH(SA). This may explain the discrepency between the
data reported by different authors [1,2,4].

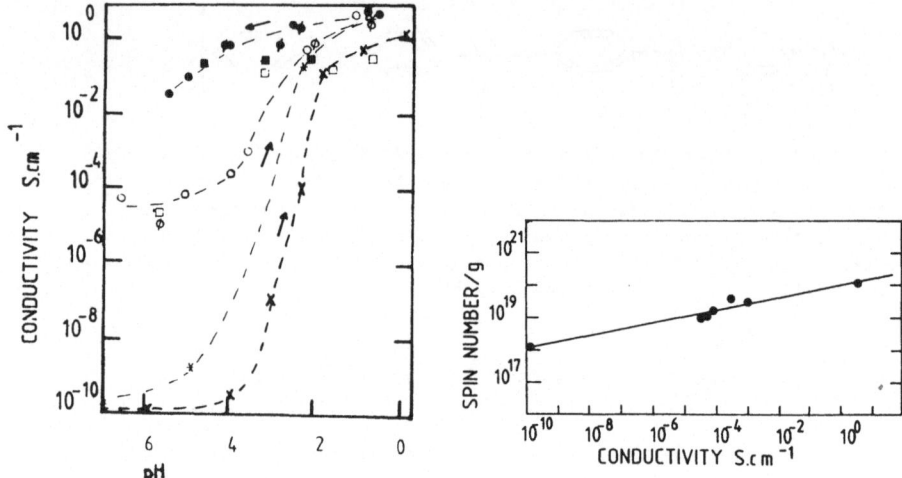

Fig.2a – Relationship between conductivity and equilibrium pH of HCl solution. For increasing pH : ● S2A ; ■ S3A ; ● S4A and decreasing pH : * S1B ; o S2B ; □ S3B ; o S4B

Fig.2b – Correlation between spin susceptibility and conductivity. The spin susceptibility is scaled in concentration of equivalent Curie spins

    In addition, we observe a spread in the conductivity value for samples of decreasing pH. Further work is in progress to elucidate the reason for this phenomenon.

    In all cases we notice a strong correlation between conductivity and spin susceptibility (Fig.2b)

A. Pron is acknowledged for critical reading of the manuscript.
This research was partly supported by a BRITE contract of the European Economic Community.

5. REFERENCES

1. M. Doriomedoff, thèse 3ème cycle, Paris ⅥI, 1974 ; M. Doriomedoff et al., J. Chem. Phys. 68, 1055 (1971)
2. J.P. Travers et al., Mol. Cryst. Liq. Cryst. 121, 195 (1985)
3. J.C. Chiang, A.G. Mac Diarmid, Synthetic Metals 13, 193 (1986)
4. A.G. MacDiarmid, J.C. Chiang, A.F. Richter, A.J. Epstein, Synthetic Metals 18, 285 (1987)
5. A.G. MacDiarmid, J.C. Chiang, A.F. Richter, N.L.D. Somasiri, to be published
6. E.M. Geniès, C. Tsintavis, J. Electroanal. Chem. 200, 127 (1986)
7. B. Villeret and M. Nechtschein, unpublished observations
8. R. de Surville, thèse 3ème cycle, Paris, 1967 ; R. de Surville et al., Electrochemica Acta 13, 1451 (1968)

Thermochromism and Acidochromism in Substituted Polyanilines

Ph. Snauwaert, R. Lazzaroni, J. Riga, and J.J. Verbist

Laboratoire Interdisciplinaire de Spectroscopie Electronique,
Facultés Universitaires Notre-Dame de la Paix,
61, rue de Bruxelles, B-5000 Namur, Belgium

1. Introduction

Conjugated polymers exhibit interesting properties as materials for rechargeable batteries [1-3], modified electrodes [4], and electrochromic display devices [5-6]. The last application involves a reversible color change depending on the redox state of the polymer; the green oxidized polyaniline films can be changed to transparent yellow by reduction in 1 M HCl in the potential range of -0.2 V to 0.6 V vs. SCE [7].

The aim of this work is to show that color changes can also take place in this kind of system under several chemical and physical treatments.

2. Experimental Observations

Electrochemical polymerization of substituted anilines on conductive glass electrodes has been carried out potentiostatically in perchloric acid solutions. Detailed experimental conditions are reported in Table I.

Poly 2,5-dimethylaniline (sample 1) is green at room temperature and undergoes an irreversible thermochromic transition (green → brown) when heated above 65°C.

Poly 2,5-dimethylaniline (sample 2) exhibits an acido-basic reactivity which also leads to color switching, strongly depending on the synthesis conditions. Addition of acetonitrile (5 %) is necessary to dissolve the monomer. When the electropolymerization has been achieved, the films turn pink when they are soaked in water. An acidic treatment (0.1 M) induces an irreversible transition from pink to green . On the other hand, a reversible transition is observed between green (pH = 7 to 1) and blue (pH = 0). It should be noted that to observe the acidochromic properties, the film has to remain in the synthesis medium for a few minutes after the polymerization potential has been switched off.

Poly 2,5-dimethoxyaniline films show a slightly different acidochromism. The films have a blue, green and purple color, respectively, at pH values of 0,7 and 14, and these can be switched reversibly.

Table 1 : Experimental conditions for electrochemical polymerisation of aniline derivatives.

Compound	Substituents (Molarity)	Potential * (Volt)	pH	Salt (Molarity)
1	2,5-dimethyl (0.1 M)	0.8	1	-
2	2,5-dimethyl (0.1 M)	0.8	2	LiClO$_4$(0.1 M)
3	2,5-dimethoxy (0.1 M)	0.45	1	-

* SCE was used as reference electrode

3. Results and Discussion

The polymers' electronic structure has been studied by UV-visible and X-ray photoelectron spectroscopies. An important point is that spectra for different forms of a given compound were taken on the same sample, which was removed from the spectrometer and treated at different temperatures or pH values before being reinserted for further study.

Thermochromism

The absorption spectra of poly 2,5-dimethylaniline at 25 and 80°C show a dramatic intensity decrease of the lowest band (775 nm) while the variation of the other bands is weak. The transition temperature can be estimated to be around 65°C from the evolution of the absorption band at 775 nm (Fig. 1).

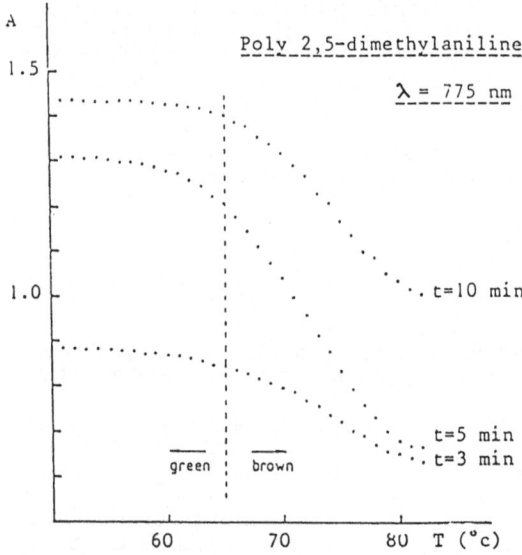

Fig. 1 : Evolution of the absorbance of poly 2,5-dimethylaniline vs. temperature, at 775 nm, for different times of polymerisation

By X-ray photoelectron spectroscopy (XPS), no modification in stoiechiometric composition is observed during the thermochromic transition. Only the shape of the N1s photoelectron line is modified. At 25°C, the N1s peak is composed of two contributions with a total width at half maximun (FWHM) of 2.1 eV. At 80°C, only one component is detected (FWHM = 1.7 eV). This modification is probably due to a geometry change of the polymer chains (allowed at high temperature) to a more stable conformation. This induces a new charge distribution (at least locally) which influences the polymer electronic structure. This thermochromic transition is observed only if the protonation level ranges from 5 to 30 %.

Acidochromism

The optical absorption spectra (Fig. 2) of poly 2,5-dimethylaniline show the influence of acidity on the electronic structure (similar behavior is observed for

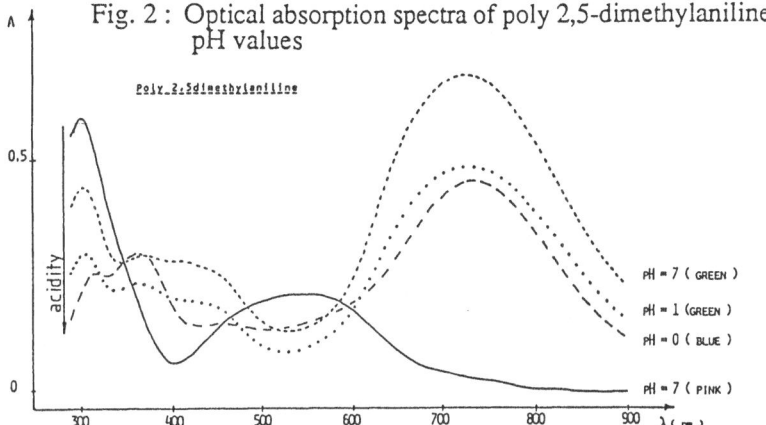

Fig. 2 : Optical absorption spectra of poly 2,5-dimethylaniline at different pH values

Poly 2,5dimethylaniline

pH = 7 (GREEN)

pH = 1 (GREEN)

pH = 0 (BLUE)

pH = 7 (PINK)

λ (nm)

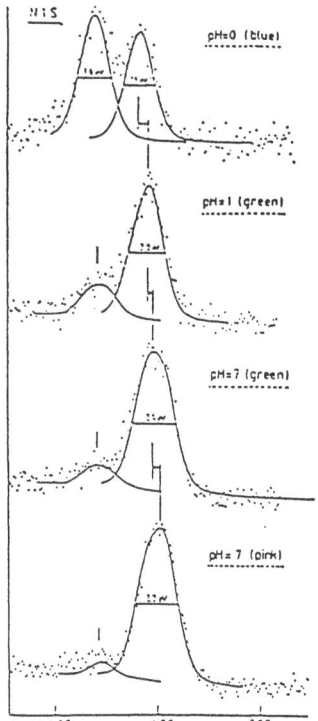

N 1 S

pH=0 (blue)

pH=1 (green)

pH=7 (green)

pH= 7 (pink)

E (eV)

Fig. 3 : Nitrogen-1s spectra of poly 2,5-dimethyl-aniline at different pH values

poly 2,5-dimethoxyaniline). The absorption band at 300 nm is generally attributed to $\pi \rightarrow \pi^*$ transitions involving molecular orbitals essentially localized on the uncharged phenyl moieties [8-9]. Its decrease at high protonation level shows the perturbation induced by the positive charges on the electronic density of the ring. The presence of a shake-up satellite, typical of the benzenic group [10], in the XPS C1s spectrum of the pink compound and its disappearance in both the blue and green polymers confirms the U.V. spectroscopic results.

251

Table II : XPS binding energies [a] (eV), FWHM (full width at half maximum) and protonation levels [b] (%) in poly 2,5-dimethylaniline.

Film color	pH	E_b C1s	E_bN1s	FWHM (N1s)	Protonation level
pink	7	285.8	399.9	2.2	2
green	7	285.7	400.2	2.4	6
green	1	285.6	400.4	2.0	31
blue	0	285.5	400.9 403.0	1.6 1.6	76

[a] Perchlorate Cl $2p_{3/2}$ line, used as common reference, was set at 208.7 eV

[b] The protonation levels have been determined from nitrogen/chlorine (N/Cl) atomic ratios.

The N1s spectra of poly 2,5-dimethylaniline (Fig. 3) also appear to be modified by the protonation level. Peak 1 grows with the acidity and peak 2 shifts towards higher binding energies (Table 2). These results indicate that positive charges are mainly localized on nitrogen sites. Contrary to the case of poly 2,5-dimethylaniline, the N1s peak of poly 2,5-dimethoxyaniline does not split into two components, probably because of the presence of methoxy groups stabilizing the positive charge and thereby allowing a better delocalization along the polymer chain. In this polymer, protonated and non-protonated nitrogen sites seem to be equivalent.

4. Acknowledgement

Ph.S. and R.L. are grateful to IRSIA (Belgium) for financial support.

5. References

1   A.G. MacDiarmid, S.L. Mu, N.L.D. Somasiri, W. Wu ; Mol. Cryst. Liq. Cryst. 121 (1985) 187.
2   A. Kitani, M. Kaya, K. Sasaki ; J. Electrochem. Soc., 133 (1986) 1069.
3   M. Kaya, A. Kitani, K. Sasaki ; Denki Kagaku, 52 (1984) 847.
4   N. Oyama, T. Ohsaka, T. Shimizu ; Anal. Chem., 57 (1985) 1526.
5   F. Garnier, G. Tourillon, M. Gazard, J.-E. Dubois ; J. Electroanal. Chem.,148 (1983) 299.
6   A. Kitani, J. Yano, K. Sasaki ; J. Electroanal. Chem., 209 (1986) 227.
7   T. Kobayashi, H. Yoneyama, H. Tamura ; J. Electroanal. Chem., 161 (1984) 419.
8   C.D. Duke, E.M. Conwell, A. Paton ; Chem. Phys. Lett., 131 (1986) 82.
9   Y. Furukawa, T. Hara, T. Hyodo, I. Harada; Synth. Metals, 16 (1986) 189.
10 J. Riga, J.J. Pireaux, J.J. Verbist; Mol.Phys., 34 (1977) 131

Spectroscopic Studies of some Model Molecules for Polyaniline

W.R. Salaneck[1], C.R. Wu[1], S. Stafström[1], M. Lindgren[1], T. Hjertberg[2], O. Wennerström[2], M. Sandberg[2], C.B. Duke[3], E. Conwell[3], and A. Paton[3]

[1]Department of Physics, IFM, Linköping University,
S-581 83 Linköping, Sweden
[2]Department of Polymer Technology, Chalmers University of Technology,
S-412 96 Göteborg, Sweden
[3]Xerox Webster Research Center, 800 Phillips Road, 0114/38D,
Webster, NY 14580, USA

The polyaniline family of conducting polymers is one of the most interesting of the newer conducting polymers [1,2]. Some details of the molecular structure of certain limiting forms of polyaniline are still uncertain. Namely, the size of the torsion angles among the benzene and quinoid rings in the reduced forms, from leucoemeraldine (LE) to pernigraniline (PNA), are under investigation [3,4]. Some earlier works are reviewed in Ref. 3. In this contribution, we report preliminary results of a spectroscopic study of certain trimer model molecules for LE and PNA, namely, N,N'-diphenyl-1,4 phenylenediamine (the "amine trimer") and N,N'-diphenyl-1,4-benzoquinoediimine (the "imine trimer") respectively. The results of our studies of preparation procedures and redox properties of those model compounds will be published separately [5]. The chemical structural formula of the amine and imine trimers, and the torsion angles ϕ_C and ϕ_e are indicated in the inserts in the figures. $\phi_e = \phi_C = 0$ defines a flat molecule, where all rings lie in the plane defined by the single and double bonds of the N-atoms. We employ ultra violet and X-ray photoelectron spectroscopy (UPS and XPS) of vapour-deposited ultra thin films, and optical absorption spectroscopy of the molecules in solution. The major issue addressed is the geometrical torsion angles, ϕ_C and ϕ_e. The results of both CNDO/S3 [3] and Valence Effective Hamiltonian (VEH) [4] quantum chemical calculations are used to access the UPS (valence electron) spectra, with the torsion angles ϕ_e and ϕ_C as parameters. The optical absorption spectra are addressed using configuration interaction (CI) treatment of the CNDO/S3 results[4]. The reader is referred to the references for further details.

In Figs. 1 and 2 are shown the experimental UPS spectra for the amine and imine molecules. Also shown are several representative electronic density-of-valence-states (DOVs) curves derived from the CNDO/S3 and VEH results by substituting a Gaussian curve for each discrete energy eigenvalue, and adjusting the width of the Gaussians such that the calculated DOVs spectrum best matches the widths of the peaks in the experimental curves. By adjusting ϕ_e and ϕ_C, the DOVs curves can be made to match the experimental UPS spectra.

Some representative optical spectra are shown in Fig. 3. An analysis will be reported separately [6]. Note the continuous evolution of the optical absorption spectra as a function of chain length. The trimer, pentamer and octamer [7] consist of alternating benzene and quinoid rings, and are all pernigraniline (like) molecules. Polyaniline base is an equal mixture of imine and amine dimers, that is, polyemeraldine [1,2].

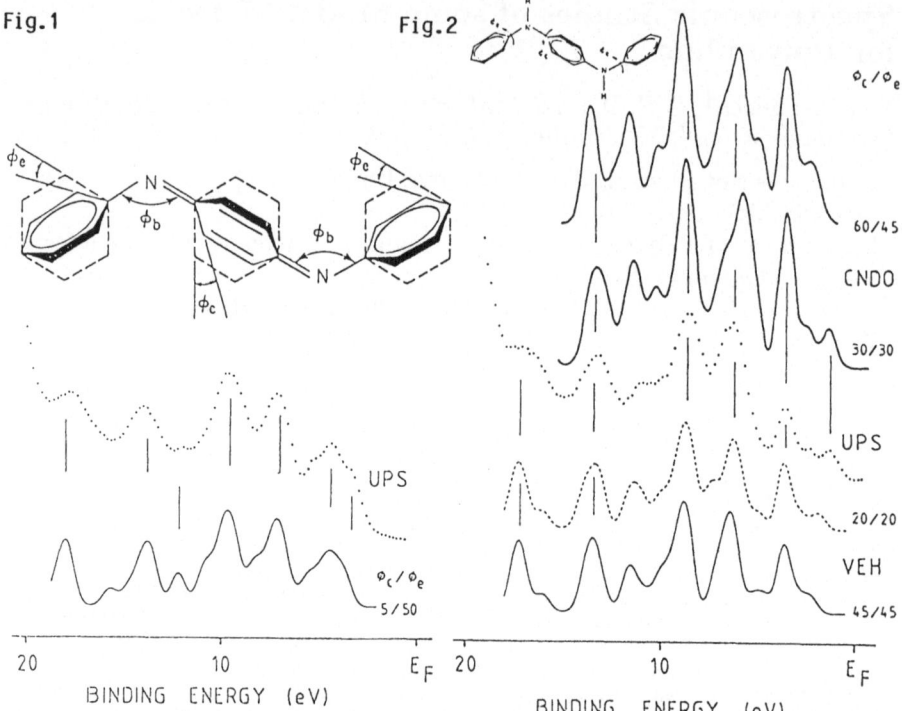

Fig. 1: The He II UPS spectra and the best-fit DOVs curve for the imine trimer shown. The torsion angles ϕ_e and ϕ_C are defined above.

Fig. 2: The He II UPS spectra and several DOVs curves for the amine trimer are shown to illustrate the sensitivity of the DOVs to ϕ_e and ϕ_C.

By varying ϕ_e and ϕ_C in both calculations, one concludes that $\phi_e \approx \phi_C \approx 30$ to $45°$ for the <u>amine</u> trimers in the molecular ion states corresponding to the UPS spectra. Since these UPS spectra were taken with He II radiation (40.8 eV photons), the spectra correspond to "sudden" transitions, and nuclear relaxation effects can be neglected (which may not be the case for when <u>lower</u> energy photons are used to study similar molecules [9]). In the <u>imine</u> case, the best values of the twist angles are near $\phi_e \approx 40$ to $50°$ and $\phi_C \approx 5°$. The quinoid group lies essentially in the plane of the molecule, while the benzoid groups are twisted out of this plane. The results of crossed polarization, magic angle spinning (CP-MAS) solid state ^{13}C NMR studies on polyemeraldine base [8] show conclusively that the quinoid groups lie in the plane of the back bone of polyemeraldine, and that the benzene rings are twisted out of the plane. The measured hyperfine splittings in recent esr/ENDOR studies of the present imine trimer are consistent with an electronic structure indicative of the geometry outlined above [5].

In CNDO/S3 + CI calculations of the optical absorption of amine and imine trimer molecules similar to those studied here (but with an $-NH_2$ group in paraposition of one of the end benzene rings), Duke et al [10] have found that the center ring (either quinoid in the imine case, or ben-

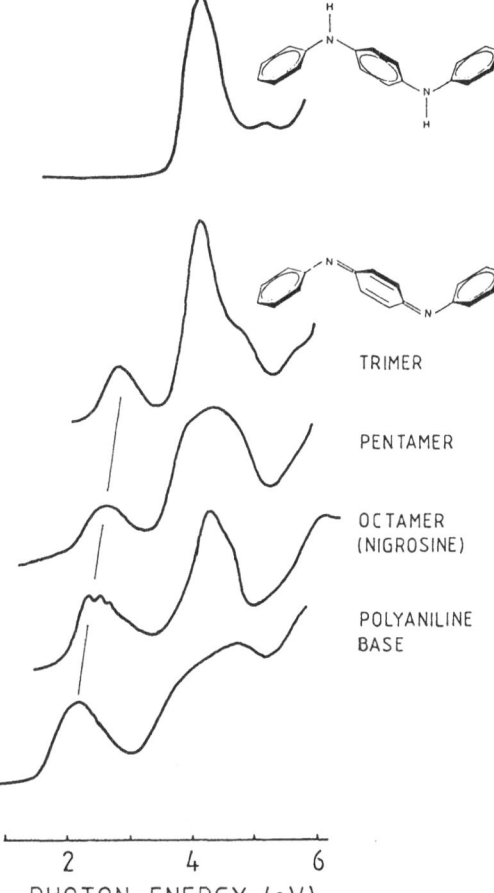

Fig. 3: Optical absorption spectra of the amine and imine trimer are compared with the spectra of an imine pentamer and that of real polyemeraldine base [7].

TRIMER

PENTAMER

OCTAMER
(NIGROSINE)

POLYANILINE
BASE

PHOTON ENERGY (eV)

zoid, in the amine case) is "twisted" out of the plane of the molecule. Nearly identical CNDO/S3 + CI results are obtained on the present amine and imine trimer molecules [6].

The geometries discussed above are physically rational, on the basis of the wave function of the frontier molecular orbital wave functions of these molecules. In the amine case, the single bonds of the N-atoms allow rotation of the center benzene ring in order to lower the total energy of the ground state of the system, by avoiding some steric interactions among hydrogen atoms on "neighboring" rings. In the imine case, the C=N double bonds are rigid and prevent the quinoid ring from rotating in ground state. In the first optically excited state, however, an electron has been removed from a state whose wave function has significant intensity on the N=C double bond, allowing this bond to become more single-bond-like, allowing a rotation of the center (quinoid) ring. The first optically excited state of the imine and amine trimers corresponds to similar, i.e., twisted, geometries [6]. These results imply that in polyemeraldine base, local geometrical distortions may accompany optical absorption at lowest energies.

In summary, the geometric structures of ground state of an imine and amine trimer, used as model molecules for polyaniline, are found to corre-

spond to both the center and end benzene rings twisted (in opposite directions) out of the plane of the molecule in the amine trimer, but that only
the end rings are twisted in the imine trimer. The quinoid ring lies in the
plane of the molecule in the imine trimer, consistent with NMR and esr/
ENDOR results as well as the rigid nature of the carbon-nitrogen double
bonds. In the optically excited state, however, both molecules are found to
exhibit a twisted geometry.

References

1.   J.-C. Chiang and A.G. MacDiarmid: Synth. Met. 13, 193 (1986).

2.   W.R. Salaneck, I. Lundström,A.G. MacDiarmid and W.S. Huang: Synth.
      Met. 13, 291 (1986).

3.   S. Stafström and J.-L. Brédas: Synth. Met. 14, 297 (1986).

4.   C.B. Duke, A Paton and E. Conwell: Chem. Phys. Lett. (submitted).

5.   M. Sandberg, T. Hjertberg, M. Lindgren and S. Stafström, J. Am. Chem.
      Soc. (submitted).

6.   C.B. Duke, E. Conwell, A. Paton, to be published.

7.   W.R. Salaneck, I. Lundström, B. Liedberg, M.A. Hasan, R.Erlandsson,
      P. Konradsson, A.G. MacDiarmid and N.L.D. Somasiri: in Electronic
      Properties of Conducting Polymers and Related Compounds, Ed. by H.
      Kuzmany, M. Mehring and S. Roth (Springer, Berlin, 1985) p. 218.

8.   T. Hjertberg, W.R. Salaneck, I. Lundström, N.L.D. Somasiri and
      A.G. MacDiarmid: Jour. Polym. Sci. Polym. Lett. Ed. 23, 503 (1985).

9.   C.B. Duke, J.W.-P. Lin, A. Paton, W.R. Salaneck, and K. Yip: Chem.
      Phys. Lett. 61, 402 (1979).

ESCA Studies of Polyaniline and Polypyrrole

H.S. Munro, D. Parker, and J.G. Eaves

Department of Chemistry, University of Durham,
South Road, Durham DH13LE, United Kingdom

1. INTRODUCTION

Electron Spectroscopy for Chemical Applications (ESCA / XPS) has been used
with considerable success in the characterisation of many different types of
polymer surface. Although there have been a few reports on the
characterisation of the surfaces of conducting polymers indicating the type of
information that can be derived from such a study, the technique is not yet
routinely employed. Apart from the elemental analysis (except H and He), the
absolute binding energy of a core level of a given element may be
characteristic of a particular functional group and/or indicative of localised
charge. With a knowledge of the relevent instrument sensitivity factors, core
level area ratios may be converted to atomic ratios from which for example
surface stoichiometries and dopant levels may be determined. By taking
examples from some polyanilines and polypyrroles produced by electrochemical
oxidation we aim to show the type of information that can be derived from a
typical ESCA study on conducting polymers.

2. Experimental

Aniline was distilled and then stored in a refrigerator for not more than one
week before use.Polyaniline films were formed by 10 minute controlled potential
electrolysis at +1.0V from aqueous solutions which were 0.1M in aniline and
either 1.0M in sulphuric acid, 1.0M in hydrochloric acid or buffered to pH7 (
approximately 0.1M in each of Na_2HPO_4 / NaH_2PO_4). For ESCA analysis, films
were prepared in a one compartment cell (100ml Pyrex beaker) using 30ml
solutions. Platinum flags (7mm x 18mm) were used as anode and cathode. After
preparation, films were washed in distilled water and then dried under
nitrogen and subsequently in vacuum.
   Samples of polypyrrole perchlorate (Sample A) and polypyrrole
tetrafluoroborate (samples B, C and D) were prepared from solutions in argon
purged, dried acetonitrile as described below.
Sample A: By controlled potential electrolysis (CPE) at +1.0V versus
Saturated Calomel Electrode (SCE) in a solution of 5 x 10 $^{-3}$ M pyrrole/ 0.1M
Tetrabutylammonium perchlorate (TBAP).Sample B: As sample A but using 0.1M
tetrabutylammonium tetrafluoroborate in place of TBAP.Sample C: As Sample B
but using 0.1M pyrrole.Sample D: As sample B but by cycling the electrode
potential between 0 V and +1.2 V versus SCE.
Coated electrodes were mounted onto a standard Kratos probe tip using double
sided Scotch tape. Core level spectra were obtained on a Kratos ES300 electron
spectrometer. Samples were positioned at an angle of 35° with respect to a
plane parallel to the location of the electron analyser slits. MgKα radiation
was used. Binding energies are quoted relative to hydrocarbon C_{1s} at 285.0 eV.
Surface elemental stoichiometries were obtained from peak area ratios
corrected with the appropriate experimentally determined elemental sensitivity
factors. An error of ± 10% may be expected in the surface stoichiometries.

257

3. POLYANILINES

The polyaniline films formed from both hydrochloric and sulphuric acid solutions were green after drying and were approximately one millimetre thick. These films adhered well to the electrode but on removal were powdery. The film in pH7 buffer solution had a bronze metallic lustre and was much thinner, probably one micron or less. In all cases the electrolysis current was observed to decrease with time, as film deposition proceeded, but for the neutral solution this decrease was much more rapid.

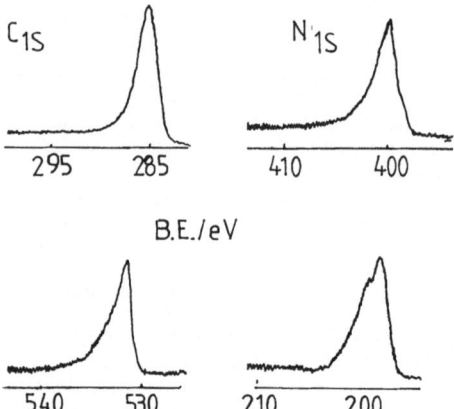

Figure 1 C_{1s}, N_{1s}, O_{1s} and Cl_{2p} core level spectra for polyaniline films grown from 1M HCl.

The C_{1s}, N_{1s}, O_{1s} and Cl_{2p} core level spectra for polyaniline films grown from 1M HCL are shown in Figure 1. The corresponding binding energies (for the main photoionisation peak in each core level) and surface elemental stoichiometries , and those for films grown in 1M H_2SO_4 and pH7 solutions are shown in Table 1.
In the films formed in sulphuric acid and phosphate buffer, the N_{1s} core levels have only a single environment. For the former this is at a binding energy 2 eV higher than in the latter (Table 1). This suggests that for the sulphuric acid grown sample, positive charge is localised on all the nitrogen atoms and, therefore, following Kitani et al [1], that the nitrogen atoms are

Table 1. Binding Energies and Surface Stoichiometries for Polyaniline Deposits.

	Binding Energy (eV)				Stoichiometry				
Electrolyte	C_{1s}	$Cl_{2p3/2}$	$S_{2p3/2}$	N_{1s}	C	N	S	Cl	O
1M H_2SO_4	285	---	169.2	402.0	6	0.64	1.5	--	4.8
1M HCl	285	199.0	---	400.0	6	1.0	--	0.6	0.6
pH7 Buffer	285	---	---	399.9	6	0.8	--	--	0.7

all protonated. For a model compound with a unit positive charge on the nitrogen atom, tetrabutyl ammonium perchlorate, the N_{1s} binding energy was found to be 402.2 eV which agrees well with the suggestion of a single positive charge localised on each nitrogen atom of the sulphuric acid grown film.

For the HCl film, the main N_{1s} photoionisation peak (Table 1) is at a binding energy of 400 eV, but has a shoulder to higher binding energy centred at ca. 402 eV. The latter constitutes 20 % of the total N_{1s} area. It has been reported [2] that the intensity of the high binding energy peak in the N_{1s} spectra of polyanilines formed in HCl is pH dependent and may be indicative of preferential protonation of the imine rather than the amino nitrogens [3].

The elemental stoichiometries (Table 1) are not consistent with " pure " doped/undoped polyanilines. For the films formed in phosphate buffer and HCl the oxygen contents are very high. The species with which the oxygen is associated has yet to be determined. The nitrogen content of the sulphuric acid film is very low, which may be due to some surface degradation process [4].As expected, for the "neutral" film, no phosphorous containing species could be detected. The levels of sulphur and chlorine in the other two films are high in comparison to the amount of nitrogen present at a binding energy of ca. 402 eV.

The binding energy of the sulphur (169.2 eV) is consistent with a sulphate environment. Some of the sulphur may be present as adsorbed sulphuric acid, consequently it is not possible to state on the grounds of stoichiometry whether the anion associated with the high binding energy nitrogen is SO_4^{2-} or HSO_4^-. The Cl_{2p} envelope has a prominent photoionisation peak at ca. 198 eV indicative of anionic chlorine. For a single environment, the Cl_{2p} level should consist of a 2:1 doublet due to spin orbit splitting ($Cl_{2p3/2}$ at lower B.E. than $Cl_{2p1/2}$). In the present case the shape of the chlorine envelope is not consistent with a single environment. As in the case for oxygen the exact nature and number of additional functionalities(anionic and/or covalent) is currently unknown.

In all the polyaniline C_{1s} spectra a step in the background level in going from the low to high binding energy side of the main photoionisation peak is observed i.e. an asymmetric peak shape. At ca 292 eV $\pi \rightarrow \pi^*$ shake up satellites are superimposed on this stepped background. The latter is indicative of the presence of localised conjugation (cf polystyrene [5]) whereas the former is associated with electron delocalisation in a disordered system [6]. The "step" is greater for the films prepared in neutral and HCl solutions than in H_2SO_4. This may be a reflection of structural differences.

4. POLYPYRROLES

Figure 2 shows curve fitted N_{1s} spectra for samples A, B and C. The N_{1s} spectrum for sample D is similar to that for Sample B and is not shown. The spectra were fitted to Gaussian components peaks as has been previously described [7].

The percentage of the total N_{1s} envelope which arises from components centred at a higher binding energy than 401 eV is given in column A of Figure 1. The percentage anion doping level, derived from the N_{1s}/Cl_{2p} (sample A) or N_{1s}/F_{1s} (samples B,C) peak area ratios corrected by experimentally determined instrumental sensitivity factors, is given in column B of Figure 2. If every oxidised nitrogen atom is associated with one singly charged dopant anion the figures in columns A and B of Figure 2 should agree. From the data

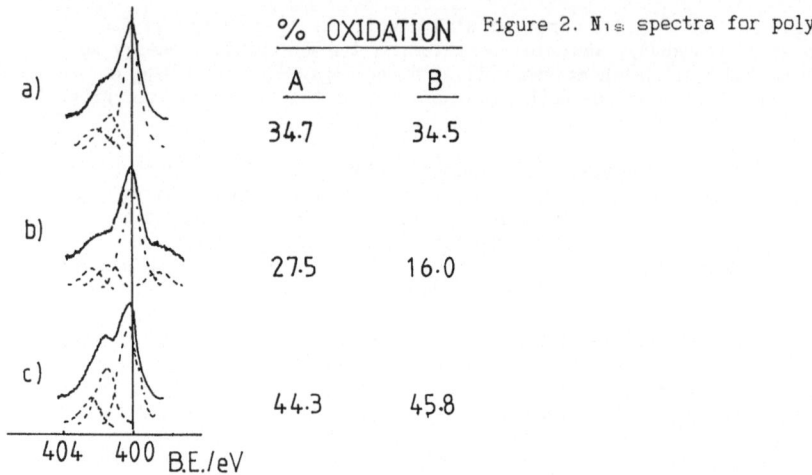

Figure 2. N_{1s} spectra for polypyrroles

	% OXIDATION	
	A	B
a)	34.7	34.5
b)	27.5	16.0
c)	44.3	45.8

404   400   B.E./eV

in Figure 2, it is clear that the extent of nitrogen oxidation increases with the percentage doping level. For sample C an unusually high anion concentration was found but this agrees very well with the N_{1s} spectrum shown in Figure 2c. Good agreement is also found for sample A (Figure 2a). For sample B the oxidation level given by the N_{1s} spectrum exceeds the level predicted by the anion concentration. The same is true for sample D where 23.2% of the total N_{1s} signal arises from oxidised components, whereas the amount of fluorine detected is only sufficient for only 16% of the pyrrole units to be associated with tetrafluoroborate anions. These results may constitute quantitative evidence for anion replacement by oxygen containing species. Low levels of oxygen, sufficient to account for the discrepancy between the two methods of calculation, were detected for these samples. The amount of oxygen detected for sample C was insignificant. Surface deficiency for tetrafluoroborate for immediately analysed samples has previously been noted [8] but not quantified in this manner.

The high binding energy component peaks used in Figure 2 were centred at 401± 0.2 eV and 402.8 ± 0.2 eV. For the percentage of oxidised nitrogen to equal the total percentage doping level, all the polypyrrole positive charge must be located on nitrogen and the average charge on the oxidised nitrogen atoms must be unity. For tetraalkylammonium cations, with a single positive charge on nitrogen, binding energies varying between 401 eV and 402.6 eV have been reported. It thus seems possible that both high binding energy components in the N_{1s} spectra of Figure 1 represent a unit positive charge, but that they may arise from nitrogen atoms which are chemically inequivalent (e.g. protonated or doubly bonded to carbon) or in structurally inequivalent sites. Supporting evidence for this hypothesis may be found in Figures 2a and c. In Figure 2a the area ratio of the 401.6 eV peak to the 402.8 eV peak is 1.6, in Figure 2c this ratio is 2.7. In both cases good agreement between the two methods of calculation of the percentage of polymer oxidation is found, this would not be expected if the two high binding energy components arose from nitrogen atoms bearing different degrees of positive charge. Using the relative component peak areas and associated fractional charges (1/6+ and 1/2+) given by Pfluger and Street [9], the percentage oxidation predicted by their lineshape analysis would be 16.7% , in poor agreement with the perchlorate doping level of approximately 33%. From C_{1s} lineshape analyses, positive charge was considered not to be associated with carbon atoms by Pfluger and Street. The C_{1s} spectra of samples A and B were found to be similar and all but superimpose when compared. This situation contrasts with the corresponding N_{1s} spectra (Figures 2a and 2b) discussed above and may be taken as evidence that positive charge is not localised on the carbon atoms. Finally it should be noted that for samples B and D (see Figure 1b) a low binding energy N_{1s} component peak was observed, which might arise from deprotonated, uncharged pyrrole nitrogen atoms.

From the above discussions on polyanilines and polypyrroles it is clear that many questions still remain to be answered. However, it should be apparent that characterisation by ESCA can provide valuable information on the nature of the polymer synthesised.

5. REFERENCES

1. A. Kitani, J. Izumi, J. Yano, Y. Hirimoto and K. Sasaki,
    Bull. Chem. Soc. Jpn., 57 (1984), 2254.
2. W.R. Salaneck. I. Lundstrom, T. Hjertberg, C.B. Duke, E. Conwell, A. Paton,
    A.G. MacDiarmid, N.L.D. Somasiri, W.S. Huang and A.F. Richter,
        Synth. Met.,18, 291 (1987).
3. J.C. Chiang and A.G. MacDiarmid, Synth. Met., 13 (1986), 193.
4. R. Hernandez, A.F. Diaz, R. Waltman and J. Bargon,
        J.Phys.Chem., 88 (1984), 3333.
5. A. Dilks, Chapter 5, p.315 in " Electron Spectroscopy: Theory Techniques
    and Applications ", Vol.4, Eds. C.R. Brundle and A.D. Baker, Acad. Press,
    London (1981).
6. D.R. Hutton, Ph.D Thesis, Durham, (1983).
7. J.G. Eaves, H.S. Munro and D. Parker, Polymer Communications,28 (1987),38.
8. R. Erlandsson, K. Inganas, I. Lundstrom and W.R. Salaneck,
    Synth. Metals, 10 (1985), 303.
9. P. Pfluger and G.B. Street, J.Chem.Phys., 80 (1984), 544.

Mass Spectroscopy of Chemical Processes in Conducting Polymers: Polyaniline

K. Uvdal[1], M.A. Hasan[1], J.O. Nilsson[1], W.R. Salaneck[1], I. Lundström[1], A.G. MacDiarmid[2], A. Ray[2], and A. Angelopoulos[2]

[1]Department of Physics (IFM), Linköping University,
 S-581 83 Linköping, Sweden
[2]Department of Chemistry, University of Pennsylvania,
 Philadelphia, PA 19104, USA

1 INTRODUCTION

We have begun a study of certain issues involved in attempting to understand (1) the the underlying chemistry and physics of the protonic acid doping process; and (2) the stability of protonic acid doped conducting polymers when exposed to environmental (e.g., H_2O, O_2) and other (e.g., NH_3, H_2O_2, etc.) gases. These issues are related, since environmental gases often influence the way in which doping induces electrical conductivity. The main analytical tools employed are mass spectroscopy, MS; temperature-dependent electrical conductivity, $\sigma(T)$; and X-ray photoelectron spectroscopy, XPS, which is also known as ESCA.

Specifically, we have begun measurements on the emeraldine chloride salt of polyaniline, or Em^+Cl^-, in the two catagories (1) and (2) above, each involving the observation of molecular species emitted from a conducting polymer sample in UHV using MS, while simultaneously monitoring the electrical conductivity, preceded and followed by analysis of the sample by XPS and $\sigma(T)$ studies. When necessary, all of these measurements can be done *in situ*, without removing the sample from the vacuum. Experimental details will be reported in a future full report [1].

2 RESULTS

Em can be "doped" to high electrical conductivity by protonic acids, which leads to the protonation of certain of the nitrogen atoms of the polymer chain [2,3]. XPS spectra of Em^+Cl^- before the treatments described below were taken with the samples in a vacuum of $1x10^{-10}$ Torr (UHV), and include C(1s) and N(1s) spectra as published previously [3]. In addition, some oxygen was always observed. In the Cl(2p) spectra (Fig.1), two peaks (actually two spin-split doublets) are observed, at binding energies of 197.9 eV and 199.6 eV, corresponding to two different types of ionic chlorine. After treatment with KOH in atmosphere, and reexamination in UHV, only a small amount of Cl(2p) remains near 200eV, corresponding to covalently bonded chlorine. The two different binding energies for the two ionically bonded chlorine atoms correspond to Cl^- in alkali chlorides (about 198 eV) and to Cl^- in metal chlorides (near 199 eV). Molecular HCl would appear near 200.5 eV [4]. These differences are readily distinguishable. Thus, we find that there exist two different Cl^- counter ion sites in Em^+Cl^-, and a small residual amount of covalently bonded chlorine, at least in the top 100 Å of the samples, as investigated via XPS.

As a sample is heated, first the conductivity increases with increasing temperature [5], as expected for "semiconductors" or in different variable range hopping models. $\sigma(T)$ increases (and is reversible) in going from 25 to about 70 °C, above which the conductivity starts to decrease slowly. HCl is released continuously from the sample during heating above about 80 °C. The Cl(2p) XPS intensity at 198 eV is reduced in the thermal process.

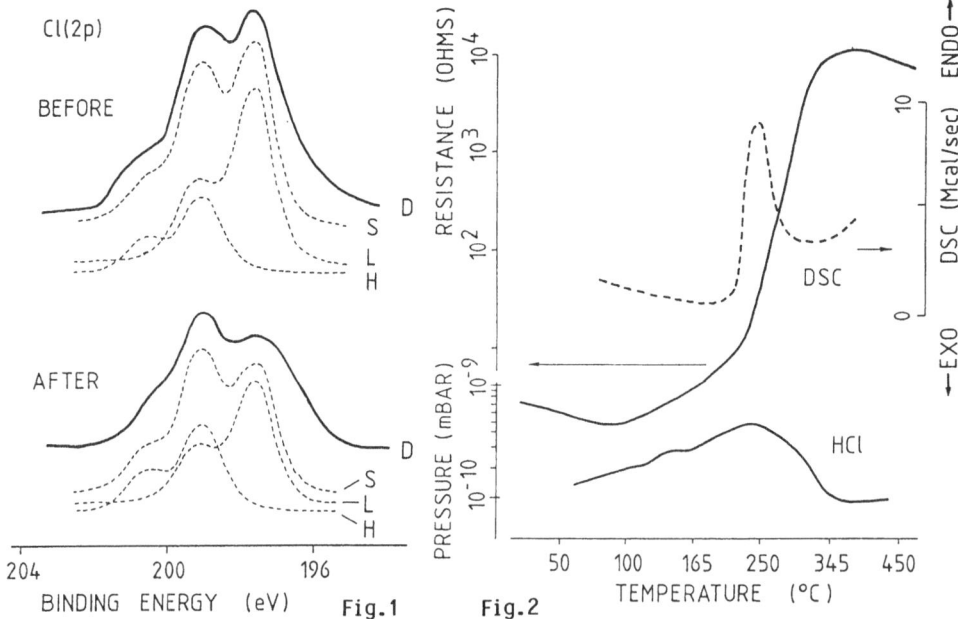

Fig.1   Fig.2

Figure 1 The Cl(2p) xps spectra before and after water vapour exposure (to about 24 mBar for two hours at 45°C) of the sample. Both measurements are done in UHV, however. Very similar spectra are obtained for heat treatment.

———— Experimental   D=Data   L=Low Binding Energy Peak
------ Simulated Cl(p) curve   S=Sum   H=High Binding Energy Peak

Figure 2 The effect of heat treatment on the electrical conductivity is shown along with a corresponding mass spectrum for M=36.

Near 200 °C the HCl intensity in the MS and the conductivity start to decrease dramatically, and well above 200 °C degradation products (molecular pieces of the polymer) appear in the MS data. In an independent study, we also have measured a large endothermic peak in the differential scanning calorimetery (DSC) spectrum [6] near 230 °C. Some of these results are collected in Fig. 2. After heating to over 400 °C, upon cooling the sample exhibits a lower conductivity (less than 10^{-1} [Ωcm]$^{-1}$ at 20°C), but a σ(T) curve indicative of metallic behavior; i.e., an increase in temperature leads to a decrease in conductivity. The thermal decomposition does not occur if the sample is heated only to near 70 °C. Examination of samples by electron microscopy, after heating past 400 °C, did not reveal any signs of melting.

As the vacuum chamber is pumped down from atmospheric pressure, the electrical conductivity decreases by about one order of magnitude. Upon sudden exposure to 24 Torr of H_2O vapor, the conductivity increases. When the water is pumped away, the conductivity returns to almost its original value, as published previously [7,8], and the 198 eV Cl(2p) intensity decreases, as shown in Fig 2. The O(1s) peak decreases slightly. At pressures below 2×10^{-4} Torr, we can do MS. Upon exposure to a low pressure of H_2O, the intensity of the HCl MS signal increases, in proportion to the H_2O pressure, as shown in Fig. 3. The conductivity increases slightly. When Em^+Cl^- is washed in distilled water, the Cl(2p) at 197.9 eV essentially disappears, while the other ionic Cl(2p) at 199.6eV remains unchanged.

263

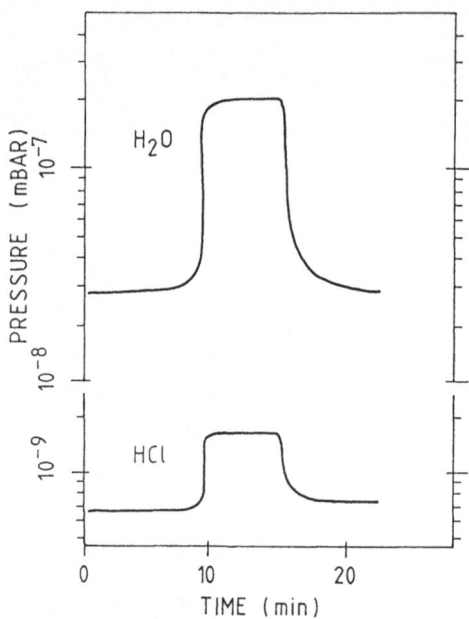

Figure 3 The HCl mass spectrum is shown in response to exposure to H_2O in the low pressure region.

3 DISCUSSION

Travers and Nechtschein (TN) [8] have observed proton exchange with the polymer when Em salts are exposed to H_2O. Unfortunately, we cannot take XPS spectra in the presence of a high vapor pressure of H_2O, but only in vacuum. We believe the elimination of a little HCl on exposure to H_2O vapor is a simple hydrolysis reaction, as expected for the hydrochloride salt of a weak base such as emeraldine.

$$\left(\begin{array}{c} \text{AMINE} \end{array} \quad \begin{array}{c} \text{IMINE} \end{array} \right) + 2\,H_2O \longrightarrow \left(\begin{array}{c} \end{array} \right) + 2\,HCl$$

The imine hydroxide would then readily lose water. We suggest that the conductivity change during exposure to water depends on the electronic screening of the Cl⁻ counter ion by the water molecules. The screening may reduce the attractive Coulomb interaction between the Cl⁻ and the protonated nitrogen, thereby facilitating the electron and proton transfer. The proton, moving free of the polymer chain, would have a certain probability of combining with a Cl⁻ in the vicinity, which would have a finite probability of being released into the vacuum. In addition, the intensity at the high binding energy side of the N(1s) spectrum is slightly reduced simultaneously with the reduction in Cl(2p) intensity, consistent with a deprotonation associated with the release of molecular HCl. We believe that the two Cl⁻ peaks may be accounted for in the following manner. There is considerable evidence [2,9] to suggest that when emeraldine base is protonated in strong aqueous HCl, only the imine nitrogens are protonated. As is well known, the relative strength of a given base to a reference acid is controlled by many factors, including the degree of hydration of the acid and base. We propose when emeraldine HCl is dried following doping with aqueous HCl, that the extent of hydration and the relative base strength of the amine and imine atoms become more similar to each other. This results in partial proton transfer from imine nitrogens to amine nitrogens as the degree of hydration decreases.

Protonation of amine nitrogens will interrupt the π-conjugation of the polymer and will reduce the conductivity. This reversible process is also qualitatively consistent with the observed changes in conductivity [2,7,8] when emeraldine HCL is pumped or exposed to H_2O vapor.

Note, that a model of "metallic-like" islands has been proposed to explain certain details of the electrical conductivity in Em-salts [10].

4 SUMMARY

We observe two distinct Cl^- counter ions in Em^+Cl^-, one of which is affected both by exposure to water vapor and by thermal treatment. We suggest a hydration effect to account for the release of HCl as well as the previously reported [7,8] increase in electrical conductivity during exposure to water vapor. Note that the local chemical equilibrium that we observe in UHV may not be the same as exists in the sample "as prepared" under ambient conditions.

ACKNOWLEDGEMENT

Work in Linköping is supported in part by grants from the Swedish National Board for Technical Development (STU), the Swedish Natural Sciences Research Council (NFR), and the New Science Group of Imperial Chemical Industries (ICI) PLC, England. Work at the University of Pennsylvania was supported by NSF-MRL Grant No. DMR-85-19059(MA) and DMR-86-15475(AR). Assistance in preparation of samples by N.L.D. Somasiri and A.P. Richter is gratefully acknowledged. Some of the results were obtained while I. Lundström was on sabbatical leave at the Moore School of Electrical Engineering, University of Pennsylvania, USA.

REFERENCES

1   K. Uvdal and J.O. Nilsson and A.G. MacDiarmid to be published.
2   J. C. Chiang and A. G. MacDiarmid, J. Synth. Met., 13, 193(1986).
3   W. R. Salaneck, I Lundström, T. Hjertberg, C. B. Duke, E. Conwell, A. Paton, A. G. MacDiarmid, N. L. D. Somasiri, W. S. Huang, and A. F. Richter, Proc. ICSM-86, Kyoto, Japan, 1986, J. Synth. Met. 13,291(1986)
4   S. Svensson, ESCA Laboratory, Uppsala University, private commun.
5   W. R. Salaneck, I. Lundström, B. Liedberg, M. A. Hasan, R. Erlandsson, P. Konradsson, A. G. MacDiarmid, and N. L. D. Somasiri, in Electronic Properties of Polymers and Related Compounds, H. Kuzmany, M. Mehring and S. Roth, Eds, (Springer-Verlag, Berlin, 1985); p-218.
6   Contracted to Kemanobel, Stockholm, Sweden.
7   M. Angelopoulos, A. Ray, A. G. MacDiarmid, and A. J. Epstein, in Proc. Intern. Conf. Elec. Proc. Cond. Polym., Vadstena, Sweden, 1986, J. Synth. Met. (in press).
8   J. P. Travers and M. Nechtschein, in Ref. 7.
9   A.G. MacDiarmid, J.C. Chiang, A.F. Richter, N.L.D. Somasiri and A.J.Epstein, to be published in conducting Polymers, Luis Alcacer ed., Reidel Pub. ,Dordrecht,Holland (1986)
10 A. J. Epstein, J. M. Ginder, F. Zuo, R. W. Bigelow, H. -S. Woo, D. B. Tanner, A. F. Richter, W. -S. Huang, and A. G. MacDiarmid, J. Synth. Met. (in press).

Synthesis and Properties of a Conducting Polymer Derived from Diphenylamine

E. Pater[1], *M. Samoc*[1], *R. Zuzok*[1], *and P. Drozdzewski*[2]

[1]Institute of Organic and Physical Chemistry,
 Technical University of Wroclaw, PL-50-370 Wroclaw, Poland
[2]Institute of Inorganic Chemistry and Metallurgy of Rare Elements,
 Technical University of Wroclaw, PL-50-370 Wroclaw, Poland

1. Introduction

Much attention has been paid in recent years to the properties of polyacetylene [1] - the archetype conducting polymer having an extended system of conjugated double bonds. Poor chemical stability of polyacetylene against oxygen and humidity lessens however the scope of its possible applications . Other conducting polymers investigated in more detail to date are, among others, polypyrrole [2,3], polythiophene [4], poly-p-phenylene [5] and polyaniline [6,7]. The last named material may be of particular interest since the oxidation of aniline leading to aniline black has been known for many years and only recently the possibility of obtaining linearly conjugated polyanilines was clearly recognized.

We present here preliminary results of investigations of electrochemical polymerization of an aniline-related compound: diphenylamine (DPA). This compound can be considered an analogue of aniline in which one hydrogen atom was replaced by the benzene ring. This substitution, however, makes unlikely the formation of a polymer structure in which all the benzene rings are coupled through nitrogens, as it is suggested for polyaniline.

It is shown that a conducting polymer can be obtained by electrochemical polymerization of DPA and we present preliminary data on its electrical conductivity and IR and Raman spectra.

2. Experimental

Electrochemical oxidative polymerization of DPA was carried out by electrolyzing solutions of DPA (purified by zone melting) in a mixture of methanole and aqueous solution of $HClO_4$. A Pt plate and a Pt wire were used as a working and a counter electrode, respectively. Layers of polymer (up to 2-3 mm thick) were deposited at constant potential of 0.65 V versus saturated calomel electrode. The colour of the solution changed during the process from the initial yellowish to dark green. Black polymer was then removed from the electrode surface, washed with methanol in a Soxhlet apparatus for about 20 hours and dried in air.

The measurements of the electrical conductivity and of Raman spectra were performed on pressed pellets. Electrical contacts for dc conductivity measurements were provided by painting with silver paste. IR spectra were taken in KBr pellets using a Specord IR 75 spectrometer.

3. Electrical conductivity

Electrical conductivity was measured for samples of the polymer obtained from solutions with various starting DPA : $HClO_4$ ratio (from 1 : 0.4 to 1 : 20). The values of the room temperature conductivity usually ranged from 10^{-6} to 10^{-2} $\Omega^{-1}cm^{-1}$ but in one case a value of appr. 1 was measured. The conductivity of samples obtained in various preparations for the same composition of the starting solution varied somewhat , however, a general trend of higher conductivity for a higher initial concentration of $HClO_4$ could be observed.

Figure 1 shows an example of a typical dependence of the dc electrical conductivity on temperature as measured in one of our samples. Thermally activated behaviour with two regions of distinct activation energies was usually observed.

These measurements concur with an early report of MALTSEV et al [8,9] who investigated electrical conductivity of the product of chemical oxidation of DPA and reported the room

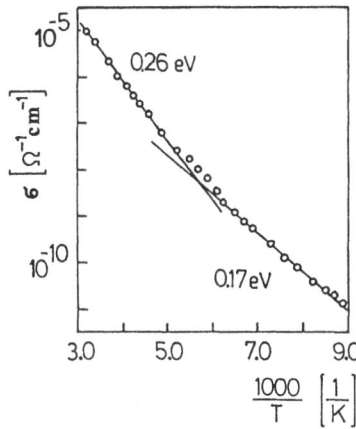

Fig. 1 Typical temperature dependence
of the electrical conductivity in
a polymer sample

temperature conductivity of the order of 10^{-4} - 10^{-3} $\Omega^{-1}cm^{-1}$. It should be also mentioned here that strong EPR signals could be observed in our samples, again similarly to the findings of the Soviet authors [8,9]. These results will be discussed in more detail elsewhere.

4. IR and Raman spectra

Figures 2 and 3 show the infrared and Raman spectra of the product of electrochemical polymerization of DPA. The polymer exhibits very low transmission in the 1650 - 4000 cm^{-1} region of the IR spectrum and the spectrum is very poorly resolved. Occasionally a weak band at about 3400 cm^{-1} could be observed - obviously due to absorption by some remnant N-H groups. The Raman spectra of the polymer bear some resemblance to the published Raman spectra of polyaniline.

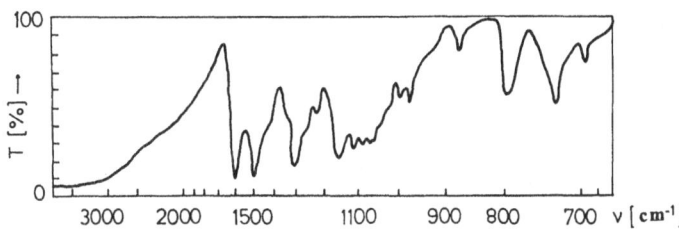

Fig.2 Infrared spectrum of the polymer

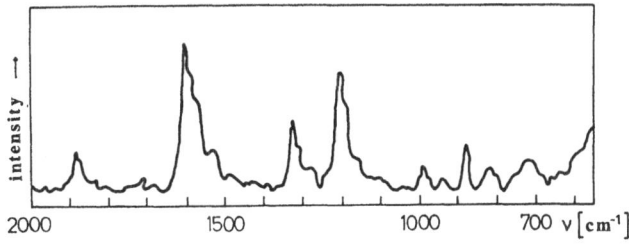

Fig.3 Raman spectrum of the polymer obtained at 5145 Å

267

The Raman spectra of polyaniline have been studied by e.g. FURAKAWA et al. [10]. They have found that an intense peak at 1521 cm⁻¹ is most probably due to a mixed mode of the -C=N- and quinone -CH=CH- stretching vibrations. Only in the spectrum of polyaniline obtained on graphite electrodes a band due to -N=N- stretching vibration has been observed (at 1404 cm⁻¹). Such behaviour of the Raman spectrum suggests predominant head-to-tail coupling, although, polyaniline can be partly polymerized through head-to-head coupling as well.

In our case the strong Raman band at 1530 cm⁻¹ can probably be explained in a similar way as in the quoted paper of FURAKAWA. There is no evidence for the -N=N- stretching mode around 1400 cm⁻¹. Therefore a quinone diimine structure for polymerized DPA may be suggested. This is in agreement with the predictions which can be made on the basis of the known first steps of electrochemical oxidation of DPA which leads to N,N'-diphenylbenzidine (see e.g. [11]).

5. Conclusions

We have shown that a conducting polymer can be obtained by electrochemical oxidation of diphenylamine in methanol in the presence of $HClO_4$. The quinone diimine structure suggested by us makes this polymer similar both to polyaniline and to poly-p-phenylene (presence of directly bound aromatic rings). In fact, one can expect the energetics of this system to be intermediate between those of these two polymers. The relatively high electrical conductivity of the polymer is most likely due to the presence of ClO_4^- anions (detected in the polymer by elemental analysis) stabilizing positive charges on the polymer backbone.

6. Acknowledgements

We are grateful to Dr. A. Jezierski for performing the preliminary ESR measurements on the DPA polymer. This work was supported by a grant from the Polish Academy of Sciences.

7. References

1. J.C.W. Chien, Polyacetylene Chemistry, Physics and Material Science, Academic Press 1984
2. A.F. Diaz, J.I. Castillo, J.A. Logan, W.Y. Lee, J.Electroanal.Chem. 129, 115 (1981)
3. R.E. Myers, J. Electronic Mat. 15, 61 (1986)
4. T.C. Chung, J.H. Kaufman, A.J. Heeger, F. Wudl, Phys.Rev. B, 30, 702 (1984)
5. M. Stamm, Mol.Cryst. Liq.Cryst. 105, 259 (1984)
6. A.G. Green, S. Wolff, Ber. 46, 33 (1913)
7. A.G. MacDiarmid, J.C. Chiang, W. Huang, B.D. Humphrey, N.L.D. Somasiri, Mol.Cryst.Liq.Cryst. 125, 309 (1985)
8. V.I. Maltsev, V.A. Itsakovich, Vysokomolekul. Soedin. 5, 1367 (1963)
9. V.I. Maltsev, V.B. Lebedev, V.A. Itsakovich, A.A. Petrov, Vysokomolekul. Soedin. 4, 848 (1962)
10. F. Furakawa, T. Hara, Y. Hyodo, I. Harada, Synth. Met. 16, 189 (1986)
11. G. Cauquis, J. Cognard, D. Serve, Tetrahedron Lett. 48, 4645 (1971)

Part VII

Polypyrrole, Polythiophene
and Polyparaphenylene

On Polaron and Bipolaron Formation in Conducting Polymers

F. Devreux, F. Genoud*, M. Nechtschein, and B. Villeret

DRF-G/SPh/DSPE-ER, CNRS 216,
Centre d'Etudes Nucléaires de Grenoble, 85 X,
F-38041 Grenoble Cedex, France

Spin concentration has been measured as a function of charge injection in polypyrrole, polyaniline, and polydithiophene. Analysis of the data in terms of polarons and bipolarons in thermodynamical equilibrium leads to polarons and bipolarons nearly degenerate for polypyrrole and polyaniline. In the case of polydithiophene the data support the bipolarons as the fundamental charged species. However, the spin susceptibility does not decrease at low temperature as expected for fundamental spinless species. We propose that the polaron pairing is hindered by potential barriers and, consequently, polarons and bipolarons are not in thermodynamical equilibrium.

1. INTRODUCTION

It is well established that several properties of conducting polymers are strongly dependent on the amount of charge injected into them in a so-called doping reaction. This doping can be achieved either chemically or electrochemically. In the case of polymers with non degenerate ground state it has been shown that, during the doping, charged states appear within the gap. The existence of these states have been explained in terms of polarons and bipolarons.

Since polarons carry spins and bipolarons are spinless it is very advantageous to use ESR technique to study the creation of these species and their mutual transformations. Several papers concerning this subject have been published to date, but the results are somewhat contradictory. Some authors have reported that very few, or no, spins are generated in the case of polypyrrole (PPy) [1], polythiophene (PT) at intermediate doping level [2-4], and poly 3 methylthiophene (P3MT) [5], whereas others have found correlations between the spins and the charges imposed into the polymer in PT [6] and polyaniline (PANI) [7], but no quantitative analysis of the data have been proposed.

In our previous work [8,9] we have demonstrated the existence of a spin/charge relationship in PPy. Interpreting the data in terms of quasi-equilibrium polaron and bipolaron populations we came to the conclusion that polarons and bipolarons are nearly degenerate. This statement was not in agreement with theoretical predictions [10]. In the present paper we extend this study to the case of polydithiophene (PDT) and PANI. The data are analyzed with a model [11] which takes into account the mechanical statistics of polarons and bipolarons in thermodynamical equilibrium in a chain. For PPy and PANI similar results are obtained,

*also member of UNIVERSITE SCIENTIFIQUE, MEDICALE ET TECHNIQUE DE GRENOBLE namely the polarons and bipolarons are nearly degenerate. However, for PDT a significantly smaller value has been observed for the absolute spin concentration, which indicates that the bipolaron state should be more favourable. Thus, the ground state being spinless, a decrease of the spin susceptibility is expected at low temperature. In order to verify this behaviour we have measured the temperature dependence of the spin susceptibility. The results are discussed below.

2. ELECTROPOLYMERISATION AND EXPERIMENTAL CONDITIONS

PPy and PDT films were grown on a platinum wire in an electrochemical cell placed inside the ESR cavity. In the case of PANI, films were obtained *ex situ* by electropolymerisation of aniline in NH_4F ; 2.35 HF and then transferred into the electrochemical cell. Experimental conditions are summarized in table 1. The films were ∼ 0.2 μm thick ; they could be repeatedly reduced and oxidized electrochemically. The redox state was controlled by a dc voltage (measured versus a reference electrode), which was incremented step by step. Steps of 10 to 50 mV were used. For each voltage values the number of Coulombs injected into the film was measured and, after an equilibrating time of 3 to 5 min., the ESR signal was recorded. The spin concentration was calculated by double integration of the ESR signal and using a calibrated reference sample placed in the same dual ESR cavity. Experimental uncertainties are estimated to be ± 30% and ± 10% for spin, and charge concentrations, respectively.

3. SPIN-CHARGE RELATIONSHIP

The evolution of the spin concentration as a function of the injected charge is given in Fig.1 for the three polymers. Spin and charge concentrations have been normalized to one ring. The three polymers

Table 1. Experimental conditions

Polymer	Composition of the solution used for electrolysis	Electrolyte used for ESR experiments	Conditions of poly-merisation
Polypyrrole	pyrrole 10^{-2}M in CH_3CN+$LiClO_4$ 10^{-1}M	CH_3CN + $LiClO_4$, 10^{-1}M	*in situ* by constant dc voltage (0.7V vs Ag/Ag')
Polydithiophene	dithiophene 10^{-2}M in CH_3CN+$LiClO_4$ 10^{-1}M	CH_3CN +$LiClO_4$, 10^{-1}M	*in situ* by constant dc current
Polyaniline	aniline 10^{-2}M in NH_4F ; 2.35 HF	HCl 1M	*ex situ* by cycling the potential from −0.2V to 0.7V vs SCE

271

qualitatively behave the same way. At first, spins -i.e. polarons- are generated at a rate close to one spin per one injected charge. Then, the rate of the spin generation decreases, the spin concentration reaches a maximum, and collapses at high doping level. For PPy and PANI the maximum spin concentration corresponds to ~ 1 spin per 12 rings. It occurs at a doping level of ~ 50% of the maximum charge. Due to the experimental uncertainties the differences between the data obtained for PPy and PANI are not significant. For PDT the maximum spin concentration is significantly lower about 1 spin per 40 rings -and it occurs at ~ 20% of the maximum charge. This gives evidence that bipolarons are more favoured in PDT than in the case of the two other polymers.

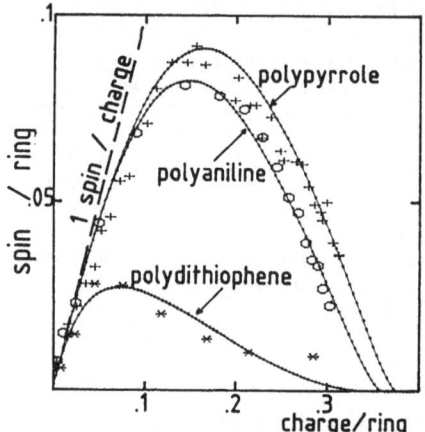

Fig.1 - Spin concentration as a function of charge injection in polypyrrole, polyaniline and polydithiophene. The theoretical curves have been obtained with the model of Ref.11 and using the parameters collected in table 2

We have used a model described elsewhere [11] to fit the data. This model consists of the complete derivation of the mechanical statistics of polarons and bipolarons in thermodynamical equilibrium in a chain. Furthermore, the model allows the lengths of polarons (L_p), bipolarons (L_b), and neutral sites (unit length) to be different. Theoretical curves are given in Fig.1 with the fit parameters collected in table 2. The uncertainties given for the values of the obtained parameters include the experimental uncertainty and the possibility of different sets of the fitting parameters.

Table 2 - Values of the polaron and bipolaron lengths and energy used in theoretical curves in Fig.1

	Polaron length L_p	Bipolaron length L_b	Energy balance $E_b - 2E_p$
Polypyrrole (11)	4.5 ± 0.2	5.3 ± 0.2	(−0.030 ± 0.03) eV
Polyaniline	5.0 ± 0.2	5.5 ± 0.2	(−0.030 ± 0.03) eV
Polydithiophene	11.0 ± 1.0	5.0 ± 0.3	(−0.075 ± 0.03) eV

For PPy and PANI slightly negative values for $E_b - 2E_p$ are obtained, which would imply fundamental bipolarons, but this result must be treated with caution taking into account the uncertainty. In the case of PDT the energy balance is more negative, with its absolute value significantly higher than the possible shift in this value due to experimental uncertainty. So, the fundamental character of bipolarons seems to be established for PDT. This is qualitatively in agreement with the conclusions of Chen et al. for polythiophene (PT) [3] and Colaneri et al. for poly 3 methylthiophene (P3MT) [5], although these authors have observed much smaller spin concentrations. The difference could be interpreted in terms of the more or less fundamental character of the bipolarons in different materials. For instance the spin concentration of 0.2 mole % reported for P3MT [5], i.e. one order of magnitude less than the value we have measured in PDT, leads to $E_b - 2E_p = -0.22$ eV (to be compared to our value of -0.075 eV for PDT). Being a subtle balance between attractive interactions and Coulomb repulsion, the value of $E_b - 2E_p$ is, of course, expected to vary, depending on the exact chemical nature of the studied samples.

4. TEMPERATURE DEPENDENCE OF THE SPIN SUSCEPTIBILITY

If bipolarons, which are spinless, correspond to the ground state ($E_b - 2E_p < 0$), the spin susceptibility is expected to vanish for $kT < |E_b - 2E_p|$. With this behaviour in view, we have performed ESR measurements of the spin susceptibility in the temperature range 4.2 - 300 K. The samples of a given oxidation state were prepared by electropolymerisation on platinum wires and introduced into an ESR quartz tube under helium gas for ensuring a good thermal contact. The temperature was controlled by an Oxford cryostat-controller system and measured with a gold-iron thermocouple.

The temperature dependence of the spin susceptibilities and the ESR linewidths is presented in Fig.2a, b, and c.

No decrease of the spin susceptibility is observed in the whole temperature range for any of the three polymers studied, in contrast to what is expected for systems with fundamental bipolarons.

In the case of PANI, as evidenced in Fig.3, the spin susceptibility can be decomposed into two contributions :

$$\chi(T) = \chi_p + \chi_c(T)$$

where χ_p is the Pauli contribution (temperature independent) and χ_c is the Curie contribution, $\chi_c(T) = N_c \mu_B^2/kT$, with N_c the number of unpaired spins and μ_B the Bohr magneton. From the data in Fig.3 we estimate that the Pauli susceptibility represents 70% of the total spin susceptibility at room temperature. Using the result of absolute spin measurements and using $\chi_p = \mu_B^2 \eta(E_F)$, we deduce a value for the density of states at the Fermi level $\eta(E_F) = 0.2$ states/ev/mole(C+N), which is in agreement with previous determination [12]. This high density of states at the Fermi level, altogether with the decrease of the ESR linewidth with decreasing temperature suggest a metallic state for PANI in its conducting form.

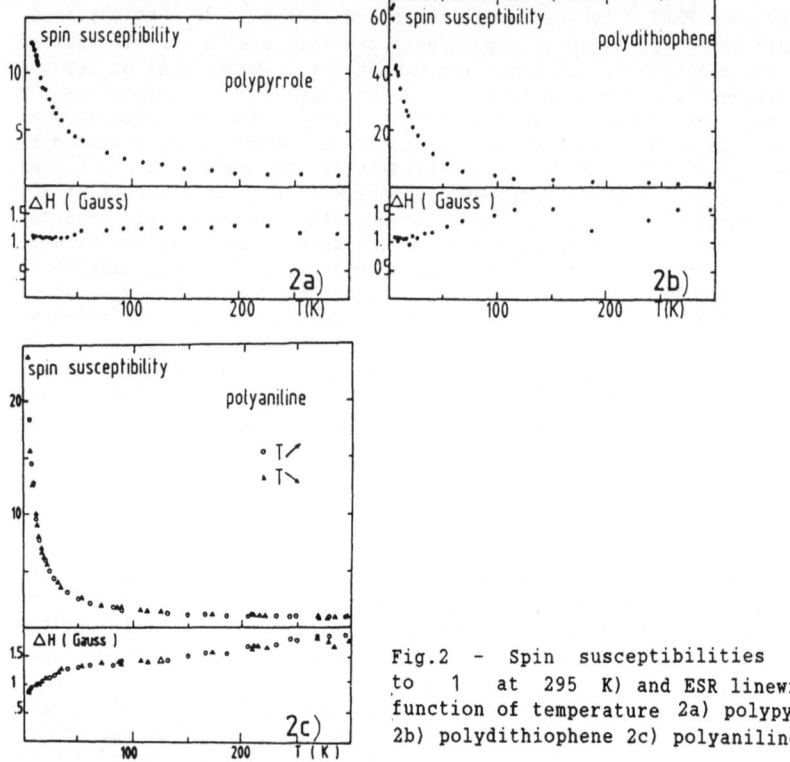

Fig.2 — Spin susceptibilities (normalized to 1 at 295 K) and ESR linewidth as a function of temperature 2a) polypyrrole 2b) polydithiophene 2c) polyaniline

For PPy and PDT no Pauli susceptibility can be extracted from the high temperature limit of the spin susceptibility. At low temperature a weak deviation from the Curie law is observed in the case of PPy, showing evidence for some spin pairing interactions. In the present study we have not observed a maximum in the spin susceptibility as we reported previously [13]. The low temperature data given in [13] might have been erroneous due to bad thermal contact between the sample and the quartz tube wall.

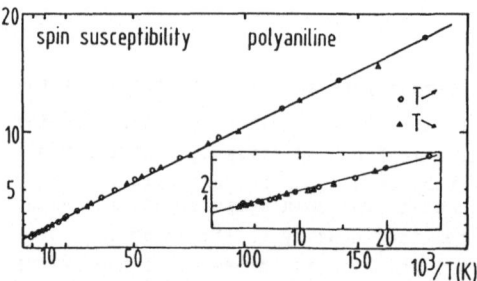

Fig.3 — Spin susceptibility (normalized to 1 at T = 295 K) versus $\frac{1}{T}$ for polyaniline

5. DISCUSSION

The remarkable point which emerges from our results is the inconsistency which apparently exists, at least for the case of PDT, between the bipolaronic character (spinless) of the charged species in their ground state ($E_b - 2E_p < 0$) on one hand, and the observed temperature dependence of the spin susceptibility on the other hand. The expected decrease of the spin susceptibility is not observed down to 4.2 K, which is a very low temperature as compared to $|E_b - 2E_p|/k = (870 \pm 350)$ K for PDT.

To explain the contradiction we propose that the polaron pairing, which should occur at low temperature, is hindered by potential barriers. Two polarons would actually gain energy by forming a bipolaron, but they have, first, to jump over a barrier. The origin of the barrier can be intrinsic : the Coulomb repulsion has a longer range than the phonon assisted attractive interaction which finally stabilizes the bipolaron state. It can also be extrinsic, for instance induced by defects and disorder. The same barriers could also make the polaron pairing upon oxidation, and de-pairing upon reduction, more difficult. The hysteresis which is observed in the charge/discharge cycles as well as the spin concentration versus potential behaviour might be connected to such barriers. Furthermore, if the barriers (or part of them) are associated with the existence of defects and disorder, several points can be explained, in particular the discrepancy in the data reported by different authors.

In conclusion, to reconcile the evolution of the spin susceptibility as a function of charge injection on one hand, and as a function of temperature on the other hand, we have proposed the existence of potential barriers which hinder the polaron pairing. Consequently, a correct description of the system should include the fact that polarons and bipolarons may not be in thermodynamical equilibrium.

Adam Pron is acknowledged for critical reading of the manuscript.

This research was supported by a BRITE contract of the European Economic Community.

REFERENCES

1. J.C. Scott, P. Pfluger, M.T. Krounbi and G.B. Street: Phys. Rev. B 28, 2140 (1983)
2. F. Moraes, D. Davidov, M. Kobayashi, T.C. Chung, J. Chen, A.J. Heeger and F. Wudl: Synth. Met. 10, 169 (1985)
3. J. Chen, A.J. Heeger and F. Wudl: Sol. St. Comm. 58, 251 (1986)
4. K. Mizoguchi, K. Misoo, K. Kume, K. Kaneto, T. Shiraishi and K. Yoshino: Synth. Met. 18, 195 (1987)
5. N. Colaneri, M. Nowak, D. Spiegel, S. Hotta and A.J. Heeger: preprint
6. K. Kaneto, Y. Kohno and K. Yoshino: Mol. Cryst. Liq. Cryst. 118, 217 (1985)
7. a) M. Kaya, A. Kitani, and K. Sasaki: Chem. Lett. 147 (1986)
   b) S.H. Glarum and J.H. Marshall: J. Phys. Chem. 90, 6076 (1986)
   c) A.G. McDiarmid, J.C. Chiang, A.F. Richter, and A.J. Epstein: Synth. Met. 18, 285 (1987)
8. F. Genoud, M. Guglielmi, M. Nechtschein, E. Geniès and M. Salmon: Phys. Rev. Lett. 55, 118 (1985)

9. M. Nechtschein, F. Devreux, F. Genoud, E. Vieil, J.M. Pernaut and E. Geniès: Synth. Met. $\underline{15}$, 59 (1986)
10. J.L. Brédas, B. Themans, J.G. Fripiat, J.M. André and R.R. Chance: Phys. Rev. B $\underline{29}$, 6761 (1984)
11. F. Devreux: Europhysics. Lett. $\underline{1}$, 233 (1986)
12. A.J. Epstein et al.: Synth. Met. $\underline{18}$, 303 (1987)
13. F. Devreux, F. Genoud, M. Nechtschein, J.P. Travers and G. Bidan: J. Phys. C3, 621 (1983)

ESR of BF$_4^-$-Doped Poly(3-methylthiophene)

M. Schärli[1], *H. Kiess*[1], *G. Harbeke*[1], *W. Berlinger*[2], *K.W. Blazey*[2], and *K.A. Müller*[2]

[1]RCA Laboratories Ltd., Badenerstr. 569,
Ch-8048 Zürich, Switzerland
[2]IBM Zürich Research Laboratory, Säumerstr. 4,
CH-8803 Rüschlikon, Switzerland

We have studied poly(3-methylthiophene) doped electrochemically with BF$_4^-$ by electron spin resonance. The spin concentration increases linearly by about one spin per injected charge up to quite high doping levels ($\approx$ 10 mol % at room temperature). The magnetic susceptibility $\chi_m(T)$ gradually changes from a Curie- towards a Pauli-type behaviour as a function of the doping concentration. Simultaneously, an increasing degree of delocalization of the corresponding spin centers is inferred from spin-lattice relaxation rates and line shape analysis. The results demonstrate that polarons are the dominant charge states in poly(3-methylthiophene) doped with BF$_4^-$.

1. Introduction

The doping-induced charge storage configurations in conjugated polymers like polythiophene (PT), poly(3-methylthiophene) (PMT), polypyrrole (PP) etc. are usually discussed using the concepts of **polarons** and **bipolarons** [1,2]. Within a single-particle theory both types of defects are associated with two localized levels positioned symmetrically about mid-gap. The **polaron** state is characterized by a charge transfer $\pm|e|$ and a spin 1/2, which gives rise to an ESR signal, while the bipolaron results from a charge transfer $\pm|e|$ and have no spin. Several ESR and optical studies have been performed on doped PT and PP in order to clarify the nature of the elementary electronic excitation [3-7]. However, these investigations have yielded quite contradictory results and up to now no coherent understanding of the doping-induced carrier generation has been established.

Two different electrochemical approaches have been used to prepare films of BF$_4^-$-doped poly(3-methylthiophene). Samples, denoted by PMT^I, have been synthesized under ambient atmosphere, while samples of the type PMT^{II} have been prepared and handled under controlled atmosphere, using a three-electrode-type electrochemical cell, equipped with a specially designed transfer module [8]. ESR studies have been performed at X-band ($\approx$ 9.4 GHz) using a rectangular cavity. Absolute spin concentrations have been determined with respect to a reference sample of Mn-doped ZnS.

2. Results and Discussion

At **low dopant concentrations** the ESR signal has been found to be a superposition of two components, namely of a Gaussian line at $g^G \approx 2.0035$ with width $\Delta H^G \approx$ 6-8 G and of a Lorentzian line at $g^L \approx 2.0029$ with width $\Delta H^L \approx 1.5$ G, respectively. This result evolves directly from the measurements of the ESR signal as a function of the incident microwave power P_W. The decomposition of the experimental ESR curve into a Gaussian and a Lorentzian has been done numerically. Measurements of the nuclear relaxation rates by NMR [9] show that the intensity of the Gaussian component remains constant as function of the dopant concentration y. The Gaussian component is attributed to the presence of localized spins, which probably stem from neutral radicals. In the present case the concentration of spins N_s^G amounts to about $3 \cdot 10^{19}$ spins/cm^3, i.e. to about 1 spin per 300 thiophene rings, and is in quantitative agreement with the value determined from NMR.

At **intermediate and high doping levels** only one symmetric line of a Lorentzian shape at $g_m \approx 2.0029$ is observed. Figure 1 shows the spin concentration $N_s^L(y)$ of the Lorentzian component for PMT^I and PMT^{II} films as a function of the dopant concentration y.

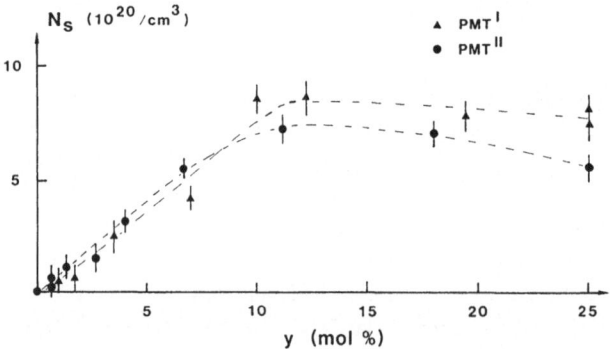

Fig. 1: Spin concentration N_s^L of the Lorentzian component as a function of the dopant concentration y at room temperature.

In both cases N_s^L increases approximately linearly by about one spin per injected charge up to y $\approx$ 10 mol % BF$_4^-$ and saturates above y $\approx$ 15 mol % BF$_4^-$. This result is a strong indication that **polarons are formed predominantly** in poly(3-methylthiophene) as a function of BF$_4^-$ concentration. At the highest doping levels, where $N_s^L(y)$ saturates or even decreases, the situation is less clear. It is tempting to infer that here the polarons start to combine into bipolarons. On the other hand the spins behave more and more like those in a metal as a function of the dopant concentration, as will be shown below, and this reduction can easily be explained by a changeover from Curie- towards Pauli-paramagnetism. This interpretation is consistent with the results obtained for the ESR linewidths and relaxation rates, as will be discussed below.

The susceptibility of the Lorentzian component, i.e. $N_s^L(T)$, gradually changes from a Curie-type behaviour at the lowest doping levels towards a Pauli-type behaviour extending down to $\approx$ 80-100 K at the highest doping levels. The temperature dependences of the linewidths, which are shown in Fig. 2, indicate that the linewidth is determined by the balance of two competing effects, namely motional narrowing and broadening by fast relaxation. The strong and approximately linear decrease of $\Delta H(y=25$ mol %) with decreasing temperature is very similar to the temperature dependence of the linewidth in metals and indicates that at high dopant concentrations the linewidth is mainly determined by spin-orbit interaction.

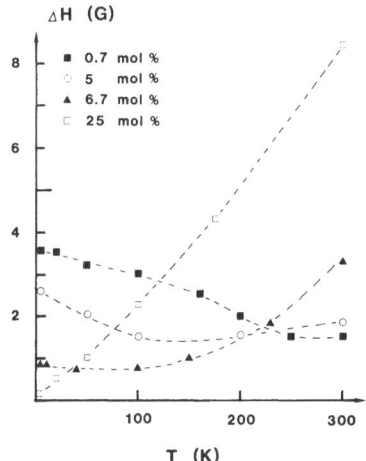

Fig. 2: Temperature dependence of the linewidths ΔH^L of samples having different dopant concentrations.

Thus, from the point of view of ESR, highly doped PMT behaves as a true metal. Further support to this statement comes from the observation of a Dysonian lineshape for sufficiently thick samples, from relaxation times and from nuclear proton relaxation as measured by NMR. By contrast, the temperature dependence of the electrical conductivity is described by an activated process up to the highest doping levels. This discrepancy could be explained by the fact that the electrical conductivity, which is a macroscopic quantity, is limited by hopping processes between segments, where the microscopic conductivity (which is probed by ESR) is much higher.

To our knowledge electronic relaxation times T_1 have not yet been reported for polythiophene. The spin-lattice relaxation time T_1 has been determined using conventional saturation technique. Figure 3 shows the spin-lattice and the spin-spin relaxation times T_1^L and T_2^L, respectively, of the Lorentzian component as a function of the dopant concentration. At the lowest dopant concentrations $(y \leq 1$ %) T_1^L is considerably longer than T_2^L. With increasing dopant concentration T_1 shortens rapidly and becomes comparable to T_2. For $y \geq 3$ mol % T_1 is always about equal to T_2, as for a metal.

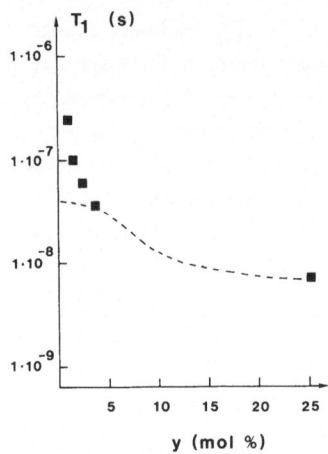

Fig. 3: Concentration dependence of the spin-lattice relaxation time T_1^L at room temperature. The dotted line represents T_2^L.

The sharp decrease of T_1^L near $y \approx 1$ mol % BF_4^- can qualitatively be related to the transport properties of this material, namely to the strong increase of the electrical conductivity by several orders of magnitude at about the same concentration.

Acknowledgement

We have benefitted from stimulating discussions with D. Brinkmann, W. Kobel, M. Mali, W. Rehwald and J. Roos. Financial support of this study by the Swiss National Science Foundation is gratefully acknowledged.

References

1 K. Fesser, A.R. Bishop and D.K. Campbell, Phys. Rev. **B 27** (1983) 4804.

2 T.C. Chung, J.H. Kaufman, A.J. Heeger, and F. Wudl, Phys. Rev. **B 30** (1984) 702.

3 J. Chen, A.J. Heeger and F. Wudl, Solid State Comm. **58** (1986) 251.

4 K. Kaneto, S. Hayashi, S. Ura, and K. Yoshino, J. Phys. Soc. Jap. **54** (1985) 1146.

5 S. Hayashi, K. Kaneto, K. Yoshino, R. Matsushita and T. Matsuyama, J. Phys. Soc. Jap. **55** (1986) 1971.

6 J.H. Kaufman, N. Colaneri, J. Scott, and G.B. Street, Phys. Rev. Lett. **53** (1984) 1005.

7 F. Genoud, M.Guglielmi, M. Nechtschein, E. Genies, and M. Salmon, Phys. Rev. Lett. **55** (1985) 118.

8 M. Schärli, to be published in J. Phys. E, Sci. Instr..

9 M. Mali et al., to be published.

Influence of the Monomer Size on the Electronic Structure of Thiophene-Like Polymers

R. Lazzaroni[1], R. Sporken[1], J. Riga[1], J.J. Verbist[1], J.L. Brédas[2;a], R. Zamboni[3], and C. Taliani[3]

[1]Laboratoire Interdépartemental de Spectroscopie Electronique, Facultés Universitaires Notre-Dame de la Paix, 61, Rue de Bruxelles, B-5000 Namur, Belgium
[2]Laboratoire de Chimie Théorique Appliquée, Facultés Universitaires Notre-Dame de la Paix, 61, Rue de Bruxelles, B-5000 Namur, Belgium
[3]Istituto di Spettroscopia Molecolare-CNR, 1, Via dè Castagnoli, I-40126 Bologna, Italy

Abstract

Recently, it has been predicted theoretically and observed experimentally that doping polythiophene to high levels can lead to a metallic-like material. In this work, we investigate the electronic structure of polythiophene and related fused-ring systems by X-ray photoelectron spectroscopy (XPS), looking for the consequences of the gap closure on the core level lineshapes. Valence band spectra of polymers prepared from thiophene and bithiophene are compared.

From both theoretical and experimental studies, it has been clearly shown that the outstanding electrical and optical properties observed in conjugated polymers are related with the presence of electronic states in the gap between the valence and conduction bands. These states, corresponding to charged defects on the polymer chains, are generally created during the doping process, but they can also be photogenerated. In systems with a nondegenerate ground state, such as polypyrrole or polythiophene, the comparison of theoretical data with optical and magnetic measurements indicates that the first step in the oxidation of the neutral polymer leads to radical-cation defects (polarons), which can recombine at higher doping levels to form spinless dications (bipolarons).

In this paper, we investigate the electronic structure of a series of sulfur-containing conducting polymers (Fig. 1), comparing the X-ray photoelectron spectra (XPS) of the oxidized and reduced (neutral) states. The second part of this work deals with the valence band spectra of polythiophene obtained from the monomer or from the dimer.

Experimental

The polymers have been prepared electrochemically in a one-compartment three-electrodes cell, using acetonitrile solutions with 10^{-1} M LiBF$_4$ or LiClO$_4$ as electrolytes. The films were grown on stainless steel or conducting glass substrates at a

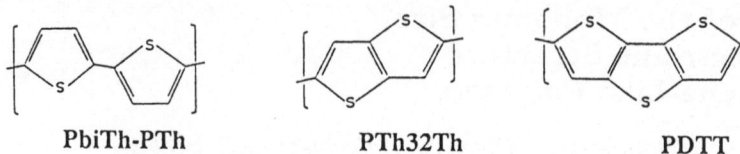

PbiTh-PTh   PTh32Th   PDTT

Fig. 1: Poly(bi)thiophene (PbiTh-PTh), Polythieno[3,2-b]thiophene (PTh32Th),
        Polydithienothiophene (PDTT).

constant potential of 1.8 (thiophene), 1.0 (bithiophene), 1.1 (Th32Th [1]) or 1.0 V
(DTT [2]) vs. the Ag/Ag+ reference electrode, whose characteristic potential is +0.3 V
vs. SCE.

XPS core level analysis

A recent XPS study of poly-3-Me-thiophene [3] focused on the modifications in the
C1s lineshapes between the oxidized and reduced states. In this paper, we mainly
analyse the evolution of the S2p level; for this line, the 2:1 intensity ratio between the
$2p_{3/2}$ and $2p_{1/2}$ components (arising from the spin-orbit coupling) limits the freedom in
the fitting procedure, restricting the number of possible hypotheses. As shown in Fig.
2a, the S2p peak in reduced PbiTh is correctly approximated by a single pair of lines
(1,2) and a small shake-up satellite (lines 3,4; separation: 2.3 eV, ratio 40:1), whereas the
same fitting pattern completely fails in reproducing the spectrum of the oxidized polymer
(Fig. 2b). In our interpretation, this discrepancy, observed throughout the series, could
be due (i) to the existence of different types of sulfur atoms or (ii) to intrinsic low-energy
electronic excitations.

In the first hypothesis, some of the sulfur atoms, located closer to anion sites, are
shifted with respect to "unaffected" atoms. This situation can be represented by a sum of
three doublets (Fig. 2c): unaffected, slightly affected (shift: 1.1 eV) and strongly affected

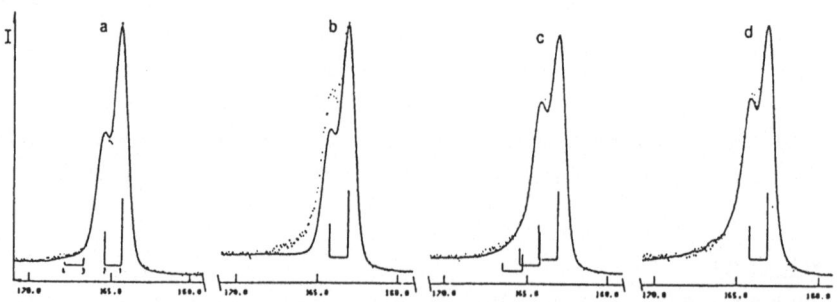

Fig. 2: S2p level: reduced PbiTh (a); oxidized PbiTh fitted with one doublet (b);
        three doublets (c); one doublet using the Doniach-Sunjic lineshape (d)

(shift: 2.1 eV) sulfur atoms. However, as already pointed out for poly-3-Me-thiophene, the intensities of the second and third components do not seem to be related to the doping level, in opposition to what would be expected.

In the second explanation, the lines would be skewed due to electronic excitations near the Fermi level screening the photogenerated core hole. In polythiophene, VEH theoretical calculations [4] have shown that the bandgap is dramatically reduced when the doping level increases and that eventually the merger of an empty bipolaronic band with the valence band could lead to a metallic-like band structure. The latter prediction has been confirmed by ESR measurements on 30%-doped PTh [5]. Since the bandgap of the neutral polymer is around 2.0 eV in this whole series of heteroaromatic compounds [6], the same evolution in the band structure can be expected upon doping. The experimental data have then been fitted with one doublet (Fig. 2d), using a Doniach-Sunjic function (asymmetry parameter $\alpha=0.2$); this lineshape, generally used for metallic compounds, is related, through the value of α, with the density of states (DOS) at the Fermi level. However, in these polymers, no DOS at the Fermi level is detected in the XPS spectra (Fig. 3), perhaps because of the intrinsically low photoemisssion cross section of the π bands. But it is to be noticed that even with UPS, a technique more sensitive to π levels, no DOS is observed below 0.2 eV [3].

On the other hand, the increase of the shake-up intensity, proposed by Jugnet et al. to explain the modification of the lines in the doped polymer, does not seem to be consistent with our results and the very small values of the bandgaps. At this point, it appears that no simple model thoroughly explains the evolution of the XPS lineshapes in sulfur-containing heteroaromatic polymers.

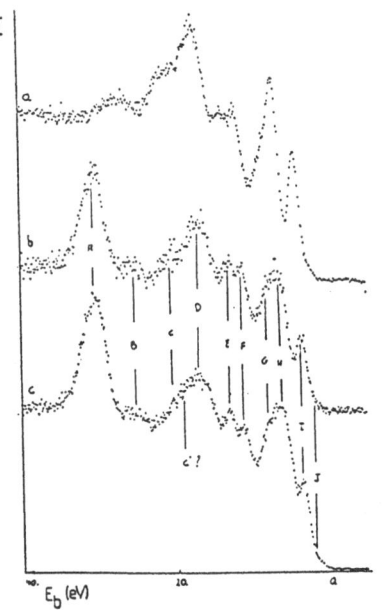

Fig. 3: XPS valence spectra:
a-reduced PbiTh,
b-oxidized PbiTh,
c-oxidized PTh

Valence band spectra

Since the polythiophene framework is believed to be constituted mainly from $\alpha-\alpha'$ linked monomer units, performing the polymerization from thiophene or $\alpha-\alpha'$ bithiophene leads to materials with very similar properties. At the XPS level, the core lines in polythiophene and polybithiophene are almost identical, but some slight differences arise in the valence spectra.

In both BF_4-doped compounds (Fig. 3b,c), the binding energies of lines A to J are 30.1, 24.8, 20.2, 16.6, 12.4, 10.8, 7.2, 5.8, 3.1 and 0.9 eV respectively. In polythiophene, line F seems somewhat lower and line C does not appear clearly, perhaps because of a third component (C') whose intensity is stronger in this polymer. These changes could be related with differences in the σ-backbone of the chains, due to the higher electrochemical potentials needed for thiophene polymerization.

It is noteworthy that modifications in the DOS related with different chain conformations have been predicted theoretically for cis-transoid and trans-cisoid polyacetylene [7]. Finally, the comparison of the spectra of oxidized and reduced PbiTh (Fig. 3a,b) points out the contributions of the BF_4^- counterions: line A corresponds to F2s-based levels and the disappearing of line G in the reduced polymer constitutes an evidence for the location of the F2p orbitals around 7.2 eV.

References
a)   Chercheur Qualifié of National Fund for Scientific Research (FNRS).
1.   R. Danieli, C. Taliani, R. Zamboni, G. Giro, M. Biserni, M. Mastragostino, A. Testoni: Synth. Met. 13, 325 (1986).
2.   P. DiMarco, M. Mastragostino, C. Taliani: Mol. Cryst. Liq. Cryst. 118, 241 (1985).
3.   Y. Jugnet, G. Tourillon, T. Minh Duc: Phys. Rev. Lett. 56, 1862 (1986).
4.   J.L. Brédas, B. Thémans, J.G. Fripiat, J.M. André, R.R. Chance: Phys. Rev. B 29, 6761 (1984).
5.   G. Tourillon, D. Gourier, F. Garnier, D. Vivien: J. Phys. Chem. 88, 1049 (1984).
6.   J. Riga, Ph. Snauwaert, A. De Prijck, R. Lazzaroni, J.P. Boutique, J.J. Verbist, J.L. Brédas, J.M. André, C. Taliani: Synth. Met., in press.
7.   J.L. Brédas, R.R. Chance, R. Silbey, G. Nicolas, Ph. Durand: J. Chem. Phys. 75, 255 (1981).

Ion Implantation of Polythiophene

H. Isotalo, H. Stubb, and J. Saarilahti

Technical Research Centre of Finland, Semiconductor Laboratory,
Otakaari 7 B, SF-02150 Espoo, Finland

Polythiophene as a free-standing film has been ion implanted
with F^+ -ions having 25 keV energy. A rapid conductivity
increase starts at doses between $5 \cdot 10^{15} \ldots 5 \cdot 10^{16}$ F^+/cm^2. The
conductivity changes by 6 orders of magnitude being $1 \cdot 10^{-2}$
$\Omega^{-1} cm^{-1}$ at a highest dose of $1 \cdot 10^{17}$ F^+/cm^2. RBS spectra show
that a $0.35 \ldots 0.40$ μm surface layer is damaged. The increasing
conductivity is accompanied by a decrease in its activation
energy and the optical energy gap at room temperature.

INTRODUCTION

Ion implantation is one way of forming a highly conducting
layer into an electrically insulating host polymer. The conductivity is often quite stable in comparison to the results
of other doping methods. Thus ion implantation can be used for
example in microelectronics when preparing conducting regions
into a polymer. Different kinds of devices or junctions can
also be processed: implanting a lightly predoped polymer with
an ion giving opposite kind of conductivity results in a stable pn - junction, which is not destroyed by diffusion as for
most other doping methods, WADA et al. /1/.

The mechanisms which produce the conducting regions are
dependent on many parameters and are not fully understood. The
energy of the incoming ion determines which type of interactions are important: electronic or nuclear. So when the ions
penetrate into the polymer the type of interaction changes. If
the ion energy is low enough and the sample thick enough the
ions stop in the polymer and may give rise to doping effects.
Many side-effects are known to occur: for example sputtering
of the surface, gas evolution and shrinkage of the polymer,
BARTKO et al. /2/.

The purpose of the present work is to study the effects of
low-energy implantation in polythiophene. This polymer is
known to be thermally stable even in doped form up to some 150
$^\circ C$, ÖSTERHOLM et al. /3/. We also wanted to compare our

earlier studies on doped polymers, for example polyparaphenylene and polythiophene to ion-implanted polythiophene, KUIVA-LAINEN et al. /4/.

1 EXPERIMENTAL

Polythiophene has been electrochemically synthesized and reduced to electrically insulating free-standing films. F^+ ions were implanted from a BF_4 source normally used for B doping of silicon. The implantation energy was kept at 25 keV and the doses in this series were $5 \cdot 10^{15}$, $5 \cdot 10^{16}$ and $1 \cdot 10^{17}$ F^+/cm^2, see table 1. The ion beam was scanned over an area corresponding to a three-inch silicon wafer, the polymer sample being some square centimeters. The maximum beam current was ~ 100 μA, which corresponds to ~ 2 μA/cm^2. To avoid serious heating effects the samples were attached firmly to an aluminum substrate by metal frames, part of the sample being behind the frame. The temperature was monitored with a test sample: polythiophene doped with $FeCl_3$ is known to be stable up to 150 $^{\circ}$C. The conductivity of our test samples did not markedly change due to implantation, which means that the effective temperature was less than 150 $^{\circ}$C.

Rutherford backscattering spectra (RBS) were measured using 2 MeV He^+ ions from a van de Graaf accelerator. To avoid destroying the samples low doses of 10^{14} He^+/cm^2 were used, NAMAVAR et al. /5/. Further, the sample was rotated so that the fixed He^+-beam drew a circle with a radius of ~ 7 mm on the polymer surface. The backscattered ions were detected from an angle of 170 $^{\circ}$.

DC-conductivity measurements were done as a function of temperature in the range 4...300 K mainly by the 2-point method, although room-temperature values were checked by 4-point measurements. Pressure or graphite contacts were used. The conductivity values were calculated assuming that the active region is the straggling layer of the F^+ ions. This layer has been calculated to be centered at a depth of R_p = 1000 Å in the polythiophene with the halfwidth of the Gaussian ion profile ΔR_p = 400 Å. When using the effective thickness ΔR_p, $2\Delta R_p$ or R_p the maximum correction factor in DC-conductivity is in any case less than 3. - A universal analytic approximation, introduced by J.P. Biersack, was used to estimate the projected range R_p and straggling ΔR_p. In this approximation the electronic and nuclear stopping powers are described by the well-known LSS theory.

2 RESULTS AND DISCUSSION

The conductivity of samples 3 and 4 are calculated using ΔR_p = 400 Å as the thickness while for samples 1 and 2 the whole sample thickness 50 μm is used. This is because the measurement current and activation energy, 380 meV, in these samples are the same. The activation energy of sample 1, 380 meV, is lower than the one normally found in undoped polythiophenes /3/. This effect may come from the fact that sample 1 was under mask during the implantation and is thus heat treated at a temperature less than 150 ^{o}C.

Table 1. Samples studied. - Series "a" are from J.-E. Österholm, Neste Oy and series "b" from our own synthesis.

Sample	Dose (F^+/cm^2)	Conductivity $(\Omega^{-1}cm^{-1})$ 300 K	Activation energy (meV) 300 K
1	———	$7 \cdot 10^{-8}$	380
2a	$5 \cdot 10^{15}$	$7 \cdot 10^{-8}$	380
2b	$5 \cdot 10^{15}$	$1 \cdot 10^{-8}$	380
3a	$5 \cdot 10^{16}$	$9 \cdot 10^{-4}$	180
3b	$5 \cdot 10^{16}$	$2 \cdot 10^{-3}$	150
4a	$1 \cdot 10^{17}$	$1 \cdot 10^{-2}$	120
4b	$1 \cdot 10^{17}$	$1 \cdot 10^{-2}$	120

DC-conductivity data are fitted in fig. 1 according to (1) and (2)

$$\sigma_{DC} = \sigma_0(T)\ e^{-(T_0/T)^{1/3}}\ , \qquad (1)$$

$$T_0 = \frac{3}{k_B \pi \alpha^2 N(E_F)}\ , \qquad (2)$$

where α is the localization length and $N(E_F)$ the density of gap states at the Fermi level E_F.

This two-dimensional hopping gives the best fit and indicates that the highly conducting ion-implanted region near the surface may be very narrow, of the order of 100 Å. The hopping conductivity can occur between dangling bonds caused by F^+

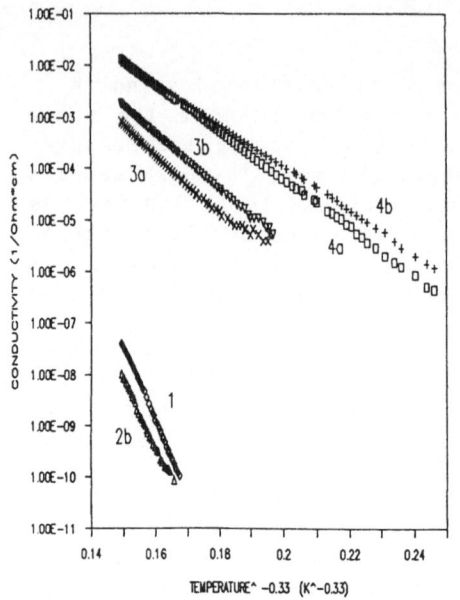

Fig 1. DC-conductivity of 25 keV
F$^+$ ion-implanted polythiophene
as a function of $T^{-1/3}$. Sample
numbers are from table 1.

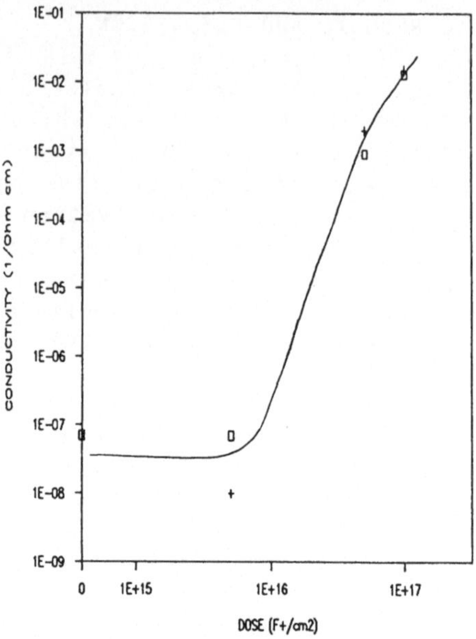

Fig 2. DC-conductivity vs
dose for ion-implanted
polythiophene.

ions. On the other hand doping effects can also have effects
on the conductivity; the highest dose corresponds to an impurity concentration of $2.5 \cdot 10^{22}$ F$^+$ ions/cm^3 or some 4...5
ions/repeat unit of polythiophene.

In fig. 2 one can clearly see the starting increase in
conductivity at a dose of $5 \cdot 10^{15}$ F$^+$/cm^2; a dose of $5 \cdot 10^{16}$ F$^+$
/cm^2 is already close to the saturation limit.

RBS spectra are seen in fig. 3. From the spectrum of sample
1 one gets a C to S ratio of 4, which corresponds to the polymer structure C_4SH_2. When comparing the highest dose samples
to the unimplanted ones the only difference is in the surface.
Using a density of 0.7 g/cm^3 the damage layer is calculated to
be 0.35...0.40 µm deep. This depth is much larger than the one
we get from a computer simulation, ΔR_p =0.1 µm, and may give an
error by a factor of 10 to our DC-conductivity data. A natural
explanation could be that the ions travel much deeper into
polythiophene because of the porous nature of the material.
The nature of the surface defects is unknown; ESCA and Raman
could be ways of studying the phenomena at the surface.

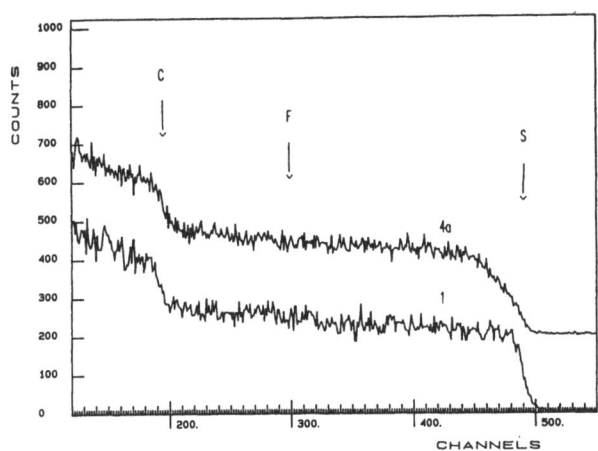

Fig 3. RBS spectra for unimplanted and heavily implanted poly-thiophenes. Sample numbers are from table 1.

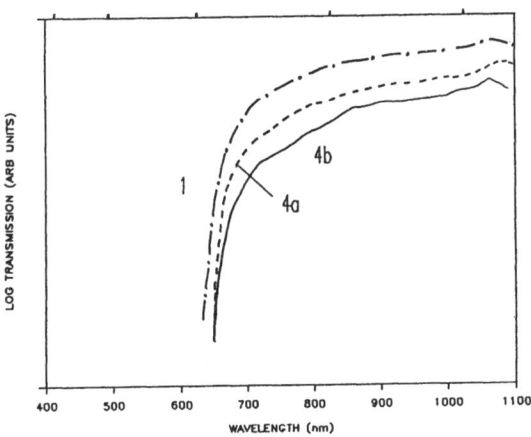

Fig 4. Optical transmission spectra. Sample numbers are from table 1.

Due to the fact that RBS sees only absolute concentrations of atoms, the implanted F^+ -ions could not be seen even for the most heavily implanted samples.

Optical transmission spectra show how the forbidden energy gap decreases upon implanting, fig. 4. The overall trans-mission decreases by a factor of 2...3. Doping-induced peaks in the transmission are not seen in this figure; on the other hand these would be difficult to distinguish because the sample thickness, 50 μm, is much larger than the active layer, 0.1 μm.

In conclusion, polythiophene can be ion implanted to give electrically conducting material. This conductivity increase is accompanied with a decrease in the optical gap and loss of sulphur at the surface. The DC conductivity mechanism is interpreted to be two-dimensional hopping in the implanted layer. Doping-induced effects could not be directly seen due to dominating effects of the host polymer. We did not see any significant difference between the two series of samples studied.

LITERATURE

1. T. Wada, A. Takeno, M. Iwaki, H. Sasabe, Y. Kobayashi: J. Chem. Soc. Chem. Commun., 1194 (1985).

2. J. Bartko, B.O. Hall, K.F. Schoch Jr.: J. Appl. Phys 59, 1111 (1986).

3. J.E. Österholm, P. Passiniemi, H. Isotalo, H. Stubb: Synth. Metals 18, 213 (1987).

4. P. Kuivalainen, H. Stubb, H. Isotalo, P. Yli-Lahti, C. Holmström: Phys. Rev. B31, 7900 (1985).

5. F. Namavar, J.I. Budnik: Nucl. Instr. and Meth. B15, 285 (1986).

Characterization of Polypyrrole-Polyvinyl Alcohol Composite Prepared by Chemical Oxidation of Pyrrole

A. Pron[1] *and K. Wojnar*[2]

[1]Centre d'Etudes Nucléaires de Grenoble, Département de Recherche Fondamentale/Service de Physique, Groupe Dynamique de Spin et Propriétés Electroniques, 85 X, F-38041 Grenoble Cedex, France; Department of Chemistry, Warsaw University of Technology, PL-00-664 Warsaw, Noakowskiego 3, Poland

[2]Department of Chemistry, Warsaw University of Technology, PL-00-664 Warsaw, Noakowskiego 3, Poland

Polyvinyl alcohol-polypyrrole conducting composites can be prepared by the exposure of solid pva films containing $FeCl_3$ or $K_3Fe(CN)_6$ to the vapours of pyrrole. It has been shown that the distribution of the oxidant within the pva matrix is crucial for the morphology and the electrical properties of the composites obtained.

In the case of pva-$FeCl_3$ the iron salt is homogeneously distributed within the pva matrix and as a result the polypyrrole film obtained on pva-$FeCl_3$ support is uniform. In the pva-$K_3Fe(CN)_6$ films $K_3Fe(CN)_6$ precipitates from the polymer matrix as a separate, crystalline phase and the resulting polypyrrole film is non-uniform and porous. The composite prepared with the use of $FeCl_3$ as oxidant shows therefore higher electrical conductivity than the corresponding composite prepared with the use of $K_3Fe(CN)_6$.

1. Introduction

Polypyrrole films are usually prepared by anodic oxidation of pyrrole in suitable aqueous or non-aqueous electrolytes [1], therefore the formation of composites consisting of polypyrrole and a non-conducting polymer is rather difficult. Few successful attempts of the application of electrochemical techniques to the formation of polypyrrole-based composites have however been reported [2]. Recently it has been discovered [3-4], that pyrrole can easily polymerise chemically on the surface of a polyvinyl alcohol film containing $FeCl_3$. The conductivity and optical transmittance of the films prepared in such a manner depend on the content of $FeCl_3$ within the pva-matrix used for the pyrrole polymerisation. For high $FeCl_3$ contents the conductivities up to 15 $ohm^{-1}cm^{-1}$ are obtained. For smaller $FeCl_3$ contents the conductivities of ca 10^{-2} $ohm^{-1}cm^{-1}$ are measured but those films exhibit high transmittance in the visible range.

In this communication we present the results of some spectroscopic, diffraction and morphological studies of pva-polypyrrole composite prepared chemically.

2. Experimental

Two types of oxidant were used for pyrrole polymerisation on pva : $FeCl_3$ and $K_3Fe(CN)_6$. In typical experiments aqueous solutions containing a mixture of pva with FeCl$_3$ in molar ratios ranging from 100 : 1 to 3 : 1 or a mixture of pva with K$_3$Fe(CN)$_6$ in molar ratios up to 100 : 1 were prepared. These solutions were then poured onto a polyester support and the solvent was evaporated in air at 323K. Films of pva-transition metal salt so formed were then suspended over a 50% V/V solution of pyrrole in ethanol and kept in a static vacuum for periods ranging from 1-7 days. Under such conditions, polymerisation of pyrrole on pva surface occurred leading to a conducting composite with good mechanical properties. The pva-transition metal salt matrix used for the polymerisation and the composite formed after the polymerisation were subjected to X-ray diffraction, Mössbauer spectroscopy, EPR spectroscopy and scanning electron microscopy studies.

3. Results and discussion

The X-ray studies of solid pva films containing up to 30% of FeCl$_3$ show that FeCl$_3$ is homogeneously distributed within the pva matrix since no phase separation between the polymer and the iron salt occurs. There are no Bragg reflection corresponding to crystalline FeCl$_3$.6H$_2$O. The only Bragg reflection at d = 4.52 A which can be observed for lower FeCl$_3$ contents is characteristic of crystalline polyvinyl alcohol. The insertion of FeCl$_3$ into the pva matrix induces increasing disorder since the reflection at d = 4.52 A gradually broadens and finally disappears with increasing FeCl$_3$ concentration. Further confirmation that no separate, crystalline phase of hydrated FeCl$_3$ is formed in the pva-FeCl$_3$ films comes from our previous ME studies [4]. The Mössbauer spectrum of pva-FeCl$_3$ recorded at 77K, although being typical of Fe(|||), is totally different from that characteristic of FeCl$_3$.6H$_2$O. Moreover, the EPR spectra of pva-FeCl$_3$ show characteristic lines at y = 4.31 and a broad shoulder in the g-region 6-10 attributable to high spin Fe(|||) but the spectrum differs significantly from the EPR spectrum of a mechanical mixture of pva and FeCl$_3$.6H$_2$O. It therefore shows that the coordination environment of iron must be different in pva-FeCl$_3$ films than in the crystalline FeCl$_3$.6H$_2$O phase.

The X-ray diffraction patterns obtained for pva-K$_3$Fe(CN)$_6$ films show that, contrary to the pva-FeCl$_3$ case, the salt in the film is present as a separate crystalline phase even at very low K$_3$Fe(CN)$_6$ concentrations. The diffraction patterns can be considered as a superposition of the reflections characteristic of crystalline K$_3$Fe(CN)$_6$ with those characteristic of crystalline pva. Moreover no influence of the increasing salt concentration on the pva reflections, is observed. Therefore the system can be regarded as a two phase heterogeneous mixture. The EPR spectrum of pva-K$_3$Fe(CN)$_6$ shows a single, very broad (ΔH_{pp} = 6000 G) line, centered at g = 2 typical of low spin Fe(|||).

The distribution of the oxidant within the pva matrix strongly influences the morphology of polypyrrole polymerised on pva. A uniform film can be obtained on pva-FeCl$_3$ as seen by scanning electron microscopy whereas the film obtained on pva-K$_3$Fe(CN)$_6$ seems porous and shows many pits. This is obviously associated with non-homogeneous distribution of the oxidant within the pva matrix.

The reduction of iron salts occurring, in pva matrix, upon the oxidative polymerisation of pyrrole is manifested in the ME [4] and in the EPR spectra. The ME spectra of pva-FeCl$_3$ and pva-K$_3$Fe(CN)$_6$ after polypyrrole deposition show the presence of high spin Fe(|||) and low spin Fe(||), respectively. Similarly in the EPR studies carried out at 77K one observes a disappearance of all lines characteristic of Fe(|||) and only a weak signal at g ≈ 2 (ΔH_{pp} ≃ 1 G), due to unpaired spins in polypyrrole chains, can be observed. No signal attributable to Fe(||) are observed since they can only be seen at temperatures <4K.

1. See e.g. G.B. Street: In <u>Handbook of Conducting Polymers</u>, Vol.1, ed. by T.Y. Skotheim (Dekker, New York, 1986) p.265
2. M.A. de Paoli, R.Y. Waltman, A.F. Diaz and J. Bargon: J. Polym. Sci. Polymer Chem. Ed. <u>23</u>, 1687 (1985)
3. T. Oho, S. Miyata: Polym. Y. <u>19</u>, 95 (1986)
4. A. Pron, W. Fabianowski, C. Budrowski, Y.B. Raynor, Z. Kucharski, J. Suwalski, S. Lefrant and G. Fatseas: Synthetic Metals <u>18</u>, 49 (1987)

Electrochemistry of Polypyrrole-Ferrocyanide Films

M. Zagórska

Institute of Inorganic Technology, Warsaw University of Technology,
PL-00-664 Warsaw, Noakowskiego 3, Poland

1. Introduction

In the majority of the studies on polypyrrole films common counterions have been used as dopants . However, recently growing interest in the incorporation of more complex species exhibiting redox properties can be noticed (e.g. /1/-/3/). It is mainly connected with the catalytic application of the polymer-modified electrodes. In addition, the presence of the electroactive doping anions makes possible to use some electrochemical methods which provide complementary information on the dopant structure and on the ion transport in the polypyrrole film.

In this work the rotating ring-disk electrode (RRDE) experiments were employed for studying polypyrrole-ferrocyanide films.

2. Experimental

The disk electrode in the ring-disk (platinum) system was covered with polypyrrole films formed potentiostatically in the solutions of 0.05 M pyrrole. Electrolytes in the polymerization process consisted of 0.2 M KNO_3 and 0.01M $K_3Fe(CN)_6$ in water or 0.1 M $(Bu_4N)_3Fe(CN)_6$ in acetonitrile. The thickness of the films obtained in aqueous solutions was determined by ellipsometric measurements "in situ" /4/. To perform rotating ring-disk experiments, RRDE with polypyrrole coated disk was transferred to 0.2 M KNO_3 solution containing neither monomer nor ferricyanide anions. The electrodes were rotated at $\omega = 170 \ s^{-1}$. A saturated calomel electrode (SCE) was used as a reference in all experiments.

3. Results and Discussion

The results of cyclic experiments performed on RRDE are presented in Fig.1. The potential of the disk covered with the polypyrrole-ferrocyanide film was swept from +0.3 V to -0.9 V and back to +0.3 V vs SCE while the platinum ring potential was kept constant at +0.5 V vs SCE.Disk currents and ring currents were recorded simultaneously during the cycles. In Fig.1-a there are two well pronounced reduction peaks of the disk current: first one connected with the ferrocyanide anions incorporated into the polymer structure and second peak which is due to the reduction of the polymer chain. Redox species leaving the polypyrrole film as a result of the polymer reduction are detected on the ring electrode (Fig.1-b). It can be seen that all ferrocyanide anions leave the film during the first sweep of the potential. In the second sweep the peak of the anion reduction current completely disappears. Similar results were reported recently /5/.

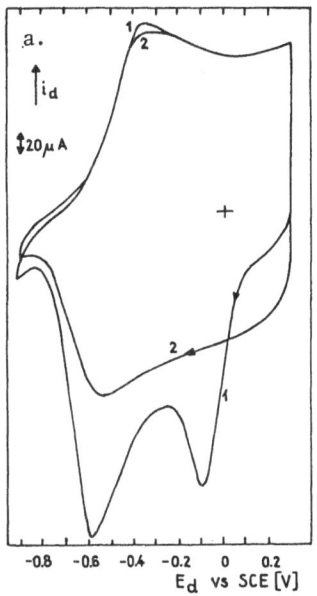

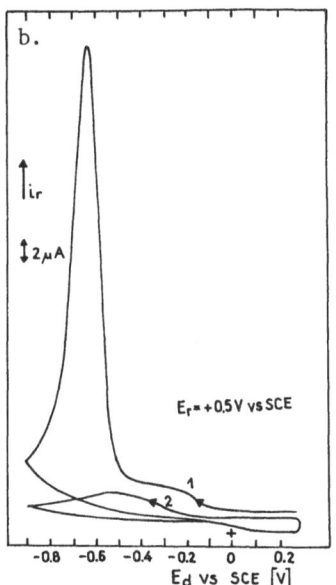

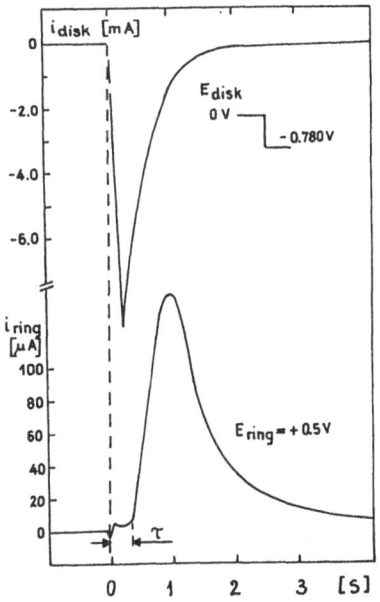

Fig. 1. Cyclic voltammograms at RRDE in 0.2 M KNO_3, ν = 50 mV/s, polypyrrole film grown on the disk in aqueous solution at $E_{polym.}$ = 0.6 V vs SCE, d = 2100 Å. a - disk current vs disk potential, b - ring current vs disk potential; 1 - first sweep, 2 - second sweep

Fig. 2. Chronoamperometric curves at RRDE in 0.2 M KNO_3, polypyrrole film grown in aqueous solution at $E_{polym.}$ = 0.6 V, d = 4250 Å

In Fig.2 chronoamperometric curves obtained by stepping the disk potential from 0 V to -0.780 V are presented. Ferrocyanide anions leaving the reduced polypyrrole arrive at the ring after the delay time (τ) (Fig.2). Assuming that it is caused by slow diffusion of dopant through the polymer film, the value of the diffusion coefficient (D) can be estimated. According to the approach proposed by DOBLHOFER et al. /6/:

Table 1. The estimated values of diffusion coefficient (D) of ferrocyanide anions in polypyrrole films

$Q_{polym.}$ (mC/cm^2)	d (Å)	τ (s)	$D \times 10^9$ (cm^2/s)
128	2100	0.1	1.4
255	4250	0.3	1.9
380	6400	0.6	2.2

$$D = d^2/\pi\tau \tag{1}$$

where d is the film thickness. The calculated values of D are given in Table1.

From the ring currents the amount of the dopant (N) leaving the polypyrrole film can be calculated:

$$N = \frac{1}{F\,K}\int_0^t i\ dt \tag{2}$$

where K is the collection efficiency. In Fig.3 the amounts of ferrocyanide anions which leave the polypyrrole film are shown. In the case of the films polymerized in aqueous solutions the dopant concentration calculated from the presented curve is 14% w/w. It is in good agreement with the value of 15% w/w obtained previously /7/ for thick films synthesized in the same conditions.

The value of N was found to be strongly dependent on the potential of polypyrrole synthesis (Table 2). The higher the polymerization potential the smaller the amount of ferrocyanide anions which leave the polymer film.

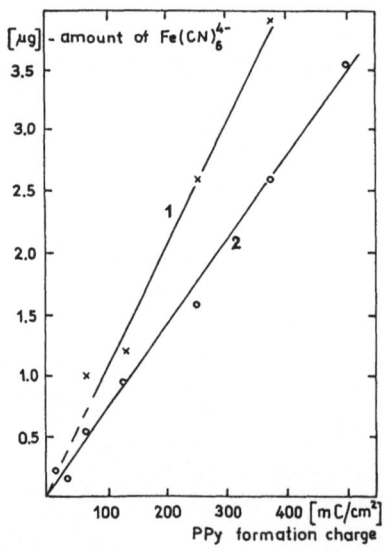

Fig. 3. Dependence of the amount of ferrocyanide anion leaving the polypyrrole film during the first redox cycle on the polymer formation charge 1 - polymerization in acetonitrile at $E_{polym.}$ = 0.7 V vs SCE, 2 - polymerization in aqueous solution at $E_{polym.}$ = 0.6 V vs SCE

Table 2. The amount of ferrocyanide anion leaving the polypyrrole film during the first reduction-oxidation cycle

$E_{polym.}$ (V)	0.60	0.75	0.90	1.25
N (%)	100	41	13	10

4. Conclusions

We have demonstrated that RRDE experiments are powerful tools to investigate the redox properties of polypyrrole films. It has been found that the apparent value of diffusion coefficient of ferrocyanide anions in the polypyrrole film is ca 10^{-9} cm^2/s. This considerably high value seems to be caused by porosity of the film and additional transport of the anions via electrolyte present in the pores.

The amount of dopants leaving the film during the reduction of polypyrrole depends strongly on the potential applied in the polymerization process. The loss of ability to release the incorporated anions may be explained by a loss of conjugation or opening of the pyrrole rings due to nucleophilic attack of water, occurring at higher potentials.Nevertheless, it cannot be totally excluded that at higher polymerization potential the structure of the polymer changes into a more compact one.

References

1. R.A. Bull, F.-R. Fan and A.J. Bard: J. Electrochem. Soc., 131, 687 (1984)
2. M. Velazquez Rosenthal, T.A. Skotheim and C.A. Linkous: Synthetic Metals, 15, 219 (1986)
3. F. Bedioui, C. Bongars, J. Devynck, C. Bied-Charreton and C. Hinnen: J. Electroanal. Chem., 207, 87 (1986)
4. M. Zagórska, K. Brudzewski, R. Mińkowski and J. Przyłuski: to be published
5. B. Zinger and L.L. Miller: in Ext. Abstr. from the International Workshop "Electrochemistry of Polymer Layers" P 2.7 (Duisburg, Sept. 15 - 17, 1986)
6. K. Doblhofer, H. Braun and R. Lange: J. Electroanal. Chem., 206, 93 (1986)
7. J. Przyłuski, M. Zagórska, A. Proń, Z. Kucharski and J. Suwalski: J. Phys. Chem. Solids, in press

Synthesis and Physical Properties of 3,4-Disubstituted Polypyrroles

J. Rühe, Ch. Kröhnke, T. Ezquerra, and G. Wegner

Max-Planck-Institut für Polymerforschung,
Jakob-Welder-Weg 11, D-6500 Mainz, Fed. Rep. of Germany

A. Introduction

The electrochemical and electrical properties of conducting polypyrrole films can be altered either by changing the counterion [1], the conformation of the counterion-chain in case of polymeric counterions [2] or by introducing substituents onto the pyrrole ring [3,4]. When the chemical structure of the polymer is altered several questions arise:
- How do the electrical properties of the polymeric film change when bulky substituents are introduced?
- Do alkyl substituents improve the solubility of the polymer?
- What information about the supermolecular structure of the polymeric film can be obtained by changing the substituent structure?

To answer these questions pyrrole derivatives substituted in the 3,4-position by a fused alkyl ring were synthesized and investigated.

B. Synthesis of the Monomers[5]

The monomers were synthesized according to eq. 1 and 2.

$$n = 2,3,4,9 \tag{1}$$

$$\tag{2}$$

The monomers were purified by flash-chromatography using silica gel and characterized by IR, ^{13}C-NMR and elemental analysis.

C. Electropolymerization

All electropolymerizations were carried out at controlled potential in single compartment cells. Thin films (100 - 1000 Å) were left attached to the electrode and used for electrochemical and spectroscopic measurements. Thick films (~ 0.1 mm)

were removed from the electrode, rinsed, dried and used for electrical measurements.
The electropolymerization is shown in eq. 3.

(3)

Cyclovoltammetric measurements on the monomers show that all oxidation reactions are completely irreversible due to polymerization. A typical voltammogram indicating the electropolymerization is shown in Fig. 2.

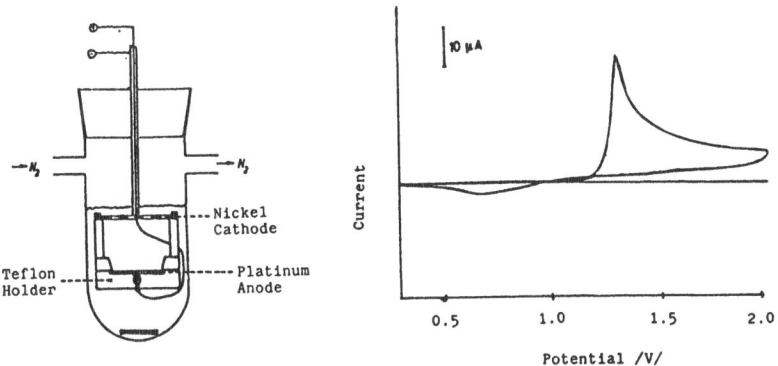

Fig.1: Electrolysis cell
Fig.2: Cyclic voltammogram of 2 (conc. 10^{-3} mol/l) in $CH_3CN/$ TBACIO$_4$ vs. Ag/Ag$^+$; sweep rate 100 mV/s

D. Electrochemical Characterization of the Polymeric Film

The galvanostatic deposited films were removed from the electrolyte, rinsed with methanol, dried in vacuum and immersed in a supporting electrolyte. Using a current of 10 µA, an electrode surface of $6.85 \cdot 10^{-3} cm^2$, and a calculated film thickness of 0.25 to 1 µm it was observed that
- the films could be reversibly reduced and reoxidized for more than 300 times within the potential window indicated by Fig. 3
- the peak separation was very small indicating that the electrochemical reaction proceeded quickly.
- the oxidation/reduction potentials shifted anodically with increasing ring size of the substituent
- application of a strongly negative potential (below -1 [V]) led to an irreversible reduction
- reduction/oxidation in water Na_2SO_4 led to an insulating film after several cycles.

Films with acceptable electrical properties could only be obtained from the bicycloalkanes 5 - 8 and not from the ketones 1 - 4 as indicated in Table 1.
Polymeric films of the azabicycloalkyl compound swell in DMSO,

Table 1: Results of cyclovoltammetric and electrical investigations

Compounds	Oxidation potential of the monomers[1]	$E_{p,c}$[2] polymer	$E_{p,a}$[2] polymer	Conductivity $[Scm^{-1}]$[3]
1	1.43	+ 0.56	+ 0.61	4)
2	1.43	+ 0.68	+ 0.93	4)
3	1.35	+ 0.68	+ 0.79	4)
4	1.46	–	–	4)
5	1.15	– 0.27	0.00	0.4
6	0.97	– 0.18	– 0.06	0.1
7	1.07	– 0.17	+ 0.26	0.01
8	1.08	+ 0.21	+ 0.55	$2 \cdot 10^{-4}$

1) all potentials in [V] vs Ag/Ag^+
2) $E_{p,c}$, $E_{p,a}$ potential of the cathodic and anodic peak resp.
3) measured at 25 °C by van der Pauw method
4) a small amount of material was deposited at the anode surface

DMF or pyridine. The electropolymerisation in THF of monomers with a fused twelve-membered alkyl ring led to the formation of purple solutions without film deposition.

E. Conductivity of the Polymers

The room temperature conductivity of all samples was measured using the four probe technique on pressed pellets of the polymers obtained as the perchlorate. The results are summarized in the last column of Table 1.
The conductivity of polypyrrole 9 synthesized under identical conditions was 1 [Scm^{-1}].
The polymers are rather stable in air; less than 0.5 % decay of the conductivity per day was observed keeping the samples in the open under laboratory conditions.
The temperature dependence of the conductivity of pressed pellets from 3,4-disubstituted polypyrroles electrodeposited from 99 % aqueous $CH_3CN/TBAClO_4$ was measured in the temperature range between 10 and 300 K. The results are shown in Fig. 4.

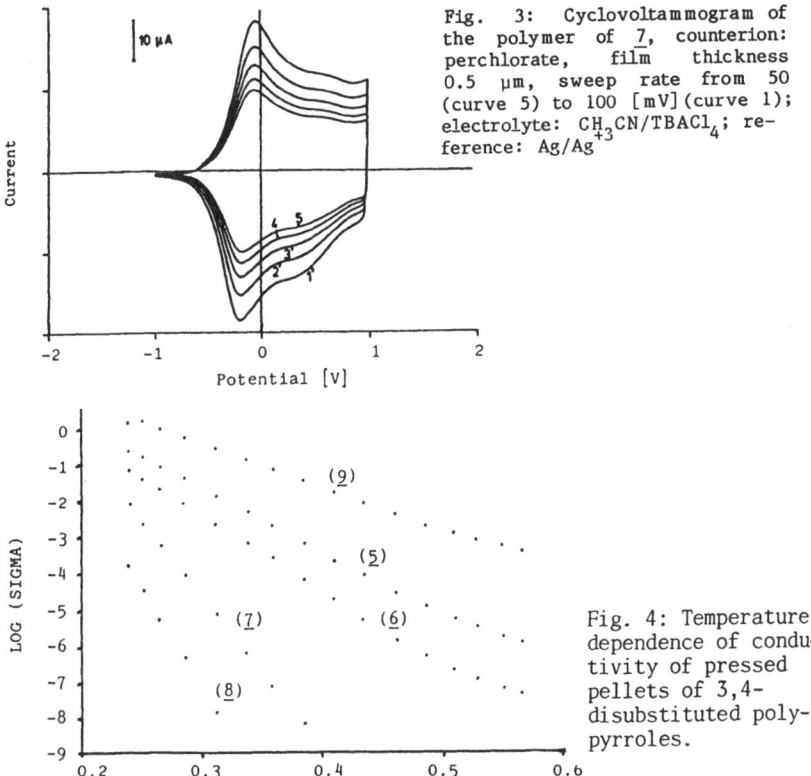

Fig. 3: Cyclovoltammogram of the polymer of $\underline{7}$, counterion: perchlorate, film thickness 0.5 μm, sweep rate from 50 (curve 5) to 100 [mV](curve 1); electrolyte: $CH_3CN/TBACl_4$; reference: Ag/Ag^+

Fig. 4: Temperature dependence of conductivity of pressed pellets of 3,4-disubstituted polypyrroles.

Both the absolute values of conductivity and the temperature coefficient increase with increasing substituent size. This is interpreted as an increase of the barrier width for the hopping of charge carriers between segments of adjacent chains. A similar behaviour was found recently in samples of unsubstituted polypyrrole having polymeric counterions, in which the barrier width could be changed by changing the counterion conformation [2].

Acknowledgement: This work was carried out in a program "Elektrisch leitfähige Polymere" supported by the BMFT through BASF, Ludwigshafen.

F. Literature
1) W. Wernet, M. Monkenbusch, G. Wegner, Mol. Liq. Cryst., 118, 193 (1985)
2) D. Glatzhofer, J. Ulański, G. Wegner, Polymer 28, 449 (1987)
3) J. M. Bureau, M. Gazard, M. Champagne, J. C. Dubois, G. Tourillon, F. Garnier, Mol. Liq. Cryst., 118, 235 (1985)
4) G. B. Street et al., J. Phys. C3, 599 (1983)
5) L. Traynor, U. S. patent 4,487,667 (1984)
6) G. Wegner et al., Synth. Met. 18, 1 (1987)

A New Polyparaphenylene Thin Film Obtained by Electroreduction: Characterization and Electrochemical Studies

G. Froyer, Y. Pelous, and G. Ollivier

C.N.E.T., LAB/ROC/TIC, BP 40, F-22301 Lannion Cedex, France

1. INTRODUCTION

Polyparaphenylene (PPP) is one of the most attractive conducting polymers as it shows a non-degenerate ground state. Much work has been done on this polymer in the last few years /1/ but it was carried out on either powder form polymer /2/ or films obtained by electrooxidation methods /3/. We have recently succeeded in forming PPP film onto various substrates by using electrochemical reduction of 1,4-dibromobiphenyl in the presence of a nickel complex /4/.

The aim of this paper is to point out some peculiarities of this PPP film grown on ITO electrodes in terms of morphology, optical absorption and electrochemical behavior.

2. SYNTHESIS SCHEME

The synthesis route is the following :
- generation (in situ) of Ni° complex (- 1.5 to - 2.0 V/Ag$^+$/Ag).
- insertion of the Ni° complex into a C-Br bond of 1,4-dibromobiphenyl.
- at a lower potential (- 2.6 V/Ag$^+$/Ag) reduction of the as-obtained compound into an anion which is the active species leading to polymerization

$$m\emptyset - Ni° \rbrack - Br \cdot \quad \xrightarrow{- 2.6\ V} \quad m\emptyset - Ni° \rbrack^-$$

$$m\emptyset - Ni° \rbrack^- + Br - \lbrack Ni° - \emptyset m \longrightarrow m\emptyset - \emptyset m + Br^- + 2\ Ni° \rbrack$$

The Ni° complex being regenerated is able to undergo new insertion which in turn gives further polymerization. Therefore one is dealing with a catalyzed electropolymerization and the knowledge of the polymerization index is not straightforward in such cases.

3. CHARACTERIZATION

The film deposit is transparent yellowish, homogeneously spread out on the electrode surface. Its thickness may be monitored up to about 1 μm. According to the potential used to grow the PPP, the material is obtained in the neutral form and shows up a porous morphology which consists of fibrils, about 500 to 1000 Å in diameter, very much like those of Kovacic PPP powder /5/.

The optical absorption spectrum is characterized (fig. 1) by a strong maximum centered at 380 - 400 nm and an absorption threshold corresponding to the interband transition at 2.8 eV. This is in agreement with FTIR data which indicate that the chains are rather long. Furthermore a shoulder at 410 - 420 nm is interpreted as coming from the contribution of longer chains according to Raman and fluorescence studies /6/.

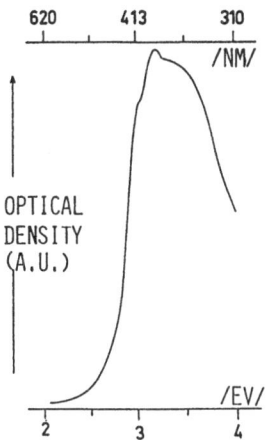

Fig. 1 : Optical absorption
spectrum of PPP film

4. RESULTS ON ELECTROCHEMICAL BEHAVIOR AND DISCUSSION

As mentioned above, we report here on the behavior of PPP films deposited on
ITO electrodes allowing both electrochemistry and spectroscopic measurements.

4.1 Cyclic Voltammetry

In order to carry out these experiments two different media were used : in the
reduction domain the electrolyte was a NBu_4BF_4 solution in THF, in the oxidation domain $LiBF_4$ in acetonitrile was employed. In both cases the counter
electrode was Pt and the pseudo reference electrode was simply a Ag wire immersed in the solution. It is interesting to note that large kinetic effects
show in the system, the internal resistance of the cell is high (organic electrolyte, ITO...) and this should be taken into account to explain the voltammogram variations as a function of sweep rate. Indeed, the best voltammograms
were recorded at 5 mV/sec.

   In Fig. 2 are reproduced the voltammograms in the oxidation (a) and reduction (b) domains. The onset of oxidation occurs at + 1.6 V/Ag wire and the
film switches from transparent yellowish to red. The reduction voltammogram is
better resolved with a reduction onset at -1.9 V/Ag wire and the corresponding
reoxidation peak appearing at about - 1.3 V/Ag wire. Under these conditions
the PPP film turns from transparent to a blue color.

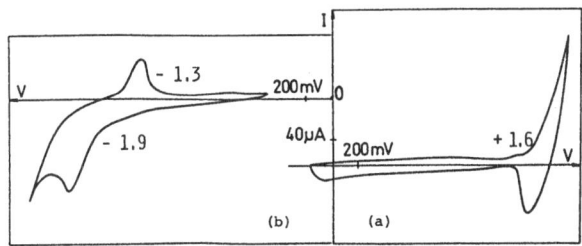

Fig. 2 : Voltammograms of PPP films (a) oxidation (b) reduction

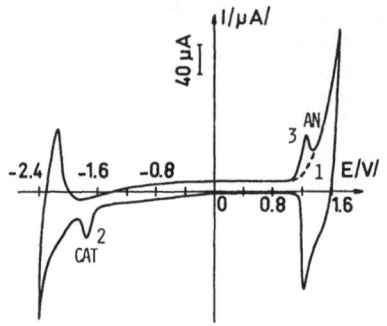

Fig. 3 : Voltammogram of
PPP film oxidized
and reduced in
the same medium,
acetonitrile-
NBu_4BF_4

We have succeeded in carrying out both experiments in the same medium. For that purpose the electrolyte was an NBu_4BF_4 solution in acetonitrile under similar conditions. Figure 3 reproduces the experimental voltammogram recorded in this experiment. Starting from the neutral film and going into the anodic domain the current intensity follows the dashed line (1) featuring the classical oxidation voltammogram. But when the potential is taken back to cathodic values a small peak (CAT) shows up at the onset of the regular reduction peak (2). If a second cycle is performed after the first one, another small peak (AN) grows on the foot of the oxidation peak (3). These new peaks do not appear as long as cycling is performed either in cathodic or anodic domain. In order to show up the film has to be taken first into the opposite domain.

This might be explained in the following way: the neutral film is fully oxidized in (1) and charges are then removed from the film at the interface electrode-film, building an insulating layer which hinders further charge removal, therefore charges far from the electrode remain in the film as long as the layer created is insulating. But when the potential is lowered to the reduction value (2) a conducting layer grows from the electrode surface through the film at the expense of the previously insulating domain. As soon as this insulating domain is consumed, a charge compensation occurs between charges - going through the film and charges + which remained in the film leading to a sharp increase in current at the very beginning of reduction (peak CAT). The same process applies for the peak AN (3) where charges + compensate the remaining charges - coming from the reduction cycle. Such a behavior was already described in the case of polyfluorenes /7/.

The mimicking peaks (AN and CAT) feature thus the very beginning of oxidation and reduction of PPP films. Their distance 2.8 V agrees well with the energy gap value obtained by optical absorption 2.8 eV.

4.2 Spectroelectrochemistry

Especially designed cells were used to perform spectro-electrochemical experiments /8/. The same media as for cyclic voltammetry were employed and the spectra were recorded at quasi-equilibrium for given potentials applied to the sample. Figure 4 shows the evolution of the absorption spectrum in the near-IR-visible range during oxidation. Two new absorptions grow at the expense of the interband transition when the potential is raised from + 0.2 V/Ag wire up to + 1.9 V/Ag wire. One is localized at 2.65 eV and the second one shifts from 0.8 eV to 1.0 eV as the potential is increased. The sketch shows that these two absorptions upon oxidation may be attributed to transitions 1' and 2' allowed by the presence of bipolarons + which generate two empty states

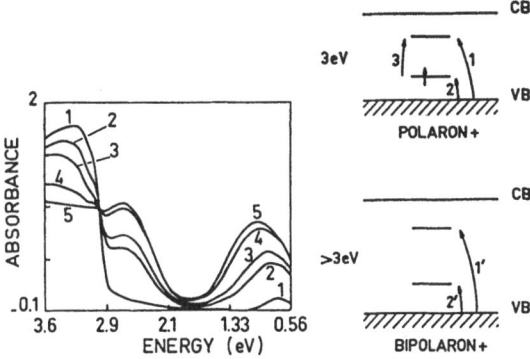

Fig. 4 : Evolution of optical absorption spectrum upon oxidation and sketch of the new transitions allowed by the presence of localized states within the gap. Applied potential/Ag wire : 1 : + 0.2 V ; 2 : + 1.5 V ; 3 : + 1.6 V ; 4 : + 1.8 V ; 5 : + 1.9 V

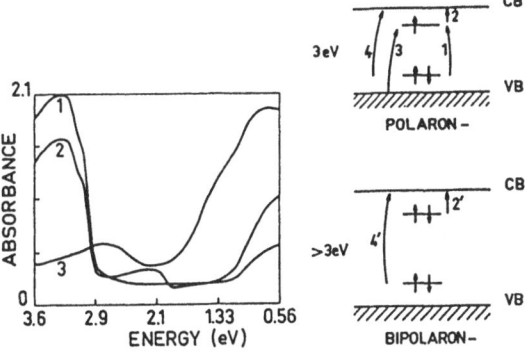

Fig. 5 : Evolution of optical absorption spectrum upon reduction of PPP film and sketch of the new transitions allowed by the presence of gap states. Applied potential/Ag wire. 1 : 0 V ; 2 : - 1.7 V ; 3 : - 1.9 V

within the band gap. It seems that even at light doping level no polarons are seen in the oxidation process. Figure 5 represents spectra corresponding to different reduction levels for the PPP film. In this case an asymmetric and weak absorption at 2.1 - 2.5 eV is generated at light reduction level as well as a huge one below 0.56 eV. As the potential is lowered down to - 1.9 V/Ag, which corresponds to the fully reduced PPP, the interband transition vanishes, the high energy absorption switches to a symmetric peak centered at 2.65 eV and the strong low-energy peak goes to 0.75 eV. Other experiments - including the kinetics of optical absorption evolution as a function of time when the film is taken suddenly from neutral to - 2.0 V - indicate that the complex absorption at 2.1 - 2.5 eV is a convolution of several absorptions which change to a single one at 2.65 eV when the sample is further reduced. The sketch indicates that during the reduction process at - 1.7 V, one may attribute the high energy absorption to transitions 1, 3 and 4 arising from new states in the band gap which correspond to polarons and the low-energy absorption to transition 2. In the fully reduced sample (- 1.9 V/Ag wire) bipolarons are created which lead to fully occupied states within the band gap, therefore transitions 2' and 4' may correspond to the absorptions at 0.75 eV and 2.65 eV.

Preliminary results obtained by in situ ESR measurements upon doping seem to corroborate the quasi absence of polarons in the lightly oxidized PPP films. Further work is underway to clarify this point.

1. See Proceedings of the ICSM'86 KYOTO in Synthetic Metals, Vol. 17-18 (1987)
2. P. Kovacic and J. Oziomek, Macromol. Syntheses, Vol. 2, J. Wiley (New-York 1966), p. 23
   T. Yamamoto, Y. Hayashi, A. Yamamoto, Bull. Chem. Soc. Japan $\underline{51}$, 2091 (1978)
3. M. Tabata, M. Satoh, K. Kaneto, K. Yoshino, J. Phys. C, Solid State Phys. $\underline{19}$, L 101 (1986)
4. J.F. Fauvarque, M.A. Petit, A. Digua, G. Froyer, Makromol. Chem., accepted for publication
5. G. Froyer, J.P. Mercier, D. Rivière, M. Le Cun and P. Auvray, Polymer $\underline{22}$, 992 (1981)
6. S. Krichene, S. Lefrant, Y. Pelous, G. Froyer, M. Petit, A. Digua, J.F. Fauvarque, Synthetic Metals, $\underline{17}$, 607 (1987)
7. J. Rault, Dissertation, Rennes (1986)
8. G. Froyer, Y. Pelous, F. Maurice, M.A. Petit, A. Digua, J.F. Fauvarque, to appear in Synthetic Metals.

Part VIII

Special Materials

Control of the State of Order in Poly(p-phenylene vinylene) and Its Effect on Iodine Doping

D.D.C. Bradley[1], *T. Hartmann*[1], *R.H. Friend*[1], *E.A. Marseglia*[1], *H. Lindenberger*[2], *and S. Roth*[2]

[1]Cavendish Laboratory, Madingley Road,
 Cambridge CB3 0HE, United Kingdom
[2]Max-Planck-Institut für Festkörperforschung,
 Heisenbergstr. 1, D-7000 Stuttgart 80, Fed. Rep. of Germany

Precursor-route syntheses of conjugated polymers are of considerable interest on account of the flexibility of sample preparation that they impart. The sulphonium polyelectrolyte route to poly(p-phenylene vinylene) [PPV] is one such synthesis that allows samples of PPV to be prepared in a variety of forms. In this paper we report the results of investigations of crystallinity at various stages in the conversion of <u>unoriented</u> films from the precursor to fully converted PPV. In partially converted unoriented samples, as is also the case for oriented films, crystallisation appears to be hindered by the presence of residual sulphur-containing moieties. However, upon full conversion the X-ray diffraction shows sharp Debye-Scherrer rings indicating an isotropic polycrystalline structure. In addition to characterisation of the state of order we have investigated its effect upon charge transfer doping. It is proposed that the difficulties experienced by other workers with iodine doping relate to the crystallinity of the samples.

1 Introduction

The sulphonium polyelectrolyte precursor route synthesis [1, 2] of PPV is shown schematically below. The conversion from the precursor to PPV is achieved by the thermal elimination of dialkyl sulphide and hydrogen halide.

$$R = CH_3, C_2H_5$$
$$X = Cl, Br$$

For elimination temperatures in excess of 300°C there is a near full conversion, as judged from infrared [3-5], elemental [6] and mass spectroscopic [7, 8] analyses, to all-trans PPV. The use of lower elimination temperatures allows the extent of conversion to be varied such that a copolymer of conjugated and non-conjugated segments is obtained in which the degree of unsaturation depends upon temperature and heating time [3-7, 9, 10]. The elimination process appears in reality to be rather complex with a number of competing steps. In particular, at moderate levels of conversion, i.e. for T > ca 150°C, the sulphonium function can undergo nucleophilic attack by the halide counterion resulting in a loss of alkyl halide and thus leaving alkyl sulphide as the side group in some of the saturated segments [7, 8, 11]. For these segments, the conversion can only proceed via elimination of alkyl mercaptan (i.e. R-S-H), a process that requires heating to temperatures of at least 300°C [6-8, 11]. It is shown below that such saturated segments can have a strong effect upon chain packing.

One point of obvious interest is whether there are any changes in the state of order corresponding to the chemical changes that occur during thermal conversion. In the case of oriented samples prepared by stretch alignment during the elimination reaction the polymer chains adopt a nematic-like packing arrangement in partially converted films whilst the fully converted polymer shows a highly crystalline structure [3, 12-14]. Reports of previous studies on unoriented films state that irrespective of their conversion the structure is always amorphous [7]. In section 2 below, we present results for combined X-ray diffraction and infrared studies on unoriented films which indicate that crystallisation may in fact be achieved by heating to 300°C.

Another point for discussion concerns the doping of PPV films using iodine. It has previously been stated that iodine will not dope PPV because the polymer's ionisation potential is too large [9, 10, 15]. MURASE et al. [6, 16] report however that the conductivity of partially converted films can be increased to a value of 2.5×10^{-3} $(\Omega \text{ cm})^{-1}$ by exposure to iodine vapour. The origin of this increase is questioned by GAGNON et al. [7] who by implication suggest that it does not relate to charge transfer involving conjugated chain segments. The results presented in Sect. 3 strongly suggest that charge transfer does occur and that the difficulties experienced with fully converted samples are associated with their crystallinity and not with their ionisation potential.

2 Characterisation of the State of Order

We have previously reported the results of investigations of the thermal conversion reaction and the control of materials properties through the use of different casting and conversion conditions [3-5, 12, 13]. These experiments and those of other workers [6, 7, 9-11, 13] clearly show that high quality samples may be readily prepared which are suitable for most experiments. Optical absorption [3, 5, 7], Raman [3], infrared [3-5] and electron energy loss [17] spectra have all been reported and the last two provide good evidence of extensive π-electron delocalisation [3, 4, 5, 17]. X-ray [12, 13, 3] and electron diffraction [14] data have also been reported with fully converted oriented films showing a high degree of crystallinity in addition to their near perfect orientation. Similar crystallinity might also be expected for unoriented films since they are likely to have a low degree of entanglement on account of inter- and intra-chain electrostatic repulsion between charged sulphonium moieties. Such behaviour would however contradict the report [7] that no crystallisation occurs.

Figure 1 shows X-ray photographs taken for samples cut from a single film at various stages in its conversion to PPV. The weakly converted precursor (Fig.1(a)) shows only a faint halo corresponding to an average spacing of ca 4.7 Å. Following heating at 80°C for 3 hours (Fig.1(b)) a further faint ring corresponding to a spacing of ca 2.1 Å could be detected in some overexposures. No change was seen following subsequent heating at 130°C for 24 hours (Fig.1(c)) but as shown in Fig.1(d), the effect of then heating at 300°C for 2 hours was to give a set of sharp rings corresponding to spacings of 4.39 Å, 4.09 Å, 3.17 Å and 2.11 Å. The form of all the diffraction patterns, i.e. consisting of Debye-

(a)

Figure 1

Scherrer rings, shows that the unoriented material is isotropic but the sharpness of the rings in Fig.1(d) indicates that crystallisation has occurred. Moreover, the Bragg-spacings of the rings match those of the first three equatorial reflections and the third meridional reflection which are the strongest reflections seen in the X-ray photographs of fully converted <u>oriented</u> samples [12, 13, 3].

Figure 2 shows the corresponding infrared spectra in the range 900 cm^{-1} to 400 cm^{-1} that were recorded on the same samples as used for the X-ray measurements. The two strongest modes, namely those at 837 cm^{-1} and 555 cm^{-1}, are assigned to the C-H out-of-plane bending and out-of-plane ring bending vibrations of the p-phenylene group. The mode at 632 cm^{-1} is, as indicated, assigned to the C-S

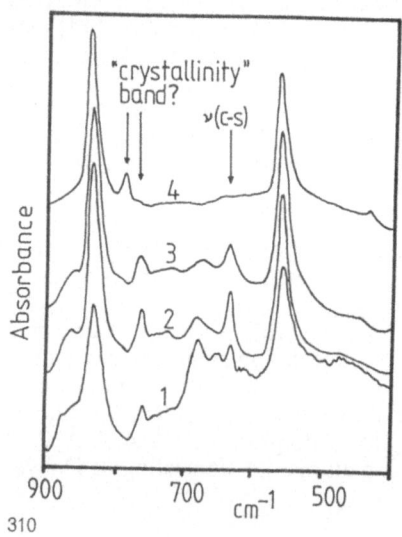

Figure 2 : Infrared spectra of unoriented samples at various stages of conversion
1: weakly converted precursor
2: 3 hours at 80°C
3: 24 hours at 130°C
4: 2 hours at 300°C

stretch mode associated with the saturated precursor segments. It is seen from the behavior of this mode that the complete elimination of sulphur, as required for full conversion, is only achieved for sample 4. The origin of the mode that appears at 763 cm^{-1} in curves 1, 2 and 3 and then apparently shifts to 784 cm^{-1} on full conversion (curve 4) is unclear. This mode is found from experiments on oriented films to be strongly polarised along the chain direction and cannot be readily assigned to a mode of either conjugated PPV or saturated precursor segments [4, 3]. One possibility [3] could be that this is a "crystallinity" band arising from a vibration involving interaction between groups on neighbouring chains and which is thus sensitive to their ordering. The upward shift in frequency would then be consistent with a stiffening of the effective force constant due to increased interactions caused by a closer packing of chains. It is emphasised that the same behaviour is seen for both oriented and unoriented films, again suggesting a strong similarity with respect to crystallinity.

In the light of these results, it is proposed that the chain packing in partially converted films, be they unoriented or otherwise, is hindered by residual sulphur-bearing side groups whose presence prevents the necessary microscopic alignment between chains in either the lateral direction or along their length. Once the elimination reaction is complete the chains are less strongly pinned relative to each other and can more readily slide into their most favourable, i.e. crystalline, packing arrangement.

3 Iodine Doping

Samples at various stages of conversion were prepared for investigations of iodine doping. The doping was attempted by exposure of free standing films to iodine vapour. Each sample was placed in a separate glass tube containing a small amount of recrystallised iodine. The tubes were then pumped to ca 10^{-4} to 10^{-5} Torr and sealed. Whereas very different behaviour is observed between those samples that have been fully converted and those of lesser degrees of conversion there was no apparent difference between oriented and unoriented films with the same degree of conversion. Samples that have previously been heated at 80°C are seen to darken rapidly and take on a black metallic lustre. Similar behaviour is observed for films previously heated at 130°C but films previously heated at 300°C showed no darkening even after being left for 24 hours. From these preliminary observations it was not immediately clear whether the darkening was due to a genuine charge-transfer reaction involving conjugated segments or whether it was simply due to a large-scale incorporation of unreacted iodine or indeed an irreversible chemical reaction. The latter can however be ruled out on account of the relative ease of undoping the films.

The infrared spectra of undoped films differ from those of the starting films only to the extent of changes that indicate a slight increase in the degree of conversion. There are no signs of reaction of the iodine with either the saturated or unsaturated segments of the chain. Confirmation of a charge-transfer reaction is obtained by recording the infrared spectra at various stages during the undoping and subsequently computing the difference spectra relative to the fully compensated film. Figure 3 shows the results for a 130°C converted, oriented ($l/l_0 = 5$) film with the infrared beam polarised parallel to the stretch direction [3]. The observed

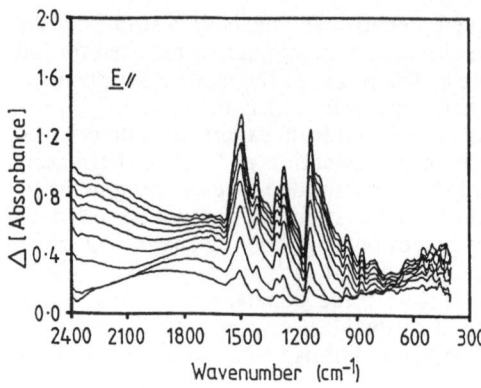

Figure 3 : Infrared difference spectra during undoping of an I_2 doped film which had been converted at 130°C. The strongest dopant induced bands appear at 1506 cm^{-1}, 1417 cm^{-1}, 1316 cm^{-1}, 1281 cm^{-1} and 1147 cm^{-1}

spectra show the same modes that are seen for AsF$_5$ doping [3, 5] and photoexcitation [3, 18] and have the expected relation to the Raman spectrum [3, 5] for modes associated with defect states created by charge transfer. Thus, it seems clear that partially converted samples of PPV do undergo a reversible charge-transfer reaction on exposure to iodine vapour.

The lack of reaction with fully converted samples would not be expected if the ionisation potential were the determining factor since ionisation potentials decrease with an increase in the length of conjugated sequences. However, as discussed above, the full conversion is also accompanied by crystallisation which would be expected to have some effect upon the doping reaction. In order to further examine the doping of the fully converted films we decided to test the effect of heating the sample under iodine vapour. Heating a fully converted sample at 90°C for two hours had no visible effect, as was also the case for heating at 180°C. Heating at 300°C, however, resulted in a shiny black film suggesting that doping had finally taken place. Examination of the latter films by infrared spectroscopy confirmed the occurrence of charge-transfer. Further work is obviously needed in order to determine the specific nature of the hinderance but a greatly reduced dopant diffusion coefficient and / or difficulties with structural rearrangement to accommodate the dopant ions are two possible explanations for the observed behaviour.

4 Summary and Discussion

We have reported measurements on the state of order in unoriented samples of precursor-route PPV at various stages of conversion. The results suggest that crystallisation is hindered in partially converted samples by the presence of residual sulphur-bearing moieties. However, upon completion of the conversion reaction crystallisation does take place. This behaviour is the same as that of oriented films [12-14, 3] implying that the crystallisation process occurs independently of orientation.

The results presented for iodine doping experiments clearly show that charge-transfer involving the conjugated segments does indeed occur for iodine vapour exposure of partially converted PPV samples and that the difficulties in doping fully converted samples are not related to their ionisation potential. It is suggested that sample crystallinity is responsible for the observed behaviour and since the same effect is seen for both oriented and unoriented samples this is an indirect confirmation that, as discussed above, the samples are very similar with regard to their crystallinity.

5 Acknowledgements

We gratefully thank Dr. I. Murase and Prof. F.E Karasz for helpful discussions and useful preprints. We also thank SERC and BP plc for financial support and the organisers of IWEPP'87 for the opportunity to present this work.

6 References

1. M. Kanbe and M. Okawara : J.Polymer Sci. A-1 6, 1058 (1968)
2. R.A. Wessling and R.G. Zimmermann : US patents #3,401,152 (1968) and #3,706,677 (1972)
3. D.D.C. Bradley : PhD Thesis, University of Cambridge (1987)
4. D.D.C. Bradley, R.H. Friend, H. Lindenberger and S. Roth : Polymer 27, 1709 (1986)
5. D.D.C. Bradley, G.P. Evans and R.H. Friend : Synthetic Metals 17, 651 (1987)
6. I. Murase, T. Ohnishi, T. Noguchi and M. Hirooka : Polymer Commun. 25, 327 (1984)
7. D.R. Gagnon, J.D. Capistran, F.E. Karasz, R.W. Lenz and S. Antoun : preprint (1987)
8. I. Murase: private communication (1986)
9. J.D. Capistran, D.R. Gagnon, S. Antoun, R.W. Lenz and F.E. Karasz : ACS Polymer Preprints 25, 282 (1984)
10. F.E. Karasz, J.D. Capistran, D.R. Gagnon and R.W. Lenz : Mol.Cryst.Liq.Cryst. 118, 327 (1985)
11. G. Montaudo, D. Vitalini and R.W. Lenz : Polymer 28, 837 (1987)
12. T. Hartmann : MPhil. Thesis, University of Cambridge (1986)
13. D.D.C. Bradley, R.H. Friend, T.Hartmann, E.A. Marseglia, M.M. Sokolowski and P.D. Townsend : Synthetic Metals 17, 473 (1987)
14. T. Granier, E.L. Thomas, D.R. Gagnon, F.E. Karasz and R.W. Lenz : J.Polymer Sci. B24, 2793 (1986)
15. J.L. Bredas, R.R. Chance and R.H. Baughman : J.Chem.Phys.76, 3673 (1982)
16. I. Murase, T. Ohnishi, T. Noguchi and M. Hirooka : Synthetic Metals 17, 639 (1987)
17. J. Fink, H. Lindenberger, B. Scheerer, A. vom Felde and S. Roth : these proceedings
18. D.D.C. Bradley, R.H. Friend, F.L. Pratt, K.S. Wong, W. Hayes, H. Lindenberger and S. Roth : these proceedings

Electronic Excitations in Polysilanes

P.R. Surjan[1], *R.A. Poirier*[2], *and H. Kuzmany*[3]

[1]CHINOIN Research Centre, P.O. Box 110, H-1325 Budapest, Hungary
[2]Memorial University of Newfoundland, St. John's, Canada
[3]Institut für Festkörperphysik der Universität Wien, and
 Ludwig Boltzmann Institut für Festkörperphysik, A-1090 Wien, Austria

Electronic transition energies were calculated for crystalline
and disordered unsubstituted polysilanes. For the calculation
the ab initio Hartree-Fock method with an STO-3G basis set is
used. The results exhibit an increase of the transition energy
of 0.7 eV by introducing disorder in the bond angle and in the
dihedral angle in agreement with an experimentally observed
shift of the optical absorption by an order-disorder phase
transition in similar materials.

1. Introduction

Recently it was observed that the electron excitation spectrum
of polysilanes exhibits certain properties similar to those of
polyacetylene in spite of the significant differences between
the electronic structures of these systems /1-4/. The similarity
concerns an increase of transition energies on introducing chain
defects by an order-disorder transition slightly above room
temperature. This transition is accompanied by a blueshift of
the optical absorption from 3.9 eV (peak) to 3.3 eV (peak) /4/.
  We report quantum mechanical excitation energy calculations on
polysilan segments in order to search for the possible structural explanation of the above effect.

2. Method and model

For the calculation the ab initio Hartree-Fock method is chosen
which accounts for electron-electron interaction in a self-
consistent manner. To describe an electronic excitation, the
single transition approximation (STA) is applied, in which the
excited state wave function corresponding to the $i \to k$
transition is taken as

$$^{1,3}\Psi_{i \to k*} = (1/\sqrt{2}) |\Psi^+_{k*\alpha} \Psi^-_{i\alpha} {}^{\pm}\Psi^+_{k*\beta} \Psi^-_{i\beta}|\Psi_o , \tag{1}$$

where Ψ^+ (Ψ^-) create (annihilate) electrons on the Hartree-Fock
orbitals with spin α or β, Ψ_o is the ground state wave function,
the + sign refers to the singlet while the - sign to the triplet
transition. The excitation energy in the STA can easily be
deduced from (1), and reads for the singlet cases:

$$^1\Delta E = \langle ^1\Psi_{i \to k*}|H|^1\Psi_{i \to k}\rangle - \langle\Psi_o|H|\Psi_o\rangle$$
$$= \varepsilon_{k*} - \varepsilon_i - (ii|k*k*) + 2(ik*|ik*) . \tag{2}$$

Note the appearance of the Coulomb and exchange interactions in ΔE which correct the orbital energy differences for the change in electron repulsion energy upon excitation.

For the evaluation of Eq.(2) we have limited ourselves to the minimum STO-3G basis set level /5/, and have studied only unsubstituted polysilane up to n=7 Si atoms. For the crystalline phase the following geometrical parameters were used: r_{Si-Si} = 2.243 Å and r_{Si-H} = 1.1423 Å. The bond- and dihedral angles are kept at their standard values 109.47° and 180.0°, respectively. The above bond lengths were obtained by geometry optimization for disilane. Hydrogens are taken in a standard tetrahedral arrangement. The numerical calculation have been performed with the MONSTERGAUSS program system /6/ with a new routine added to evaluate transition energies in the single transition approximation /7/.

The evaluated lowest singlet transition energies were found to follow a n^{-2}-law in agreement with experimental results /8/. Extrapolation to infinity yielded a scaling factor of 0.3 if compared quantitatively with the experiments.

3. Disorder in bond angles

We performed calculations by introducing small perturbations (up to 10°) around the standard value α_o = 109.47° corresponding to a tetrahedral arrangement. While in the crystalline case the value E = 12.0 eV was found, we obtained ΔE values in the range 11.7 to 12.7 eV on introducing a maximum disorder of 10° in α. A decrease in the skeleton bond angles leads to an increase in ΔE by 0.7 eV (Tab.1). Seemingly, this blue shift could explain the experimental findings during the order-disorder phase transition, but it must be noted that a change in α involves a strong increase of the ground state energy (Tab.1). This change seems to be too high for the corresponding states to be activated around room temperature (kT ≈ 0.6 kcal/mol).

Tab.1. Relative ground state energies and excitation energies in the Si_7H_{16} chain varying the Si-Si-Si bond angle in the middle of the chain

Geometry α	Relative ground state energy /kcal/mol/	Excitation energy /eV/
crystalline: 109.47°	0.0	12.0
bent:		
105.0°	4.5	12.4
100.0°	13.7	12.7

4. Effect of conformation

The crystalline geometry considered gives a planar zig-zag arrangement in the all-trans staggered conformation. To study the role of conformational defects of the Si backbone, we picked up a single dihedral angle δ in the middle of the chain and evaluated the STA transition energies as a function of δ. The result is given in Fig.1. A blue shift of 0.6 eV in ΔE occurs upon introducing a twist towards the cis-type conformation. The excitation energy curve is flat around 180° and around 0°. The corresponding increase in the total ground state energy was found to be only 2.4 kcal/mol, which is small enough to allow the cis-type structure to be thermally activated under the experimental conditions.

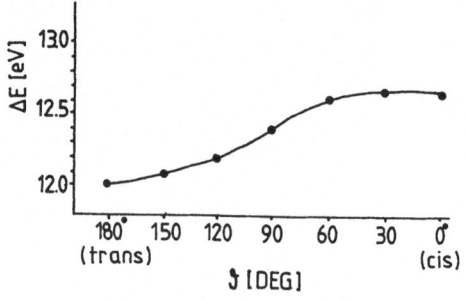

Figure 1
The first singlet transition energy of Si_7H_{16} as a function of a dihedral angle in the middle of the chain

5. Discussion and Conclusion

In light of the above results we may conclude that the experimentally observed blue shift in polysilanes is very probably due to conformational changes in the Si skeleton. The theoretically calculated increase of 0.6 or 0.7 eV in the first transition energy has to be scaled down by a factor of 0.3. Nevertheless qualitative agreement with the experiments is obtained. The total experimentally observed blueshift of the absorption could result from multiple defects on the same chain.

The calculated change in transition energy was caused by a single cis-type defect in an otherwise trans-conformational chain. Disorder in bond length can also lead to a blue shift in electronic transitions, but the associated ground state energy differences are too high and such type of defects can be excluded under the experimental circumstances. On the other hand, the change in the ground state energy upon rotation from the trans-like to the cis-like conformation is only 2.4 kcal/mol, less than the rotation barrier in ethane (≈ 3 kcal/mol). We emphasize that no side chains were considered in our calculation because the Si backbone serves as a chromophore of the polymer for the transition studied. The conformational disorder of the Si chain may explain the UV-VIS spectrum of polysilanes, though the details of this mechanism are yet to be established.

Acknowledgement

This work was supported by the Stiftung Volkswagenwerk. One of us (PRS) acknowledges the Natural Sciences and Engineering Research Council of Canada for supporting a research visit to the University of Newfoundland.

References

1. R.D. Miller, D. Hoffer, J. Rabolt and G.N. Fickes: J. Am. Chem. Soc. 107, 2172 (1985)
2. H. Bock, W.Ensslin, F. Fehér and R.Freund: J. Am. Chem. Soc. 98, 668 (1976)
3. K. Takeda and N. Matsumoto, J. Phys. C. 18, 6121 (1985)
4. H. Kuzmany, J.F. Rabolt, B.L. Farmer and R.D. Miller: IBM Research Reports, Solid State Phys., 9/15/86 San Jose, California, 1986
5. W.J. Hehre, R.F. Stewart and J.A. Pople: J. Chem. Phys. 51, 2657 (1969)
6. M.R. Peterson and R.A. Poirier: Program MONSTERGAUSS, Dept. Chem. University of Toronto, Toronto, Canada, and Dept. Chem., Memorial Univ. of Newfoundland, St. Johns, Canada.
7. P.R. Surján: STA routine to MONSTERGAUS, Dept. Chem., Memorial Univ. of Newfoundland, St. Johns, Canada, 1986
8. C.G. Pitt, M.M. Bursey, and P.F. Rogerson: J. Am. Chem. Soc. 92, 519 (1970)

Doping Experiments
with μ-(Pyrazine)phthalocyaninatoiron(II)

M. Hanack and *A. Leverenz*

Institut für Organische Chemie, Lehrstuhl für Organische Chemie II
der Universität Tübingen, Auf der Morgenstelle 18,
D-7400 Tübingen, Fed. Rep. of Germany

I. Introduction

Phthalocyaninatoiron(II) and -ruthenium(II) compounds $[PcML]_n$
(M = Fe, Ru) with different bridging ligands L = 1,4-diisocyano-
benzene (1,4-dib), 1,4-dicyanobenzene (dcb), 4,4'-diisocyano-
biphenyl (dibph) and 4,4'-bipyridylacetylene (bpyac), 2,3,5,6-
tetramethyldiisocyanobenzene (Me_4dib), 2,3,5,6-tetrachlorodiiso-
cyanobenzene (Cl_4dib) and 1,3-diisocyanobenzene (1,3-dib) were
reacted with iodine using different doping procedures. All the
compounds $[PcMLI_y]_n$ which were formed were characterized by
TG measurements, infrared-, resonance Raman- and [57]Fe-Mößbauer
spectroscopy /1/. We also report on electrochemical doping
experiments with (μ-pyrazine)-phthalocyaninatoiron(II) $[PcFe(pyz)]_n$
using a variety of different conditions and counterions. It was
intended to study a) the influence of various counterions upon
the conductivity and stability of the compound $[PcFe(pyz)]_n$ and
b) to investigate if oxidation occurs at the macrocycle or at
the central metal atom.

II. Electrochemical doping experiments

The samples described were electrolysed galvanostatically. The
composition of the doped $[PcFe(pyz)]_n$ was established by elemental analyses and TG/DTA. The theoretical doping level was calculated by the equation Q = J x t. The influence of current
density (J) on the specific conductivity σ_{RT} of the doped samples
is given in Table 1.
As Table 1 shows, $[PcFe(pyz)]_n$ doped with a current density of
120 $\mu A/cm^2$ had a maximum conductivity of 4.7 x 10^{-2} S/cm.
Samples of $[PcFe(pyz)]_n$ doped with current densities >120 $\mu A/cm^2$
revealed a decrease in conductivity. The conductivity of the
doped $[PcFe(pyz)]_n$ also depends on the degree of oxidation.

Table 1   Conductivity of doped samples of $[PcFe(pyz)]_n$ as a function of current density.

Current density $[\mu A/cm^2]$	Conductivity[a] $[S/cm]$	max. potential V
90	3.2×10^{-2}	1.7
120	4.7×10^{-2}	1.9
160	3.5×10^{-2}	1.9
200	2.3×10^{-2}	2.3
300	4.3×10^{-4}	10 (decomp)

$[PcFe(pyz)]_n$ was electrolysed in 0.1 m solution of $TBABF_4$ to a charge quantity of 0.5 F (TBA = tetrabutylammonium).
a) Pressed pellets $p = 10^8$ Pa by four-probe method.

Figure 1 shows the dependence of conductivity on the doping level from y = 0.0 to y = 1.0 with BF_4^- as the counterion. A maximum conductivity is obtained for $[PcFe(pyz)(BF_4)_{0.45}]_n$ at the doping level y = 0.5. Further oxidation leads to a decrease in conductivity. The solvent has very little effect on the conductivities.

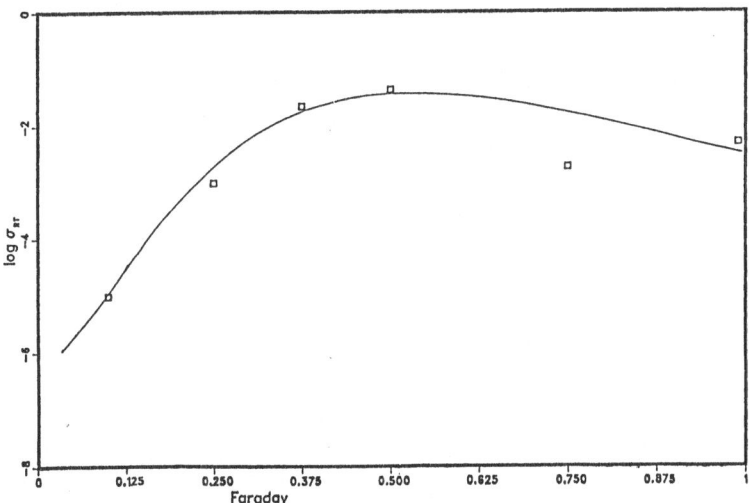

Figure 1   Conductivity[a] of doped $[PcFe(pyz)]_n$[b] vs. doping level[c]

a) By four-probe van der Pauw technique. b) Doped with 0.1 m $TBABF_4$ in CH_2Cl_2 with $J = 120\ \mu A/cm^2$. c) Found by charge quantity measurements.

The influence of different counterions on the conductivities is shown in Table 2. $[PcFe(pyz)]_n$ can be doped electrochemically with all the counterions, except BPh_4^-, which has no noticeable effect. In all other cases stoichiometric compounds of the compositions given in Table 2 are formed. In the cases with BF_4^- and PF_6^- as the counterion the compounds are thermally stable up to 140°C. From Table 2 it can be concluded that the oxidation potential of the supporting electrolytes $TBAHSO_4$, TBASCN and $TBAClO_4$ is smaller than the oxidation potential of $[PcFe(pyz)]_n$. Therefore electrochemical oxidation partly leads to decomposition of these supporting electrolytes and not exclusively to doping of the macrocycle.

Table 2   Conductivity and stoichiometry of doped $[PcFe(pyz)]_n$ (all samples were electrolysed up to a current quantity of 0.5 F per unit).

Supporting electrolyte	Solvent	Conductivity[a]	Stoichiometry[b]
Bu_4NBF_4	CH_2Cl_2	4.7×10^{-2}	$[PcFe(pyz)(BF_4)_{0.45}]_n$
Bu_4NPF_6	CH_2Cl_2	3.5×10^{-2}	$[PcFe(pyz)(PF_6)_{0.5}]_n$
Bu_4NBPh_4	CH_2Cl_2	2×10^{-6}	$[PcFe(pyz)]_n$
Bu_4NHSO_4	CH_2Cl_2	1×10^{-5}	$[PcFe(pyz)(HSO_4)_{0.4}]_n$
Bu_4NSCN	CH_2Cl_2	1.5×10^{-6}	$[PcFe(pyz)(SCN)_{0.3}]_n$
$LiClO_4$	H_2O	3×10^{-3}	$[PcFe(pyz)(ClO_4)_{0.3}]_n$

a) Measured by four-probe van der Pauw technique, $P = 10^8$ Pa.
b) Found by elemental analysis. All samples C_6H_5Cl per Pc unit.

From IR spectra it can be seen that independent of the counterion an increase in conductivity of the samples is accompanied by a progressive growth of a broad electronic absorption, which is indicative of an oxidation process occurring in the macrocycle forming a radical cation. Similar effects were observed in a number of other doped $[PcMO]_n$ and $[PcML]_n$ compounds /2a, 2b, 2c/.

That the anodic oxidation process takes place mainly at the macrocycle and not at the central metal atom is also shown by ^{57}Fe-Mößbauer spectra of $[PcFe(pyz)(BF_4)_{0.5}]_n$. The data are given in /3/.

From the described experiments it follows that electrochemical doping of $[PcFe(pyz)]_n$ with several counterions results in an increase of conductivity by up to four orders of magnitude. If the counterion is both small enough to enter the channels between the stacks and is not destroyed at the applied oxidizing potential good conductivities are obtained.

Literature

1. a) M. Hanack, U. Keppeler, H.-J. Schulze, Synth. Met., in press.
   b) U. Keppeler, O. Schneider, W. Stöffler, M. Hanack, Tetrahedron Lett. 3679 (1984).
2. a) B.N. Diel, T. Inabe, J.W. Lyding, K.F. Schoch, T.J. Marks, C.R. Kannewurf, J. Am. Chem. Soc. 105, 1551 (1983).
   b) B.N. Diel, T. Inabe, N.K. Jaggi, J.W. Lyding, O. Schneider, M. Hanack, C.R. Kannewurf, T.J. Marks, L.H. Schwartz, J. Am. Chem. Soc. 106, 3207 (1984).
3. M. Hanack, A. Leverenz, Synth. Met., in press.

Optical and Electrical Properties of a New Conductive Polyheterocycle: Poly(1,4-di(2-thienyl))benzene

C. Taliani[1], R. Danieli[1], R. Zamboni[1], G. Ruani[2], and P. Ostoja[3]

[1]CNR-Istituto ISM, Via Castagnoli 1, I-40126 Bologna, Italy
[2]Università di Parma, Dipartimento di Fisica,
  Via M. D'Azeglio 85, I-43100 Parma, Italy
[3]CNR-Istituto LAMEL, Via Castagnoli 1, I-40126 Bologna, Italy

A new conjugated conductive polymer with non-degenerate ground state is obtained by electrochemical polymerization of 1,4-di(2-thienyl)benzene. The polymer backbone consists of bithiophene segments separated by para-phenylene units attached in α position. The electrical specific conductivity for the ClO_4^- doped polymer is of the order of 1×10^{-4} S cm^{-1}.

The comparison of the vibrational spectra of the doped and of the undoped polymers shows the drastic increase of the doping-induced bands in the 1600-1000 cm^{-1} spectral range which is characteristic of the presence of IR activated charge defects in the polymer backbone.

The evolution of the midgap absorption bands at 2 and 1 eV respectively is consistent with the formation of two symmetric bipolaron bands in the gap which for the undoped species is at 3 eV.

1 INTRODUCTION

The exploratory research for new conductive polymers which is taking place in recent years, is guided by the quest for new materials and, at the same time, by research on the relations between the molecular structure and the charge transport properties. Following this line we have prepared a new conductive polymer with a modified backbone structure /1/ (Fig.1).

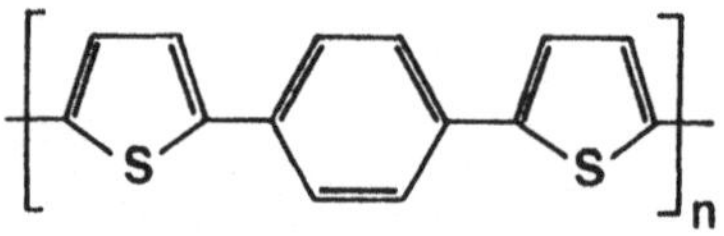

Fig. 1

The "molecular design" of the monomer of 1,4-di(2-thienyl)benzene is such that the π electron conjugation scheme of the resulting electropolymerized polymer possesses the mixed character of polythiophene (PT) and polyparaphenylene (PPP).

2 EXPERIMENTAL

The electropolymerization was performed in a two compartment cell equipped with Pt electrodes with the monomer (1.4×10^{-3} M/1) and $LiClO_4$ dissolved in acetonitrile at constant current density of 1 mA cm^{-2}. Under these conditions the oxidation potential vs SCE was 1.06 V.

Conductivity was measured both by the four probe compaction method and with a Van der Pauw test pattern realized on a high-resistivity silicon substrate which was previously used as anode in the electropolymerization. Silicon was made conductive by photogeneration of carriers. "In situ" NIR-VIS-UV analysis was performed on a modified electrochemical cell by applying a controlled current to a SnO_2 coated quartz electrode covered by the polymer in the presence of $LiClO_4$ in propylene carbonate vs a Pt counterelectrode. "As grown" IR spectra were recorded with a Bruker FT-IR (113v) spectrophotometer on the polymer grown on a conductive silicon substrate. Raman scattering is measured with a pulsed excitation at 532 nm.

3 RESULTS AND DISCUSSION

The "as grown" FT-IR spectra of the doped and undoped polymers for the same deposit is shown in Fig.2. At first instance we notice that the doped polymers possess a much stronger absorption intensity compared to the undoped polymer. The main spectral features of the undoped polymer are similar to those shown by PT /2/. In fact the 791 cm^{-1} band, which is assigned to the C-H out-of-plane vibration of the 2-5 disubstituted thiophene according to /3/, is only slightly displaced from the frequency shown in PT (787 cm^{-1}). The two strong bands at 1492 and 1431 cm^{-1} are assigned to the antisymmetric and symmetric C=C stretching, respectively. The doped

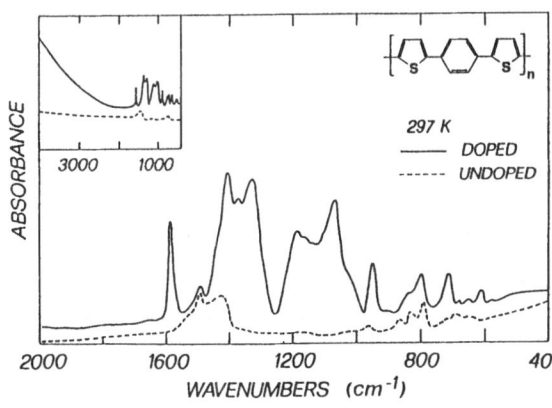

Fig.2 FT-IR spectra of poly (1,4-di(2-thienyl)) benzene "as grown" on conductive Si single crystal.

spectrum, on the other hand, shows a number of doping-induced bands (DIB) which are very similar in intensity and shape to PT, apart from the strong line at 1581 cm^{-1} which we assign to the C=C symmetric stretching vibration of the benzene ring.

A confirmation of this assignment is given by the Raman spectrum which is shown in Fig.3.

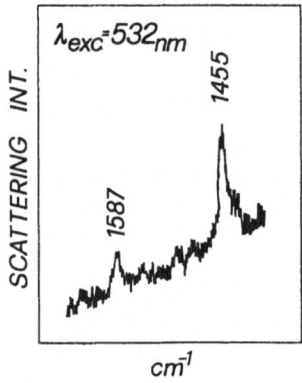

Fig.3 Pulsed Raman spectrum
of the undoped poly(1,4-di
(2-thienyl))benzene.

The Raman scattering signal is strongly disturbed by fluorescence emission. The 1587 cm^{-1} is evidently due to the symmetrical C=C stretching vibration which becomes allowed in the IR by the breaking of the selection rules due to the formation of charged defects in the polymer backbone.

The evolution of the optical absorption upon doping, in the UV-VIS-NIR spectral range, is reported in Fig.4.

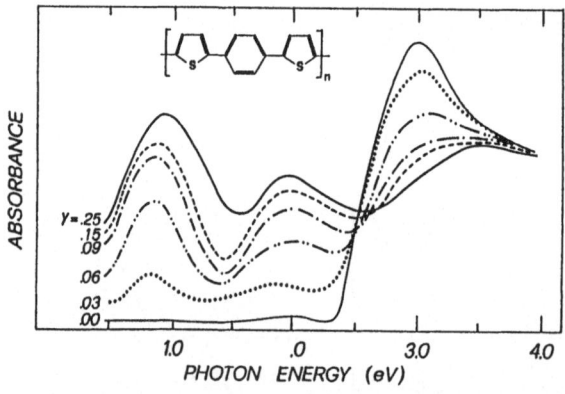

Fig.4 "In situ" UV-VIS- NIR spectra at different doping levels. Y indicates the dopant (ClO$_4^-$) concentration in moles per monomer mole.

The intensity of the $\pi - \pi^*$ electronic transition at 3 eV decreases upon doping and two new bands emerge at lower energy. For the fully doped polymer such bands are observed at 1 and 2 eV.

4 CONCLUSIONS

The mixed character of the conjugated backbone structure of poly(1,4-di(2-thienyl))benzene leads to an increase of the energy of the lowest $\pi - \pi^*$ electronic transition (3 eV) compared to PT (2.67 eV) and becomes almost equal to PPP-Kovacic /4/. The benzene ring is involved in the charge transport as shown by the strong 1581 cm^{-1} band in the doped polymer IR spectrum. The DIB marks the presence of charge defects which couple effi-ciently with the doping induced electronic absorption in the NIR; this behaviour is typical of conductive polymers like PT and polypirrole. Assuming that the number of charge carriers is proportional to the inten-sity of the DIB, since the relative intensity of the DIB is similar to PT, there should be approximately the same concentration of carriers in the polymer backbone. The new electronic transition in the NIR are perfectly symmetric with respect to the Fermi level according to the bipolaron model.

REFERENCES

1. R.Danieli, P.Ostoja, M.Tiecco, R.Zamboni and C.Taliani: J.Chem.Soc.Chem. Commun. 1476 (1986)
2. C.Taliani, R.Danieli, R.Zamboni, P.Ostoja and W.Porzio: Synth.Met. 18, 177 (1987)
3. M.Akimoto, Y.Fukuyama, H.Takouchi, I.Harada, Y.Soma and M.Soma: Synth. Met. 15, 353 (1986)
4. A.Heim and G.Leising: Molec.Cryst. and Liq.Cryst. 118, 309 (1985)

Preparation, Spectroscopic and Electrical Characterization of a New Polyheterocycle: Poly-benzo(1, 2 − b; 4, 3 − b′)dithiophene

C. Taliani[1], *R. Danieli*[1], *R. Zamboni*[1], *G. Giro*[2], and *F. Sannicolò*[3]

[1]Istituto Spettroscopia Molecolare, CNR,
 Via Castagnoli 1, I-40126 Bologna, Italy
[2]Istituto FRAE, CNR, Via Castagnoli 1, I-40126 Bologna, Italy
[3]Dipartimento Chimica Org. e Industriale, Università,
 Via Golgi 19, I-20133 Milano, Italy

A new polymer has been electrochemically prepared starting from the monomer mole cule: benzo(1,2-b;4,3-b')dithiophene (Fig. 1). The optical properties of the "as grown" polymer film have been investigated both in the UV-VIS and IR for the doped and undoped forms. The low electrical conductivity, 1×10^{-11} S cm^{-1}, is in agreement with the absence of doping-induced bands (DIB) in the mid-IR spectra and with the high energy (2 eV) of the doping-induced electronic absorption.

We suggest that the low conductivity is related to the lack of extended π electron conjugation on the polymer backbone due to a high degree of disorder induced by α-β' and β-β' linkages.

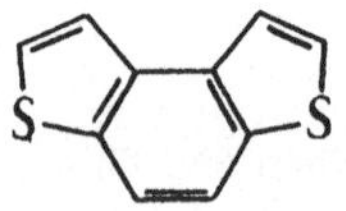

Fig. 1. Benzo(1,2-b;4,3-b')dithiophene

1. INTRODUCTION

In recent years a certain number of heterocycles have been polymerized in order to obtain materials which in the doped form show interesting charge transport properties. We have recently considered large molecules formed by condensed thiophene rings as candidates for the electropolymerization of conducting polymers. Along this line we have prepared for the first time the poly-dithieno(3,2-b;2',3'-d)thiophene /1/, the poly-thieno(3,2-b)thiophene /2/ and the poly-(1,4-di(2-thienyl)benzene) /3/.

In the present work we report on the electrical and spectroscopic properties of a new polymer obtained from benzo(1,2-b;4,3-b')dithiophene hereafter referred to as BDT.

2. EXPERIMENTAL

The BDT molecule was synthesized following the method described in Ref. /3/. The electropolymerization was performed in a two-compartment cell equipped with platinum electrodes with BDT (0.01 M/I) and Bu$_4$NClO$_4$ (0.06 M/I) dissolved in dichloromethane at constant current density of 1 mA cm^{-2}. The oxidation potential is 1.46V vs. SCE. Conductivity measurements were performed both with four points and two points configurations on pressed pellets of the black brittle deposit scratched off the electrode.

The electrical conductivity at room temperature is 1×10^{-11} S cm^{-1} for pellets ranging in thickness from 0.01 to 0.5 mm.

Vibrational spectra were obtained on "as grown" polymer layers electropolymerized on low resistivity silicon monocrystals as well as on KBr pellets using a Bruker FT-IR (IFS-113v). UV-VIS spectra were performed on films grown on tin oxide coated quartz electrodes.

3. RESULTS AND DISCUSSION

Typical elemental composition and empirical formula of poly-BDT is listed in Table 1 which shows an average doping level of one anion unit every eight monomer units.

Table 1

Elemental composition (percentage)				Empirical Formula			
C	H	S	Cl	C_{10}	$H_{4.78}$	$S_{1.995}$	$Cl_{0.125}$
54.27	2.16	28.90	2.01				

FT-IR spectra of the chemically undoped poly-BDT and the BDT monomer in KBr pellets are shown in Fig. 2.

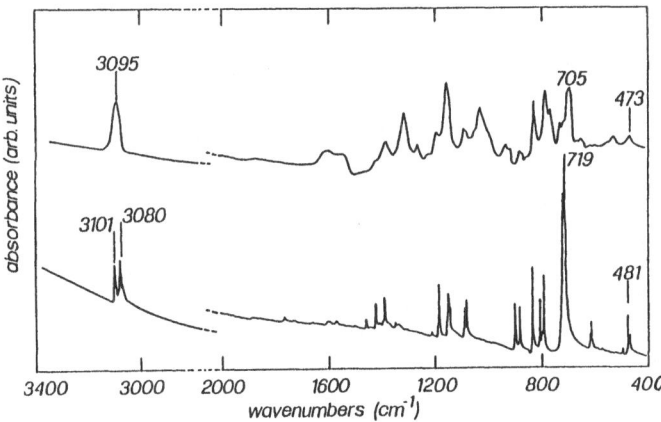

Fig. 2. FT-IR spectra of the BDT monomer (lower trace) and poly-BDT (upper trace) in KBr pellets.

We notice that the neutral polymer spectrum shows a remarkable broadening of the bands with fwhm ranging from 30 to 40 cm^{-1}, which is a definite sign of a high degree of amorphous disorder. Among the many IR bands of the monomer in the low-frequency range we may tentatively assign the in-plane ring deformation of the thiophene ring at 719 cm^{-1} as well as the out-of-plane deformation of the thiophene ring at 481 cm^{-1}. Both bands are observed in the neutral polymer at lower frequency, 705 and 473 cm^{-1} respectively. The monomer spectrum shows two bands at 3101 and 3080 cm^{-1} which are assigned to the CH stretching of the benzene and the thiophene rings respectively, while the polymer shows only one broad band centered at about 3095 cm^{-1}.

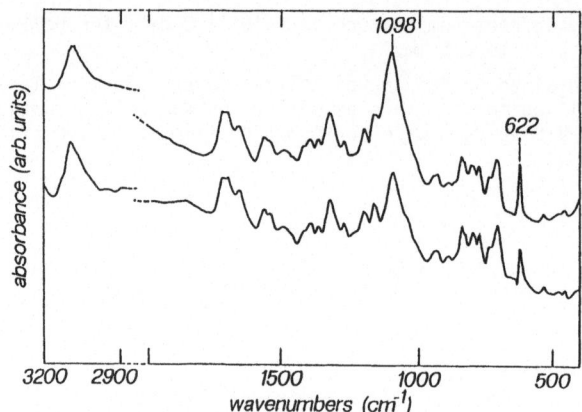

Fig. 3 FT-IR spectra (upper trace) of doped (ClO_4^-) poly-BDT polymer film "as grown" after electropolymerization on conductive silicon monocrystal used as electrode in the preparation. The lower trace shows the same film after treatment with hydrazine used as undoping agent.

The "as grown" FT-IR spectra of the doped and chemically undoped poly-BDT, electropolymerized on conductive silicon single crystals, are shown in Fig.3.

Both spectra, apart from a slight reduction of the 622 and 1098 cm^{-1} band, assigned to the ClO_4^- dopant group, are almost identical. In contrast with poly thiophene and other fused thiophene ring based polymers /5/ the doped poly-BDT does not show any doping-induced band as well as any increase in the absorption intensity on the high-energy side of the IR spectrum.

The electronic spectrum of the "as grown" polymer electropolymerized on SnO_2 coated quartz is shown in Fig. 4.

The $\Pi-\Pi^*$ transition is observed at 2.9 eV (428 nm) as a very broad (fwhm=0.8 eV) absorption envelope and on the lower energy side we observe a pronounced shoulder peaked at about 2.0 eV. The $\Pi-\Pi^*$ absorption is slightly reduced in intensity by undoping while the 2.0 eV band is strongly reduced.

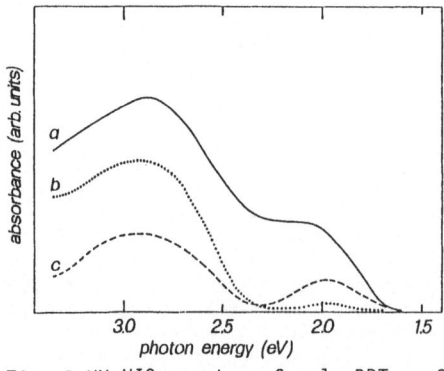

Fig. 4 UV-VIS spectra of poly-BDT on SnO_2 coated quartz: a)"as grown" doped (ClO_4^-) poly-BDT; b) chemically (hydrazine) undoped poly-BDT; c) after subsequent iodine doping.

Since the subsequent doping with I_2 vapour causes a decrease of the $\Pi-\Pi^*$ absorption band and an increase of the 2.0 eV band, we assume that the latter marks the presence of a doping-induced charge transfer transition. We notice that the chemical undoping causes a deterioration of the film which is reflected by a decreased intensity of the $\Pi-\Pi^*$ transition. Further doping and undoping cycles cause the same effects shown in the first cycle apart from the progressive deterioration of the film.

Despite the fact that the doped polymer is easily formed at the electrode and new polymer continues to grow on it, proving that the deposit is conductive enough to make the electropolymerization proceed, we notice that the electrical conductivity is extremely low compared to other doped poly-heterocycles /5/.

The low value of the electrical conductivity agrees well with:
a) the absence of any low-energy doping induced electronic absorption below 1.5 eV attributable to mobile carriers absorption
b) the lack of any new IR band induced by doping which reflects the absence of IR activated charge defects in the polymer backbone
c) the relative high energy of the first $\Pi-\Pi^*$ electronic transition (2.9 eV) which reflects a low degree of conjugation in the polymer backbone
d) the low doping level of one anion for every eight monomer units.

We suggest that, because of the relatively high oxidation potential of the monomer, as well as the relatively high reactivity of the carbon atoms in the β position with respect to the sulphur atoms, the resulting electropolymerized polymer is affected by disorder due to $\alpha-\beta'$ and $\beta-\beta'$ linkages which reduce the extension of the Π electron conjugation.

ACKNOWLEDGMENTS

We thank Mr. G. Tasini for technical assistance. This work was partly supported by a grant from P.S. "Nuovi Materiali" of CNR, Italy.

1. P.Di Marco, M.Mastragostino, C.Taliani: Mol.Cryst.Liq.Cryst. 118, 241 (1985)
2. R.Danieli, C.Taliani, R.Zamboni, G.Giro, M.Biserni, M.Mastragostino, A.Testoni: Synth. Met. 13, 325 (1986)
3. R.Danieli, P.Ostoja, M.Tiecco, R.Zamboni, C.Taliani: J.C.S. Chem. Comm. 1473, (1986)
4. R.M.Kellogg, M.B.Groen, H.Wynberg: J. Org. Chem. 32, 3093 (1967)
5. C.Taliani, R.Danieli, R.Zamboni, P.Ostoja, W.Porzio: Synth. Met. 18, 177 (1987)

Investigation of Undoped and Alkali-Metal Doped Poly(p-phenylene Selenide)

W. Czerwiński*, J. Fink, and N. Nücker

Kernforschungszentrum Karlsruhe, Institut für Nukleare Festkörperphysik, P.O. Box 3640, D-7500 Karlsruhe, Fed. Rep. of Germany

1. INTRODUCTION

In the last few years, conducting polymers with non-degenerate ground state and with conductivity mechanisms somewhat different from the case of trans-polyacetylene have attracted strong interest. Polaron-type defects in these polymers (Polyparaphenylene (PPP), Polypyrrole (PP), Polythiophene (PT), Polyparaphenylene sulfide (PPS)) leading to the presence of quinoid structure segments, have been theoretically predicted [1] and are expected to depend on the presence of impurities [2] as well as on the chain length of the polymer [3]. Another polymer with a non-degenerate ground state is poly-paraphenylene selenide (PPSe) which was investigated up to now only incidentally [4,5]. Moreover, investigations on the electronic structure of PPSe have not been reported up to date. It is known [6] that polyparaphenylene sulfide doped by strong p-dopant molecules has a rather high conductivity ($\sim$1 S/cm). Substitution of sulfur by selenium may be favourable for high conductivity due to the increased metallic character of selenium and due to a decrease in electronegativity. Moreover, there is experimental evidence [7,8] that doped selenium compounds have higher conductivities than similar sulfur systems. However, also opposite results were reported [4]. In this contribution we have used electron energy loss spectroscopy and electron diffraction to investigate the electronic and crystal structure of undoped and doped PPSe.

2. EXPERIMENTAL

PPSe was synthesized from p-dibromobenzene and sodium selenide prepared in situ directly from its elements in N,N-dimethyl-formamide in argon atmosphere at a temperature of $130\pm2°C$. The reaction time t was varied from t = 6h to t = 24h in order to obtain polymers with different chain length. After removal of oligomers with THF (72h extraction in Soxhlet apparatus) a yellow (t = 6h) to dark green (t = 24h) powder was dried in vacuum (p > 10^{-5} Torr) at 180°C during 4h. The resulting polymers were characterized by infra-red spectroscopy, mass spectroscopy, X-ray diffraction, EPR and differential scanning calorimetry [9]. Chemical analysis was used to calculate the average degree of polymerization (n) by the content of bromine sitting at the ends of the chains in the final product. Doping was achieved by exposure of the samples to alkali metal vapour at room temperature. Electron energy loss spectroscopy (EELS) and electron diffraction measurements were performed with a spectrometer described elsewhere [10].

---

*on leave from Institute of Chemistry, University N. Copernicus, Torun, Poland

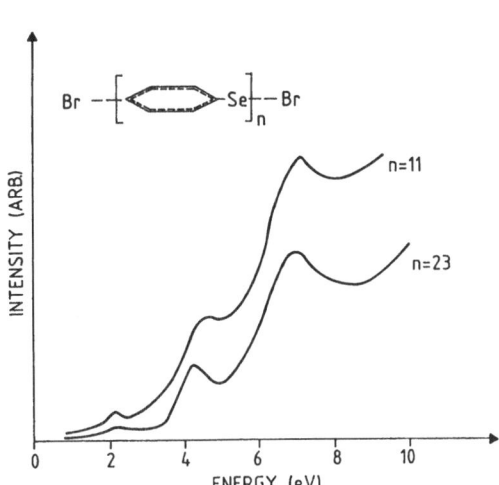

Fig. 1 Energy-loss spectra of the valence excitation region of PPSe with different chain lengths.

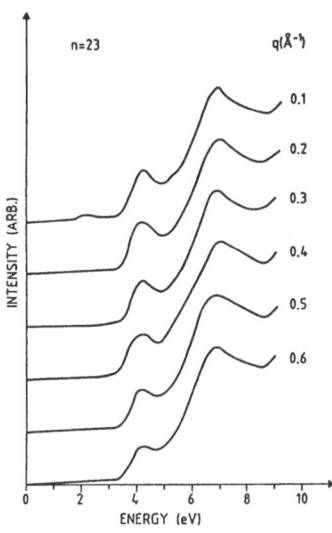

Fig. 2 Momentum dependence of the loss function of undoped PPSe.

3. RESULTS AND DISCUSSION

Typical loss spectra of the valence-excitations of PPSe having different chain length are shown in Fig. 1. The most pronounced structure in these spectra is a peak at 7 eV, the energy of which is independent of the number of repeat units (n) in the polymer chain and does not change upon variation of the momentum transfer q. The electrons participating in this excitation are localized in the benzene rings, leading to rather flat π and π^* bands. Similar transitions have been observed in PPP and PPS [11]. In addition, an excitation is observed at 4.3-4.5 eV above the onset of the absorption at 3.5 eV corresponding to excitations from the highest occupied π band to the lowest unoccupied π^* band. For PPSe with longer chains this peak is shifted to slightly lower energy. Fig. 2 shows the momentum dependence of the valence excitations in PPSe. The 7 eV transition exhibits no dispersion which is typical of a transition between flat bands. Contrary to observations in PPP and PPS also the plasmon at 4.3 eV shows almost no dispersion. This indicates that the highest occupied π band and the lowest unoccupied π^* band is rather narrow. Therefore, the localization of the π electrons near the Fermi level increases going from PPP via PPS to PPSe. The gap of PPSe is comparable to those of PPP and PPS. For all samples, a small structure near 2.2 eV appears in the gap. The intensity of this excitation decreases slowly with increasing chain length. In Fig. 3 these changes are shown in more detail as a function of the average number n of repeat units in the chain. At higher momentum transfers this excitation with low intensity is strongly decreasing, indicating a rather localized state, probably connected with Br atoms at the end of the chains. Recent results on PPP prepared by the Yamamoto method indicate an excitation in the gap at the same energy.

In Fig. 4 we compare the loss spectra of pure PPSe (n=23) with those of Cs doped PPSe. Upon doping, the π plasmon at 4.3 eV is shifted to 2 eV, while the excitation at 7 eV is broadened and there is an indication of a splitt-

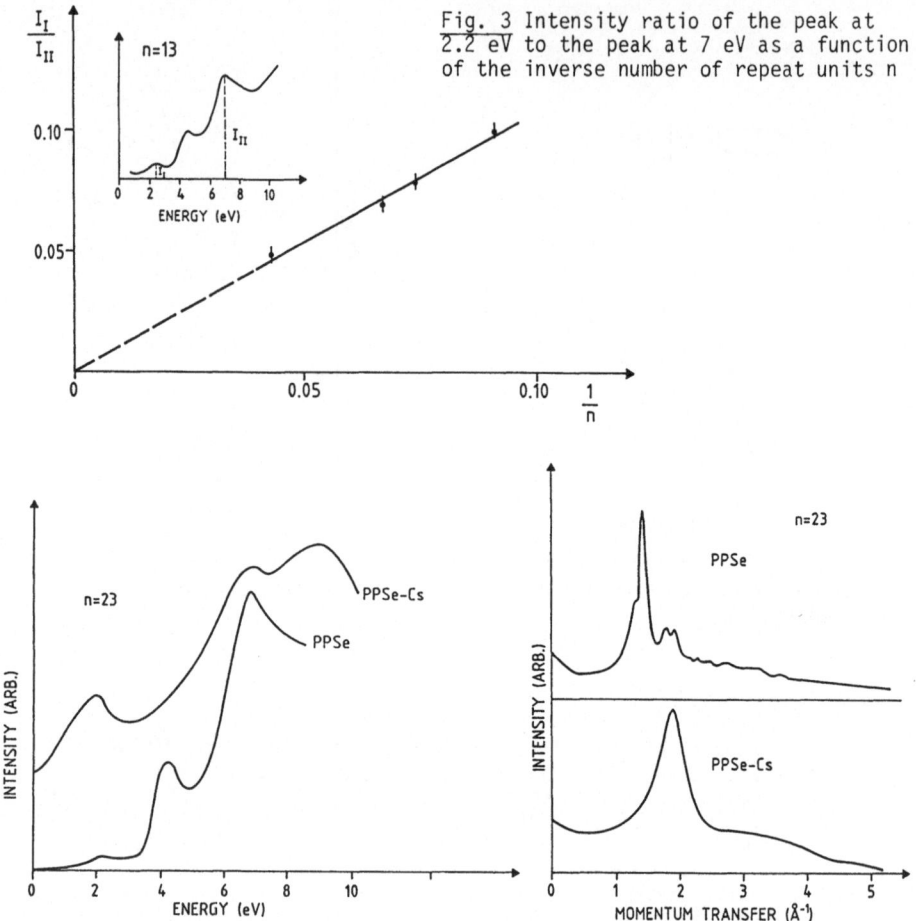

Fig. 3 Intensity ratio of the peak at 2.2 eV to the peak at 7 eV as a function of the inverse number of repeat units n

Fig. 4 Loss spectra of Cs-doped PPSe and pristine PPSe

Fig. 5 Elastic electron scattering spectra of pristine PPSe and PPSe-doped with Cs.

ing. Similar results were observed for PPP doped with Li [12] where the gap is closed too, and the lowest π plasmon is shifted to lower energies. However, the splitting of the 7 eV π plasmon may indicate the decomposition of the aromatic structure by a chemical reaction of the carbon atoms with Cs, or possibly an intra- or interchain cross-linking.

The latter interpretation is supported by the electron diffraction data of undoped and Cs doped PPSe shown in Fig. 5. For the undoped PPSe we find pronounced peaks corresponding to d values 4.85 Å, 4.47 Å, 3.61 Å, and 3.35 Å which can be assigned to the 110, 200, 112, 211 reflections assuming an orthorhombic structure. Upon doping with Cs, the diffraction pattern changes from a spectrum typical of crystalline structure to that of a disordered material. The changes of the diffraction data may again indicate a possible reaction of the Cs atoms with the polymer. Further work is necessary to clarify this question.

REFERENCES

1. J.L. Brédas, B. Thémans, J.G. Fripiat, and J.M. André: Phys. Rev. B 29, 6761 (1984)
2. K. Sanechika, T. Yamamoto, and A. Yamamoto: J. Polym. Sci. Polym. Lett. Ed. 20, 365 (1982)
3. L.W. Shacklette, H. Eckhardt, R.R. Chance, G.G. Miller, D.M. Ivory, and R.H. Baughman: In Polymer Science and Technology, ed. by R.B. Seymour (Plenum Press, New York 1981) p. 115
4. S. Tanaka, M. Sato, K. Kaeriyama, H. Kanetsuna, M.Kato, Y. Suda: Macromol. Chem. Rapid Commun. 4, 231 (1983)
5. D.J. Sandman, M. Rubner, and L. Samuelson: J. Chem. Soc. Chem. Commun. 1133 (1982)
6. J.F. Rabolt, T.C. Clarke, K.K. Kanazawa, J.R. Reynolds, and G.B. Street: J. Chem. Soc. Chem. Commun. 348 (1980)
7. J.H. Perlstein: Angew. Chem. Int. Ed. Engl. 16, 519 (1977)
8. H. Guenther, M.D. Bezoari, P. Kovacic, S. Gronowitz, A.B. Hörnfeldt: J. Polym. Sci. Polym. Lett. Ed. 22, 65 (1984)
9. W. Czerwiński, S. Nielek, M. Wełniak: to be published in "Angew. Makromol. Chem."
10. J. Fink: Z. Phys. B - Condensed Matter 61, 463 (1985)
11. G. Crecelius, J. Fink, J.J. Ritsko, M. Stamm, H.-J. Freund and H. Gonska: Phys. Rev. B 28, 1802 (1983)
12. J. Fink, B. Scheerer, M. Stamm, B. Tieke, B. Kanellakopulos, and E. Dornberger: Phys. Rev. B, Rap. Commun. 30, 4867 (1984).

Structural Characterization of Polymers Prepared from Iodophenylacetylene by Heating at Different Temperatures

M. Rotti, H. Krikor, and P. Nagels*

Rijksuniversitair Centrum, University of Antwerp,
B-2020 Antwerpen, Belgium

Information on the structural evolution of polymers prepared by heating iodophenylacetylene at 150, 400 and 600°C was acquired from chemical analysis, IR and UV spectroscopy. Polymerization is accompanied by iodine and hydrogen loss. The reactions involve a polycondensation of dimers yielding the formation of linear chains (150°C). At 400°C intramolecular cyclo-addition of phenylethynyl side groups occurs. With increase in temperature to 600°C the aromatic sections of the polymer grow, resulting in a graphite-like structure.

1. INTRODUCTION

Phenylacetylene is readily polymerized by using Ziegler-Natta catalysts as well as by heat. The resulting polymers exhibit a high electrical resistivity which is much less affected by doping with electron attracting or donating molecules as compared to polyacetylene [1]. In many doping experiments, molecular I_2 is incorporated into the material by exposure to its vapour phase. It seems interesting to study the effect of a direct substitution of iodine for hydrogen on the β position in the acetylene group. This paper discusses the results of a structural investigation based on IR-UV spectroscopy and chemical analysis of polymers prepared by heating the iodophenylacetylene monomer at three different temperatures.

2. EXPERIMENTAL PROCEDURES

Iodophenylacetylene (IPA) was synthesized by adding a solution of iodine in liquid ammonia to the acetylene derivative. The mixture was allowed to stand for 12 hours. The compound was precipitated by adding small portions of water. After filtering, it was washed with an aqueous solution of $Na_2S_2O_3$ in order to remove free iodine. A first thermal polymerization was carried out in argon filled quartz ampoules at 150°C during a short time (5 min). By extracting the brown viscous mass with benzene/methanol mixtures of different volume ratios, four fractions were separated with molecular masses equal to 460, 1280, 1950 and 2400. The iodine content of each fraction was determined by chemical analysis and yielded 53.9, 17.2, 6.2 and 3.0 wt % I, respectively. The fraction with $\bar{M}$ = 1950, present in the highest amount, was subjected to a further heat treatment at 400°C for 4 hours under the same experimental conditions. Black insoluble powders were obtained. These polymers were heated at 600°C for 4 hours.

The polymerization of the monomer, using a $Ti(O\ Bu)_4$-$AlEt_3$ catalyst system, proved to be unsuccessful. The presence of free iodine, formed

---

*Also at Physics Department, S.C.K./C.E.N., B-2400 Mol (Belgium)

during polymerization, probably poisoned the active species of the Ziegler-Natta catalyst.

3. RESULTS AND DISCUSSION

The IR spectrum of the iodophenylacetylene monomer is shown in Fig. 1(a). Among the multiple absorption bands typical of stretching and deformation vibrations of C=C and =C-H bonds of the benzene ring, two characteristic bands lying at 2165 and 525 cm^{-1} are observed. They can be assigned to stretching vibrations of -C≡C- and -C-I groups. As can be seen from Fig. 1(b) the intensity of these bands strongly decreases in the IR spectrum of the fraction with $\bar{M}$ = 1950 isolated from the material, which was obtained after heating at 150°C.

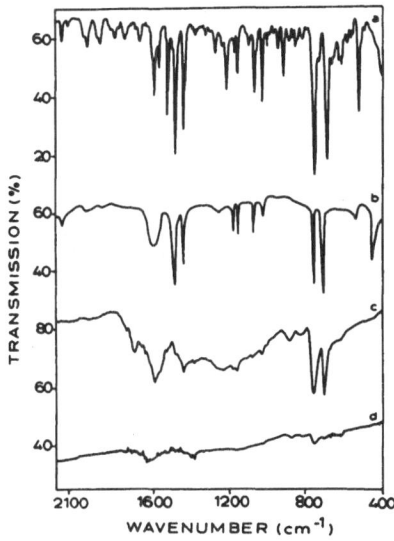

Fig. 1. Infrared spectra of (a) iodophenylacetylene monomer and of its polymers prepared by thermal polymerization at (b) 150°C (5 min); (c) 400°C (4 hrs); (d) 600°C (4 hrs)

The presence of double bonds in the polymer chain was evidenced by the possibility of bromine addition at the sp^2 C-atoms. In the IR spectrum of the brominated material a decrease in absorption band intensity at 1600 cm^{-1} was observed, indicating a reduction in the number of double bonds. Three new bands appeared at 2950, 2875 and 620 cm^{-1}, the former two attributable to saturated hydrocarbons, the latter one to vibrations of aliphatic C-Br bonds. From the bromination experiment it also follows that the broad absorption band at 1600 cm^{-1}, observed in the IR spectrum of fraction $\bar{M}$ = 1950, is a superposition of two bands associated on the one hand with the quadrant stretching of the benzene ring and on the other hand with the stretching vibration of the C=C bond present in the chain.

The IR spectrum of PIPA, polymerized at 400°C for 4 hours, exhibits much less structure (Fig. 1(c)). The bands at 2165 and 525 cm^{-1} have completely disappeared. In addition to this, the intensities of the absorption bands at 1180, 1155, 1070 and 1025 cm^{-1}, typical of monosubstituted benzenes, have strongly decreased. Two new bands located at 830 and 880 cm^{-1} appeared in this spectrum. They can be assigned to out-of-plane deformation of the =C-H group in benzene derivatives of varying degree of substitution. The IR spectrum of the polymer heated at 600°C shows a structureless absorption pattern, which is characteristic for the presence of condensed aromatics. One of the major features of the IR data is the gradual decrease of the C-I band intensity upon heating, indicating a loss of iodine. This is confirmed by chemical analysis (Table 1).

Table 1. Chemical analysis (wt %) of iodophenylacetylene polymerized at 150, 400 and 600°C

	C	H	I
PIPA-150	88.6	4.6	6.2
PIPA-400	95.3	3.2	0.3
PIPA-600	95.6	1.5	0.16

The molar extinction coefficient ε calculated from optical absorption data in the UV range (250-500 nm) is represented in Fig. 2(a) for the fraction $\bar{M}$ = 1950 polymerized at 150°C. A broad peak displaying a maximum at 340 nm appeared in the UV spectrum. It is known that the conjugated molecules benzene and phenylacetylene exhibit a sharp π-π^* absorption at 254 and 285 nm, respectively. The shift of the maximum to a higher wavelength (340 nm) in the polymer can be related to a coupling of the phenyl ring to segments of the polymer chain containing some number of conjugated double bonds. The appearance of the broad maximum followed by a tail extending to 500 nm points to a varying degree of conjugation.

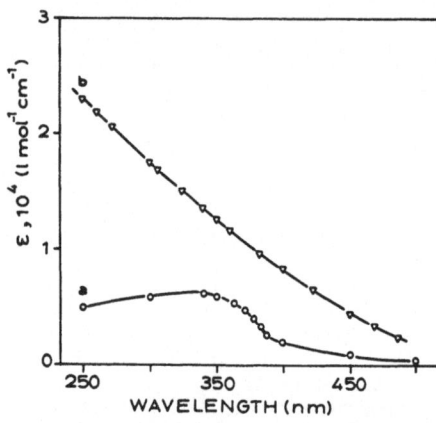

Fig. 2. Molar extinction coefficient ε versus wavelength in the UV range (250-500 nm) of iodophenylacetylene polymerized at (a) 150°C (5 min); (b) 400°C (2 hrs)

The UV spectrum of the polymer prepared at 400 and 600°C could not be recorded because of the insolubility of the materials in the common solvents. A fraction soluble in chloroform with a molecular mass $\bar{M}$ = 2100 could be isolated when the polymerization at 400°C was stopped after 2 hours. In the whole wavelength range from 250 to 500 nm this polymer shows a higher optical absorption than the fraction with approximately the same molecular weight ($\bar{M}$ = 1950) but polymerized at 150°C. The molar extinction coefficient continuously decreases, which might give some evidence for the presence of a polyaromatic system of varying length.

A structural analysis of thermally polymerized iodophenylacetylene using the same techniques was carried out by Cherkashin et al. [2]. They concluded that the initial step in the polymerization is the formation of a dimer involving the displacement of an iodine atom. The first fraction extracted from the mass obtained upon heating at 150°C, has a molecular weight $\bar{M}$ = 460 and an iodine content of 53.9 wt %, which corresponds to the composition of the iodophenylacetylene dimer. The IR and UV spectra, together with the ease of bromination, indicate that the propagation leads to the formation of a polymer backbone containing some sections of double bonds, to which phenylethynyl groups and phenyl rings are attached alternatively. During this polycondensation reaction, molecular iodine will be split off, as is experimentally observed. One iodine atom remains at each end of the chain.

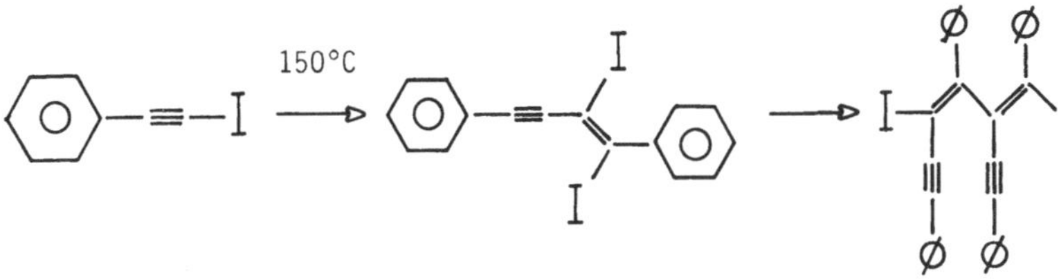

This reaction scheme differs from the one proposed for polyphenyl-acetylene which involves the formation of a conjugated linear chain with phenyl side groups. Heating these structures at 400°C gives rise to intramolecular cyclo-addition reactions of the phenylethynyl side groups. As a result, a polyaromatic structure is formed.

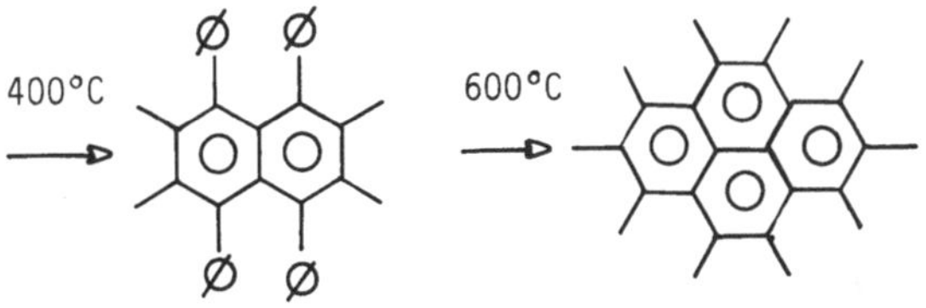

Upon heating to 600°C hydrogen is eliminated which might be a consequence of further cyclization. As shown by the structureless IR spectrum, the formation of a graphite-like structure can be assumed.

References

1. J.R. Ferraro, K. Martin, A. Furlani and M.V. Russo: Appl. Spectroscopy 38, 267 (1984)
2. M.I. Cherkashin, P.P. Kisilitsa, O.G. Sel'skaya and A.A. Berlin: Vysokomol. soyed A10, 196 (1968)

337

Metallic Coordination Polymers Using CS₂ as Starting Material

H.J. Keller[1], *T. Klutz*[2], *H. Münstedt*[3], *G. Renner*[1], *and D. Schweitzer*[2]

[1]Anorganisch-Chemisches Institut der Universität Heidelberg,
D-6900 Heidelberg, Fed.Rep. of Germany
[2]MPI, Abteilung Molekulare Physik,
D-6900 Heidelberg, Fed.Rep. of Germany
[3]AWETA der BASF, D-6700 Ludwigshafen, Fed.Rep. of Germany

Organic polymers with "metallic" properties have found widespread interest during the last few years /1,2/. Acetylene and aniline as well as different nitrogen and sulfur heterocycles have been used as starting materials. One main problem hampering the technical application of these solids up to now is their environmental and thermal instability. Therefore, we introduced metal ions to stabilize polymeric backbones with high electrical conductivity. Because of the enormous coordination ability of sulfur to many transition metal ions we decided to use a polymeric carbon-sulfur backbone. In view of a future technical application one has to start from less expensive and readily available chemicals like CS_2 e.g., in order to obtain a carbon-sulfur polymer. Since CS_2 cannot be readily polymerized to highly conducting solids a different approach has to be found. As reported earlier /3,4,5/ CS_2 can be reacted to thiapendione which can be converted into ethylenetetrathiolate (TT) by cleavage with strong chemical bases. Additionally thiapendione can be "dimerized" to bis(1.3-dithiole-2-one)tetrathiafulvalene which can be converted into tetrathiolatotetrathiafulvalene (TTF-TT) again by a cleavage reaction with strong bases. Both tetrathiolates react with transition metal ions like nickel(II) or copper(II) to coordination polymers which are remarkably inert chemically as well as thermally /6/. The reactions and their final products are schematically summarized in figure 1.

In the case of nickel (M = Ni) m = 0 and p = 0. The polynuclear compound contains formally nickel(IV) without counterions. Using copper(II) as metal ion a Cu(III)/Cu(I) mixed valence solid with m = 1 and p = 1 is obtained.

The compounds which are obtained as powders can be pressed to compact pellets which show conductivities up to 150 S/cm. The electrical conductivity of some of these materials varies only slightly in the temperature range between 4.2 K and 400 K (figures 2 and 3). Thermopower measurements (figure 4) indicate a metallic character of these solids. Principally the conductivity of the isolated TT derivatives is somewhat lower (2 - 7 S/cm) compared with the TTF-TT compounds (40 - 150 S/cm). A comparison between equivalent nickel and copper compounds shows that the copper species usually conduct better than the nickel derivatives which are much more stable environmentally compared to the copper compounds. The higher stability of the nickel species may be explained by the surprising fact that these materials are obtained in an electrically neutral "undoped" state.

SYNTHESIS

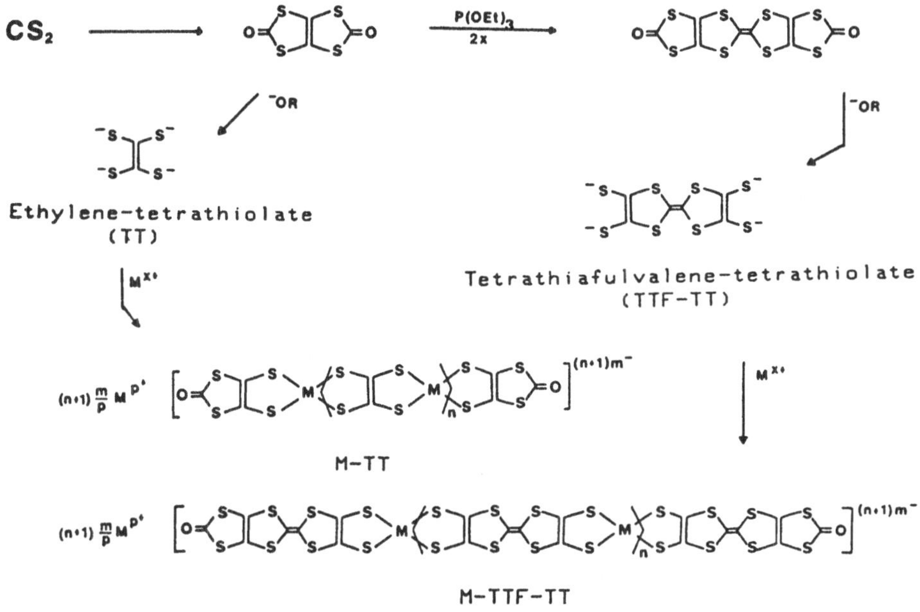

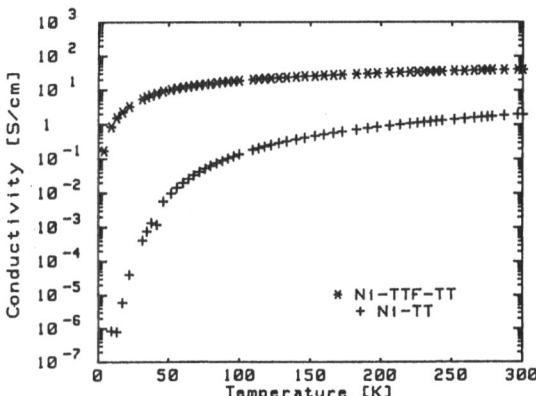

CS_2 ⟶

Ethylene-tetrathiolate
(TT)

Tetrathiafulvalene-tetrathiolate
(TTF-TT)

M-TT

M-TTF-TT

Fig. 1   Schematic description of the syntheses of "polymer"
          metal tetrathiolates starting from CS_2 (M = Ni or Cu)

Fig. 2   Electrical conductivity of Ni-TT (crosses) and
          Ni-TTF-TT (stars) as pressed pellets in the temperature
          range between 4.2 K and 300 K

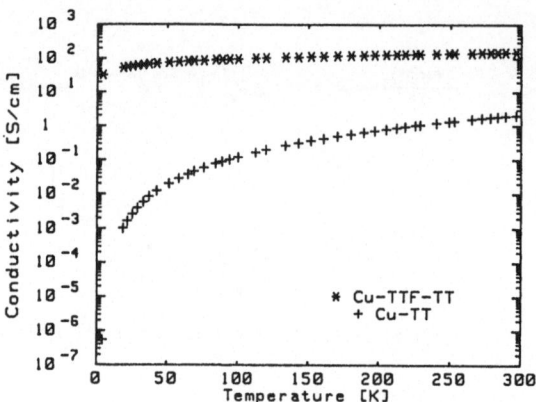

Fig. 3 Electrical conductivity of Cu-TT (crosses) and
Cu-TTF-TT (stars) as pressed pellets in the temperature
range between 4.2 K and 300 K

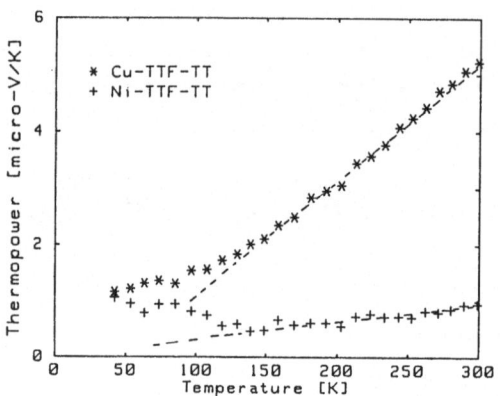

Fig. 4 Thermopower of pressed pellets of Ni-TTF-TT (crosses)
and Cu-TTF-TT (stars) in the temperature range between
40 K and 300 K

Literature

1. Proceedings Int. Conf. Science and Technol. of Synthetic Metals, Synth. Met., Vol. 19 (1987)
2. Proceedings Yamada Conf. Phys. Chem. Quasi One-dimensional
   Conductors, Physica B, (1987)
3. H. Poleschner, W. John, G. Kempe, E. Hoyer, E. Fanghänel:
   Z. Chem. 15, 345 (1978)
4. G. E. Holdcroft, A. E. Underhill: Synth. Met. 10, 427 (1985)
5. R. Vicente, J. Ribas, P. Cassoux, L. Valade: Synth. Met. 13,
   265 (1986)
6. G. Renner: Dissertation, University of Heidelberg, (Heidelberg 1987)

$(QP)_4(SbF_6)_3$: A Model Compound for Doped Polymers?

W. Grauf[1], J.U. von Schütz[1], H.P. Werner[1], H.C. Wolf[1], K. Göckelmann[2], V. Enkelmann[2], and G. Wegner[2]

[1]3. Physikalisches Institut, Universität Stuttgart,
Pfaffenwaldring 57, D-7000 Stuttgart 80, Fed. Rep. of Germany
[2]Max-Planck-Institut für Polymerforschung,
Jakob-Welder-Weg 11, D-6500 Mainz, Fed. Rep. of Germany

The p-quaterphenylene (QP) can be considered as a finite segment of a poly-p-phenylene (ppp) chain. Therefore, the interaction between the molecules in the radical cation salt of QP can be regarded as a model for the interchain interaction in doped polymers. In reality, the conductivity of $(QP)_4(SbF_6)_3$ is only 10^{-1} S/cm at room temperature (three orders of magnitude lower than that of doped ppp) and thermally activated. Magnetic resonance experiments (χ, ESR and NMR) yield, however, high spin concentration and narrow ESR-lines, as expected for mobile electrons. This discrepancy can be explained by the crystal structure which favourizes a 1:1 charge transfer between the stacks, thus leading to localized electrons (with spin exchange) and to a semiconductive state.

1. Introduction

The quaterphenylene-molecule (QP) consists of four benzene rings, σ-linked linearly, resembling short pieces of a polyparaphenylene (ppp) chain. Oxidized QP-molecules might be compared with doped ppp having chopped polymer segments in the solid state. The radical cation salt (rcs) of QP was therefore assumed to be a model compound for the examination of the interstack interactions of conducting polymers with high doping concentration. Under these aspects we have performed ESR-, NMR- and conductivity experiments on $(QP)_4(SbF_6)_3$. Whereas the magnetic properties were very similar to those of (fluoranthene)$_2$AsF$_6$ [1], an organic metal with high conductivity, the conductivity of our salt was rather poor and semiconductive.

Such behaviour can only be understood if spin and charge carriers are considered to be decoupled. The reason for it is seen in the crystal structure which can lead to a groundstate in which exactly one electron sits on each site of the stack.

2. The Crystal

The stable radical cation salt crystals were grown by anodic oxidation of the QP's in the presence of the SbF$_6$-anions as described in [2,3].

The main features of the crystal structure are seen in Fig. 1: There are triads of QP's in the stack, one uncharged QP per unit is incorporated in the anion sheet, resulting in the composition of $(QP)_3QP(SbF_6)_3$.

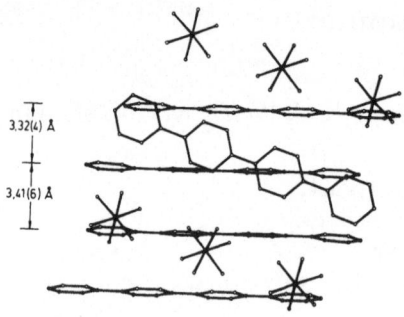

Fig. 1
The crystal structure of
quaterphenylenehexafluoro-
antimonate

3. Experimental Results

a) Conductivity

The conductivity is rather poor ($\sigma \approx 10^{-1} \ldots 10^{-3}$ Scm^{-1} at 300 K) and semiconductive, with an activation of about 0.1 eV.

b) ESR

The ESR lines are Lorentzian shaped, about 50 mG broad at 300 K. Their homogeneity was proved by ESR pulse experiments. With decreasing temperature ΔBpp decreases first till about 35 mG at 180 K to increase after that reaching 3 G at 4.2 K. The ESR intensity, drawn in fig. 2 by stars, is only weakly temperature dependent in the range of narrow ESR lines (300 K - 180 K).

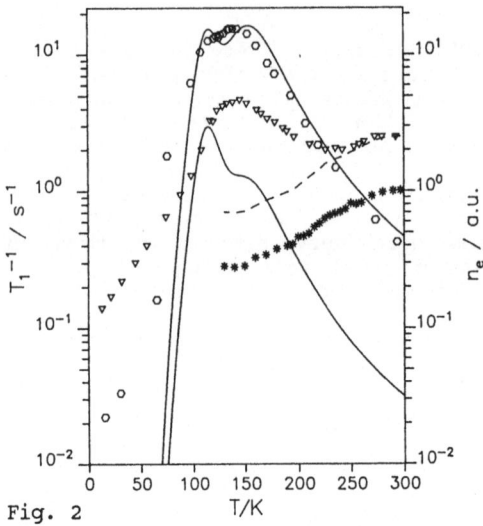

Fig. 2

The proton (∇) and fluorine ($\circ$) relaxation rates T_{1H}^{-1} and T_{1F}^{-1} as function of T.

The solid lines are the calculated relaxation rates due to reorientational motion of the SbF$_6$ molecules (see discussion). The electron spin density ($*$) is deduced from ESR intensities. It has to be scaled with the amplitude factor of the hyperfine interaction to get the relaxation rate (---).

c) NMR

The proton and fluorine relaxation rates were measured as a function of temperature at a frequency of 44 MHz (Fig. 2).

At room temperature T_{1H}^{-1} exceeds T_{1F}^{-1} by a factor of 6, stays more or less constant, whereas the fluorine relaxation exhibits a strong temperature dependence. Both relaxation curves have maxima at about 150 K to drop considerably below. Further we have measured the frequency dependence of T_{1H}^{-1} at 300 K which follows the relation $T_{1H}^{-1} \alpha \omega^{-0.4}$ between 0.1 and 180 MHz.

4. Discussion

As mentioned above, the magnetic properties of the QP salts are very similar to those of (fluoranthene)$_2$XF$_6$ (X = P, As, Sb) [1]. There, the narrow ESR-lines are explained by mobile electrons with averaged fine- and hyperfine interaction; the conformity between the proton relaxation T_{1H}^{-1}(T) and ESR-intensity by the presence of the electrons on the cation-stack solely.

The fluorine relaxation could be calculated quantitatively via FF- and FH-interaction modulated by the thermally activated reorientational motion of the SbF$_6$-octahedra. The results of such calculations are drawn in Fig. 2 as solid lines for T_{1F}^{-1} and T_{1H}^{-1}. The total proton relaxation is the sum of electron-proton (broken line) and proton-fluorine interaction (solid line).

In contrast to the fluoranthene salts which exhibit metal-like conductivities in the order of 10^3 Scm^{-1} [4], σ of our system is rather poor and semiconducting.

First attempts to attribute this discrepancy to crystal imperfections, small undistorted zones separated by insulating zone boundaries, could be ruled out by microwave conductivity measurements which coincided with the DC-results [2].

Taking into account the stoichiometry and structure of the crystal, we should have a half-filled band, if each molecule in the cation stack is charged as assumed. If there exists strong electron-electron repulsion, however, (the cost in energy of putting two electrons on one site is too great) only one electron can be placed on each site (state) and the band is, in effect, full [5]! Consequently spin and charge transportation is decoupled. The "spin" mobility via exchange averages the fine structure and relaxes the nuclei, their preferential 1-d-exchange being reflected in the frequency dependence of $T_{1H}^{-1} \alpha 1/\sqrt{\omega}$ [6].

5. Conclusions

The evaluation of the mechanisms of conductivity in polymers via the properties of perfectly ordered crystalline radical-cation-salts may fail just because of the well-defined stoichiometry. Although there can exist large transfer integrals, full bands or pinned charge density waves can mask the conductivity totally, as demonstrated on our system.

343

Acknowledgement

This work was supported by the Stiftung Volkswagenwerk. We would like to thank R. Kimmich and E. Schauer, University of Ulm, for the field cycling experiments.

References

[1] W. Höptner, M. Mehring, J.U. von Schütz, H.C. Wolf, B.S. Morra, V. Enkelmann and G. Wegner
Chem. Phys. 73, 252 (1982), Mol. Cryst. Liq. Cryst. 93, 395 (1983)
[2] K. Göckelmann, V. Enkelmann and H.W. Helberg, to be published
[3] V. Enkelmann and K. Göckelmann, G. Wieners and M. Monkenbusch
Mol. Cryst. Liq. Cryst. 120, 195 (1985)
[4] W. Stöcklein, B. Bail and M. Schwoerer:
In Electronic Excitations and Interaction Processes in Molecular Aggregates, ed. P. Reineker, H. Haken and H.C. Wolf
Springer Series in Solid State Sciences (1983)
[5] L. Coleman, J.A. Cohen, A.F. Garito and J. Heeger
Phys. Rev. B7, 2122 (1973)
[6] G. Soda, D. Jérome, M. Weger, J. Alizon, J. Gallice, H. Robert, J. Fabre and L. Giral
J. Phys. 38, 931 (1977)

344

Part IX

Related Topics

Supermolecular Structures Based on Langmuir-Blodgett Films

M.C. Petty

Molecular Electronics Research Group, School of Engineering and Applied Science, University of Durham, Science Laboratories, South Road, Durham, DH1 3LE, United Kingdom

1. Introduction

The Langmuir-Blodgett (LB) technique is now a well established method for producing ultra-thin films of organic materials on solid surfaces. These layers can be exploited in electronic and optoelectronic devices[1]. Recent advances in the film deposition equipment allow alternate-layers of different molecules to be built-up. This control of the molecular architecture of organic crystals is unique to the LB technique and permits the fabrication of superlattices with precisely defined symmetry properties. Such molecular assemblies can exhibit pyroelectric, piezoelectric and non-linear optical phenomena.

In this paper we shall briefly review the LB deposition process. Examples of the use of this technique to produce non-centrosymmetric structures exhibiting pyroelectric or non-linear optical effects will then be given.

2. Langmuir-Blodgett Technique

LB films are prepared by first depositing a small quantity of an amphiphilic material, dissolved in a volatile solvent, on the surface of carefully purified water. When the solvent has evaporated, the organic molecules may be compressed to form a floating "two-dimensional" solid. The hydrophilic and hydrophobic parts of the amphiphilic molecules ensure that, during this process, the individual molecules are all aligned in the same way. In the condensed state monolayers may be conveniently removed from the water surface by dipping and raising a suitably prepared solid substrate through the monolayer/air interface. Careful control of the surface pressure of the monolayer is required during this process. A number of different deposition modes are possible [1]; for the most common "Y-type" deposition, the monolayer is transferred from the water surface on both the upward and downward movement of the substrate.

The development of alternate-layer troughs [2,3] has enabled simple organic superlattice structures to be conveniently fabricated using the LB technique. In one such system [2] a single PTFE glass fibre loop is fed around PTFE rollers so as to define two surface compartments; a solid barrier separates these two areas. A cylinder, to which a substrate is attached, is an integral part of this barrier. If the cylinder is rotated continuously in one direction so that the substrate is dipped first through one monolayer (material A) and then withdrawn through the other (material B), alternate layers with the sequence ABABABA ..... are produced.

3. Pyroelectric LB Films

Pyroelectric detectors have become widely used in recent years for the thermal detection of infrared radiation because they are simple, rugged, low cost and do not require cooling. Significant advantages can accrue from the use of pyroelectric materials in thin film form (< 1μm). These include the relatively low thermal mass of the detector element and a reduction in the high - frequency noise equivalent power. The LB process allows the fabrication of the necessary polar structures for the observation of pyroelectricity. Moreover, the technique produces films that do not require poling, and that are much thinner than are usually attainable by the more conventional means. Figure 1 shows how alternate-layers of long chain fatty acids and long chain amines can be built-up to produce structures with an overall polarization [4-7].

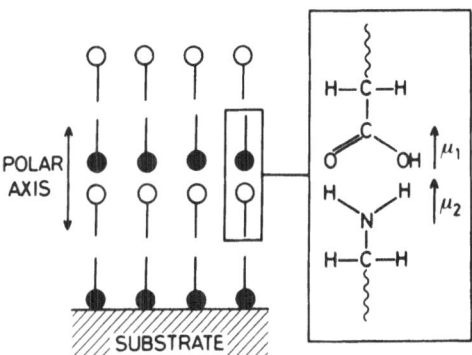

Fig. 1 Alternate-layer structure of fatty acid/fatty amine

The pyroelectric activity in such molecular assemblies can be measured using either a static or a dynamic method. In both cases a device structure is produced by sandwiching the acid/amine multilayer assembly between two metallic electrodes. For the static measurement technique the resulting device is mounted on a copper block and heated or cooled at a constant rate by a thermoelectric element. The pyroelectric current, i, is then monitored and the pyroelectric coefficient, p, is calculated from the equation

$$i = pA \left(\frac{dT}{dt} \right),\qquad(1)$$

where $\frac{dT}{dt}$ is the rate of change of temperature, and A is the device area.

The static pyroelectric current obtained from a 99-layer film of ω-tricosenoic acid alternated with docosylamine is shown in Fig. 2 [7]. It is apparent that the current flows in opposite senses when the sample is heated and when it is cooled, and it can be seen that the current is proportional to the rate of change of temperature, as predicted by (1). The dynamic pyroelectric method of CHYNOWETH [8] involves the sample being heated by a black-body source modulated by a mechanical chopper; the pyroelectric voltage can be measured by a phase-sensitive detection system, synchronized to the modulation frequency. The resulting signal may be used to calculate the pyroelectric coefficient using a theory developed by CHRISTIE et al [6]. Dynamic pyroelectric coefficients of the order 1 nC

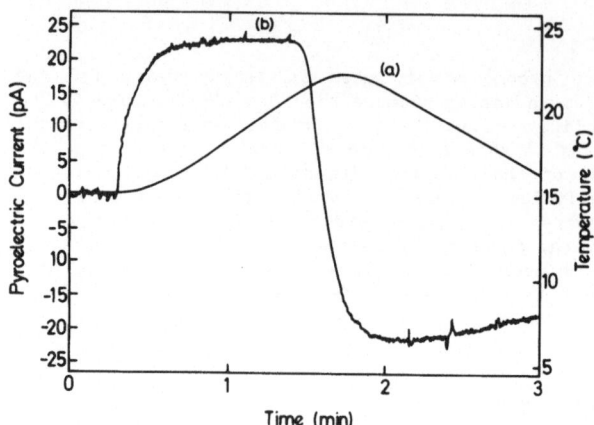

Fig. 2   Time variation of (a) the temperature and (b) the static
         pyroelectric current for a 99 layer acid/amine LB device during a
         heat-cool cycle

cm^{-2} K^{-1} have now been obtained for multilayer assemblies of simple fatty
acids and fatty amines [5-7]. The origin of the temperature dependent
polarization still remains somewhat uncertain, although the dependence of p
on the expansion coefficient of the substrate indicates a significant
secondary contribution (i.e. a thermally induced piezoelectric polarization)
[7].

More recently JONES et al [9] have reported on multilayer arrays based
on a fatty acid and a long chain aniline derivative; pyroelectric
coefficients approaching 2 nC cm^{-2} K^{-1} were demonstrated. Although such
values are still small compared to other materials (e.g. ceramics), one
material figure of merit for a pyroelectric element (applicable, for
instance, to a pyroelectric vidicon) is given by p/ε', where ε' is the real
part of the permittivity [10]. The very low values of ε' for LB films
(typically <5) therefore make the pyroelectric multilayer arrays discussed
here look quite attractive for practical devices.

4.   LB Films for Non-Linear Optics

Non-centrosymmetric LB arrays can also form the basis of non-linear optical
signal processing elements. The polarization, P, induced in a medium by an
external field is given by[11]

$$P = P_0 + \chi^{(1)}E + \chi^{(2)}E^2 + \chi^{(3)} E^3 + \ldots, \tag{2}$$

where P_0 is a constant, E is the electric (optical) field and $\chi^{(n)}$ is the

n'th order susceptibility tensor. On the molecular level, a molecule placed
in an electric field experiences a polarizing effect measured through an
induced dipole moment, $\Delta\mu$; its relation to E is expressed in terms of the
linear polarizability α (which is the origin of refractive index), and of
the higher order hyperpolarizabilities β, γ, ...   Thus

$$\Delta\mu = \mu - \mu_0 = \alpha E + \beta E^2 + \gamma E^3 + \ldots, \tag{3}$$

where μ_0 is a constant.

348

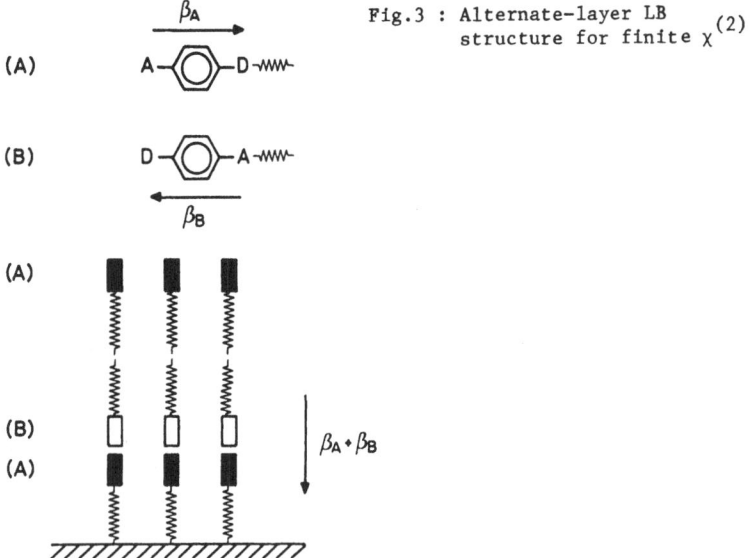

Fig.3 : Alternate-layer LB structure for finite $\chi^{(2)}$

Some organic substances with π-electron systems exhibit the largest known susceptibility coefficients, often considerably larger than those of the more conventional inorganic dielectrics [11]. For a finite second-order $\chi^{(2)}$ susceptibility, a non-centrosymmetric structure is required. This is of course possible with single monolayers deposited by the LB technique, and there are now a number of reports of second harmonic generation from such systems [12-14]. Figure 3 illustrates how a multilayer LB film may be built-up with a finite $\chi^{(2)}$.

The molecular array consists of alternate layers of materials A and B. Both molecules consist of a delocalized π-electron system between a donor and acceptor group (required for a high β hyperpolarizability value), plus an aliphatic tail (required for monolayer formation). However, in one of the molecules the donor group is next to the hydrocarbon tail, while in the other the acceptor group is in this position. Thus the β values for alternate-layers of the two molecules should be additive, producing a thin film possessing an overall $\chi^{(2)}$ value. This effect has recently been demonstrated by NEAL et al. [15,16]; one material was a hemicyamine dye [17], while the other was a nitrostilbene derivative [18]. It is interesting to note that the second-order optical nonlinearity for a bilayer of these materials was found to be much greater than expected by simply adding the polarizabilities for the individual dye molecules.

Acknowledgements

The author wishes to thank Professor G. G. Roberts (Oxford University) and Professor W. J. Feast (Durham University) for their contributions to this work. And to past and present members of the Durham Molecular Electronics Group, whose work is presented here: J. P. Lloyd, P. Christie, B. Holcroft, D. B. Neal and C. A. Jones.

References

1. M. C. Petty: In <u>Polymer Surfaces and Interfaces</u>, ed. by W. J. Feast and
    H. S. Munro (John Wiley & Sons Ltd., 1987) p.163
2. B. Holcroft, M. C. Petty, G. G. Roberts and G. J. Russell : Thin Solid
    Films, <u>134</u>, 83 (1985)
3. M. F. Daniel, J. C. Dolphin, A. J. Grant, K. E. N. Kerr and G. W. Smith
    : Thin Solid Films, <u>133</u>, 235 (1985)
4. G. W. Smith, M. F. Daniel, J. W. Barton and N. Ratcliffe : Thin Solid
    Films, <u>132</u>, 125 (1985)
5. P. Christie, G. G. Roberts and M. C. Petty : App. Phys. Lett., <u>48</u>, 1101
    (1986)
6. P. Christie, C. A. Jones, M. C. Petty and G. G. Roberts : J. Phys. D. :
    App. Phys., <u>19</u>, L167 (1986)
7. C. A. Jones, M. C. Petty and G. G. Roberts : Proc. IEEE 6th Int. Symp.
    Applications of Ferroelectrics, Bethlehem PA, 195 (1986)
8. A. G. Chynoweth : J. App. Phys., <u>27</u>, 78 (1956)
9. C. A. Jones, M. C. Petty, G. G. Roberts, G. H. Davies, J. Yarwood, N.
    M. Ratcliffe and J. W. Barton : Thin Solid Films, in press
10. A. M. Glass, J. S. Patel, J. W. Goodby, D. H. Olson and J. M. Geary :
    J. Appl. Phys., <u>60</u>, 2778 (1986)
11. J. Zyss : J. Non-Cryst. Solids <u>47</u>, 211 (1982)
12. O. A. Aktsipetrov, N. N. Akhmediev, E. D. Mishina and V. R. Novak :
    JETP Lett., <u>37</u>, 207 (1983)
13 I. R. Girling, N. A. Cade, P. V. Kolinsky and C. M. Montgomery : Elec.
    Lett., <u>21</u>, 169 (1985)
14. I. R. Girling, N. A. Cade, P. V. Kolinsky, R. J. Jones, I. R. Peterson,
    M. M. Ahmad, D. B. Neal, M. C. Petty, G. G. Roberts and W. J. Feast :
    J. Opt. Soc. Am. B, in press
15. D. B. Neal, M. C. Petty, G. G. Roberts, M. M. Ahmad, W. J. Feast, I. R.
    Girling, N. A. Cade, P. V. Kolinsky and I. R. Peterson: Elec. Lett.,
    <u>22</u>, 460 (1986)
16. D. B. Neal, M. C. Petty, G. G. Roberts, M. M. Ahmad, W. J. Feast, I. R.
    Girling, N. A. Cade, P. V. Kolinsky and I. R. Peterson : Proc. 6th IEEE
    Int. Symp. Applications of Ferroelectrics, Bethlehem PA, 89 (1986)
17. I. R. Girling, N. A. Cade, P. V. Kolinsky, J. D. Earls, G. H. Cross and
    I. R. Peterson : Thin Solid Films, <u>132</u>, 101 (1985)
18. M. M. Ahmad, W. J. Feast, D. B. Neal, M. C. Petty and G. G. Roberts :
    J. Molec. Elec., in press

Preparation and Structure
of Langmuir-Blodgett Films of Polymers

C. Bubeck

Max-Planck-Institut für Polymerforschung,
Postfach 3148, D-6500 Mainz, Fed. Rep. of Germany

1. Introduction

The electronic properties of polymers are strongly influenced by their structure, molecular packing and morphology. Therefore a large interest exists in an improved structural control of polymer thin films. This contribution will provide a brief introduction and a survey of the possibilities and the inherent problems of the Langmuir-Blodgett technique. This technique has been extensively described in the literature [1,2] and the progress in this field is summarized in some special volumes [3]. The simplified preparation consists of the following steps:

o A solution of amphiphilic molecules is spread at the air-water interface of a Langmuir-trough
o The solvent evaporates
o The area of the monolayer is continuously reduced by means of movable barriers
o The surface pressure can be adjusted to a defined value below the collapse pressure by electronic control of the barrier position
o The monolayer can be transferred to solid substrates provided that their surface is either hydrophobic or hydrophilic
o Multilayers can be built up by consecutive dipping and withdrawing of the substrate

The film thickness is exactly given by the number of dippings multiplied by the thickness of the individual layer. Reproducible multilayer production is only possible if all preparation parameters are carefully controlled [1]. Especially, extreme cleanliness of all parts and chemicals is necessary.

To obtain polymer thin films two preparation routes will now be described and compared: the first way starts with the preparation of monolayers of amphiphilic monomers that are polymerized in the layer environment, the second way begins with preformed polymers that are spread at the water surface and are subsequently transferred to solid substrates.

2. Multilayers of Monomers and their Polymerization in the Solid State

At present LB-multilayers of polydiacetylenes are the only example where the polymer backbone consists of a fully conjugated π-electron system. The structure of a monomeric and polymeric bilayer is shown schematically in Fig.1. The polymerization is initiated either by UV-irradiation in

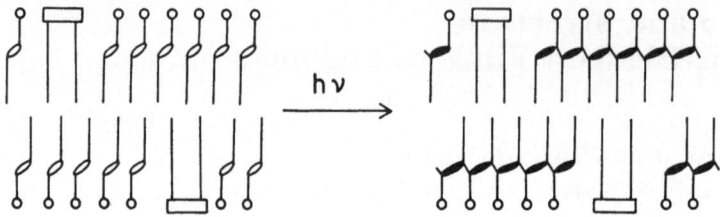

Fig.1 Simplified structural model of a monomeric and a polymeric diacetylene bilayer with embedded surface-active dyes. The length of the
alkyl chains is typically given by n=8 and m=12.

the range of the absorption band of the monomer /4,5/, or it can be
sensitized for visible light by the incorporation and optical excitation
of surface-active dyes /6/. The range of the sensitization of the dye
is not only restricted to the layer plane where the dye is located. In
addition, up to three neighbouring layers can be polymerized, as studied
by an evaluation of the polymer absorption band at 640 nm. The question
whether this is an electron transfer process over more than 90 Å or
simply due to a diffusion process of dye molecules during the layer
transfer to solid supports is hard to answer and is currently investigated /7/.

This also shows how important it is to know the morphology and the
real structure of molecular aggregates whenever electronic properties
are discussed. For example, the LB-layers of diacetylenes consist of
domains of two-dimensional crystallites that have grown already on the
water surface /5,8/. At present it seems that the formation of domains
is a general phenomenon of low molecular weight compounds that form a
solid analog phase at the water surface. This problem is hard to overcome
and is a severe drawback for all optical applications of LB-films,
because the domain boundaries act as scattering centres.

The positive and negative properties of LB-multilayers built up by
monomer units and subsequent polymerization in the solid state can be
summarized in the following manner:

+ Controlled thickness
+ Reasonably well defined structure
+ Increased stability
- Domain structure leads to light scattering at grain boundaries
- Side reactions such as photodecomposition can lead to
  a reduced conversion
- Steric constraints for the solid-state polymerizability

352

3. Multilayers from Preformed Polymers

If macromolecules are spread at the water surface, the cross-over of the polymer chains very often prevents the formation of real monolayers /9/. Nevertheless it was recently shown that multilayers can be built up from preformed polymers /10-13/. Using this way the restrictions due to steric constraints of solid-state polymerizability and the domain problem could be overcome. The advantages and the disadvantages of LB-multilayers of preformed polymers are briefly summarized:

+ The number of processing steps for the film preparation
  is reduced
+ No constraints with respect to solid-state reactions
- Low molecular order due to chain crossings
- The structural and thickness control is harder to achieve

4. Structural Investigations by FTIR-Spectroscopy

A manifold of structural information can be derived from Fourier Transform Infrared Spectroscopy. The principles of grazing incidence reflectance were derived by Greenler /14/ and recently demonstrated with LB-films /15/. Improvements in the spectrometer performance and the measurement technique have allowed to get high-quality IR-spectra of monolayers on evaporated metallic mirrors /15,16/. It turns out that the spectrum of the first monolayer on solid supports is very different with respect to the subsequent layers. This demonstrates strong influences of the substrate surface to the packing and orientation of the first layer.

5. Literature

1. H. Kuhn, D. Möbius, H. Bücher: In Physical Methods of Organic Chemistry, ed. by A. Weissberger, P. Rossiter, Vol. 1 part 3B, p.577 (Wiley Interscience 1972)
2. G.G. Roberts: Adv. in Physics 34, 475 (1985)
3. Special issues on LB-films: Thin Solid Films, Volumes 68 (1980), 99 (1983), 132-134 (1985)
4. B. Tieke, H.-J. Graf, G. Wegner, B. Naegele, H. Ringsdorf, A. Banerjie, D. Day, J.B. Lando: Colloid Polym. Sci. 225, 521 (1977)
5. B. Tieke: Adv. Polym. Sci. 71, 79 (1985)
6. C. Bubeck, B. Tieke, G. Wegner: Ber. Bunsenges. Phys. Chem. 86, 499 (1982)
7. G. Duda, C. Bubeck: to be published
8. G. Lieser, B. Tieke, G. Wegner: Thin Solid Films 68, 77 (1980)
9. F.H. Müller: Zeitschr. f. Elektrochem. 59, 312 (1955).
10. R.H. Tredgold, C.S. Winter: J. Phys. D 15, L 55 (1982)
11. A.J. Vickers, R.H. Tredgold, R. Hodge, E. Khoshdel, I. Girling: Thin Solid Films 134, 43 (1985)
12. S.J. Mumby, J.D. Swalen, J.F. Rabolt: Macromol. 19, 1054 (1986)
13. E. Orthmann, G. Wegner: Angew. Chem. 98, 1114 (1986)
14. R.G. Greenler: J. Chem. Phys. 44, 310 (1966)
15. J.D. Swalen, J.F. Rabolt: In Fourier Transform Infrared Spectroscopy, ed. by Ferraro, Basile Vol. 4, 283 (Acad. Press 1985)
16. T. Arndt, G. Bubeck: to be published

BEDT-TTF Radical Salts:
Organic Metals and Superconductors

*D. Schweitzer**

3. Physikalisches Institut der Universität Stuttgart,
Pfaffenwaldring 57, D-7000 Stuttgart 80, Fed. Rep. of Germany

A review of the structural, electronic and superconducting properties of
some BEDT-TTF-radical salts is given.

Introduction

After the first discovery of superconductivity in an organic metal - the radical salts of TMTSF (tetramethyltetraselenafulvalene) - under pressure [1] and
ambient pressure [2] soon it was clear that in addition to the usual intrastack
contacts between the donor molecules, intermolecular contacts between molecules
in neighbouring stacks are important in such materials. These interstack contacts result in a less pronounced one-dimensional electronic behaviour leading
to a stabilization of the metallic character down to low temperatures.

Therefore, SAITO et al. [3] had the idea that the somewhat nonplanar structure of BEDT-TTF (bis-ethylenedithiolotetrathiofulvalene) might amplify the interstack contacts in radical salts of this donor resulting in an even more isotropic electric behaviour. Electrochemically prepared crystals of $(BEDT-TTF)_2 \cdot$
$ClO_4(TCE)_{0.5}$ showed in fact the typical electrical behaviour of a quasi two-
dimensional organic metal [3, 4] down to low temperatures, but did not become
superconducting. In a similar salt $(BEDT-TTF)_4(ReO_4)_2$ the two-dimensionality
is somewhat less marked but PARKIN et al. [5,6] did find a superconducting
state near 2 K under an isotropic pressure of about 4 kbar, whereby the isotropic pressure was necessary to suppress a metal-insulator transition at 80 K and
ambient pressure. This was an important result because for the first time in an
organic sulphur donor system, superconductivity was observed. Nevertheless, the
electrochemical preparation of this radical salt is not easy due to the fact
that a large number of radical salts with different stoichiometries and crystal
structures might grow simultaneously. For that reason we started to prepare
radical salts of BEDT-TTF with I_3 counterions in THF-solutions [7-9] leading
to two radical salts with identical stoichiometries but different structures
(see fig. 1), the so-called α- and β-$(BEDT-TTF)_2I_3$ phases. Both types of crystals can be easily distinguished by eye because α-phase crystals have usually
a plate-like shape while ß-phase crystals are canted rhombohedrons. In addition the room temperature ESR linewidth of the conduction electrons can be
used to discriminate unequivocal between both salts (70 to 110 Gauss for
α-phase crystals and 20 to 25 Gauss for ß-phase crystals depending on the direction of the crystals with respect to the magnetic field [10, 11]).

α-$(BEDT-TTF)_2I_3$ crystals are two-dimensional organic metals with a nearly
isotropic electrical conductivity ($\sigma_{300} \approx$ 60-250 S/cm) within the ab-plane, a
1000 times smaller conductivity in the c*-direction and a metal insulator transition at 135 K [8, 9]. ß-$(BEDT-TTF)_2I_3$ crystals are two-dimensional metals as

---

* on leave from Max-Planck-Institut für Med. Forschung, Abt. Molekulare Physik,
Jahnstr. 29, 6900 Heidelberg, Fed. Rep. Germany

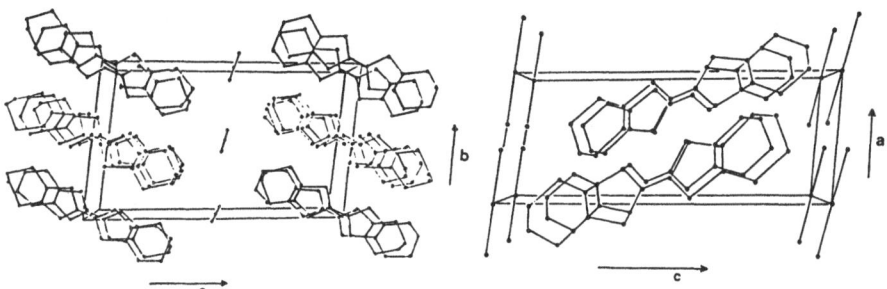

Figure 1 Stereoscopic projections of the structure of α-(BEDT-TTF)$_2$I$_3$ (left) and ß-(BEDT-TTF)$_2$I$_3$ (right) along the stacking axes.

well ($\sigma_{300} \approx 10$-50 S/cm) but stay metallic down to low temperature and - as was shown first by SCHEGOLEV et al./12/ - become superconducting at 1.3 K and ambient pressure.

In the following a review of the structural and physical properties of α- and ß-(BEDT-TTF)$_2$I$_3$ shall be given and the conditions will be described under which superconductivity at ambient pressure and 8 K in both types of crystals is observed.

ß-(BEDT-TTF)$_2$I$_3$

The discovery of superconductivity in ß-(BEDT-TTF)$_2$I$_3$ at T_c =1.3 K and ambient pressure /12/ was confirmed by WILLIAMS et al. /13/ and Meissner-effect measurements /14/ demonstrated the bulk property of the superconductivity in these crystals. A short time later it was found that under a pressure of 1.3 kbar the superconducting transition in ß-phase crystals can be raised to 7.5 K /15,16/. After a particular pressure-temperature cycling procedure - pressurization up to 1.5 kbar at room temperature and a release of the helium gas pressure at temperatures below 125 K - superconductivity at T_c = 8 K and ambient pressure was found /17-19/. The observed sharp superconducting transition - as measured by the resistivity of the crystal - as well as the suppression of the superconducting state by a magnetic field along the c*-axis at various temperatures is shown in fig. 2. The confirmation of bulk superconductivity at 8 K as well was obtained by Meissner-effect /20/ and ac-susceptibility measurements /18/. However, this specially prepared superconducting state at 8 K and ambient pressure

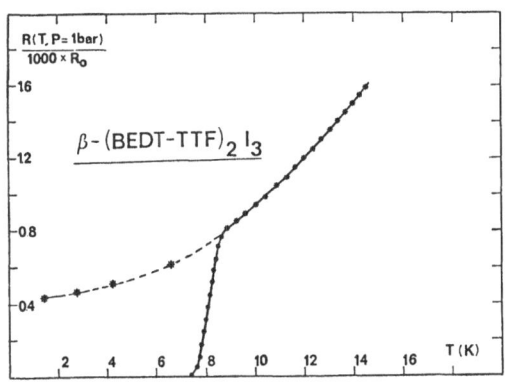

Figure 2
A superconducting transition at 8 K and ambient pressure in ß-(BEDT-TTF)$_2$I$_3$ after a special pressure-temperature cycling procedure /17/ as well as the suppression by a magnetic field along the c*-axis at various temperatures.

is meta-stable because superconductivity can only be obtained as long as the crystal temperature does not exceed 125 K. At this temperature a metal-metal phase transition was observed by thermopower measurements /10/. Recently it was shown that this meta-stable 8 K superconducting state can be obtained as well by an electronic excitation with laser light at temperatures below 125 K /21/. In order to understand these somewhat surprising facts, the structural properties of the ß-(BEDT-TTF)$_2$I$_3$ crystals under the different conditions have to be discussed. At room temperature ß-crystals are triclinic /22, 23/ (a=15.243, b=9.070, c=6.597 A; α=109.73, ß=95.56; γ=94.33°; V=848.9 A³) and the structure is very similar to the Bechgaard-salts. Below 195 K at ambient pressure an incommensurate structural modulation exists /24-26/ and the origin of this modulation is connected with an anion-cation interaction. Below 125 K a commensurate superstructure with a unit cell of about three times as large as at room temperature was observed /27/ (a=18.269, b=21.04, c=6.543 A; α=93.56, ß=94.84, γ=99.86°; V=2461 A³). The basic structural change below 125 K with respect to the room temperature structure lies in a pronounced distortion of the triodide chains and especially in a change of the linear and symmetric I$_3$-anions at room temperature into non-linear and asymmetric anions. This finding was confirmed by resonance Raman-investigations /21/. Therefore, it was assumed /27/ that the symmetric linear structure of the I$_3$-anions at room temperature is stabilized down to low temperatures by the special temperature cycling procedure /17,18/ and this more symmetric structure results in the high T$_c$-superconducting transition at 8 K. In fact, this assumption was confirmed recently by neutron diffraction experiments /28, 29/ and it was shown that no incommensurate modulated structure at 4.5 K exists, but furthermore, that the terminal ethylene-groups of the BEDT-TTF-molecules in the stacks are all ordered, in contrast to the low T$_c$-superconducting structure where only on one side the ethylene groups in the stacks are ordered, while on the other side they occupy the two possible conformations statistically /28, 29/.

It was shown /21/ that an electronic excitation of the ß-phase crystals at temperatures below 125 K by laser light can induce - at least in a thin layer on the surface - the structural transformation from the low temperature T$_c$=1.3K structure into the more ordered and symmetric high temperature T$_c$= 8 K structure. This transformation is observed in the resonance Raman-spectrum by a disappearance of the splitting of the symmetric stretching mode of the I$_3$-anions with time at constant laser power (see fig. 3) or immediately at high light intensity ($\approx$ 50 mW) /21/. Therefore, at temperatures between 1.5 and 8 K at least the surface of the crystal - in the moment it is not clear whether IR-radiation might even switch the whole volume of the crystal - can be switched optically from the normal conducting into the superconducting state. But only those parts of the surface are switched which were irridiated by light (optical storage !). However, this transformation is again only stable as long as the crystal temperature does not exceed 125 K /21/.

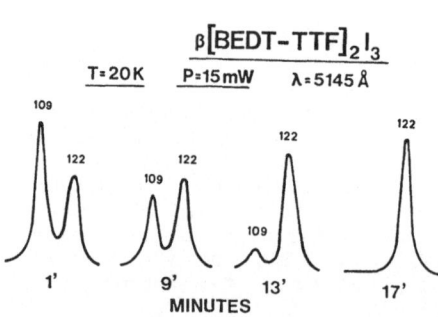

β[BEDT-TTF]$_2$I$_3$

T=20K P=15mW λ=5145 Å

MINUTES

Figure 3
Structural transformation from the low into the high temperature superconducting phase in ß-(BEDT-TTF)$_2$I$_3$ by irradiation with light /21/ as observed by the change of the resonance Raman-spectrum of the symmetric stretching mode (109 and 122 cm^{-1}) of the I$_3$-anions with time at constant laser power (15 mW, at 5145 Å).

α - and α_t-(BEDT-TTF)$_2$I$_3$

The unit cell of α-(BEDT-TTF)$_2$I$_3$ crystals (triclinic: a=9.211, b=10.85, c=17.488 A; α=96.95, ß=97.97, γ=90.75, V=1717 A^3 /9/) at room temperature is twice as large as the unit cell of the ß-phase. In contrast to the ß-phase crystals in α-(BEDT-TTF)$_2$I$_3$ two crystallographically different stacks occur and there exist large dihedral angles between the molecular planes (59.4 and 70.4°) of neighbouring donor molecules. The I$_3$-anions are linear and all the terminal ethylene-groups of the BEDT-TTF-donor molecules in both crystallographically non-equivalent stacks are ordered. Under ambient pressure the α-(BEDT-TTF)$_2$I$_3$ undergoes a metal-insulator phase transition at 135 K/7-9/, which can be suppressed by an isotropic pressure of >12 kbar, but no sign of superconductivity could be observed down to 100 mK /30/. It was claimed /31/ that doping the α-phase crystals with iodine results in a metallic state below the insulator phase transition at 135 K followed by a transition to supercon- ductivity at around 3.2 K. Similar results as in ref. /31/ for the resistivity for such iodine-doped α-crystals could be obtained by GOGU /32/ as well, but microwave conductivity /33/ as well as ac-susceptibility measurements /32/ showed that this metallic state is not a bulk effect in these crystals.

Recently, BARAM et al./34/reported a structural transformation of α-phase crystals into ß-phase crystals by tempering the crystals at a temperature of 70-100°C for about 10 to 20 hours. This is a quite surprising result because of the essential differences in the structures of both phases but the struc- tural transformation was confirmed by Weisenberg-pictures /34/. The most sur- prising fact after the structural transformation was that these crystals show a sharp decrease in resistivity at around 8 K, indicating a superconducting tran- sition at this temperature, but proof by a simultanously applied magnetic field was not reported. Further it was not clear whether or not the eventual super- conducting transition is a bulk effect in these tempered crystals.

Another open question was what the differences in the structures between the normal grown ß-(BEDT-TTF)$_2$I$_3$ and the tempered α_t-(BEDT-TTF)$_2$I$_3$ crystals are. (In the following tempered α-phase crystals are called α_t-(BEDT-TTF)$_2$I$_3$). This is an important question since ß-(BEDT-TTF)$_2$I$_3$ becomes superconducting under

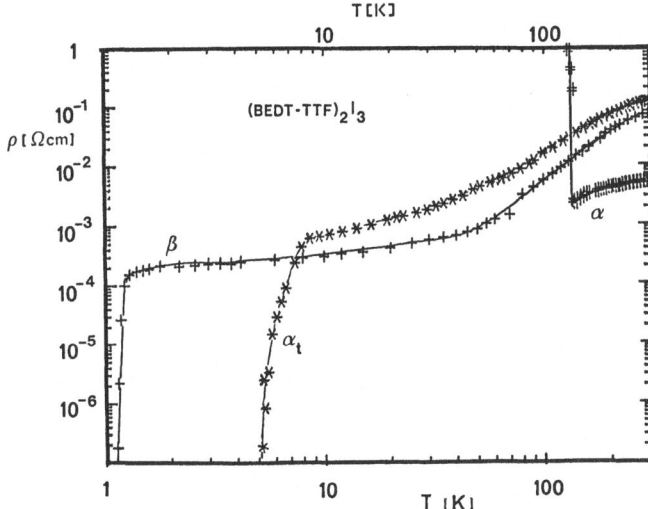

Figure 4   Resistivity versus temperature (logarithmic scales) for
α -, α_t- and ß-(BEDT-TTF)$_2$I$_3$.

normal conditions at 1.3 K while in α_t-(BEDT-TTF)$_2$I$_3$ this transition seems to be at around 8 K, whereas the ß-(BEDT-TTF)$_2$I$_3$ shows superconductivity only after special treatment (temperature pressure cycling or electronic excitation by light).

A systematic study of the temperature-dependence of the resistivity (fig.4,5) and ac-susceptibility (fig. 6) both with and without applying a magnetic field as well as ESR- (fig. 7), NMR- (fig. 8), resonance Raman- (fig.9) and thermo-power investigations on α_t-(BEDT-TTF)$_2$I$_3$ /35/ has shown that bulk superconductivity at 8 K and ambient pressure exist in such α_t-crystals. In contrast to the specially prepared 8 K meta-stable superconducting state in ß-crystals here in α_t-(BEDT-TTF)$_2$I$_3$ the superconducting state is stable and entirely repro-ducable for several temperature cycles up to 380 K.

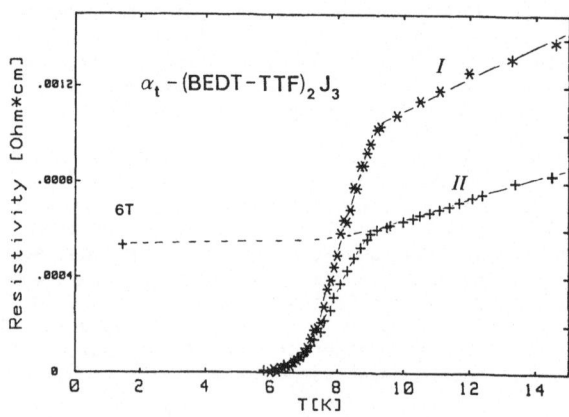

Figure 5

Resistivity of two α_t-(BEDT-TTF)$_2$I$_3$ crystals(I and II) in the temperature region between 0 and 15 K /35/. The suppression of superconductivity in sample II by applying a magnetic field of 6 T at 1.3 K paral-lel to the c*-axis is shown as well.

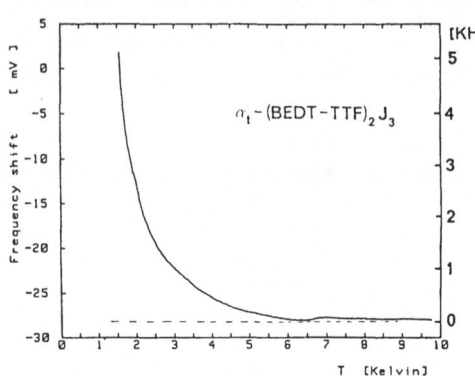

Figure 6

Increase in resonance fre-quency of a LC-circuit due to exclusion of the rf-field by diamagnetic shielding currents (ac-susceptibility) in α_t(BEDT-TTF)$_2$I$_3$ (sample II of fig. 5) by lowering the tempe-rature /35/.

^{13}C-NMR solid state investigations /35/ (fig. 8) have shown that by tempe-ring α-crystals above 70 °C for several days in air all crystals used in the experiment (≈ 0.3 g) were totally converted into the new α_t-(BEDT-TTF)$_2$I$_3$-phase.

The ESR-linewidth measurements (fig. 7) as well as the resonance Raman-spectra (fig. 9) indicate the phase transition after tempering as well /35/. The upper critical fields H$_{c2}$ (fig. 10)/35/ are anisotropic but with 2.5 to 11 T (depen-ding on the direction of the magnetic field with respect to the crystal axes)

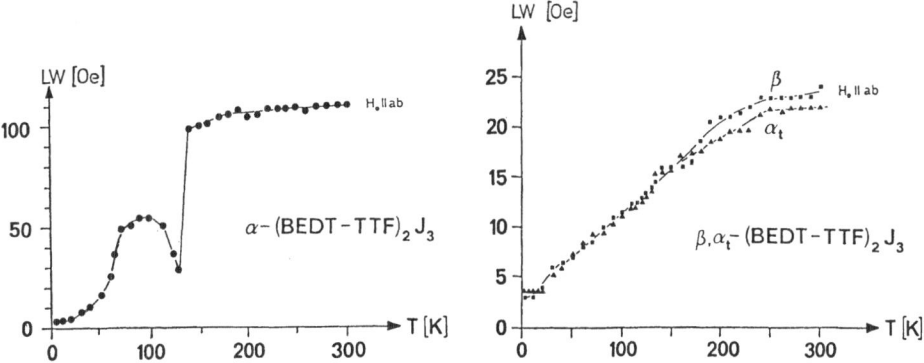

LW [Oe]

Figure 7   Temperature dependence of the ESR-linewidth for α-(BEDT-TTF)$_2$I$_3$
(left) as well as for ß-(■) and α_t-(BEDT-TTF)$_2$I$_3$ (▲)(right)/35/.

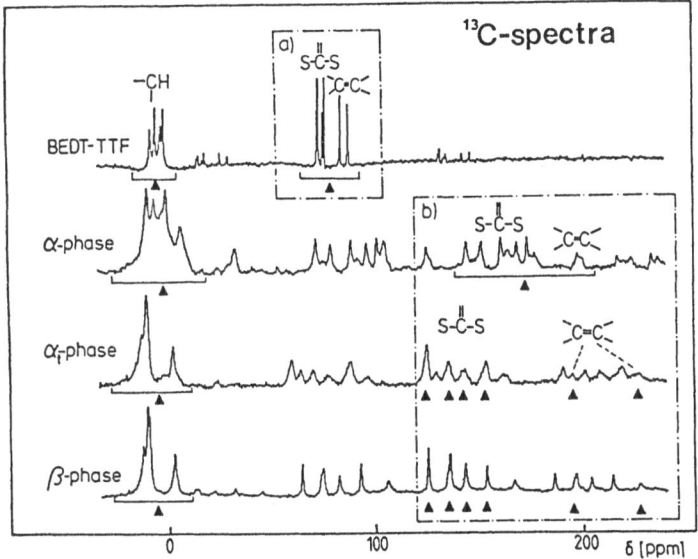

Figure 8   Magic angle spinning (MAS) ^{13}C-NMR spectra at 300 K of BEDT-TTF
and of the organic metals α-, ß- and α_t-(BEDT-TTF)$_2$I$_3$. The center bands are
marked by ▲ in the framed parts as well as in the region of the ethylene
groups (ν = 68 MHz, spinning frequency between 4-5 KHz)/35/.

relatively high. All the results from ref. /34, 35/ indicate that the α_t-crystal
structure at room temperature is identical with the one of the high T$_c$-super-
conducting phase of ß-crystals. This is probably due to the fact that in the
α-phase crystals at room temperature (and probably at 75 °C as well) all terminal ethylene-groups of the BEDT-TTF donor molecules are ordered as in the
high T$_c$-superconducting ß-phase and therefore the phase transition at 75 °C
results in this structure. To clarify the situation, further X-ray structure
investigations are necessary.

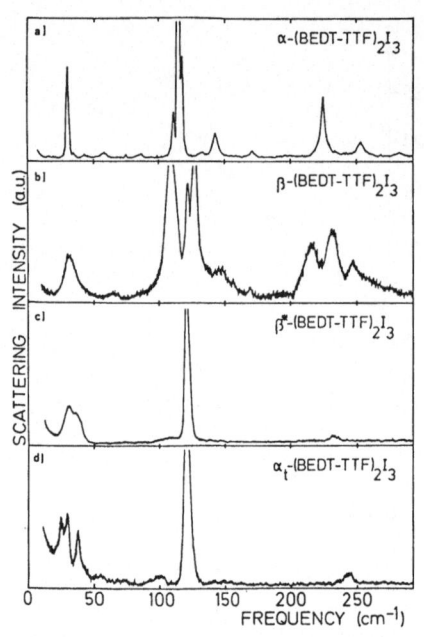

Figure 9

Low energetic parts of the resonance Raman-spectra at 2 K of
a)α-(BEDT-TTF)$_2$I$_3$
b)β-(BEDT-TTF)$_2$I$_3$ (low temperature
superconducting (1.3 K) phase)
c)β-(BEDT-TTF)$_2$I$_3$ (high temperature
superconducting (8 K) phase, as
prepared by optical excitation/21/).
d)α_t-(BEDT-TTF)$_2$I$_3$
(excitation wave length 4880 A,
10 mW) /35/.

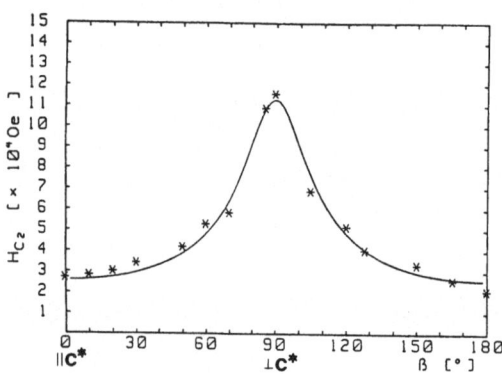

Figure 10

Upper critical fields H$_{c2}$ at
a temperature of 1.3 K for
α_t-(BEDT-TTF)$_2$I$_3$ as evaluated
by the mid-transition of the
resistivity curve for several
magnetic field directions/35/.

BCS superconductivity

An important question is whether or not the usual electron phonon coupling is
responsible for the superconductivity in the radical salts of BEDT-TTF. Tunneling experiments in the normal metallic /36/ as well as in the superconducting
state /37, 38/ were carried out. Tunneling experiments of HAWLEY et al. /37/
on β-(BEDT-TTF)$_2$IAuI-crystals (superconducting transition at ambient pressure
at 4.5 K /39, 40/) did show a superconducting gap Δ_T in the ab-plane about
5 times larger than the expected BCS-value. More recent tunneling experiments
/38/ on β-(BEDT-TTF)$_2$I$_3$ as well as on β-(BEDT-TTF)$_2$IAuI in the ab-plane as well
show values of Δ_T only about 15 % larger than the BCS weak coupling value and
are in very good accordance with the conventional electron phonon theory of
superconductivity. However, the superconducting gap might be quite anisotropic
and in fact preliminary band calculations /41/ indicate that the Fermi-surface

has several separate sheets, which obviously may possess different superconducting gaps. In such a situation the average gap may correspond to the BCS-value
and the maximum value of the gap may be considerably higher /38/. Therefore
more experimental data for tunneling contacts with different orientations in
the ab-plane are needed.

Conclusions

Bulk superconductivity at 8 K and ambient pressure exists in β-(BEDT-TTF)$_2$I$_3$
and α_t-(BEDT-TTF)$_2$I$_3$ crystals. While in the ß-crystals the superconducting
state has to be prepared under special conditions (temperature pressure cycling
procedure or electronic excitation) and is only meta-stable, in α_t-(BEDT-TTF)$_2$I$_3$
this superconducting state at 8 K is stable and entirely reproducable for many
temperature cycles up to 380 K.

Acknowledgement

I would like to thank all my colleagues and co-workers (see references) for
very effective cooperation during all the investigations in recent years. My
special thanks to Professors H.J. Keller, H. Endres, M. Weger and Dr. Swietlik
for many helpful discussions and efficient teamwork in Heidelberg.

Literature

1. D. Jérome, A. Mazaud, M. Ribault and K. Bechgaard: J. Phys. Lett. $\underline{4}$,
   L 95 (1980)
2. K. Bechgaard, K. Caneiro, M. Olsen, F.B. Rasmussen and C.S. Jacobsen:
   Phys. Rev. Lett. $\underline{46}$, 852 (1981).
3. G. Saito, T. Enoki, K. Toriumi and H. Inokuchi: Solid State Comm. $\underline{42}$,
   557 (1982).
4. G. Saito, T. Enoki, H. Inokuchi and H. Kobayashi: Journal de Physique
   $\underline{44}$, C 3, 1215 (1983).
5. S.S.P. Parkin, E.M. Engler, R.R. Schumaker, R. Lagier, V.Y. Lee, J.C.
   Scott and R.L. Greene: Phys. Rev. Lett. $\underline{50}$, 270 (1983).
6. S.S.P. Parkin, E.M. Engler, R.R. Schumaker, R. Lagier, V.Y. Lee, J.
   Voiron, K. Carneiro, J.C. Scott and R.L. Greene: Journal des Physique
   $\underline{44}$, C 3, 791 (1983).
7. D. Schweitzer, I. Hennig, K. Bender, K. Dietz, H. Endres and H.J.Keller:
   In Verhandlungen d. Deutschen Physikalischen Gesellschaft, Physikerta-
   gung Münster, März 1984, MO 32, p. 584 (1984).
8. K. Bender, K. Dietz, H. Endres, H.W. Helberg, I. Hennig, H.J. Keller,
   H.W. Schäfer and D. Schweitzer: Mol. Cryst. Liq. Cryst. $\underline{107}$, 45 (1984).
9. K. Bender, I. Hennig, D. Schweitzer, K. Dietz, H. Endres and H.J.Keller:
   Mol. Cryst. Liq. Cryst. $\underline{108}$, 359 (1984).
10. I. Hennig, K. Bender, D. Schweitzer, K. Dietz, H. Endres, H.J. Keller,
    A. Gleitz and H.J. Helberg: Mol. Cryst. Liq. Cryst. $\underline{119}$, 337 (1985).
11. R. Rothaemel, L. Forró, J.R. Cooper, J.S. Schilling, M. Weger, P.Bele,
    H. Brunner, D. Schweitzer and H.J. Keller: Phys. Rev. B$\underline{34}$, 704 (1986).
12. E.B. Yagubskii, I.F. Schegolev, V.N. Laukhin, P.A. Kononovich, M.V.
    Kartsovnik, A.V. Zvarykina and L.I. Buravov: Sov. Phys. JETP Lett. $\underline{39}$,
    12 (1984).
13. J.M.Williams, T.J. Emge, H.H. Wang, M.A. Beno, P.T. Copps, L.N. Hall,
    K.D. Carlson and G.W. Crabtree: Inorg. Chem. $\underline{23}$, 2558 (1984).
14. H. Schwenk, C.P. Heidmann, F. Gross, E.Hess, K. Andres, D. Schweitzer
    and H.J. Keller: Phys. Rev. B$\underline{31}$, 3138 (1985).

15. K. Murata, M. Tokumoto, H. Anzai, H. Bando, G. Saito, K. Jajimura and
    T. Ishiguro: J. Phys. Soc. Japan 54, 1236 (1985).
16. V.N. Laukhin, E.E. Kostyuchenko, Yu.V. Sushko, I.F. Schegolev and E.B.
    Yagubskii: Soviet Physics JETP Lett. 41, 81 (1985).
17. F. Creuzet, G. Creuzet, D. Jérome, D. Schweitzer and H.J. Keller: J.
    Physique Lett. 46, L-1079 (1985).
18. F. Creuzet, D. Jérome, D. Schweitzer and H.J. Keller: Europhys.Lettr.
    1, 461 (1986).
19. F. Creuzet, C. Bourbonnais, D. Jérome, D. Schweitzer and H.J. Keller:
    Europhys.Lett. 1, 467 (1986).
20. H. Veith, C.P. Heidmann, F. Gross, A. Lerf, K. Andres and D. Schweitzer:
    Solid State Commun. 56, 1015 (1985).
21. R. Swietlik, D. Schweitzer and H.J. Keller: submitted to Phys.Rev. B.
22. T. Mori, A. Kobayashi, Y. Sasaki, H. Kobayashi, G. Saito and H.Inokuchi:
    Chem. Lett. 957 (1984).
23. V.F. Kaminskii, T.G. Prokhorova, R.P. Shibaeva and E.B. Yagubskii:
    JETP Lett. 39, 17 (1984).
24. P.C.W. Leung, T.J. Emge, M.A. Beno, H.H. Wang and J.M. Williams: J.Am.
    Chem. Soc. 106, 7644 (1984).
25. T.J. Emge, P.C.W. Leung, M.A. Beno, A.J. Schultz, H.H. Wang, L.M. Sowa
    and J.M. Williams: Phys. Rev. B. 30, 6780 (1984).
26. P.C.W. Leung, T.J. Emge, M.A. Beno, H.H. Wang, J.M. Williams, V. Petri-
    cek and P. Coppens: J. Am. Chem. Soc. 107, 6184 (1985).
27. H. Endres, H.J. Keller, R. Swietlik, D. Schweitzer, K. Angermund and
    C. Krüger: Z. Naturforsch. 41a , 1319 (1986).
28. A.J. Schultz, M.A. Beno, H.H. Wang and J.M. Willaims: Phys.Rev. B33,
    7823 (1986).
29. A.J. Schultz, H.H. Wang, J.M. Williams and A. Filhol: J.Am. Chem. Soc.
    108, 7853 (1986).
30. H. Schwenk, F. Gross, C.P. Heidmann, K. Andres, D. Schweitzer and H.J.
    Keller: Mol. Cryst. Liq. Cryst. 119, 329 (1985).
31. E.B. Yagubskii, I.F. Schegolev, V.N. Laukhin, R.R. Shibaeva, E.E. Kosty-
    uchenko, A.G. Khomenko, Yu.V. Sushko, A.V. Zvarykina, Pis'ma Zh. Eksp.
    Theor. Fiz. 40, 387 (1984).
32. E. Gogu: private communication.
33. M. Przybylski, H.W. Helberg, D. Schweitzer and H.J. Keller: Synthetic
    Metals 19, 191 (1987).
34. G.O. Baram, L.I. Buravov, L.C. Degtariev, M.E. Kozlov, V.N. Laukhin,
    E.E. Laukhina, V.G. Orischenko, K.I. Pokhodnia, M.K. Scheinkmann, R.P.
    Shibaeva and E.B. Yagubskii, JETP Lett. 44, 293 (1986).
35. D. Schweitzer, P. Bele, H. Brunner, E. Gogu, U. Haeberlen, I. Hennig,
    T. Klutz, R. Swietlik and H.J. Keller: submitted to Zeitschr.f. Physik B.
36. A. Nowack, M. Weger, D. Schweitzer, H.J. Keller: Solid State Comm. 60,
    199 (1986).
37. M.E. Hawley, K.E. Gray, B.D. Terris, H.H. Wang, K.D. Carlson, J.M. Wil-
    liams: Phys. Rev. Lett. 57, 629 (1986).
38. A. Nowack, U. Poppe, M. Weger, D.Schweitzer and H. Schwenk: submitted
    to Zeitschr. f. Physik B.
39. H.H. Wang, M.W. Beno, U. Geiser, M.A. Firestone, K.S. Webb, L. Nunez,
    G.W. Crabtree, K.D. Carlson, J.M. Williams, L.J. Azevedo, J.F. Kwak,
    J.E. Schirber: Inorg. Chem. 24, 2466 (1985).
40. E. Amberger, H. Fuchs and K. Polborn: Angew. Chem. 97, 968 (1985).
41. J. Kübler, M. Weger, C.B. Sommers: submitted to Sol. State Comm.

Electrical Conducting Molecular Crystals in a Polymer Matrix

G. Heywang

Zentrale Forschung, BAYER AG,
D-5090 Leverkusen, Fed. Rep. of Germany

Summary

Films containing only 0.4 - 0.8 % of a radical salt based on tetracyanoquinodimethane (TCNQ) are transparent and permanently antistatic. The charge-transfer complexes form a three-dimensional network. The specific conductivity is 10^{-6} to 10^{-7} S/cm. The formation of the conducting network depends on the viscosity, temperature, solvent and evaporation speed.

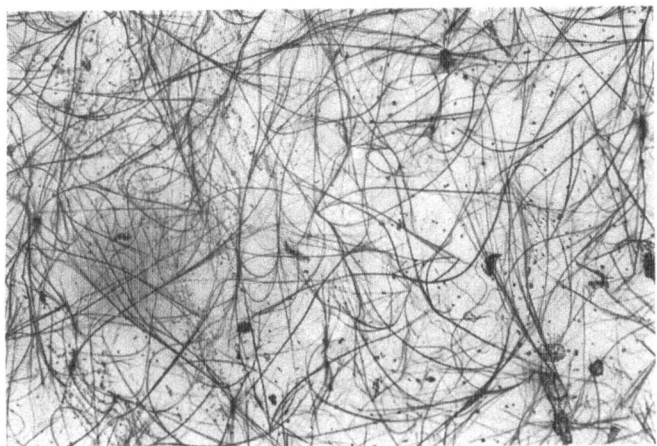

Film of polycarbonate with 0.8 % N-methylquinolinium-TCNQ$_2$

This talk has been published by
J. Hocker, F. Jonas, H.-K. Müller in
Die Angewandte Makromolekulare Chemie 145/146 (1986) 191-210.

Part X

Applications

Grafting, Ionomer Composites, and Auto-doping of Conductive Polymers

P. Audebert[1], *G. Bidan*[1;*], *M. Lapkowski*[2], *and D. Limosin*[3]

[1]Equipe d'Electrochimie Moléculaire, Laboratoires de Chimie,
  Département de Recherche Fondamentale, Centre d'Etudes Nucléaires
  de Grenoble, 85 X, F-38041 Grenoble Cedex, France
[2]Institut of Inorganic Chemistry and Technology,
  Silesian Technical University, PL-Gliwice, Poland
[3]Laboratoire d'Electrochimie Organique et Photochimie Rédox,
  UA CNRS No. 1210, Université Scientifique, Technologique et
  Médicale de Grenoble, Domaine Universitaire, BP 68,
  F-38402 Saint Martin d'Heres Cedex, France

The easy electrodeposition of poly(pyrrole) even when starting from
N-derived pyrrole monomers has opened up a new field in the coating of
electrodes by various specific molecules (electroactive, photo-sensitive,
catalytic centres ...). In order to overcome some difficulties due to the
poly(pyrrole) matrix (instability upon cycling, inclusion of protons ...)
this method is now being extended to poly(derived-thiophenes). But new
requirements have arisen specifically due to the severe conditions of
electropolymerization of 3-derived thiophenes. The doping process of
conductive polymers has been exploited to form interpenetrated composites
in which the second polymer component is an ionomer acting as a
macro-dopant of the conductive matrix. The quasi-immobilisation of the
anionic dopants in these structures allows them to function as
cation-exchangeable polymers with new applications in battery and water
deionization. In connection with the immobilization of the dopant we
present a new approach, the "auto-doped" polymers, in which the sulphonate
dopant is grafted to the polymer via an alkyl linkage.

INTRODUCTION

The use of molecular reagents to manipulate the properties of electrode
surfaces appears to be one of the most promising approaches for the
development of applications such as electrochemical synthesis, energy
conversion and storage, electrochromic displays, sensors and new kinds of
microelectronic devices /1a-c/. The confinement of specific molecules (SM)

---

* Author to whom correspondence should be addressed

/2/ at the interface electrode-electrolyte has quickly moved towards immobilization inside a polymer film.

This film deposited at the surface of the electrode must meet at least the following requirements :
- it must allow the electronic accessibility of the SM from the electrode,
- it must be porous to the substrates which have to react with the SM,
- it must maintain good adhesion and chemical stability at the electrochemical potentials and at the contact with the chemical reagents needed to operate the electrode.

Electroconductive polymers are of growing interest as an improved matrix to confined SM at the surface of an electrode. As a matter of fact, these polymers possess an intrinsic electronic conductivity and an ionic permeability due to the doping process. In addition, most of them such as poly(pyrrole) /3/, poly(aniline) /4/ and poly(thiophene) /5/ can be easily electrodeposited as films of adequate thickness. This thickness is difficult to monitor when the classical "dip coating" method is used for the deposition of functionalized poly(derived-vinyl) film.

This article focusses on the method of covalent binding of a SM to the conductive polymer matrix, a procedure we have developed for the functionalization of the poly(pyrrole). The advantages, limitations and improvements of this approach are emphasized mainly through a review of our previous results concerning the electrochemical behaviour of poly(pyrrole) films grafted with viologen groups /6/, polypyridinyl complexes of ruthenium /7/, ferrocene groups /8/, nitroxide functions /9/ and anthraquinone groups /10/, so that the coverage is not intended to be a critical assessment of the entire field. Functionalization of conductive polymers via modification of the doping anions will not be presented here.

Some salient aspects of our alkyl chain binding method are illustrated by new results presented here with their corresponding experimental parts : the association of viologen radical cations inside poly(pyrrole) films is evidenced by a spectroelectrochemical study, and depends on the alkyl chain length. The interest in copolymerization of a derived pyrrole monomer with pyrrole to reduce the steric hindrance of the SM and to increase the electronic conductivity of the polymer is exemplified with

367

the nitroxide function. In order to overcome the limitations due to the "proton sponge" behaviour of the poly(pyrrole) matrix we report here a preliminary study concerning the functionalization of poly(thiophene) with ferrocene and viologen groups.

Recent aspects of the preparation of interpenetrated composites of conductive polymers with ionomers are illustrated with ionomer composites and a new approach, "the auto-doped" polymers, in which the sulphonate dopants are bound to the polymer matrix. Applications of these very promising new materials are presented from literature results.

EXPERIMENTAL PART

1 Chemical Synthesis

Chemical analyses (elemental and high resolution mass spectroscopies) of the new compounds presented here were correct.

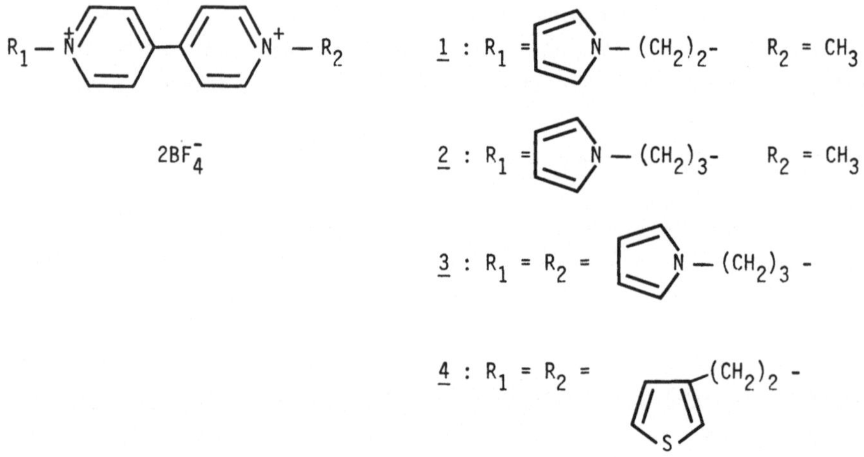

The synthesis of compound 3 has been previously described /11/. The synthesis of compound 2 has been briefly reported /6/, 1 was prepared in a similar way : the first step was a monoquaternization in benzene of a fourfold excess of 4,4'-bipyridine with the N-(tosylethyl)pyrrole /12/. The resulting white precipitate was filtered off and quaternized by refluxing in ethanol with a fivefold excess of methyl iodide. The resulting yellow precipitate was then dissolved in ethanol and passed through an ion exchange column (Amberlite IRA-93 in BF_4^- form). Elimination of the solvent gave the N-methyl, N'-(2-pyrrol-1-yl-ethyl)-4,4'-bipyridinium ditetrafluoroborate 1 in a 35 % global yield related to the initial

N-(tosylethyl)pyrrole. mp = 170°C decomposition. The structure was verified on the basis of ^{1}H-NMR data in CD_3 CN. δ (ppm) vs. TMS = 4.38 (s, 3H), 4.43 (t, 6Hz, 2H), 4.88 (t, 6 Hz, 2H), 5.97 (t, 2.2 Hz, 2H), 6.52 (t, 2.2 Hz, 2H), 8.34 (m-d, 7.5 Hz, 6H), 8.82 (d, 4.8 Hz, 2H).

N,N'-bis-(2-thien-3-yl-ethyl)-4,4'-bipyridinium ditetrafluoroborate 4.

The 4,4'-bipyridine was quaternized in two steps. The first monoquaternization was done using the 2-(3-thienyl)-ethyltosylate and the second quaternization using the 2-(3-thienyl)-ethylbromide which is more reactive for the addition on the deactivated second nitrogen of the 4-pyridine-4'-pyridinium group. A typical procedure is the following :

- In a dry box, 25 cm^3 (4 x 10^{-2} mol) of n-butyllithium (Janssen, 1.6 M in hexane) is slowly added to a solution of 5.13 g (4 x 10^{-2} mol) of 2-(3-thienyl)-ethanol (Janssen) in 50 cm^3 of dry THF. The solution was then added to an equimolar amount of tosylchloride (7.62 g) dissolved in 20 cm^3 of dry THF. After 5 min stirring, the mixture is taken out of the box, evaporated to dryness, and extracted with a mixture of diethyloxide/ water-Na_2CO_3. Recrystallisation from a mixture of ethanol/petroleum ether yielded 10.73 g (95 %) of the 2-(3-thienyl)-ethyltosylate as white crystals. This tosylate is transformed in the bromide derivative by adaptation of a general two-phase procedure /13/ : a mixture of 12 g of KBr in 20 cm^3 H_2O, 5.65 g (2 x 10^{-2} mol) of the tosylate and 500 mg of dicyclohexyl-18-crown-6 was refluxed for two hours. After cooling, extraction with petroleum ether yielded 3.6 g (94 %) of the 2-(3-thienyl)-ethylbromide of high enough purity for its direct use in quaternization.

2.82 g (10^{-2} mol) of the 2-(3-thienyl)-ethyltosylate and a twofold excess of 4,4'-bipyridine (3.1 g) in a 50 cm^3 solution of benzene were refluxed for three days. The resulting white precipitate was filtered off, yielding 4.18 g (95 %) of N-(2-thien-3-yl-ethyl)-4-(4'-pyridyl)- pyridinium tosylate.

2.2 g (5 x 10^{-3} mol) of this pyridylpyridinium were refluxed for two days in 10 cm^3 ethanol in the presence of a fourfold excess (3.6 g) of 2-(3thienyl)-ethylbromide. The resulting yellow precipitate was filtered off, then washed with cold ethanol and yielded 1.18 g (43 %) of the monobromide monotosylate salt of 4. This salt was dissolved in a 50/50 mixture ethanol-water and passed through the ion exchange column. Evaporation of the solvent yielded 900 mg (88 %) of slightly yellow-green colored crystals of the ditetrafluoroborate 4. mp = 300°C decomposition. IR (KBr) :

$\nu(cm^{-1})$ = 3140-3080 (4,4'-bipyridinium and thiophenic C-H), 1645 (bipyridinium rings), 1120-1040 (BF_4^-).

^{1}H-NMR (DMSO-D_6) data are as follows : δ (ppm) = 3.3 (t, 7.5 Hz, 4H), 4.9 (t, 7.5 Hz, 4H), 7.02 (d, 5Hz, 2H), 7.2 (d,3Hz, 2H), 7.45 (d-d, 5Hz and 3Hz, 2H), 8.68 (d, 7.5 Hz, 4H), 9.18 (d, 7.5 Hz, 4H).

The new compounds 5 and 6 were synthesized by addition of the suitable lithium alcoholate to the 1,1'-bis-chlorocarbonyl-ferrocene (Ventron). A typical procedure, exemplified by the preparation of 6 is :

- In a dry box, 1.65 g (1.3 x 10^{-2} mol) of 2-(3-thienyl)-ethanol is added to 20 cm^3 of absolutely anhydrous THF and converted to the lithium salt by 8 cm^3 (1.3 x 10^{-2} mol) of n-butyl lithium (1.6 M in hexane). 2 g (0.64 x 10^{-2} mol) of 1,1'-bis-chlorocarbonylferrocene is then directly added. After 10 min stirring, the mixture is taken out of the dry box, evaporated to dryness, triturated with diethyloxide and filtered to remove the insoluble lithium chloride. Recrystallization in the mixture diethyloxide/hexane yielded 2.6 g (82 %) of 6 as orange crystals. The compound 5 was obtained in a 78 % yield.

Bis [1,3-bis(pyrrol-1-yl)-prop-2-yl] -ferrocenyl-1,1'-dicarboxylate 5.
mp = 132°C. IR(KBr) : ν (cm^{-1}) = 3105 (pyrrolic and ferrocenic C-H), 2940 (alkyl CH and CH_2), 1700 (C=O), 1140 (C-O). ^{1}H-NMR (CDCl$_3$) data are as follows : δ (ppm) = 3.95 (d, 6Hz, 8H), 4.18 (t, 2Hz, 4 H), 4.65 (t, 2Hz, 4H), 5.33 (qt, 5.7 Hz, 2H), 6.13 (t,2.2 Hz, 8H), 6.63 (t, 2.2 Hz, 8H).

bis(2-thien-3-yl-ethyl)-ferrocenyl-1,1'-dicarboxylate 6.
mp = 78°C. IR(KBr) : ν(cm^{-1}) = 3100 (thiophenic and ferrocenic C-H), 2920 and 2980 (alkyl CH_2), 1700 (C=O), 1150 (C-O). ^{1}H-NMR (CDCl$_3$) data are as follows : δ (ppm) = 3.03 (t, 6.5 Hz, 4 H), 4.24 (s, 4 H), 4.4 (t, 6.5 Hz, 4 H), 4.73 (s, 4H), 7.03 (m, 4H), 7.2 (m, 2H).

7

8

1-(pyrrol-1-yl)-but-4-yl-sulphonate, tetrabutylammonium salt 7

The potassium salt of 7 was prepared in an identical way to the 1-(pyrrol-1-yl)prop-3-yl-sulphonate potassium salt we have previously described /14/ and with the same 94 % yield.

The potassium salt of 7 was readily converted into the tetrabutyl-ammonium salt by stirring with an equimolar amount of $N(Bu)_4$ OH (Fluka purum 40 % by weight in water) in a two-phase dichloromethane/water solution. Evaporation of the dichloromethane phase, and then drying under vacuum over night yielded 87 % of 7 as an oil which slowly crystallized. mp = 93°C. IR(KBr) : $\nu(cm^{-1})$ = 3100 (pyrrolic C-H), 2960 (alkyl CH_2 and CH_3), 1200 (sulphonate group). ^{1}H-NMR (CD_3 CN) data are as follows :
δ (ppm) = 1-1.8 (m, 32H), 2.53 (t, 7.2 Hz, 2H), 3.17 (m, 8H), 3.87 (t, 6.8 Hz, 2H), 6 (t, 2.2 Hz, 2H), 6.66 (t, 2.2 Hz, 2H).

1-(1-thien-3-yl-ethyl-2-oxy)-but-4-yl-sulphonate tetrabutylammonium salt 8

In the dry box, 10 g (7.8 x 10^{-2} mol) of 2-(3-thienyl)-ethanol in 100 cm^3 of absolutely anhydrous THF is converted to the lithium salt by addition of 48.7 cm^3 (7.8 x 10^{-2} mol) of n-butyllithium. The solution is then added to a 12 cm^3 (10 % excess) solution of 1,4-butane sultone (Aldrich) in a mixture of 50 cm^3 of THF and 50 cm^3 of DMSO. After evapora-tion of the THF, the remaining solution was stirred for one hour at 90°C, taken out of the dry box, cooled and poured into a vigorously stirred dichloromethane/diethyloxide solution. After successive washings of the resulting precipitate with CH_3CN and CH_2Cl_2, drying under vacuum over night yielded 14 g (66 %) of a white-cream precipitate. This potassium salt of 8 was converted into the tetrabutylammonium salt as described for 7, with a 57 % yield giving 15 g of a viscous oil.
IR(KBr) : $\nu(cm^{-1})$ = 3080 (thiophenic C-H), 2960 (alkyl CH_2 and CH_3) 1200 (sulphonate group). ^{1}H-NMR (CD_3CN) data are as follows : δ(ppm) = 1-1.7 (m, 32 H), 2.5 (t, 7.5 Hz, 2H), 2.83 (t, 6.8 Hz, 2H), 3.16 (m, 8H), 3.4 (t, 6Hz, 2H), 3.58 (t, 6.8 Hz, 2H), 6.9-7 (m, 2H), 7.27 (m, 1 H).

2 Electrochemical apparatus

Analytical studies were made using a 3 mm^2 Pt disk electrode in a two-compartment cell. Acetonitrile, AN, (BDH, HiPerSolv) and propylene carbonate, PC, (Fluka purum) were distilled inside the dry box on P_2O_5 and sodium, respectively. Tetraethylammonium tetrafluoroborate, TEAFB (Fluka) and $LiClO_4$ (G.F. Smith Chem. Co.) were dried at 100°C and 150°C respectively for two days. Potentials were referred to an $Ag/10^{-2}M$ Ag^+ electrode in the same solvent as the cell.

3 Electrosynthesis and electrochemical studies of the polymers

All experiments were performed in the dry box. Data for the electrodeposition are gathered in Table 1.

After synthesis, film-coated electrodes were transferred after thorough rinsing to clean electrolyte (without monomer) in a similar cell to that used for the electrodeposition. The electrolytes were the same, except for compounds 7 and 8 for which 0.1 M $LiClO_4$ was used in AN and PC, respectively.

4 Spectroelectrochemical apparatus

The spectroelectrochemical cell has already been described /15/. The optically transparent electrodes were of I.T.O. coated glass, S = 1 cm^2. The solvents, electrolytes and concentrations of the monomer were the same as in the electroanalytical experiments. The reference electrode was

Table 1

	Monomer	Concentration (mol/litre)	Solvent	Electrolyte	E(mV) of electrosynthesis
Pyrrole derivatives	1	2×10^{-3}	AN	TEAFB	+ 650
	2	2×10^{-3}	"	"	"
	3	2×10^{-3}	"	"	"
	5	2×10^{-3}	"	$LiClO_4$	+ 850
	7	10^{-2} + Pyrrole 10^{-1}	"	- *	+ 570
Thiophene derivatives	4	2×10^{-2}	PC	TEAFB	+2000
	6	2×10^{-2}	"	$LiClO_4$	+1300
	8	2×10^{-1} + 3-methyl thiophene 5×10^{-1}	"	- *	+1250

* 7 and 8 act as the electrolyte salt

$Ag/10^{-2}$ M Ag^+ in CH_3CN. The electrochemical apparatus was the same as in the analytical studies. Spectrophotometric data were obtained from an optical multi-channel analyser OMA III model 1460 equipped with a high performance gateable intensified diode array detector model 1420 B and a monochromator model 1229. The spectra and chronoabsorptometric curves were plotted on a HP 7474 A plotter. Cell filling and film preparations were effected inside the dry box.

5 Conductivity

Conductivity measurements were performed using the four-probe technique on free-standing films (when possible) synthesized on a 2 x 2 cm^2 I.T.O. coated glass electrode, with the same electrochemical conditions mentioned above.

RESULTS AND DISCUSSION

When a derived pyrrole monomer is subjected to electrooxidation a film of poly(derived-pyrrole) is formed at the surface of the electrode, the electroactivity of which is the sum of the electroactivity of the pending specific molecule (SM) and that of the polypyrrolic matrix /6-10/.

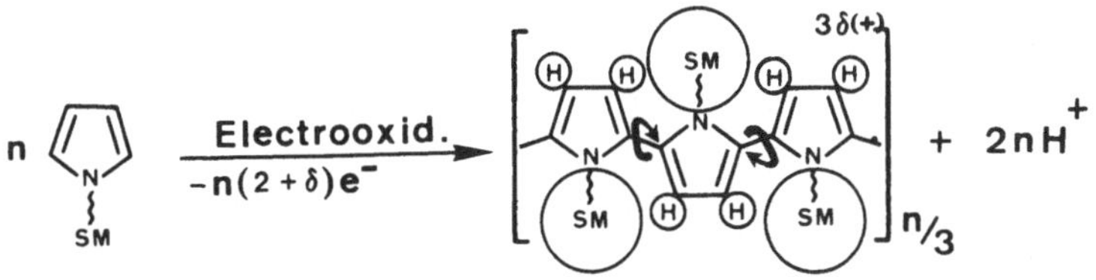

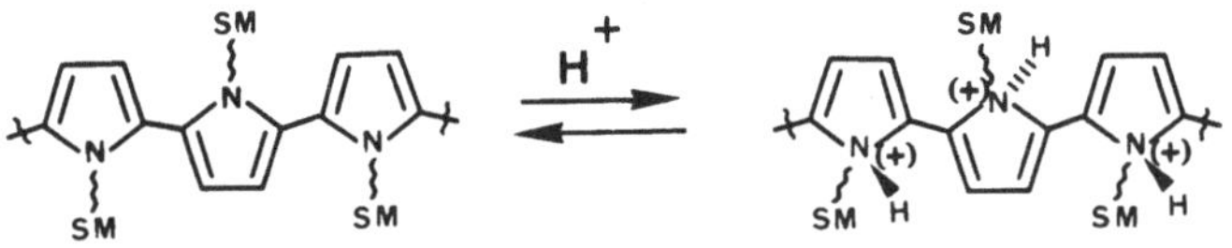

Scheme 1

The interactions beetween the SM and the polypyrrole matrix are mainly of steric or chemical nature (Scheme 1, the wavy line ∼ symbolizes the alkyl linkage, δ is the doping level).

Steric interactions influence the polymerisability of the monomer (i.e. its ability to form electroactive polypyrrole chains) and lead to a decrease in conductivity. This latter effect is due to the relative twisting of the pyrrole rings to minimize these interactions. A linear C_6 alkyl chain or an isopropyl chain with two pyrroles for one SM appear to be the more suitable bindings /12/ for groups of usual bulkinesses /8-10/. In the case of very bulky groups such as polypyridinium complexes of ruthenium S. COSNIER et al. /16/ have shown that the use of several pyrrole rings attached to the complex on different ligands allows a better stability of the polymer upon electrochemical cycling.

Another approach to minimizing the effect of the bulkiness of the SM is to reduce the steric hindrance by copolymerization with the underived pyrrole.

The electropolymerization at 0.8 V of the monomer $\underline{9}$ in 0.1 M LiClO$_4$ CH$_3$CN, using 0.3 C/cm^2 gives a free-standing film in a 55 % deposition yield and with a conductivity value $\sigma = 1.3 \ 10^{-4} \ \Omega^{-1} \ cm^{-1}$ /9/. If, instead of pure $\underline{9}$, one starts with a mixture of 2 x 10^{-3} M of $\underline{9}$ and 4 x 10^{-3} M of pyrrole, electropolymerization in the same conditions provides a free-standing film in a 89 % deposition yield and with a conductivity value $\sigma = 1.3 \ \Omega^{-1} \ cm^{-1}$. Compared with the homopolymer (Fig. 1a), the copolymer (Fig. 1b) exhibits an electroactivity of the polypyrrole matrix shifted to negative potentials due to the coinsertion along the polypyrrole chains of pyrrole rings with N-derived pyrrole rings. Moreover, from the voltammograms shown in Fig. 1a,b, integration of the

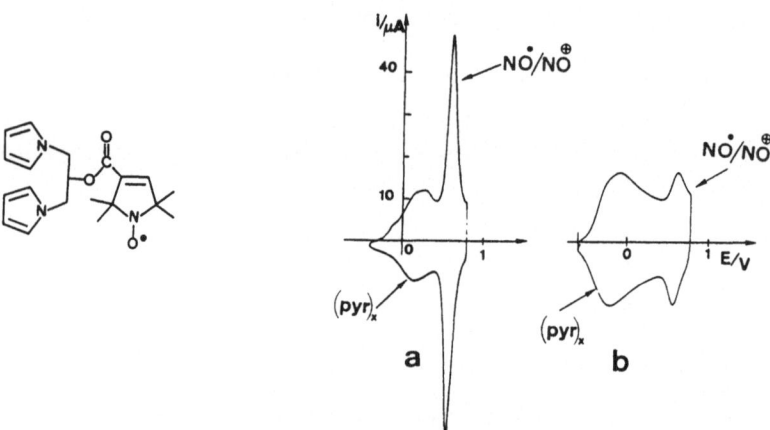

Fig. 1 : Cyclic voltammograms in AN 0.1 M LiClO$_4$, v = 10 mVs^{-1} of a) film of poly ($\underline{9}$) and b) film of copoly ($\underline{9}$ + pyrrole) deposited (Q = 0.3 C cm^{-2}) on Pt electrodes (3 x 10^{-2} cm^2) in a) a 2 x 10^{-3} M solution of $\underline{9}$ and b) a mixture of 2 x 10^{-3} M of $\underline{9}$ and 4 x 10^{-3} M of pyrrole both in AN 0.1 M LiCO$_4$.

nitroxide and of the polypyrrole electroactivities allows one to deduce that the nitroxide group has been diluted about four times. Thus an appreciable gain in the deposition yield and a four orders of magnitude increase in the conductivity value have been obtained with only a weak dilution of the SM inside the film.

Now a question arises : is the conductivity a key point for the use of these modified electrodes ? It seems not ! A part of the answer is already given by the fact that well-defined electroactivities are exhibited by viologen groups /6/, ruthenium complexes /16/, or anthraquinone groups /10/ when included in poly(pyrrole), although the latter is in its semiconductive state ($\sigma = 10^{-8} \, \Omega^{-1} \, cm^{-1}$) at the redox potentials of these SM. In such cases it would be concluded that the poly(derived-pyrrole) film acts as a redox pending group polymer in which charge transport is accomplished by electron hopping from one redox site to another. This hypothesis involves interactions between the pending groups. For instance, the decrease in electronic interactions between the nitroxide groups is evidenced by the narrowing of the EPR signal of the copolymer film compared with the one of the homopolymer film (Fig. 2).

$\Delta H_{PP} = 16 \, G$ $\Delta H_{PP} = 6.6 \, G$

a   b

Fig. 2 : EPR signal recorded at 9.5 GHz, with a modulation amplitude of 1 G of a) film of poly (9) b) film of copoly (9 + pyrrole) prepared according to Fig. 1 caption.

Another proof of the interactions between pending groups in poly(pyrrole) is exemplified by our optical study of poly(pyrrole) film containing various viologen groups. The dimeric association of viologen radical cations in solution /17/ or in polymer film /18/ is well known and is generally characterized by a blue shift (around 530 nm) of the band located around 610 nm /18b/. The poly (1), poly (2) and poly (3) differ only by the length of the alkyl chain linkage and by the number (one or two) of pyrrole rings attached to the viologen group. The optical spectra of these films ($10 \times 10^{-3} \, C \, cm^{-2}$) when maintained at -0.8 V, show an important dimeric association band for the poly (2) (see curve b, Fig. 3) and, conversely, a weaker association band for poly (3) (see curve c, Fig. 3) whose absorption spectrum is comparable to the one of a 10^{-3} M diluted solution of N,N'-dimethyl viologen (see curve d, Fig. 3). So it

375

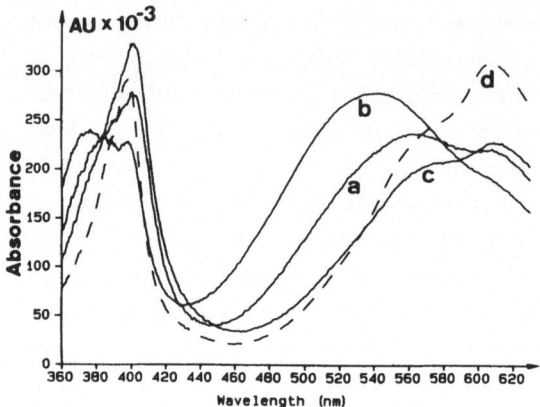

Fig. 3 : Absorption spectra at -0.8 V (vs. Ag/Ag$^+$) of a) poly ($\underline{1}$), b) poly ($\underline{2}$), c) poly ($\underline{3}$) films (Q = 10^{-2} C) deposited on a 1 cm^2 I.T.O. glass in AN 0.1 M LiClO$_4$; d) 10^{-3} M solution of the N,N'-dimethylviologen radical cation.

appears that a trimethylene linkage increases the mobility of the viologen group compared with a dimethylene one, and that for the same trimethylene chain length two attachments to the polymeric skeleton dramatically decrease the dimeric association.

<u>Chemical interactions</u> between the SM and the polymer matrix are mainly due to the protons tightly bonded by the poly(pyrrole), which acts as a "proton sponge". These protons come from the electrochemical condensation of the pyrrole rings during electropolymerization (see Scheme 1, paths 1 and 2).

We have previously shown that during the electrochemical cycling of poly(pyrrole) films containing nitroxide groups (-NO$^\cdot$) or anthraquinone groups (-AQ), these groups were slowly transformed into the electroinactive NOH$^{\cdot +}$ /9/ and the dihydroanthraquinone AQH$_2$ /10b/ forms respectively, according to the following reactions (Scheme 2) :

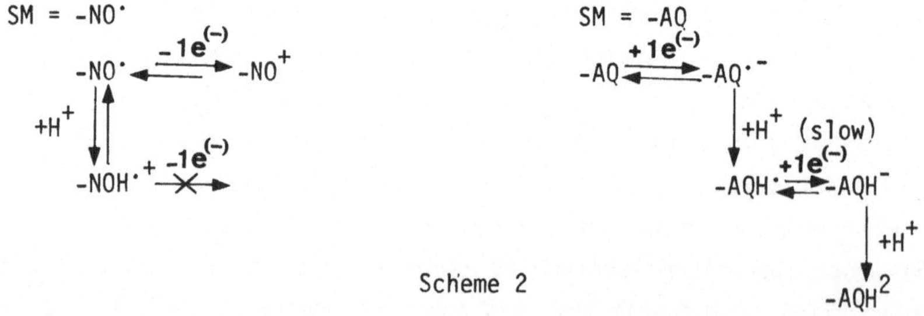

SM = -NO$^\cdot$

$$-NO^\cdot \underset{}{\overset{-1e^{(-)}}{\rightleftarrows}} -NO^+$$

$$+H^+ \updownarrow$$

$$-NOH^{\cdot +} \xrightarrow{+ -1e^{(-)}} \times$$

SM = -AQ

$$-AQ \underset{}{\overset{+1e^{(-)}}{\rightleftarrows}} -AQ^{\cdot -}$$

$$\downarrow +H^+ \text{ (slow)}$$

$$-AQH^\cdot \underset{}{\overset{+1e^{(-)}}{\rightleftarrows}} -AQH^-$$

$$\downarrow +H^+$$

$$-AQH^2$$

Scheme 2

376

After cycling, dipping these films in a 2,4,6-collidine solution regenerates the nitroxide electroactivity, but does not reactivate the anthraquinone activity. However, if before cycling the poly(pyrrole-anthraquinone) film is dipped in the collidine solution the protonation process is considerably slowed : after 4000 cycles we observe only a 50 % decrease of the signal of the AQ/AQ$^{\cdot-}$ couple instead of the complete disappearance in the case without the collidine pretreatment /10b/.

More recently, in the case of a poly(pyrrole) film containing 1,10-phenanthroline complexing cavities, we have observed competitive complexation reactions between the protons included in the film and various cations (Cu^{I} or Ni^{II}) /19/ in solution. Replacing the poly(pyrrole) matrix by the poly(thiophene) one appears to be a very attractive way to overcome this "proton sponge" disadvantage. Moreover, when compared with poly(pyrroles), poly(thiophenes) are often considered as more adherent and more stable to electrochemical cycling.

Consequently, as a first step, we have synthesized the derived thiophene monomers 4 and 6 in order to compare their electrochemical behaviour with these of the analogous derived pyrrole monomers 3 and 5. To our knowledge we report here the first results concerning the functionalization of poly(thiophene). As a matter of fact, previously reported poly(3-alkyl thiophenes) /20/ do not include specific molecules.
The electropolymerization of thiophene monomers compared with pyrrole monomers requires more severe experimental conditions. The solvent and the electrolyte salt must be perfectly dried /21a/ and the solution of the thiophene monomer must be ten to one hundred times more concentrated than the analogous pyrrole one /21b/. This latter condition can be very limiting when the grafted specific molecule is weakly soluble.

Electropolymerisation of thiophene-viologen 4 and thiophene-ferrocene 6 has been performed using a 2×10^{-2} M solution in propylene carbonate (see the experimental part). Compared with usual thiophene derivatives the potential of electropolymerization of 4 is very high (2 V vs. Ag/Ag$^+$) and only thin films ($\Gamma = 5 \times 10^{-9}$ mol cm^{-2}) /22/ can be electrodeposited with a low deposition yield.

Strong steric interactions and the existence of charge transfer complexes between thiophene and viologen may be invoked to explain the rise of the oxidation potential of the thiophene ring. At 2 V electrolysis the poly (thiophene) so formed begins to be degraded and the electrode is quickly passivated. After rinsing the film is transferred to a clean electrolyte and gives a cyclic voltammogram in agreement with the presence

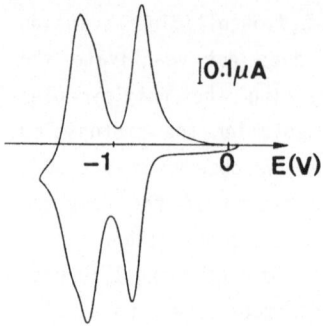

Fig. 4 : Cyclic voltammogram in AN 0.1 M TEAFB of a 5 x 10^{-9} mol cm^{-2} film of poly ($\underline{4}$) deposited (Q = 2.5 x 10^{-3} C) at 2 V vs.Ag/Ag$^+$ on a Pt electrode (3 x 10^{-2} cm^2), v = 10 mVs^{-1}.

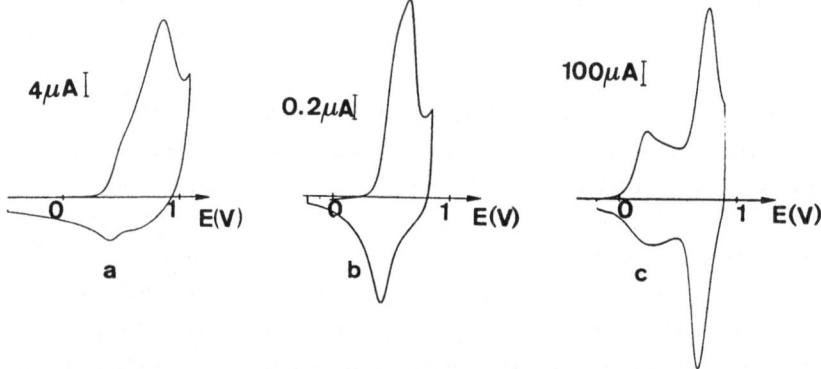

Fig. 5 : Cyclic voltammogram in AN 0.1 M LiClO$_4$ of Pt electrodes (3 x 10^{-2} cm^2) coated with a) a 8 x 10^{-8} molcm^{-2} film of poly ($\underline{6}$) deposited (Q = 3.5 x 10^{-3} C) at 1.3 V vs.Ag/Ag$^+$, v = 20 mVs^{-1}, b) v = 1 mV s^{-1}, c) film of poly ($\underline{5}$) deposited (Q = 8 x 10^{-2} C) at 0.8 V vs.Ag/Ag$^+$, v = 50 mVs^{-1}.

of viologen species confined at the electrode surface (see Fig. 4). The cycling stability of the viologen group in this poly ($\underline{4}$) is comparable with that of those included in a polypyrrole matrix /6/.

Electropolymerization of the thiophene-ferrocene $\underline{6}$ is easier (see the experimental part) and film of poly ($\underline{6}$) with Γ= 8 x 10^{-8} molcm^{-2} can be obtained, the cyclic voltammogram of which is shown in Fig. 5. Compared with the well-defined cyclic voltammogram of the poly(pyrrole-ferrocene), poly ($\underline{5}$) (see Fig. 6), the expression of the electroactivity of the ferrocene group inside poly ($\underline{6}$) is only appreciable at low sweep rate (Fig. 5b). These preliminary results on functionalized poly(thiophene) are somewhat disappointing, however, we are currently investigating thiophene derivatives via a longer 3-alkyl linkage, which would appear to be easier to polymerize.

Among the groups which can be immobilized into a conductive polymer, substituted sulphonate anions acting as the dopant play a central part. Exploiting the doping property of conductive polymers in this way, various specific functions bearing sulphonate anions have been introduced into the polypyrrole /23a,b/. The same approach has been used to produce composites of poly(pyrrole) with poly(styrene sulphonate) /23b;24/, poly(vinyl sulphate) /23b.25a-c/ or Nafion /26a-d/, a perfluorosulphonated polymer. In these cases the ionomeric matrix plays the role of a "macro-dopant" of the polypyrrole matrix. These composites are interesting because of their mechanical properties and melt processability /24/, which are introduced by the second non-conductive component of the system, but the most striking properties come from the immobilization of the anionic dopant, and consequently the reduction (dedoping) of these composites is accompanied by the movement of cations (M^+), which have been called cathodic "pseudo-dopants" /25b/ (see the reaction (1), Scheme 3, the wavy line of $SO_3\sim$ symbolizes the ionomer matrix, δ is the doping level).

Reduction of a composite polypyrrole/polysulphonate.

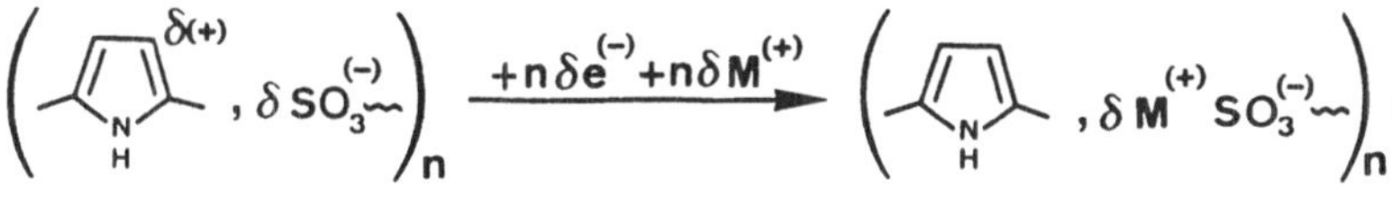

$$\left(\includegraphics{pyrrole}\,,\delta\,SO_3^{(-)}\!\sim\right)_n \xrightarrow{+n\delta e^{(-)}+n\delta M^{(+)}} \left(\includegraphics{pyrrole}\,,\delta\,M^{(+)}SO_3^{(-)}\!\sim\right)_n \quad (1)$$

Oxidation of a classical polypyrrole

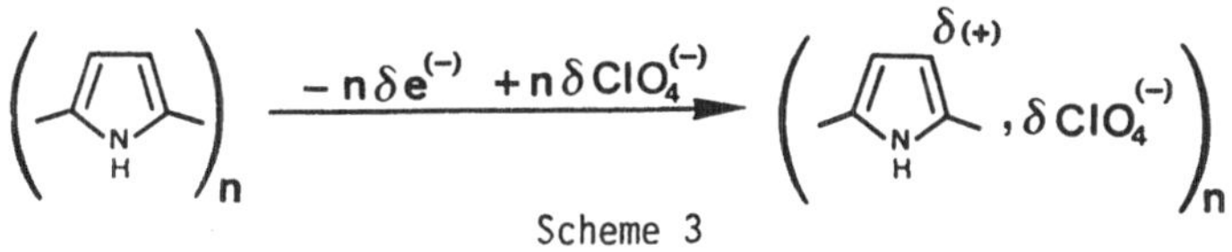

$$\left(\includegraphics{pyrrole}\right)_n \xrightarrow{-n\delta e^{(-)}+n\delta ClO_4^{(-)}} \left(\includegraphics{pyrrole}\,,\delta\,ClO_4^{(-)}\right)_n \quad (2)$$

Scheme 3

The association in an electrochemical device of two electrodes coated with the two types of polypyrrole, the cation-exchangeable and the anion-exchangeable ones (see Scheme 3), provides a novel type of polymer battery /26b/ or electrochemical deionization system /26a/. Another interesting aspect of the cation-exchange ability of these composites of poly(pyrrole) and ionomer lies in the association with a lithium anode in a battery. In such a case the amount of electrolyte may be reduced as much as possible since it is not consumed during the electrochemical cycling (see Scheme 4), and the specific energy density of the system is increased.

polypyrrole/Li battery

Charging process (classical anion exchangeable polypyrrole)

Cathode $\left(\underset{H}{\bigvee_{N}}\right)_n$ $+ n\delta ClO_4^{(-)}$ $\xrightarrow{-n\delta e^{(-)}}$ $\left(\underset{H}{\bigvee_{N}}{}^{\delta(+)}, \delta ClO_4^{(-)}\right)_n$

$ClO_4^{(-)}$
$Li^{(+)}$ (electrolyte)

Anode $n\delta Li^{(+)}$ $+ n\delta e^{(-)}$ $\longrightarrow$ Li

Charging process (cation exchangeable poly(pyrrole))

Cathode $\left(\underset{H}{\bigvee_{N}}, \delta Li^{(+)} SO_3^{(-)}\sim\right)_n$ $\xrightarrow{-n\delta e^{(-)}}$ $\left(\underset{H}{\bigvee_{N}}{}^{\delta(+)}, \delta SO_3^{(-)}\sim\right)_n$

Anode $n\delta Li^{(+)}$ $\xrightarrow{+ n\delta e^{(-)}}$ Li

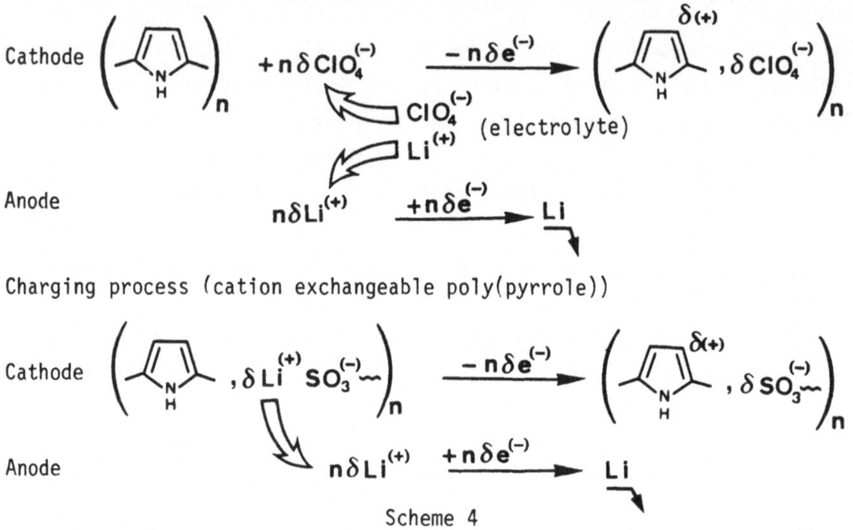

Scheme 4

However, a severe limitation on the use of such a polypyrrole/ionomer composite for battery applications is the weight of the polymer skeleton of the ionomer matrix, which penalizes the specific energy density. To overcome this difficulty we have developed another approach, similar to the method of functionalization, in which the sulphonate anion is bound to the poly(pyrrole) through an alkyl chain linkage.

1-(pyrrol-1-yl)prop-3-yl (PPS) and 1-(pyrrol-1-yl)but-4-yl (PBS) sulphonates are easy to synthesize /14/ but cannot be directly electropolymerized in their pure forms /27/, even when starting from the acetonitrile soluble tetrabutylammonium salt (see the experimental part) instead of the potassium salt. This phenomenon is confirmed the recent results of F. WUDL et al. on similar derivatives of thiophene /29/ and of REYNOLDS et al. on PPS /30/. However, if these salts are used as electrolyte during the electropolymerization of underived pyrrole, copolymers of pyrrole and PPS or PBS are obtained. For instance, electrooxidation at constant current (0.4 mA cm^{-2}) of a solution of pyrrole (1M) and PPS potassium salt (0.1 M) in pure DMSO or in aqueous (10%) acetonitrile gives free-standing films (16 C cm^2) in electrochemical yields of 97 % and 89 % /32/, with conductivities of $\sigma = 3 \times 10^{-2}$ Ω^{-1} cm^{-1} and $\sigma = 2 \times 10^{-1}$ Ω^{-1} cm^{-1} respectively. Elemental analysis data correspond to the following formulas for DMSO 10 or aqueous acetonitrile 11 prepared films.

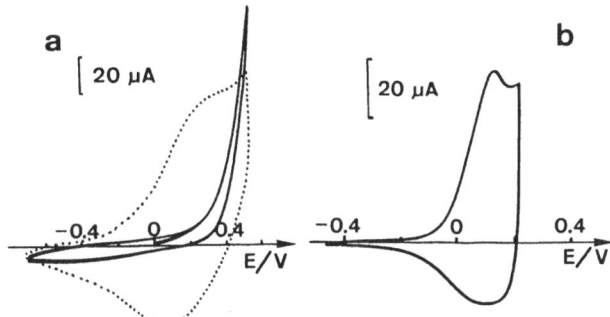

The observed temperature dependence of the conductivity of the sulphonate auto-doped polypyrrole 10 follows the variable range hopping process (the log of the conductivity is linear vs. $T^{-1/4}$) /31/.

Now the question that arises is : are sulphonate-derived pyrrole rings included along the polypyrrole chains or not ? When a mixture of pyrrole and 7 in the form of its <u>tetrabutylammonium salt</u> is electrooxidized in AN (see the experimental part) the electrodeposited film <u>does not exhibit any electroactivity</u> in this solution of electrosynthesis. In fact, during electrodeposition by scanning there is no change in the successive voltammograms (Fig. 6a, solid line). Classical anion exchangeable poly-pyrrole is electroactive in tetrabutylammonium perchlorate or tetrabutyl-ammonium p-toluene sulfonate, so the difference in behaviour may be explained by the suppression of the anion movement. If the film is transferred, after thorough rinsing, to a clean electrolyte containing LiClO$_4$ the electroactivity slowly appears and reaches a maximum value after 50 scans (Fig. 6a, dotted line) accounting for the slow diffusion of

Fig. 6 : a) Solid line : cyclic voltammetry on Pt electrode (3×10^{-2} cm^2) of an AN solution 10^{-2} M 7 and 0.1 M pyrrole, v = 20 mVs^{-1} (first 5 scans), dotted line : 50th cyclic voltammogram of the 5-scan electrodeposited copoly (7 + pyrrole) film (Q = 2×10^{-3} C) transferred into AN 0.1 LiClO$_4$; b) cyclic voltammogram of a copoly (8 + 3-methylthiophene) film (Q = 1×10^{-3} C) in PC 0.2 M 8 and 0.5 M 3-methylthiophene.

the lithium cations inside the film. It is noticeable that when synthesized in anhydrous AN using the tetrabutylammonium salt, free-standing films are more difficult to obtain than in aqueous AN, and their conductivities are dramatically low ($\sigma = 10^{-6} \Omega^{-1}$ cm^{-1}).

Similar experiments have been performed with the tetrabutylammonium salt of the thiophene derivative $\underline{8}$. Preliminary results show that electrodeposited films of copoly ($\underline{8}$ + 3-methylthiophene) appear to be electroactive even in the solution of electrosynthesis (Fig. 6b). This difference with the auto-doped copoly ($\underline{7}$ + pyrrole) may be accounted for by a more porous structure due to the long linkage (6 CH_2 and one oxo-connection), which allows the diffusion of the tetrabutylammonium cation.

CONCLUSION

Various functions have been grafted on electrode surfaces via covalent binding to a poly(pyrrole) matrix. The main problems arising in the electrochemical behaviour of these modified electrodes come from steric hindrance and proton interactions with the grafted specific molecules. The steric hindrance may be reduced by adjusting the nature or the length of the alkyl linkage, or by copolymerization of the N-substituted pyrrole with the underived pyrrole. Treatment of a functionalized poly(pyrrole) film with a basic non-nucleophilic base, the 2,4,6-collidine, greatly diminishes the proton content of the film, but does not totally suppress the "proton sponge" effect. Replacing the poly(pyrrole) matrix by the poly(thiophene) one to overcome this disadvantage is not so easy, owing to the severe conditions on the concentration of the starting monomer and the electrolyte dryness required for the electropolymerization of 3-derived thiophene monomers.

The immobilization of the doping anions inside the polymer allows the doping process to be reversed from an anionic exchange to a cationic one. This immobilization can be achieved either by using a macro dopant such as an ionomer or by the covalent binding of the sulphonate anions to the so-called "auto-doped" polymers. These cation-exchangeable conductive polymers are new materials which appear very promising in applications such as charge controllable membranes, batteries, water deionization, sensors ...

ACKNOWLEDGEMENTS
We are thankful to A. DERONZIER and J.-C. MOUTET for fruitful discussions concerning the spectrophotometric behaviour of viologen derivatives, and

to B. GALLAND for giving us samples of the compounds 1, 2 and 3. The authors thank CNRS, PIRSEM and AFME (ATP "Préparations électrochimiques et générateurs électrochimiques") for partial financial support.

Bibliographie

1. a) L.R. Faulkner: Chem. and Eng. News, Feb. 27, 28 (1984) ; b) J. Schreurs, E. Barendrecht: Red. Trav. Chim., Pays-Bas, 103, 205 (1984) ; c) M.S. Wrighton: Science, 231, 32 (1986)

2. By "specific molecules" we mean molecules which exhibit a particular function such as a redox center, an electrocatalyst, an electrochrom, a photosensitizer ...

3. A.F. Diaz, J.J. Castillo, J.A. Logan, W.-Y. Lee: J. Electrochem. Soc., 129, 115 (1981)

4. E. Genies, C. Tsintavis: J. Electroanal. Chem., 195, 109 (1985)

5. G. Tourillon, F. Garnier: J. Electroanal. Chem., 135, 173 (1982)

6. G. Bidan, A. Deronzier, J.-C. Moutet: J. Chem. Soc., Chem. Commun., 1185 (1984)

7. G. Bidan, A. Deronzier, J.-C. Moutet: Nouv. J. Chim., 8, 501 (1984)

8. G. Bidan: Extended Abstracts of the Spring Meetings of the Electrochemical Society, Toronto, Canada, May 12-17, 1985, p. 105

9. G. Bidan, D. Limosin: Ann. Phys., (PARIS), 11, 5 (1986)

10. a) P. Audebert, G. Bidan, M. Lapkowski: J. Chem. Soc., Chem. Commun., 887 (1986) ; b) P. Audebert, G. Bidan, M. Lapkowski: J. Electroanal. Chem., 219, 165 (1987)

11. L. Coche, A. Deronzier, J.-C. Moutet: J. Electroanal. Chem., 198, 187 (1986)

12. G. Bidan, M. Guglielmi: Synt. Met., 15, 49 (1986)

13. D. Landini, F. Montanari: J. Chem. Soc., Chem. Commun., 879 (1974)

14. P. Auric, G. Bidan: accepted in J. Polym. Science, Polymer Physics Ed.

15. E.M. Genies, C. Santier: French patent No 85 1 6805, Nov. 14th. 1985

16. S. Cosnier, A. Deronzier, J.-C. Moutet: J. Electroanal. Chem., 193, 193 (1985)

17. A. Deronzier, B. Galland, M. Vieira: Nouv. J. Chim., 6, 97 (1982)

18. a) J.G. Gaudiello, P.K. Ghosh, A.J. Bard: J. Am. Chem. Soc., 107, 3027 (1985) ; b) M. Furue, S. Yamanaka, L. Phat, S. Nozakura: J. Polym. Science : Polym. Chem. Ed., 19, 2635 (1981) ; c) Y. Nambu, K. Yamamoto, T. Endo: J. Chem. Soc., Chem. Commun., 574 (1986)

19. G. Bidan, B. Divisia-Blohorn, J.M. Kern, J.-P. Sauvage: in preparation

20. a) M. Sato, S. Tanaka, K. Kaeriyama: J. Chem. Soc., Chem. Commun., 873 (1986) ; b) R.L. Elsenbaumer, K.Y. Jen, R. Oboodi: Synt. Met., 15, 169 (1986)

21. a) A.J. Downard, D. Pletcher: J. Electroanal. Chem., 206, 147 (1986) ; b) M. Sato, S. Tanaka, K. Kaeriyama: J. Chem. Soc., Chem. Commun., 713 (1985)

22. The apparent surface concentration of an electroactive species is determined from the charge under the anodic peak associated with the first electronic transfer of the viologen group assuming 1 Faraday per mol

23. a) M.V. Rosenthal, T.A. Skotheim, C.A. Linkous: Synt. Met., 15, 219 (1986) ; b) T. Shimidzu, A. Ohtani, T. Iyoda, H. Honda: J. Chem. Soc., Chem. Commun., 1414 (1986)

24. N. Bates, M. Cross, R. Lines, D. Walton: J. Chem. Soc., Chem. Commun., 871 (1985)

25. a) T. Shimidzu, A. Ohtani, T. Iyoda, K. Honda: J. Chem. Soc., Chem. Commun., 1415 (1986) ; b) T. Iyoda, A. Ohtani, T. Shimidzu, K. Honda: Chem. Let., 687 (1986) ; c) T. Shimidzu, A. Ohtani, T. Iyoda, K. Honda: J. Chem. Soc., Chem. Commun., 327 (1987)

26. a) F.-R. Fan, A.J. Bard: J. Electrochem. Soc., 133, 301 (1986) ; b) R. Penner, C.R. Martin: J. Electrochem. Soc., 133, 310 (1986) ; c) G. Nagasubramanian, S. Di Stefano, J. Moacanin: J. Phys. Chem., 90, 4447 (1986) ; d) P. Aldebert, P. Audebert, M. Armand, G. Bidan, M. Pineri: J. Chem. Soc., Chem. Commun., 1636 (1985)

27. On the other hand they were chemically polymerized by $Fe(ClO_4)_3$ /14/, which we have used previously in the chemical synthesis of various poly(pyrroles) /28/.

28. a) G. Bidan: Report CEA-R-5321, 18 p., September 1985 ; b) P. Audebert, G. Bidan: Synt. Met., 14, 71 (1986).

29. A.O. Patil, Y. Ikenoue, F. Wudl, A.J. Heeger: J. Am. Chem. Soc., 109, 1858 (1987)

30. N.S. Sundaresan, S. Basak, M. Pomerantz, J.R. Reynolds: to be published in J. Chem. Soc., Chem. Commun.

31. J.-P. Travers, P. Audebert, G. Bidan, Mol. Cryst. Liq. Cryst., 118, 149 (1985). In this paper, the 3-pyrazoline cycle in Table 1 p. 150 must be corrected to a pyrrole cycle.

32. The electrochemical yields have been calculated on the basis of the elemental analysis and by weighing the electrodes, assuming that pyrrole rings bearing the sulphonate group participate in the electro-oxidation process.

On the Charge Storage Mechanism of Conducting Polymers

J. Heinze, J. Mortensen, and M. Störzbach

Institut für Physikalische Chemie der Universität Freiburg,
Albertstr. 21, D-7800 Freiburg, Fed. Rep. of Germany

1 Introduction

Most conducting polymers are made conductive by doping. However, ever since such systems have been studied it has been recognized that this doping is not comparable to the classical doping of typical inorganic semiconductors. Obviously, the p-doping corresponds to an oxidation and the n-doping to a reduction process. One may either use chemical electron donors or acceptors as redox reagents or steer the redox process electrochemically. Nevertheless, although conducting polymers can be charged and discharged reversibly, the basic principles of the charge storage mechanism were not properly understood. Thus, electrochemists have interpreted typical cyclic voltammograms of conventional conducting polymers such as polypyrrole (PPy) or polythiophene (PTh) as a combination of faradaic and capacitive charging processes, thereby assuming that the redox potential of all electroactive segments in the polymer is simply given by the mean value of the anodic and cathodic peak potentials E_{pa} and E_{pc} in the forward and reverse scan /1-3/. On the other hand, in the literature on physics the bipolaron model was proposed /4-6/ which ideally postulates the formation of multiple thermodynamically stable diion states (=bipolarons) associated with local geometric distortions of the chain. Theoretical calculations as well as spectroscopic results suggest that this stabilization, involving the formation of chinoid-like structures, starts in the monoionic state, but increases considerably in the diionic bipolaron state. In addition, it is assumed that the locally distorted bipolaron state comprises only four or five monomeric units of a polymer segment and that the energy gain, in comparison to two polaron states, amounts to approximately 0.4 eV. Up to now, all electrochemical measurements give only indirect evidence for the existence of bipolaron states. In recent experiments NECHTSCHEIN et al. /7,8/ concluded from ESR and electrochemical data that the bipolaron state is not noticeably more stable than the polaron state.

One of the difficulties in the interpretation of the electrochemical findings results from the fact that polymeric systems, in contrast to monomeric molecules with one or two redox sites, often possess an irregular structure and have a more or less broad molecular weight distribution. Therefore, voltammograms of polymers prepared under only slightly different conditions may exhibit quite different charging/discharging characteristics. However, if one takes care to form a regularly structured material the voltammograms will be reproducible and

very typical. This may be achieved by the electropolymerization
of defined oligomers such as di- or quaterthienyl /9/ or by
using mild preparation conditions so as to avoid side reactions. Given these prerequisites one obtains cyclic voltammograms which have a nearly identical "structure", even if different types of monomers are polymerized /10/. Apparently, all
chain-like conjugated polymers with the exception of polyaniline (PANI) undergo qualitatively similar charging/discharging
processes.

Until now, the main barrier for a conclusive interpretation
of voltammetric data on conjugated polymers is the lack of
controlled measurements on well-defined model systems such as
oligomeric p-phenylenes and, consequently, the lack of unequivocal evidence on the principles of charge storage in a conjugated system with multiple electrophoric groups. Therefore, to
clarify the situation and to test the applicability of the
bipolaron model we have studied the reductive charging/dis-
charging behaviour of several newly synthesized soluble oligomers of the p-phenylene-vinylene type /11/ as well as of some
oligomers – up to the tetramer – belonging to the p-phenylene
series.

2 Results and Discussion

The cyclic voltammograms, given in Fig. 1., represent reduction
experiments within the homologous series of oligo-p-phenylenes-
vinylenes. To avoid additional side reactions the experiments
were carried out under superdry conditions in THF/NaBPh$_4$ /12/.
The results provide the first uneqivocal evidence of the cha-

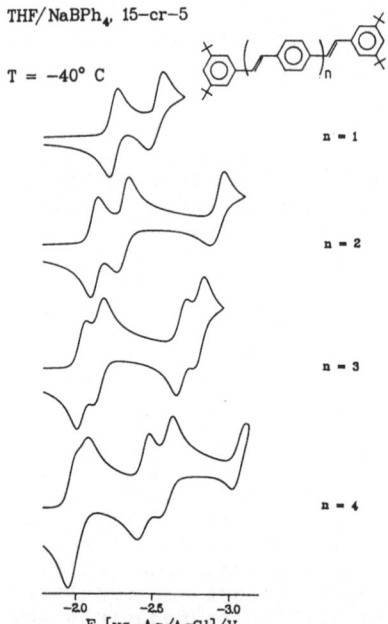

THF/NaBPh$_4$, 15-cr-5

T = -40° C

n = 1

n = 2

n = 3

n = 4

−2.0   −2.5   −3.0
E [vs. Ag/AgCl]/V

Fig. 1.:
Cyclic voltammograms (corrected
with respect to background currents) for the reduction of oligo
-p-phenylenevinylenes
(v = 100 mV/s)

racter of the charge storage mechanism in such conjugated oligomers. As can be seen, the number of redox steps increases with increasing chain length of the oligomer. While the basic compound with n = 1 gives rise to a dianion, the homologue with n = 4 transforms into a stable pentaanion. Three effects are especially significant. Firstly, the potentials of already existing redox states shift to less negative values when the next higher homologue is reduced. Obviously, the redox energies of different states gradually approach a common convergence limit with increasing chain length. Secondly, the redox states degenerate pairwise with increasing chain length, but no "two-electron" transfers are observed as predicted for energetically stabilized diion states. Thirdly, adding successive monomeric subunits in the molecular chain enlarges the number of accessible redox states in agreement with expectations. However, the energetic gap increases strongly between the lowest and the highest charged states. In the case of the pentaanion it amounts to more than 1V.

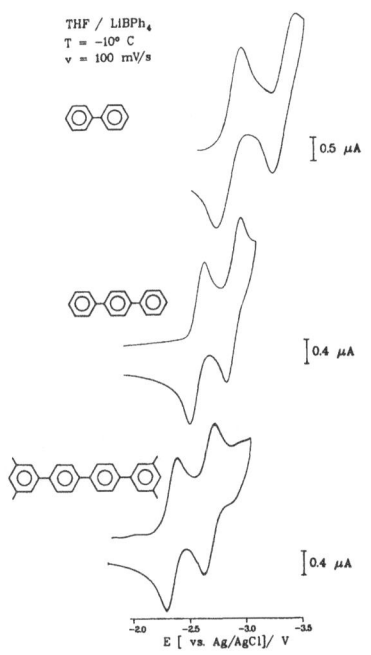

THF / LiBPh$_4$
T = –10° C
v = 100 mV/s

0.5 µA

0.4 µA

0.4 µA

-2.0   -2.5   -3.0   -3.5
E [vs. Ag/AgCl]/ V

Fig. 2.:
Cyclic voltammograms for the reduction of oligo-p-phenylenes, limit for supporting electrolyte reduction -3.4 /V/ vs. Ag/AgCl

Similar results are obtained when oligo-p-phenylenes of defined structure are reduced (Fig. 2). As a phenylene unit is smaller than a phenylenevinylene unit, the reduction potentials for the oligophenylenes are more negative than those of the oligophenylene-vinylenes. Again, with increasing chain length the redox states approach a common convergence limit. On the other hand, the energetic gap between the lowest and highest charged states increases as a function of chain length. In the case of quaterphenyl the potential separation between the first and the third redox states is at least of the order of 1.2 /V/.

The exact value cannot be given because the supporting electrolyte will be electrolyzed at -3.4 /V/, i.e. before the third redox wave appears. The enhanced energetic gaps between the redox states of oligo-*p*-phenylenes, in comparison to those of an oligo-*p*-phenylenevinylene with the same number of monomeric units, are due to the small size of a monomeric phenylene unit, which causes a strong coulomb interaction of the charges in the chain.

Several conclusions on the redox properties of conducting polymers, especially of poly-*p*-phenylenevinylene and poly-*p*-phenylene, can be drawn from all these experimental data. By extrapolation it can be deduced that the interaction energy between the charges in a diionic state decrease to a minimum only when the chain contains at least eight to nine monomeric units. This number may be even higher when conducting polymers are oxidized because the screening of the charges at the polymeric chain is weaker in the presence of large counterions such as Li^+ perchlorate or hexafluorophosphate than in the presence of Li^+ or Na^+. Hence, it follows that in a corresponding polymer with the degree of polymerization m approximately m/4 charges can be stored which interact only by spin pairing. As all these states are almost energetically degenerate only "one" thermodynamic redox potential can be observed. In terms of the bipolaron model, one bipolaron or spinless diion state comprises a segment of eight to ten monomeric units in the polymer. However, as the experimental data also reveal, in contrast to the predictions of the bipolaron model /6/, no geometric stabilization effects of the diionic species are observed in the solution. In addition, for charge numbers greater than m/4 in a polymeric chain the increasing coulomb repulsion between the charges shifts all the successive redox states to higher energies. This means that outside of the degenerated redox states the ionization energy increases in the case of oxidation, and the electron affinity decreases in the case of reduction.

All these data obtained with oligomers of defined chain length prepare the way for the construction of a model for the charge storage mechanism in conducting polymers. To achieve this we may assume that a reasonable number of redox states in such systems are energetically degenerated, followed by redox states of higher energy. From the fact that the number of degenerated states is small in comparison to such of higher energy - for instance, in polypyrrole under favourable conditions only 20-25% of the charge is located in the peak-shaped wave - it must be concluded that the chain lengths in these polymers are "finite" and relatively short. However, it is unclear if our model, which is deduced from the redox properties of defined soluble oligomers, can be unconditionally correlated with the experimental results of conventional conducting polymers. The main objection to the application of the model may be that typical charging/discharging curves of PPy or PTh, which exhibit a broad current plateau in the anodic tail after the peak-shaped wave, are interpreted as a combination of faradaic and capacitive charging and discharging processes. In particular, it has been suggested that the current in the range of the anodic tail is mainly produced by capacitive effects /2/. However, recent measurements with newly electropolymerized

PPy films reveal that the major part of the charging/dischar-
ging processes in the whole accessible potential range is
faradaic /10/. Given these preconditions, we have simulated
voltammetric current voltage curves for n-step electron transfer processes which involve degenerated redox states as well as
such of increasing energy /13/. Although the simulated voltammograms were similar to experimental findings, showing two
equally high peak-shaped waves in both scan directions and a
broad current plateau, there was still a significant difference
between experiment and theory. Contrary to calculations, in
experiments the peak-shaped wave in the reverse scan is shifted
to lower potentials by approximately 400-500 /mV/ and the
corresponding peak current is only half the value of the anodic
peak. As has been showed independently with other systems /14/
such effects are due to conformational changes of the system
during the charging and the discharging process, provided that
all the conformers have different redox potentials. In the case
of chain-like conjugated polymers we assume that after the
formation of monoionic redox states (=polarons) in the chain,
the polymer changes its structure from a twisted to a planar
form, whereby the monoionic and the higher charged states are
energetically simultaneously stabilized. As a consequence of
these processes, a single sharp wave again appears at the
beginning of the voltammetric response. In the reverse scan the
planar form of the chain exists up to the end of the discharging processes and then regenerates into the starting system.
As the planar form is energetically stabilized, the redox
potentials for the discharging steps are now lower than those
for the charging steps. Therefore, asymmetrical charging/dis-
charging curves result. We have simulated this mechanism for
the polypyrrole system and have obtained a very reasonable
agreement between theory and experiment (Fig.3). Obviously, our
model fits the experimental data which give in situ the ionization energies or the electron affinities of the polymers stu-

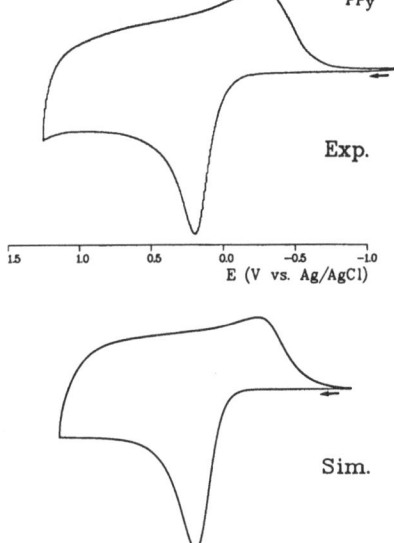

Fig. 3.:
Experimental and simulated cyclic
voltammograms for the oxidation
of PPy
The simulation was done with the
assumption that 10 redox states
are forming the anodic peak-
shaped wave. In the reverse scan
the energetic stabilization of
these states amounts to 300 /mV/.

died. Thus, in terms of the solid state models our results favour the polaron concept rather than the bipolaron model.

Acknowledgement

Financial support by the Stiftung Volkswagenwerk is gratefully acknowledged.

References

1. F.A., Diaz, J.I., Castillo, J.A., Logan, W.-Y. Lee: J. Electroanal. Chem. 129, 115 (1982)
2. S.W. Feldberg: J. Am. Chem. Soc. 106, 4671 (1984).
3. R.J. Waltman, J. Bargon, A.F., Diaz: J. Phys. Chem. 87, 1459 (1983)
4. J.L. Brédas, R.R. Chance, R. Silbey: Phys. Rev. B26, 5843 (1982)
5. J.L. Brédas, B. Themans, J.G. Fripiant, J.M. André, R.R., Chance: Phys. Rev. B28, 6761 (1984)
6. J.L. Brédas, G.B. Scott: Acc. Chem. Res. 18, 308 (1985)
7. M. Nechtschein, F. Devreux, F. Genoud, E. Vieil, J.M. Pernaut, E. Geniès: Synth. Metals 15, 59 (1986)
8. F. Genoud, M. Guglielmi, M. Nechtschein, E. Geniès, M. Salmon: Phys. Rev. Lett. 55, 118 (1985)
9. J. Heinze, K. Hinkelmann, M. Dietrich, J. Mortensen: DECHEMA Monographien 102, 209 (1986)
10. J. Heinze, M. Dietrich, J. Mortensen: Makromol. Chem., Makromol. Symp. 8, in press
11. J. Heinze, J. Mortensen, K. Müllen, R. Schenk: J. Chem. Soc., Chem. Commun. 1987, in press
12. J. Heinze: Angew. Chem. 96, 823 (1984); Angew. Chem. Int. Ed. Engl. 23, 831 (1984)
13. J. Heinze, J. Mortensen, M. Störzbach: Ber. Bunsenges. Phys. Chem., in press
14. M. Dietrich, J. Heinze, H. Fischer, F.A. Neugebauer: Angew. Chem. 98, 999 (1986); Angew. Chem. Int. Ed. Engl. 25, 1021 (1986)

Electrochemical Investigations and Neutron Activation Analysis of Polyacetylene

G. Nagele[1], *G.E. Nauer*[1], *and H. Kuzmany*[2]

[1]Institut für Physikalische Chemie der Universität Wien,
Währingerstraße 42, A-1090 Wien, Austria
[2]Institut für Festkörperphysik der Universität Wien and
Ludwig-Boltzmann-Institut für Festkörperphysik,
Strudlhofgasse 4, A-1090 Wien, Austria

From galvanostatic pulse- and impedance-measurements combined with data of neutron activation analysis the following conclusions could be drawn: 1. The values of the double layer capacitance are related to the macroscopic $(CH)_x$-film surface and do not seize the fibrillar interior surface under the given conditions. 2. Lithium is inserted together with solvent molecules during n-doping. 3. At the beginning of the undoping process an insertion of counterions takes place.

1. Introduction:

Galvanostatic pulse- and Fast-Fourier-transform (FFT) impedance measurements can provide rapid information about electrode kinetics. The relatively simple models used for calculation of pulse- and impedance-measurements are deduced for ideally flat, compact and equally conducting electrodes. An additional condition is semiinfinite diffusion in the electrolyte solution /1,2/. So it is not obvious that the applied methods also hold for a system like polyacetylene, which is far away from ideal behaviour. $(CH)_x$ is an intercalation-type electrode with porous structure and changing conductivity during doping. Diffusion processes are not restricted to the electrolyte but also play an important role in the polymer-electrode itself. It could be shown, however, that impedance- and pulse-measurements can be applied to a system like polyacetylene under certain conditions /3,4/. They give at least qualitative or semiquantitative information about the kinetics of doping, undoping or selfdischarge behaviour.

Combining the electrochemical methods with neutron activation analysis, additional information about the doping/undoping-process, side-reactions or efficiency of doping can be drawn.

2. Experimental:

$(CH)_x$-films were prepared by the SHIRAKAWA method and thermally isomerized. The $(CH)_x$-samples were either contacted at only one point of the film by a Pt-pressure contact (point contact) or over the whole backside area (backside-contact). Further experimental details for n- and p-doping have been reported previously /5/. After electrochemical investigation the samples were washed with pure solvent and subsequently dried under vacuum for at least 12 hours.

The amount of Al, Ti, Cl, Li and F in the doped or partially dedoped $(CH)_x$-samples was determined by neutron activation analysis (NAA).

3. Results:

The Nyquist plots of n-doped and partially dedoped $(CH)_x$-electrodes and their corresponding equivalent cells are given in figure 1. From the shape of the curves one can see that during doping, charge transfer is rate determining at high, and diffusion is rate determining at low frequencies /1,5/. The Nyquist-plot of partially dedoped $(CH)_x$ indicates that the reaction rate is determined by diffusion over the whole investigated frequency range.

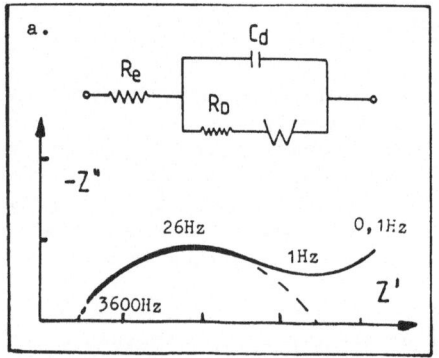 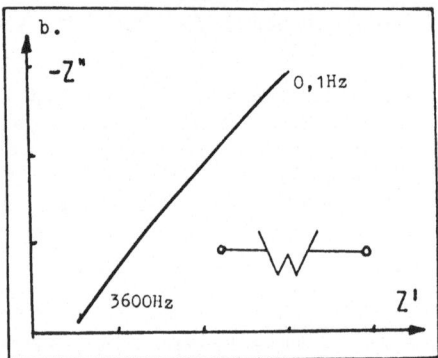

C_d...double layer capacitance
R_D...charge transfer resistance
R_e...electrolyte resistance

-W-...Warburg impedance /1/
Z.....impedance

Fig.1: Nyquist plots of $(CH)_x$ in different states with corresponding equivalent cells: a. after 6 mole-% doping at 0.7V in $LiClO_4$/THF, and b. after partially undoping at 2.8V vs. Li/Li^+ in $LiClO_4$/THF. Similar diagrams have been obtained for n- and p-doping with $LiBF_4$ as salt and sulfolane as solvent.

The magnitude of the capacitance of the electrical double layer (C_d) of p- and n-doped $(CH)_x$ in different electrolytes, determined from galvanostatic pulse measurements, is given in table 1. C_d is hardly changed by the doping-undoping process /4/. The values of C_d of $(CH)_x$ received from pulse- and impedance-measurements are in reasonable agreement to each other as shown elsewhere /5/.

The results of neutron activation analysis show inhomogeneities of doping at $(CH)_x$-samples which were contacted at only one point.

Table 1: Double-layer capacitance of n- and p-doped $(CH)_x$-electrodes in different electrolytes determined after 50 µs of pulse duration:

doping ion	electrolyte	$C_d(\mu F/cm^2)$
BF_4^-	$LiBF_4/SL^*$	¦2.3
Li^+	$LiBF_4/SL$	0.64
Li^+	$LiClO_4/THF^{**}$	1.23

*...saturated $LiBF_4$ solution in sulfolane
**...0.1m $LiClO_4$ solution in tetrahydrofurane

Samples which were contacted at the whole backside of the film, were doped homogeneously. However, no significant difference in the impedance measurements was observed for the two cases of contacting.

For samples which were partially dedoped after n-doping in a $LiClO_4$/THF-electrolyte a Cl-concentration of up to 0.6 mole-% was found, which could arise from inserted ClO_4^--counterions.

Table 2 shows a comparison between doping degrees determined by coulometric and activation analytical measurements. One can see that the highest doping degree up to which $(CH)_x$ can be doped in $LiClO_4$/THF with only low coulometric losses is 18 mole-%. However, at this high doping level the samples could not be undoped completely. The maximum reversible doping degree is 8 mole-% /10/.

Table 2: Doping degrees and doping efficiency of n-doped $(CH)_x$ (electrolyte: $LiClO_4$/THF)

coulometry	NAA	doping efficiency
3.8 mole-%	3.5 mole-%	92 %
18.8 mole-%	18.1 mole-%	96 %
27.1 mole-%	17.8 mole-%	65 %

4. Discussion:

Regarding the surface electrode/electrolyte as a condensor, an important quantity given by both electrochemical methods is the double-layer-capacitance (C_d). It is proportional to the electrochemically active surface and has to be distinguished from the intrinsic capacitance (C_I) discussed in Ref. /6/ and /7/. The latter is some orders of magnitude higher and related to the total charge storage capacity of the polymer.

It was shown by WILL /3/ that only the exterior surface of the $(CH)_x$-film is responsible for the magnitude of the double-layer capacitance, whereas the interior fibrous structure has no influence under the given conditions. This was explained by the degree of wetting: The high viscosity of the used electrolyte ($LiBF_4$/sulfolane) caused a wetting of only the exterior $(CH)_x$-film. The interior surface, which is more than three orders of magnitude higher, remains unwetted. Following this explanation one should expect a huge increase of the wetted surface-area for an electrolyte with low viscosity like THF which is able to penetrate into the interior of the microporous structure of the polymer. The double-layer-capacitance should increase proportional to the degree of wetting.

As can be seen from table 1, however, the magnitude of C_d increases by only a factor of two if THF is used instead of SL as solvent. So another explanation seems to be more likely: It is known from impedance- and galvanostatic pulse-studies of porous metals and metaloxides /8,9/ that C_d decreases to a minimum value with increasing frequency or decreasing pulse-duration. This value corresponds to the outer macroscopic surface of the samples. The smaller the pore-diameters the lower is the frequency where C_d is determined by only the exterior surface. This explanation can be transformed to studies of poly-acetylene-electrodes. According to the very small interfibrillar space for pulse durations of 50 μs, the fibrillar $(CH)_x$-electrode definitely behaves like a plain and compact one.

Table 1 shows that the double-layer-capacitance of Li-doped $(CH)_x$ is much lower than the one for BF_4-doped. This can be explained by solvent coinsertion which was also detected by other methods /10,11/. Strongly solvated Li^+ has a higher radius than lightly or unsolvated BF_4^-. The higher radius causes an increase of the Helmholtz-layer and a decrease of C_d.

The Nyquist-plots for doping and undoping in figure 1 have a very different shape. They indicate that during undoping, diffusion is hindered to a much higher extent than during doping. This can be explained by a counter-movement of ions with opposite charges and possibly an intercalation of counterions: The deintercalation of inserted ions seems to be strongly hindered. Under these conditions an intercalation of counterions simulaneous to deintercalation of previously inserted ions should take place during undoping to achieve electroneutrality in the polymer. Affirmation for this assumption is given by data from neutron-activation analysis: They showed that partially dedoped samples contained up to 0.6 mole-% of counterions.

5. Conclusion:

The following conclusions could be drawn:
1. Double-layer capacitances calculated from galvanostatic pulse- and impedance measurements are related to the exterior surface of the $(CH)_x$-films but do not seize the interior microfibrillar structure. This can be explained rather by the frequency dependence of porous electrodes than by the degree of wetting of the surface.
2. Li-doped $(CH)_x$-samples which were partially undoped contained up to 0,6 mole-% of chlorine from ClO_4^--counterions which were intercalated simultaneously with deintercalation of previously inserted Li^+-ions.
3. From the C_d-values of $(CH)_x$ doped with different ions, confirmation for solvent-coinsertion with Li^+ could be found.

Acknowledgements:

We are indebted to Doz. F. Grass from the Österreichisches Atom-institut for the neutron activation analytical measurements. We would also like to thank J. Kürti for helpful discussions. This work was supported by the Fonds zur Förderung der wissenschaftlischen Forschung in Österreich.

References:

1. R.D. Armstrong, M.F. Bell, A.A. Metcalfe, Electrochemistry Vol. 6, 98, The Chem. Soc. 1978, Oxford
2. T. Berzins, P. Delahay, J. Am. Soc. 77, 6448 (1955)
3. F.G Will, J. Electrochem. Soc 132, 2351 (1985)
4. T. Osaka, T. Kitai, Bull. Chem. Soc. Jpn. 57, 759 (1984)
5. G. Nagele, G.E. Nauer, H. Kuzmany, J. Kürti, Synth. Met., in press

6. J. Tanguy, N. Mermillod, M. Hoclet, J. Electrochem. Soc.
    134, 795 (1987)
7. J. Tanguy, N. Mermillod, this volume
8. N.M. Beekmans, L. Heyne, Electrochimica Acta 21, 303 (1976)
9. K.J. Euler, Electrochimica Acta 17, 619 (1972)
10. L.W. Shacklette, J.E. Toth, N.S. Murthy, R.H. Baughman,
    J. Eletrochem. Soc. 132 (1985)
11. J.J. Andre, M. Bernard, B. Francois, C. Mathis, Journal de
    Physique C3-199 (1983)

Low Frequency Impedance Measurements on Polypyrrole

J. Tanguy and M. Slama

D-LETI/DEIN-LERA/CEN. SACLAY,
F-91191 Gif-sur-Yvette Cedex, France

From impedance results obtained on polypyrrole we have obtained evidence for three different doping mechanisms. Two ionic relaxation processes characterised by two distinct and relatively short relaxation times are responsible for the capacitance effect. Another very low frequency process is connected with ionic diffusion and mass transport in the thick samples. This effect is responsible for the difference observed between the total charge and the capacitive one.

INTRODUCTION

The measurement of a.c. impedance has become a powerful technique for the analysis of complex redox mechanisms observed in some conducting polymers. In previous papers [1-3] we demonstrated a doping process in polypyrrole governed by an ionic relaxation mechanism characterised by a relatively short relaxation time. In this paper we have tried to separate clearly the diffusion process which appears at a very low frequency (10^{-4}Hz) from the relaxation one (10^{-2}Hz – 10^{-1}Hz).

EXPERIMENTAL RESULTS

Measurements were performed on chemically synthesized polypyrrole in the form of thin pellets of pressed powder. The polypyrrole was synthesized by direct oxidation of the monomer by ferric perchlorate in water [1]. The electrochemical cell used for impedance measurements has been described elsewhere [1-3].

In Fig. 1 we have plotted an admittance diagram obtained for a 4 mg polypyrrole sample. The smaller high frequency semicircle (a) is related to the network formed by the electrolyte resistance R_o, the transfer resistance R_T and the double layer capacitance C_d. The low frequency semi circle (b) can be related to a complex relaxation phenomenon [3] represented by two capacitances C_1 and C_2 in series with two resistances R_1 and R_2. The parallel impedance Z takes into account a low frequency diffusion process. We will discuss these different mechanisms in detail.

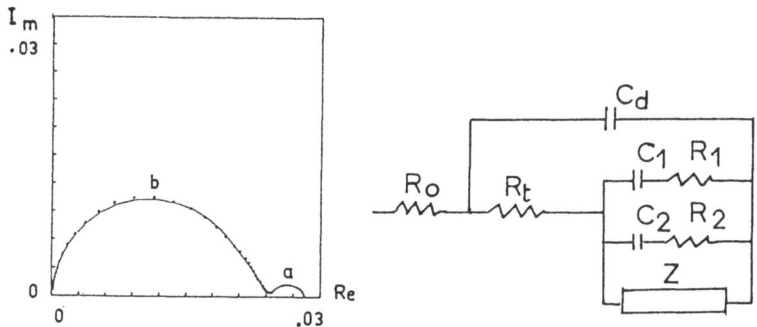

Fig.1 - A.c. admittance diagram for a 4 mg polypyrrole sample and its electrical equivalent circuit

THE CAPACITANCE EFFECT AS FUNCTION OF THE FREQUENCY

After about 30 cycles the voltamograms becom very stable and the a.c. impedance diagram appears totally reproducible. In Fig. 2a the capacitance curve has a plateau and the conductance curve exhibits a maximum characteristic of a relaxation process. We were able to decompose these curves into two relaxation processes corresponding to two capacitance plateaus C_1 and C_2 and two relaxation times T_1 and T_2 (Fig. 2b). The very low frequency part of the diagram (d) is attributed to the diffusion process. In studying the dependence of the two capacitances C_1 and C_2 on the electrode potential it appears in Fig. 3 that when the electrode potential increases the capacitance C_2 corresponding to the long relaxation time T_2 decreases continuously . On the other hand the capacitance C_1 connected to the short

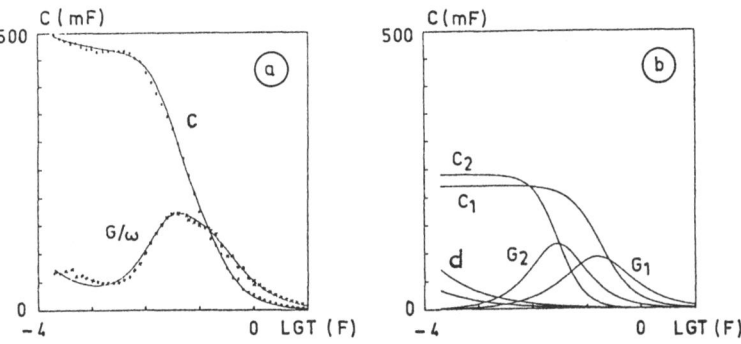

Fig.2 - Fit of the low frequency capacitance and conductance at the end of the 35th charge (2a). In (2b) these curves are decomposed into three parts which represent two distinct relaxation processes (C_1, G_1) and (C_2, G_2) and a diffusion process (d) at a very low frequency.

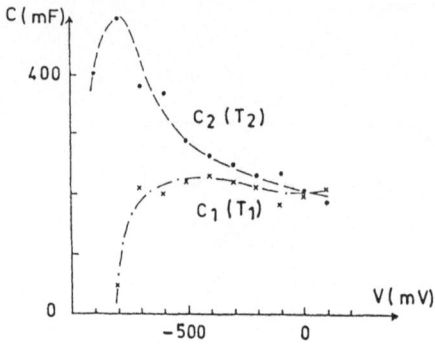

Fig.3 - The two capacitances C_2 and C_1 as a function of the electrode potential

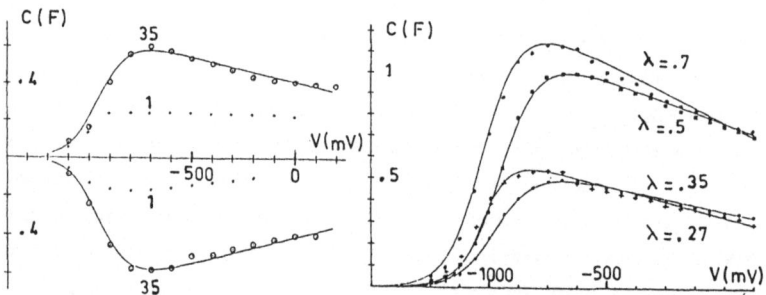

Figs.4 and 5 - Fit of the capacitance assuming a capacitance value proportional to the total charge

relaxation time T_1 increases and becomes stable at a high potential. This evolution can be related to the decrease of the polaron density with increasing potential which has been reported by different groups working on polypyrrole [6,7].

FIT OF THE CAPACITANCE AS A FUNCTION OF THE POTENTIAL

The low frequency capacitance values derived from the plateaus in the capacitance curves obtained at different potentials have been plotted as a function of the potential in Fig. 4. We have tried to fit the capacitance variation assuming that the capacitance effect is proportional to the total charge [4,5]. This charge is assumed to follow the NERNST relation ; then we have :

$$C(E) = \frac{Q_o}{V_o} \exp\left[A \frac{E - E_o}{KT}\right] \left[1 + \exp A \frac{E - E_o}{KT}\right] - \lambda (E - E_o) \qquad (1)$$

398

where E is the electrode potential, E_O is the redox potential, A is a coefficient taking into account the Coulombic interactions between the charges and Q_O is the total charge.

The correction term $\lambda(E - E_O)$ is introduced in order to fit the decreasing part of the capacitance due to hindrance between the ions when the total charge is increasing. In Fig. 5 we report the capacitances obtained for different polypyrrole samples having different total charges. It appears that the expression (1) fits the experimental results well and we found that λ increases regularly with the total charge, which confirms the hypothesis of the hindrance effect increasing with the ionic charge.

DISCUSSION AND CONCLUSION

Impedance results can be summarized as due to three distinct doping mechanisms : two relaxation processes arising at moderately low frequencies ($10^{-1} - 10^{-2}$ Hz) and a diffusion and mass transport phenomenon at very low frequencies (10^{-4}Hz). The capacitance effect due to the two relaxation processes can be fitted by a Nernstian law with a correction term due to interaction between the charges at high doping levels (λ term). Considering the variation of the amplitude of the two relaxation phenomena (C_1 and C_2) and the fact that the C_2 value continuously decreases as the electrode potential increases, we tend to assume that the charges responsible for the C_2 capacitance may be connected with the existence of polarons in polypyrrole. In fact different workers have shown that the number of polarons is continuously decreasing when the potential is increasing [6,7]. The difference observed between the two relaxation times T_1 and T_2 is compatible with an energy difference of about 20 meV measured between polaron and bipolaron species [7]. However, we need a direct comparison with EPR measurements in order to confirm this hypothesis.

[1] N. MERMILLIOD, J. TANGUY, F. PETIOT, J. Electrochem. Soc. 133 (6)(1986)1073.

[2] J. TANGUY, N. MERMILLIOD, M. HOCLET, Synth. Metals 18(1987) 7-12.

[3] J. TANGUY, N. MERMILLIOD, M. HOCLET, J. Electrochem. Soc. 134(4)(1987) 795.

[4] S.W. FELDBERG, J. Am. Chem. 106 4671 (1984)

[5] J.F. OUDARD, PhD Thesis INPG Grenoble France (1986)

[6] J.C. SCOTT, J.L. BREDAS, K. YAKUSHI, P. PFLUGER and G.B. STREET, Synth. Metals 9,2, 165 (1984)

[7] NECHTSCHEIN, F. DEVREUX, F. GENOUD, E. VIEIL, J.M. PERNAUD, E. GENIES, Synth. Metals 15 1986 59-78

This work was partly supported by a CEC BRITE contract n° R11B—0109—D(B)

Poly(alkyl thiophenes) and Poly(substituted heteroaromatic vinylenes): Versatile, Highly Conductive, Processible Polymers with Tunable Properties

R.L. Elsenbaumer, Kwan-Yue Jen, G.G. Miller, H. Eckhardt, L.W. Shacklette, and R. Jow

Polymer Laboratory, Corporate Research, Allied-Signal, Inc., Morristown, NJ 07960, USA

1. Introduction

Conductive polymers derived from heteroaromatics such as thiophenes, furans, and pyrroles have been found to be a highly versatile conductive polymers. Unlike other conducting polymers, such as polyacetylene and polyphenylene, the polyheteroaromatics can be substituted with a variety of substituents to give polymers with widely varied physical and optical properties while maintaining high levels of conductivity. Especially versatile are polymers based on thiophenes. Here, long-chain alkyl substituents render the rigid polythiophenes solution processible in both the neutral and conductive forms. Further, strong electron donating or electron withdrawing groups can dramatically influence both polymer ionization potential and band gap. Such ready substitution allows one to fine tune the oxidative and reductive potentials of the polymer as well as manipulate the optical properties of the neutral and conductive forms. Adding a vinyl linkage between heteroaromatic rings reduces both the band gap and ionization potential of the parent polymer while providing a simplified route to their preparation and fabrication via the sulphonium polyelectrolyte precursor route. Herein we describe the preparation and properties of these novel materials.

2. Poly(alkyl thiophenes)

Early experiences taught that substitution onto the rigid conjugated polymer chains may improve polymer processibility but at the expense of many orders of magnitude in conductivity [1-3]. A desire to find conductive polymers suitable for wide scale applications and the willingness to sacrifice some conductivity for improved processibility, led us to investigate the properties of long-chain alkyl substituted polythiophenes.

Alkyl thiophenes were prepared according to the procedure of TAMAO et al [4] and diiodinated by the method of BARKER et al [5]. The diiodothiophenes were polymerized by the nickel-catalyzed Grignard coupling method similar to the one used by KOBAYASHI et al [6] using 2-methyl tetrahydrofuran as the reaction solvent [7]. Random copolymers were prepared by polymerizing a mixture of monomers.

Undoped polymer solubility increased with increasing chain length of the substituent in the order n-butyl > ethyl >> methyl. The soluble polymers were readily characterized by i.r., nmr, and UV spectroscopy establishing that the polymers have a regular linear, highly conjugated structure.

Copolymers of long-chain alkyl thiophenes with 3-methylthiophene gave soluble, amorphous polymers with higher glass transition temperatures and higher conductivities than the homopolymers.

The conductivities of various doped poly(alkylthiophenes) are given in Table 1. For this class of polymers it appears that both the size of the substituent and the nature of the dopant have only a minor influence on

Table 1. Conductivities of Chemically Prepared Poly(alkylthiophene) Films

Polymer	Dopant	Conductivity
Poly(3-methylthiophene) (Tg 145 °C)	NOSbF$_6$	5 S/cm
Poly(3-butylthiophene) (Tg 45 °C)	NOSbF$_6$	6
	FeCl$_3$	15
Poly(3-butylthiophene-co-3-methyl-thiophene) (Tg 58 °C)	NOSbF$_6$	20
	FeCl$_3$	50
Poly(3-octylthiophene) (Tg 25 °C)	FeCl$_3$	1
Poly(3-octylthiophene-co-3-methyl-thiophene) (Tg 40 °C)	NOSbF$_6$	10
	FeCl$_3$	20

conductivity. For a random copolymer of 3-butylthiophene and 3-methylthiophene (50:50) a maximum in the conductivity was reached at a doping level of 25 mol% as determined by *in situ* electrochemical doping. At this doping level, the polymer's potential relative to a silver electrode was +1 volt which puts the doped polymer near the threshold of stability to water vapor at room temperature (environmental stability). At elevated temperatures a much reduced level of stability would be expected in humid environments. In fact, from accelerated thermal testing, the conductivity half-life of a FeCl$_3$ doped butylthiophene methylthiophene copolymer at 25 °C in normal environments is about 8.1 years but at 100 °C the half-life is only 42 hr.

Films of the neutral poly(alkylthiophenes) show an absorption maxima at 460-480 nm (deep red color in transmission). On doping, this absorption band disappears with the growth of a broad (free electron) absorption that extends from the visible to the near IR (850-5000nm) (deep blue color in transmission and reflection). Alkyl substitution increases the band gap of the parent polymer by only 0.1 eV (E$_g$ 2-2.2 eV) possibly because of a slight increase in steric interactions between adjacent thiophene rings. High molecular weight homopolymers of 3-butylthiophene doped with SbF$_6^-$ or PF$_6^-$ are soluble in moderately polar solvents such as nitrobenzene or benzonitrile giving homogeneous blue solutions. Optical absorption spectra of the solutions of doped polymers resemble the solid state spectra, and have been interpreted as indicative of the polymer chains predominately supporting polarons [8].

The alkylated polythiophenes are not totally immune to steric effects. In fact, subtle steric effects are evident; the size and number of alkyl substituents , the method of polymerization and the regiospecificity of polymerization all play an important role in polymer properties (electrical and optical). Steric effects are especially evident with dialkyl substitution on the thiophene rings. Polymerization of 3,4-dimethylthiophene by the Grignard coupling method gives a lemon yellow polymer which is less extensively conjugated and has a larger band-gap compared to the monosubstituted polymers. Less effective conjugation also raises its ionization potential such that it is not doped by mild doping agents such as iodine. More aggressive dopants (NOSbF$_6$) give a complex with a conductivity of 0.5 S/cm (pressed pellet). Electrochemically prepared poly(3,4-dimethylthiophene) reportedly has a conductivity of 10 S/cm (film) [9].

Differences in polymer properties as a function of polymerization method are further exemplified by comparison of our results on chemically prepared poly(alkyl thiophenes) with those of KAYERIMA et al [10] on electrically prepared polymers. Electrochemically prepared poly(3-butylthiophene) is reported to have a conductivity of 110 S/cm while we observe a conductivity of only 15 S/cm for the chemically prepared polymer. The 1H nmr spectrum of neutral, chemically coupled poly(3-butylthiophene) shows two resonance for the aryl methylene, one at 2.6 and the other at 2.85 ppm (TMS) in the ratio of 21:79. We attribute the high field resonance to "head-to-head" coupled butylthiophene (containing 3,3'-dibutyl-2,2'-bithiophene units) and the low field resonance to "head- to-tail" coupled butylthiophene (containing 3,4'-dibutyl-2,2'-bithiophene units). Because of increased steric interactions, we suspect that the presence of "head-to-head" coupling gives rise to a polymer with reduced conductivity.

To test the hypothesis that regiospecificity plays a role in polymer conductivity, we compared the conductivity obtained on a random copolymer of butylthiophene and methylthiophene (50:50) with the conductivity obtained on a polymer prepared from the regiospecific dimer 4-butyl-3'-methyl-2,2'-bithiophene (see figure 1). The random copolymer was amorphous with a conductivity of 50 S/cm (FeCl$_3$ doped) whereas the more regiospecific copolymer was partially crystalline and had a much higher conductivity of 140 S/cm. However, the regiospecificity of the Grignard coupling of the diiododimer is not high, the ratio of head-to-head to head-to-tail coupling of the dimer is 37:63. It is to be expected that even more regiospecific polymers should give even more highly conductive polythiophenes.

Random Copolymer
σ = 50 S/cm
Amorphous.

Regiospecific Copolymer
σ = 140 S/cm
Partially Crystalline.

3. Poly(heteroaromatic vinylenes)

The difficulties associated with the synthesis of high molecular weight poly(alkylthiophenes) by chemical polymerization led us to investigate alternate methods of polymerization. Given the simplicity of preparation of poly(phenylene vinylene) (PPV) via the sulfonium polyelectrolyte precursor route and the ease of synthesis of the starting monomers, we chose to investigate the applicability of this route to the general preparation of poly(heteroarylene vinylenes).
    Our first target was poly(2,5-thienylene vinylene) (PTV). In 1970, KOSSMEHL [11] reported the synthesis of PTV using a Wittig condensation reaction. But this procedure formed intractable, low molecular weight powders with low conductivities. Based on the conductivity results, it was generally believed that this class of polymers would not be highly conductive owing to the apparent mismatch in chemical potential of vinylene and thiophene units [12]. Upon reinvestigation, we have now found that high molecular weight, freestanding films of PTV can be conveniently prepared via a modified KANBE [13] and WESSLING [14] procedure and that PTV prepared by this route gives high conductivities on doping.
    Thiophene was chloromethylated using formaldehyde and hydrochloric acid. Then the bischloromethyl derivative was converted into a bis-sulfonium salt by treatment with dimethylsulfide in methanol. The noncrystalline sulfonium salt was treated with one equivalent of base in water at 0°C which resulted in formation of a viscous solution of the polyelectrolyte. Films of the

402

polyelectrolyte could be cast from this solution at 0°C, but elimination of HCl and dimethylsulfide to give the insoluble conjugated polymer occurred at temperatures much above this. Gentle warming to about 60°C nearly completed the elimination and formation of the conjugated structure as evidenced by the films change in color from light yellow to orange to red to purple and finally to a golden lusterous color.

Higher molecular weight polymer was prepared by using 2,5-bis(tetrahydro-thiophenonium methyl) thiophene chloride in place of the dimethyl sulfonium salt. The PTV film analyzed for $(C_{5.98}H_{4.1}S)_x$. Rapid heating of the cast prepolymer results in low density foams of PTV while gentle warming with a slow increase in temperature results in more dense shiny gold films.

Optical absorption spectra of PTV were recorded on thin films cast onto quartz substrates. A broad absorption peak is observed which begins at 710 nm and extends down into the UV. The absorption maximum occurs at about 600 nm. The absorption onset suggests that the optical bandgap is 1.74 eV, which is very close to that predicted by BREDAS et al [15] (1.6 eV). By looking at a series of oligomers and extrapolating to high polymer, KOSSMEHL [16] predicted that the absorption maximum for PTV should be 530 nm. The observation of the maximum at 600 nm (solid state) suggests that PTV prepared by this technique is extensively conjugated and of high quality.

Infrared spectra recorded on freestanding films of PTV show a strong absorption at 800 cm^{-1} for a 2,5-disubstituted thiophene ring, and a peak at 935 cm^{-1} for a trans olefin. A strong broad absorption which extends from the near-IR to 2000 cm^{-1} is probably due to some free electron absorption, and is consistent with the observation that PTV is slightly doped by air (oxygen).

A remarkably rapid conductivity increase was observed on exposure of 25 micron thick films of PTV (density 0.2 g/cc) to different oxidizing dopants. As shown in Table 2, conductivities as high as 62 S/cm were measured on doped polymers; average values were around 45 S/cm. Correcting for the low density of these films, conductivities as high as 200 S/cm might be expected for dense

Table 2. Conductivities of Doped Poly(heteroaromatic vinylenes)

Poly(thienylene vinylene) film[a]		Poly(furylene vinylene) film	
Dopant	Conductivity	Dopant	Conductivity
Iodine	62 S/cm	iodine	36 S/cm
FeCl$_3$	56	FeCl$_3$	25
NOSbF$_6$	25	Electrochemical[b]	4
Electrochemical[b]	37	Air	10^{-4}
Air	10^{-4}		

[a] low density film
[b] NaPF$_6$ in dimethoxyethane

doped PTV films. Undoped films exhibited a strong ESR signal (g-factor = 2.0024, Hpp=3.3G) and exhibited low levels of conductivity (10^{-6}-10^{-5} S/cm) much like *trans*-polyacetylene [17]. Air exposure caused a significant rise in conductivity (to 10^{-4} to 10^{-3} S/cm), presumably due to air (oxygen) doping. But unlike polyacetylene, PTV showed no signs of degradation on long-term air exposure; PTV films remained golden and flexible even after 1 year in air.

To study what effect substituents would have on the properties of PTV, a series of oligomers and low molecular weight polymers were prepared and characterized. Initial attempts to prepare high molecular weight substituted thienylene vinylene polymers by the quinodimethane route were unsuccessful due to the high reactivity of the bischloromethyl derivatives of thiophenes substituted with electron donating groups. Thus, for initial screening purposes we chose to prepare these polymers by coupling the 5,5'-dilithio derivatives of 3,3'-disubstituted dithienoethylene with 1,2-dichloroethylene. Average degrees of polymerization were on the order of 20 which is sufficiently high to assess a polymer's electronic properties.

A methyl substituent on the thiophene rings in PTV causes little change in polymer properties. But alkyl substituents the size of propyl or greater result in polymers with much increased solubility in organic solvents at room temperature. In fact, even the doped form of PTV bearing butyl groups appears to be soluble. The presence of the alkyl groups has little effect on the conductive properties of the polymers; methyl, ethyl, and butyl substituted thienylene vinylene polymers all reached conductivities around 1-5 S/cm on doping (pressed powder pellets).

From model studies on oligomers, the presence of strong electron donating substituents was found to have a pronounced effect on the band gap. The strong electron donating methoxy substituent was found to close the gap by 0.3 eV. Such a large effect on a small molecule warranted the investigation of such substitution on the electronic properties of the high polymer.

Poly(3-methoxythienylene vinylene) (PMOXTV) was prepared and the optical spectrum of a cast thin film shows two significant features: the absorption edge for the pi to pi* transition occurs at 800 nm which is significantly higher than that observed for PTV, and a second, weaker, very broad peak appears at 1000 nm which extends out into the near IR. The first peak clearly indicates that the band gap has been reduced by at least 0.2 eV as a result of the methoxy substituent on the thiophene rings. The broad absorption in the near IR region most likely is due to a weak, free electron absorption which suggests that the polymer has been slightly oxygen doped. Apparently, this is the case since we have observed a significant rise in conductvity on exposure of virgin PMOXTV films to air.

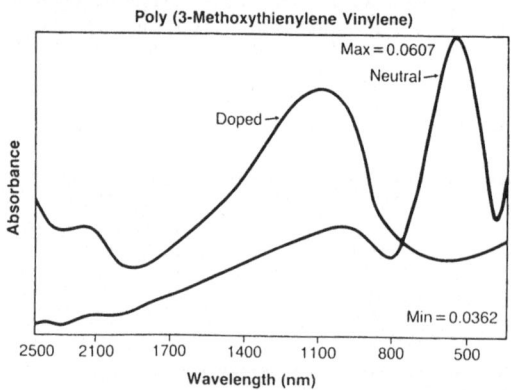

Fig. 1 Absorption spectra of undoped and doped poly(3-methoxythienylene)

On further doping PMOXTV with strong acceptor dopants such as ferric chloride or nitrosonium salts, the near IR absorption increases at the expense of the absorption in the visible. In fact, the absorption in the visible is weak and thin films of the doped polymer on glass substrates appear nearly colorless.

We also envisioned a facile synthesis of poly(furanyl vinylene) (PFV) by a similar route to that used to prepare PTV. The bischloromethyl derivative of furan is more reactive and more unstable than the thiophene analog. We have found an alternate route to bis chloromethyl furan starting from readily available dimethanol furan. Conversion to the bissulfonium salt proceeded smoothly to give a crystalline adduct. Polymerization to the polyelectrolyte proceeded as with the thiophene analog. But, unlike the thiophene analog, the polyelectrolyte precursor to PFV is more stable (less prone to elimination) at room temperature. Casting and fabrication of coatings is easier with the furan system than the thiophene system. Elimination from the precursor polymer begins at about 40°C and proceeds rapidly at 60°C. Nearly complete elimination is achieved by final heat treatment at 150°C as evidenced by elemental analyses. The resulting polymer has a golden lustrous color and looks identical to that of PTV.

The absorption edge of thin PFV film occurs at 700 nm which equates to an optical band gap of 1.76 eV. Absorption maxima are observed at 633 and 530 nm. On doping, a strong absorption appears in the near IR which peaks at about 1700 nm, and a weaker absorption is evident at 1000 nm. The latter peak may be due to polaron absorption, similar to what is observed with polythiophene [18,19,20].

Treatment of PFV with a variety of dopants results in highly conductive compositions; typical conductivities are on the order of 30 S/cm (see Table 2).

4. Conclusions

Extensive spectroscopic and electrochemical characterization of a variety of substituted polythiophenes and poly(heteroaromatic vinylenes) provides a detailed description of structure/property relationships within these classes of polymers. A summary of substituents and their influence on the electronic properties of the parent polymer is given in figure 2. One sees that these substituent effects are additive and allows one to fine tune polymer properties in a predictable way.

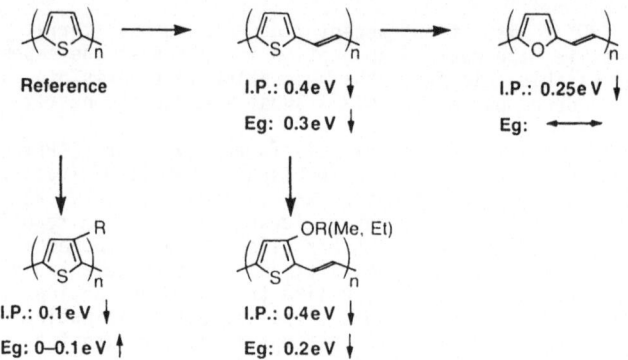

Fig. 2 Substituent effects on the band gap and ionization potentials for substituted polythiophenes and poly(heteroaromatic vinylenes).

5. References

1. W. Deits, P. Cukor, M. Rubner, and Jopson, Ind. Eng. Chem. Prod. Res. Dev., 20, 696 (1981).
2. J. Bargon, S. Mohmand and R.J. Waltman, IBM J. Res. Develop., 27, 330 (1983).
3. G. Tourillon and F. Garnier, J. Electroanal. Chem., 161, 407 (1984).
4. K. Tamao, S. Losama, I. Nakajima, M. Kunada, A. Minato, and K. Suzuki, Tetrahedron, 38, 3347 (1982).
5. J. Barker, P.R. Huddleston and M.L. Wood, Syn. Commun., 5, 59 (1975).
6. M. Kobayashi, J. Chen, T.-C. Moraes, A.J. Heeger, and F.Wudl, Synth. Met., 9, 77 (1984).
7. R.L. Elsenbaumer, K.Y. Jen, G.G. Miller, and L.W. Shacklette, Synth. Met., 18, 277 (1987).
8. S. Hotta, S.D.D.V. Rughootuputh, A.J.Heeger, and F. Wudl, Macromolecules, 20, 212 (1987).
9. G. Tourillon, D. Gourier, P. Garnier and D. Vivien, J. Phys. Chem., 88, 1049 (1984).
10. M.Sato, S. Tanaka and K. Kaeriyama, Synth. Met., 18, 229 (1987).
11. G. Koßmehl, M. Hartel, and G. Manecke, Macromol. Chem., 131, 15 (1970).
12. R.H. Baughman, J.L. Brédas, R.R. Chance, R.L. Elsenbaumer, and L.W. Shacklette, Chem. Rev., 82, 209 (1982).
13. M. Kanbe and M. Okawara, J. Polym. Sci. Part A-1, 6, 1058 (1968).
14. R.A. Wessling and R.G. Zimmerman, U.S. Pat. #3,401,152 and U.S. Pat #3,706,677.
15. J.L. Brédas, R.L. Elsenbaumer. R.R. Chance, and R. Silbey, J. Chem. Phys. , 78 (9), 5656 (1983).
16. G. Koßmehl, Ber. Bunsenges. Phys. Chem. , 83, 417 (1979), and references therein.
17. A.G. MacDiarmid and A.J. Heeger, Synth. Met., 1, 101 (1979).
18. M. Sato, S. Tanaka, and K. Kaeriyama, Synth. Met., 14, 279 (1986).
19. G. Harbeke, E. Meier, W. Kobel, M. Egli, H. Kiess and E. Tosatti, Solid State Commun., 55, 419 (1985).
20. K. Kaneto, S. Hayashi, S. Ura, and K. Yoshino, J. Phys. Soc. Jpn., 54, 1146 (1985).

Electrical Conductivity in Heterogeneous Polymer Systems

B. Wessling

Zipperling Kessler & Co, Kornkamp 50,
D-2070 Ahrensburg, Fed. Rep. of Germany

1. Introduction

It has been known for decades that it is possible to change the electrical
conductivity of insulating polymers by orders of magnitude by adding conductive fillers. It began with the use of carbon black for field smoothing
at cables, transport belts and so on. Today, antistatic and moderately conductive compounds have developed to become specialities which are used in
(in addition to the previously mentioned application areas) the packaging
and handling of electronic devices, explosives etc.; antistatic floors,
pipes, and housings; self-regulating resistance heaters or overvoltage
protection (using the so-called PTC effect); and - most recently - electrolytic processes as anodes /15/.

With the discovery of intrinsically conductive polymers (ICPs) a new
perspective in the electrical conductivity of polymeric materials has
opened: Recently we were able to show for the first time that, in contrast
to earlier predictions, this new class of polymers can be processed /18/.
This would allow control of the morphology and structure of ICP and therefore also their electrical properties, which could also lead to a deeper
understanding of their practical use.

2. Carbon Black as a Conductive Filler

Numerous different carbon blacks are now available as antistatic additives
for plastics. They differ predominantly in their specific surface area
(100 to 1200 m^2/g depending on type) and their porosity (dibutylphthalate
(DBP) absorption, 100 - 400 ml/100 g depending on type) /1-3/. The factor
responsible for these characteristic values of the carbon structure is
generally agreed to be their morphology: the smallest structural units are
semi-crystalline (graphitic) and form spheroidal primary particles with
10 - 100 nm diameter; these combine into larger, more or less hollow
(measurable with the aid of DBP absorption) secondary agglomerates with
different structures, which are up to 1 μm in size. To these is usually
attributed the conductivity of compounds /1-8/. It is generally assumed
that the primary particles have contact to each other by broad material
bridges, as were found for reinforcing carbon black /5/.

It is known that the conductivity of carbon black filled compounds does
not increase linearly with the content of conductive carbon black. Rather,
at a characteristic proportion there is a sudden increase in conductivity,
whilst it increases only slightly with extra added carbon. This phenomenon
is known as percolation. Figure 1 shows the percolation behaviour in poly-
propylene of carbons with various specific surface areas /9/. Carbon blacks
with larger surface areas or greater porosities (DBP absorption) have a
distinctly lower percolation threshold.

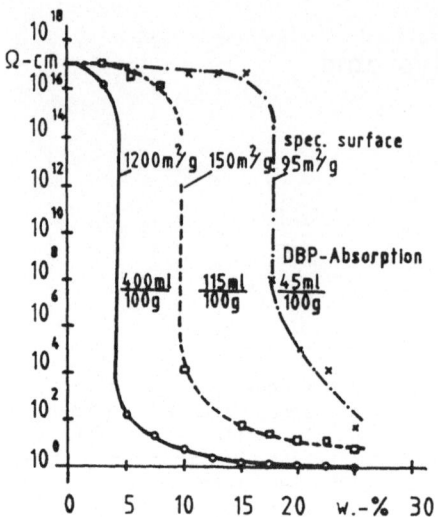

Fig. 1. Percolation behaviour of different types of carbon black

It is assumed that the conductivity of a compound depends on the structured agglomerates being sufficiently close to each other or in direct contact above the percolation point and on the continuous current pathways created thereby.

A model is nowadays assumed, especially by the carbon black manufacturers, according to which the approximately 30 nm large primary particles arrange themselves linearly into a (secondary) agglomerate or aggregate, whose structure is retained, at least partially, even during compounding into the polymer /4,5/. The concept has therefore become established that the carbon structure is degraded during compounding /6/, and "overprocessing" /1/ of the carbon black during compounding and subsequent processing of the compounds (e.g. in injection moulding) should be avoided.

3. Post Polymerisation Processing of ICP

Since the start of broad world-wide research in the field of ICP (like polyacetylene, PAc; polypyrrole, PPy; and the like) their improcessability after polymerisation implies a strong drawback. So until now, research has concentrated on films, thin layers or powders, as they are available directly from polymerisation.

Recent results /11,12,18/ indicate that pure ICPs show flow properties under high shear conditions, which may lead to highly crystalline and oriented products. In these cases, the original morphology and crystal structure of the ICP is dramatically changed /18/, as seen from SEM /12/ and X-ray diffraction (Fig.2). We expect the resulting crystals to behave differently from the native ICP (e.g. conductivity-temperature dependence).

4. Polymer Blends with Intrinsically Conductive Polymers (ICPs)

In contrast to processing of the pure ICPs, the preparation of ICP blends with conventional polymers was quite quickly a generally accepted idea.

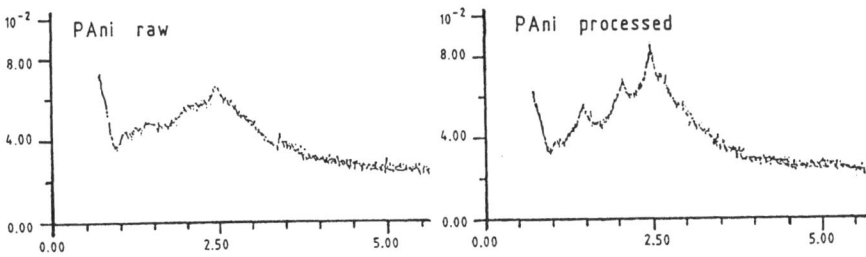

Fig. 2. X-ray diffraction curves of PAni

Several groups tried to polymerize ICPs inside a PS-, PB- or PVOH-matrix
or to prepare block or graft copolymers /16/. In all cases, 2-phase systems
develop due to the incompatibility (insolubility) of the ICP with the
matrix; these blends or composites have not been successfully processed.

In 1982 we began to work on a concept according to which ICP powders
should be polymerized and afterwards incorporated into a (thermoplastic)
matrix. This demands a knowledge of the morphology after completed polymerisation and after incorporation. What morphology does an ICP show after
polymerisation? The result of our work /13,11/ is that we consider a
spherical form of primary particles to be the general form in all ICPs,
when polymerized in suspension. Every other morphology, e.g. fibrils, is
secondary. This morphology appears to be similar to that of carbon black,
so it was often concluded that ICP in blends should be treated like carbon
black compounds: highly structured agglomerates of ICPs were prepared or
conserved and these were distributed in the matrix. That was why HOCKER
et al./14/ tried to polymerize highly structured PAc, as they are assumed
to occur in highly conductive carbon black (Ketjenblack EC).

We in contrast followed a different idea: assuming that the primary
particles in ICPs would only have point-like contacts and are agglomerated
only by (rather strong!) physical forces without chemical bonds as in
"material bridges", we tried to disperse the ICPs in polymer matrices.
Dispersion involves disagglomeration and wetting of the surface of the
primary particles by the matrix. This proved to be possible 3 years ago
/11,18/. SEM analysis revealed that the primary particles were isolated
from each other. True dispersions (in contrast to suspensions) of ICPs in
various solvents (which are stable to settling) or polymer matrices are
always intensely coloured. PAc is blue, PPy violet, PAni green, etc./11/.
This also showed that PAc and other ICP do not have a "structure" as is
assumed to occur in conductive carbon black.

How can dispersed, non-structured ICP particles percolate? Would not
this concept lead to very high percolation thresholds like 50 or 80 %?
Last year we were able to show for the first time that, by dispersing PAni
and increasing its concentration, percolation occurs at 9 - 10 vol.%.
Saturation (that is: the blend shows the same conductivity as the pure
raw PAni powder, compressed) is now observed between 40 and 50 vol.%
(Fig.3, curve 1) /18, 15/.

Recently we succeeded in dispersing PPy almost completely, which led to
a comparable percolation curve (Fig.3, curve 2) /19/.

These blends are thermoplastically processable. First experiments show
that they behave chemically and in respect to conductivity essentially like

409

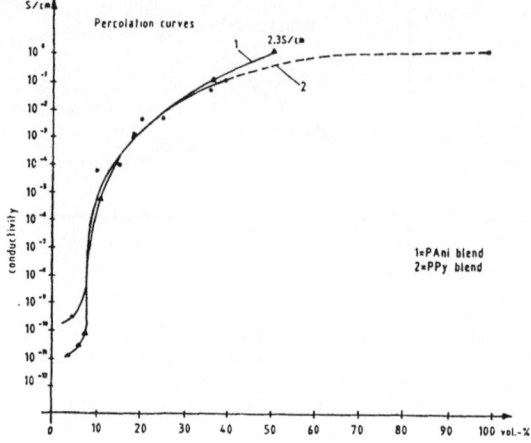

Fig. 3. Percolation curves of PAni and PPy

pure films of PAni or PPy or PAc: they are chemically and electrochemically reversibly reducible and oxidizable /19,20/; their temperature-conductivity dependence seems to be the same as is found for pure ICP films and differs significantly from carbon black compounds /21/, see Figs. 4 and 5.

At least we can conclude that percolating systems from previously dispersed (=isolated) particles (spheres) do not differ significantly from the pure films, which are also percolating systems. We would also like to propose that the structure of the percolated system does not differ significantly in both systems: spheres agglomerate to chain-like linear structures which build up networks. This is what we assumed we saw in fibrils of pure films /13/ during our morphology studies.

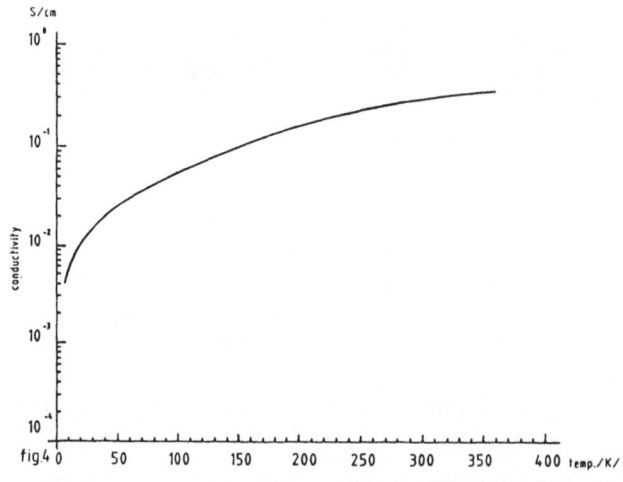

Fig. 4. Conductivity temperature relationship in an PAni blend

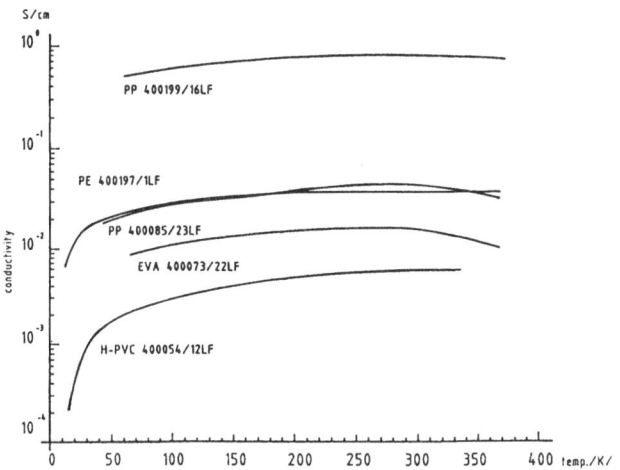

Fig. 5. Conductivity temperature relationship in carbon black compounds

5. Two Different Percolation Mechanisms?

We saw that in carbon black compounds percolation is assumed to occur for
statistical and geometric reasons: irregular, highly structured agglomerates
of carbon black are said to contact each other with a certain probability.
How can an ICP, which shows no "structure" in its agglomerates and is fully
dispersed, percolate at that low volume content (10 %)?

We propose now /15/ that the driving percolation force is the interfacial
energy. In these 2-phase systems, two processes occur and compete: dispersion
(= wetting of the dispersed phase by the matrix polymer: adhesion) and
agglomeration (= point-like contacts of dispersed particles, chaining:
cohesion). At the percolation point, adhesion and cohesion are in balance
- this is our new hypothesis, which is currently being investigated further.
We propose that this process takes place both in fully dispersed carbon
black compounds and ICP blends. The difference in properties between carbon
black compounds and ICP blends results from their different basic properties.

Literature

1. Ketjenblack EC, data sheet,(Akzo Ltd., Chicago, 1984)
2. Philblack XE-2, data sheet,(Philips Petroleum, Botlek, 1985)
3. Pigments, publication no. 69,(Degussa AG, Frankfurt/M., 1977)
4. Kleinheins, G.H.: Anwendungen strahlungsvernetzter leitfähiger Kunststoffe, in: Elektrisch leitende Kunststoffe; H.J. Mair, S. Roth (Hrg.),
   Carl Hanser Verlag, München, Wien, 1986, p. 77/91
5. Sichel, E.K. (ed.): Carbon Black Polymer Composites;(M.Decker, New York,
   1982)
6. Probst, N.: Conductive Compounds on the Move; Eur.Rubber Journal, reprint,
   1984
7. Jansen, J.: The Critical Conductive Filler Loading in Antistatic Composites, J.Appl.Phys. 46 (1975), p. 966/969
8. Verhelst, W.F.: Conductive Carbon Black; Paper, SRC, Kopenhagen, 1985
9. Gilg, R.G.: Ruß für leitfähige Kunststoffe, in: Elektrisch leitende
   Kunststoffe, H.J. Mair, S. Roth (Hrg.),(Carl Hanser Verlag, München,
   Wien, 1986) p. 55/76

10. <u>Semiconductive Compounds</u>, data sheets, Zipperling Kessler & Co, Ahrensburg, 1986
11. Wessling, B.; Volk, H.: <u>Post Polymerisation Processing of Conductive Polymers (II)</u>, Workshop on Synthetic Metals, Brookhaven, N.Y./USA, Synthetic Metals, <u>15</u> (1986) 183-193
12. Wessling, B.; Volk, H.: <u>Thermoplastic Conversion of Doped Polyaniline from the Amorphous to the Partially Crystalline State</u>, Synthetic Metals <u>16</u> (1986) 127/131
13. Wessling, B.: <u>Contribution to the Discussion on the Morphology of Polyacetylene (PAc)</u>, Makromol.Chem. <u>185</u> (1984) p.1265/1275
14. J. Hocker u.a. EP-OS 62211 (1982)
15. Wessling, B.: <u>Electrically Conductive Polymers</u>, Kunststoffe/ German Plastics <u>76</u> (1986) 10, p.930/936
16. a) F. Bates, G. Baker: Makromolecules 16 (1983) p.704/707
    b) M. Galwin, G. Wnek: J.Polym.Sci., Polym.Chem. Ed. 21 (1983) p.2727/2737
    c) M. Aldissi: Synthetic Metals 15 (1986) p.141/153
    d) A. Pron et al: Synthetic Metals 18 (1987) p.49/52
17. S. Miyata, T. Ojio, Y.E. Whang: Abstracts of International Conference on Science and Technology of Synthetic Metals, p.427
    O. Nowa, M. Hikita, T. Tamamura: Synthetic Metals 18 (1987) p.677/682
18. B. Wessling, H. Volk : Synthetic Metals 18 (1987), p.671/676
19. B. Wessling, H. Volk, R. Mathew, L. Campbell, V. Kulkarni: unpublished results
20. J. Heinze, private communication (cyclovoltammetry)
21. D. Schaefer-Siebert, S. Roth, B. Wessling, H. Volk: to be published

Organic and Polymeric Non-linear Optical Materials: Properties and Applications

*D. Bloor**

GEC Research Limited, Marconi Research Centre, Chelmsford,
Essex, CM2 8HN, United Kingdom

1. Introduction

As communication and data transfer systems increasingly employ optical
fibres there is a growing requirement for electro-optic and optical devices
with improved performance. This requires the preparation and assessment of
materials with larger non-linear optical coefficients and the development
of techniques for the fabrication of practical devices from these
materials. The latter is far from simple, since the main requirement is for
integrated optical devices which can be easily connected to mono-mode
optical fibres. Indeed, it has taken many years to establish the technology
required to produce the lithium niobate devices that are just entering the
market.

Oxides, such as lithium niobate, have been investigated for many years
in view of the large optical non-linearities displayed by ferroelectric
materials. More recently it has been shown that organic materials could,
because of their highly polarisable electron systems, exhibit even larger
non-linear effects. Furthermore, quantum chemical theoretical models of
organic molecules developed to predict structure and linear optical
properties can be extended to provide a sound basis for the molecular
design of non-linear optical materials [1,2]. Problems remain in both
predicting the values of the non-linearities for crystals and achieving the
necessary crystal structures. The former problem arises since the
polarisation of the molecule is determined by the local field (E_ℓ),
Equation 1, while that of the medium depends on the externally applied
field (E), Equation 2:

$$p = \alpha \, E_\ell + \beta \, E_\ell{}^2 + \gamma \, E_\ell{}^3 + \_\_\_ \quad , \tag{1}$$

$$P = \chi^{(1)} \, E + \chi^{(2)} \, E^2 + \chi^{(3)} \, E^3 + \_\_\_ \quad . \tag{2}$$

Thus proper evaluation of $\chi^{(2)}$ and $\chi^{(3)}$ depends on evaluation of the
local field [3,4]. Crystal symmetry places restrictions on the values of
the even terms in Equation (2). Since in a centrosymmetric medium change
of sign of the applied field must result in a reversal of the polarisation,
the even terms must be zero. Materials displaying second-order
non-linearity must, therefore, be non-centrosymmetric while no restrictions
are placed on the symmetry of the medium for third-order non-linearity.
One consequence of this is that the production of a molecule with a high β
is no guarantee of the production of a crystal with high $\chi^{(2)}$.

Despite these problems, considerable success has been achieved in
producing organic crystals with both good non-linear optical properties and

---

* Permanent address: Dept. of Physics, Queen Mary College, Mile End Road,
London, E1 4NS, U.K.

413

considerable resistance to radiation damage [5,6]. There are, however, serious problems in the fabrication of the thin single crystal films, with the correct crystallographic orientation necessary for the development of integrated optic devices. In contrast, the ease of fabrication of thin polymer films has resulted in a growing interest in their application to integrated optics. In the following sections some recent developments of organic and polymeric materials with large second- and third-order non-linearities will be reviewed. An example of an integrated optic device is presented in each case.

2. Second-Order Non-Linear Materials

The ground rules for the production of a high molecular hyperpolarizability, β, are simple. The molecule should possess a large dipole moment in either the ground or excited state. Most materials studied fall into the former category. This requirement is met by an extended, linear or planar, molecule with a strong electron donating group attached through a conjugated structure to a strong electron accepting group. The best known example is nitroaniline in which the donor (amine-group) is connected to the acceptor (nitro-group) via the conjugated (phenyl) ring. The strong influence of crystal packing is shown by the absence of any macroscopic second-order non-linearity for para- and ortho-nitroaniline but its presence for meta-nitroaniline. It should be noted that extension of the conjugated section of the molecule, which at first sight would appear to be beneficial, is not necessarily so. This follows since the important quantity is β per unit volume not β per molecule. It has been shown that β per unit volume is a maximum for polyphenyls for three repeat units, while for polyenes a maximum is reached at about 20 repeat units [2]. Thus compact molecules can have near optimum molecular properties and be much easier to obtain as optical quality single crystals. An example of this is DAN, Fig. 1, which has been shown to exhibit efficient second harmonic generation [7].

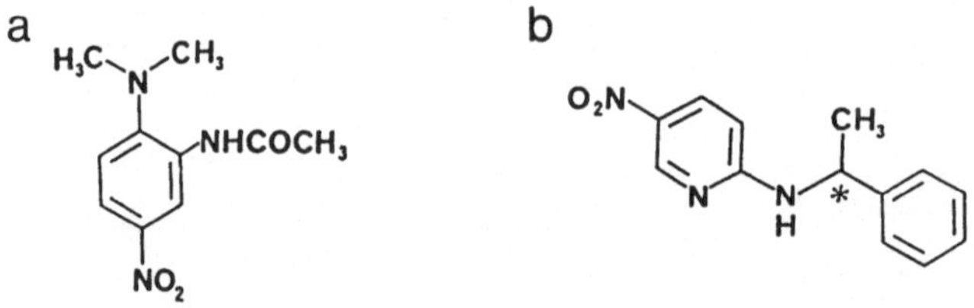

Figure 1 Molecular structures of
      (a) 2-(N, N-dimethylamino)-2-nitroacetaniline (DAN) and
      (b) R-(+)-5-nitro-2-[N-(1-phenylethyl)-amino] pyridine (MBA-NP).

The fabrication of integrated optic structures from organic crystals is not simple [6,8]. In contrast, polymeric optical waveguides can be formed with relative ease. If amorphous polymers, such as poly(methyl methacrylate), are utilised the optical loss can be low. Values of 0.15 dB/cm have been reported for planar PMMA guides [9] and as low as 10^{-4} dB/cm for optical fibres fabricated from deuterated PMMA [10]. In addition, techniques exist for the definition of channel guides in polymer films [11,12]. Polymer films with finite second-order non-linearities can be produced in a number of ways. An inactive, amorphous polymer can be loaded with active molecules, such as nitroaniline [13,14]. Such loaded systems are isotropic and anisotropy must be impressed by heating the loaded polymer above its glass transition temperature (T_g), applying an electric field to align the molecular dipoles and cooling below T_g to freeze in the molecular alignment. This process is analogous to the method of electric field-induced second harmonic generation (EFISH) used to determine β for molecules in solution [15]. Second harmonic generation has been observed in electrically polarised, loaded, films of PMMA [13,14]. Alternatively, the active groups can be attached to the polymer backbone as pendant groups. In both cases electrically polarised films are unstable and tend to relax back to an isotropic structure. This problem can be overcome by incorporating non-linear groups into liquid crystal polymers so that order is imposed by the inter- and intra-molecular interactions. High order parameters can be obtained but the liquid crystal phases display characteristic, strongly scattering textures [14]. It is necessary to apply an electric field to obtain a uniformly orientated film that is optically transparent [14].

Though scattering is a serious problem in optical waveguides their use eases other problems. In crystals phase matching of interacting waves can occur only along a few special directions, if at all. This problem is virtually eliminated in optical waveguides since the matching of propagation velocity of fundamental and harmonic guided waves can be achieved for a specific thickness of the guiding layer [6]. Thus optimum second harmonic generation can be achieved by varying guide thickness, e.g. by using a wedged rather than a parallel sided film [8]. High conversion efficiency is possible for a moderate second-order non-linearity provided the propagation losses are low. The latter requirement can be met, as noted above. The former requires a high density of active groups though some reduction in β per unit volume is inevitable due to the presence of inactive constituents of the polymer. In general the proportion of active groups in loaded polymers, typically 10% maximum, is lower than can be achieved by bonding the active group to the polymer chain. An exception to this general rule has been found for the system MBA-NP (Fig. 1) loaded in polycarbonate [16]. Glassy films with loadings as high as 65% MBA-NP have been obtained by casting from solution. Guiding has been observed in these films with losses of 2.4 and 6.2 dB/cm for the TE0 and TE1 modes respectively. These values indicate that the loss is principally due to scattering by irregularities at the film surface rather than in the bulk of the film.

Current organic and polymeric materials compare favourably with inorganic materials for second harmonic generation and optical parametric devices. The advantages for electro-optic devices are less clear since the organic materials lack the ferroelectric (atomic displacement) contribution which gives rise to large effects in inorganic materials. In addition, much work remains to be done to match the present sophistication of lithium niobate integrated optic devices.

3. Third-Order Non-Linear Materials

A high third-order non-linearity occurs for long conjugated molecules. This was established from studies of β-carotene and related polyenic molecules [17,18]. An early extension to poly-diacetylene (PDA) crystals [19] attracted little attention despite the observations of $\chi^{(3)}$ values larger than those of Ge and GaAs. Recently, interest has revived and extensive studies have revealed a large, fast third-order non-linearity in the infra-red which is resonantly enhanced in the vicinity of the main absorption peak of the polymer [20]. Larger values have been reported for polyacetylene [21], however, the fibrous morphology renders it unsuitable for integrated optics. PDAs can be obtained as thin, high quality crystals of large area [22] allowing the possibility of low loss optical guiding. A potential application is in optical logic devices based on the optical Kerr effect. An unequal path, channel waveguide, interferometer will display an intensity-dependent output. For such a device PDA films offer performance falling between that of GaAs and GaAs/AlGaAs superlattices [23].

A number of attempts have been made to calculate the third-order susceptibilities of PDAs. AGRAWAL et al [24] assumed that the intense optical absorption was an interband transition and obtained expression appropriate for a one-dimensional semiconductor. Subsequent studies of photoconduction [25] and electro-absorption [26] have shown that the interband transition lies well above the optical absorption which is due to a transition to an excitonic state. KAJZAR and MESSIER [27] have calculated $\chi^{(3)}$ invoking a two-photon resonance with the first excited ^1A_g state. From observations of a negative value of $\chi^{(3)}$ at 1.06 μm they found it necessary to place the two-photon state above the allowed one-photon state energy. Recently TOKURA et al [28] have reinterpreted the electro-absorption as due to the ^1A_g level. Though no non-linear data were recorded for the particular PDA studied,this interpretation appears inconsistent with the observation of negative $\chi^{(3)}$ at 1.06 μm in other PDAs. Thus, there remains the need for a full theoretical study of PDAs.

Rapid assessment of PDAs is difficult since the third harmonic of 1.06 μm radiation is strongly absorbed. Langmuir-Blodgett films are thin enough for this not to be a problem, but the difficulty cannot be avoided in single crystals. Second harmonic generation will occur when an electric field is applied to distort the electron distribution in the polymer chain [15]. Such solid-state EFISH has been reported for Langmuir-Blodgett and polycrystalline PDA films [29]. The results show deleterious effects due to charge injection and trapping, i.e. signals decreasing after prolonged pumping and persistence of SHG after the E-field is removed. These effects can be avoided by the use of mixed crystals in which the predominant defects are chain ends [30]. The system toluene sulphonate (TS)/fluorobenzene sulphonate (FBS) diacetylene forms good solid solutions with polymerization kinetics showing increasing chain-end defect density with increasing FBS content [31]. Mixed crystals in the composition range 0% to 50% FBS only show SHG when the d.c. and optical E-fields are both parallel to the polymer chain direction. Signal intensity increases with FBS content. Typically SHG decay curves after the E-field is turned off are shown in Figure 2. Two decay components are observed, an initial fast

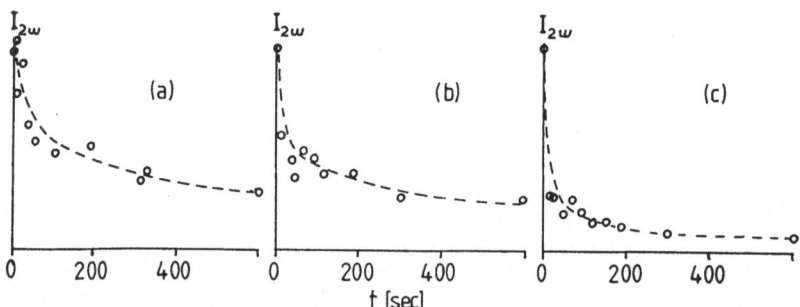

Figure 2   Decay of SHG after removal of E-field for TS/FBS crystals with FBS contents of (a) 5%, (b) 10% and (c) 20%. Second harmonic intensities normalised at t = 0.

and a final slow decay. The faster decay dominates at high FBS concentration, at 50% FBS the SHG has decayed essentially to zero after two probe pulses. These results are consistent with the distribution of trapped charge throughout the sample as FBS content increases and the existence of two charge recombination mechanisms, one involving little charge migration, e.g. tunnelling, the other involving charge transport along the polymer chain to recombination centres.

As for second-order materials, much work remains to be done to construct practical devices from third-order polymeric materials. In both areas progress will require strong interaction between the technologist and the pure scientist.

Acknowledgements

The SERC and Royal Society are thanked for an Industrial Fellowship, support under the Joint Opto-Electronic Research Scheme of the SERC and DTI is also acknowledged. British Drug Houses are thanked for supplying DAN and FBS.

References

1. S.J. Lalama, A.F. Garito: Phys. Rev. B. 20, 1179 (1979)
2. S. Allen, J.O. Morley, D. Pugh, V.J. Docherty: Proc. SPIE 682, 20 (1986)
3. M. Hurst, R.W. Munn: J. Mol. Electron. 2, 35 (1986)
4. M. Hurst, R.W. Munn: J. Mol. Electron. 2, 43 (1986)
5. J.C. Williams, ed.: Non-linear Optical Properties of Organic and Polymeric Materials (Amer. Chem. Soc., Washington D.C. 1983)
6. J. Zyss: J. Mol. Electron. 1, 25 (1985)
7. P.A. Norman, D. Bloor, J.S. Obhi, S.A. Karaulov, M.B. Hursthouse, P.V. Kolinsky, R.J. Jones, S.R. Hall: J. Opt. Soc. Amer. B in press
8. H. Sasaki, T. Kinoshita, N. Karasawa: Appl. Phys. Lett. 45, 333 (1984)
9. R. Schriever, H. Franke, H.G. Festl, E. Kratzig: Polymer 26, 1423 (1985)
10. T. Kaino, K. Jinguji, S. Nara: Appl. Phys. Lett. 42, 567 (1983)
11. M.D. Lechner: In Electronic Properties of Polymers and Related Compounds, ed. by H. Kuzmany, M. Mehring, S. Roth, Springer Ser. Solid-state Sci., Vol. 63, 301 (Springer, Berlin, Heidelberg, 1985)
12. W. Driemeier, A. Brockmeyer: Appl. Opt. 25, 2960 (1986)
13. K.D. Singer, J.E. Sohn, S.J. Lalama: Appl. Phys. Lett. 49, 248 (1986)
14. J.B. Stamatoff, A. Buckley, G. Calundann, E.W. Choe, R. De Martino, G. Khanarian, T. Leslie, G. Nelson, D. Stuetz, C.C. Teng, H.N. Yoon: Proc. SPIE 682, 85 (1986)
15. S. Kielich: IEEE Jour. Quant. Electr. QE-5, 562 (1960)
16. A.R. Oldroyd, G. Elliott, S. Mann, N.J. Parsons: Proc SPIE 800 in press.
17. J.P. Hermann, J. Ducuing: J. Appl. Phys. 45, 5100 (1974)
18. K.C. Rustagi, J. Ducuing: Opt. Commun. 10, 258 (1974)
19. C. Sauteret, J.P. Hermann, R. Frey, F. Pradère, J. Ducuing, R.R. Chance, R.H. Baughman: Phys. Rev. Lett. 36, 956 (1976)
20. G.M. Carter, J.V. Hryniewicz, M.K. Thakur, Y.J. Chen, S.E. Meyler: Appl. Phys. Lett. 49, 998 (1986)
21. F. Kajzer, S. Etemad, G.L. Baker, J. Messier: Phys. Rev. Lett. in press.
22. M. Thakur, S. Meyler: Macromols. 18, 2341 (1985)
23. H. Kawaguchi: Opt. Lett. 10, 411 (1985)
24. G.P. Agrawal, C. Cojan, C. Flytzanis: Phys. Rev. B 17, 776 (1978)
25. H. Bassler: In Polydiacetylenes, ed. by D. Bloor, R.R. Chance (Martinus Nijhoff, Dordrecht, 1985) p. 135

26. L. Sebastian, G. Weiser: Phys. Rev. Lett. 46, 1156 (1981)
27. F. Kajzar, J. Messier: In Polydiacetylenes, ed. by D. Bloor,
    R.R. Chance (Martinus Nijhoff, Dordrecht, 1985) p. 325
28. Y. Tokura, T. Koda, A. Itsubo, M. Miyabayashi, K. Okuhara, A. Ueda:
    J. Chem. Phys. 85, 99 (1986)
29. P.A. Chollet, F. Kajzar, J. Messier: In Polydiacetylenes, ed. by
    D. Bloor, R.R. Chance (Martinus Nijhoff, Dordrecht, 1985) p. 317
30. J.S. Obhi, D. Bloor, D.J. Ando, P.A. Norman, P.V. Kolinsky,
    B. Movaghar: GEC J. Res. 4, 256 (1986)
31. V. Enkelmann: Die. Makromol. Chem. 184, 1945 (1983)

Preparation of Conducting Polymers by UHV-Compatible Methods

W.R. Salaneck, I. Lundström, A. Mohammadi, S. Akbar*, and A. Platau

Department of Physics (IFM), Linköping University,
S-58183 Linköping, Sweden

I: Introduction:

In future potential applications of electrically conducting organic polymers as active or passive components in integrated electronic circuits of various sorts, it could be of enabling importance to have fabrication methods which are compatible with present-day and especially projected future semiconductor processing technology. The chemical, electrochemical and catalytic methods presently used may be difficult to apply in such a context. Despite the fact that we believe that the "best" applications of conducting polymers have yet to be discovered, we recently proposed that vacuum processing is a potentially attractive method for preparing thin films of conducting polymers with controlled electronic and optical properties [1]. One can envision the incorporation of conducting polymer thin films into what might be considered "integrated electronics". In this context, e.g., thin conducting polymer films could possibly find use as the active components in integrated (bio)gas sensors with built-in logic on an (V)LSI chip.

There are (at least) three general catagories of vacuum processing of polymers that can be envisioned. First, monomer molecules can be condensed upon a cooled substrate and subsequently polymerized in situ via radiation polymerization. All types of radiation can, in principle, be considered, including electrons, photons (IR, UV, soft X-rays, etc.), ions, and others. Second, radiation treatment of initially insulating polymers can be used to modify their (surface) chemical structure to form "conducting polymers" (which can be doped), using radiation as described above. In both cases, radiation treatment would be followed by exposure to gases of "dopant" molecules. Third, a substrate in UHV can be exposed to individual streams of monomer, dopant, oxidant, and catylist molecules simultaneously in order to accomplish a chemical vapor deposition (CVD) of a conducting polymer film directly. Although other possibilities exist, some work carried out in our laboratories in these areas will be outlined below.

II: Radiation Polymerization:

Several years ago [2] we reported that pyrrole could be polymerized, in the form of ultra-thin films condensed upon a metal substrate in UHV, using soft X-ray radiation. In that work, pyrrole was condensed at temperatures well below that necessary to just condense the molecules onto a cooled substrate in the vacuum chamber. The use of too low temperatures led to the formation of polymer films with a relatively high degree of disorder. The resultant radiation-polymerized ultra-thin films were studied by ultra violet photoelectron spectroscopy (UPS) and X-ray photoelectron spectroscopy (XPS, which is also known as ESCA).

---

* Permenent address: Department of Chemistry, University of Baluchistan, Sariab Road, Quetta, Pakistan.

Recently, we have extended our work to the case of thiophene [1]. In UHV, the thiophene monomer condenses at -130 C. Upon exposure to the soft X-radiation from one of the typical laboratory sources used in XPS, in our case Mg K$_\alpha$ at 1253.6 eV, thiophene polymerizes slowly enough that the evolution of the XPS C(1s) and S(2p) core level spectra can be observed as a function of time. Comparison can be made with the corresponding spectra of electro-chemically made polythiophene [3], and the binding energies as well as the relative intensities are found to be equal. In the UPS and XPS valence band spectra, not only is good agreement between the spectra of the radiation polymerized and electro-chemically prepared material obtained, but through comparison with model Valence Effective Hamiltonian (VEH) calculations [4], it is clear that in these radiation polymerized films the monomer molecules are joined via α-α linkages. The π-band seen at lowest binding energy is the definite indication no significant population of β-β or α-β linkages occurs [1,3,5]. The relative high quality of these polythiophene films may be ascribed to the use of as high a temperature as possible in condensing the monomer molecules (-130 C), which allows maximum molecular mobility, enabling the molecules to reorient themselves as necessary during the polymerization process. For spectroscopic purposes, however, these films were formed upon metal substrates, precluding the possibility of studying the electrical conductivity, had doping been attempted.

For this conference, we have attempted to polymerize thiophene using hard X-ray radiation. We exposed thiophene, sandwiched between two ultra thin sapphire plates and sealed from the atmosphere, to hard (relative to Mg K$_\alpha$ radiation) 40 KeV tungsten X-radiation in a conventional X-ray diffractometer. Following exposures of various lengths of time, the optical absorption spectra of the contents between the two sapphire plates could be recorded. Since no metal substrate or electrode was involved, this system had the potential of allowing doping of any resulting films, so that electrical conductivities could be studied as well. Unfortunately, the thiophene polymerizes very slowly. The rate of polymerization is extremely small as compared with the rate for Mg K$_\alpha$ radiation, even after estimating the attenuation in the sapphire plate and interaction efficiency of the hard X-rays with the liquid thiophene. There were, however, some weak indications of optical absorption in the irradiated samples near the photon energy corresponding to the π-π transitions in electro-chemically polymerized polythiophene [6]. The onset of optical absorption was at shorter wave lengths in the radiation polymerized samples than in electro-chemically made material, indicating relatively short conjugation lengths in those "oligomers" that were formed in the hard X-ray irradiation process. Since these observations are preliminary, and the radiation process was not able to be carried to completion, the spectra are not presented in these proceedings.

We view the aromatic heterocyclic molecules, of which pyrrole and thiophene are representative, as a potential source of materials of which solid-state radiation polymerization may be utilized in order to form highly crystalline, or at least highly ordered, conducting polymers. One can envision ordered overlayers of, e.g., thiophene adsorbed upon single-crystal metal surfaces, polymerized by irradiation, with subsequent layers grown epitaxially upon the first. Alternatively, one can envision single crystals of the monomers grown at low temperatures, and then polymerized via irradiation, analogous to the highly developed techniques used today in the preparation of single crystals of the polydiacetylenes [7]. The development of such techniques is obviously many years away. The fact that the aromatic heterocycles do polymerize under soft X-radiation, however, may possibly serve as fuel in this area of research involving conducting polymers.

III: Chemical Vapor Deposition (CVD):

Another general method for the preparation of conducting polymers that might be compatible with vacuum processing is chemical vapor deposition (CVD). We have carried out a survey of what might be termed the "type of instrumentation" that might be suitable in the preparation of conducting polymers by a CVD process [8-10]. Initial studies on polypyrroles were carried out in UHV [8], using pyrrole, $FeCl_3$, and H_2O as the reaction species. It was necessary to expose these gases to a substrate cooled to between -20 C and 0 C, in order to increase the residence time of the molecules (but not condense them) on the substrate. Pressures near one Torr (total) were finally found to be necessary in order to achieve film formation. In these experiments, the properties of the polypyrrole films formed were independent of the identity of the substrate, presumably because of the use of $FeCl_3$ in the vapor. Highly disordered, cross-linked polymer films with poor electrical conductivities were produced in these early experiments [8]. Moving to a mild vacuum system, and using H_2O_2 instead of the $FeCl_3$, better polymer films were formed, which had optical absorption and infra red spectra very much like electro-chemically prepared polypyrrole. Most important, however, was that electrical conductivities as high as about 10 % of those for conventional electro-chemically prepared materials were achieved in the CVD materials. Finally, recent experiments [11] in a carrier-gas system indicate that very shiny, uniform-looking thin films can be formed, when Cl_2 is used as an oxidant. We note in passing that the optimization of conditions has not been carried out yet in any of the types of CVD system that we have studied. In a sense, we have used the type of apparatus as the variable in the first round of these exploratory studies. The next phase [11] involves the optimization of the CVD conditions with one system, which we have chosen to be the carrier-gas system.

IV: Summary:

We have suggested that vacuum compatible, vapor phase methods for the preparation of thin films of electrically conducting organic polymers may possibly find applications in future "integrated electronics" devices. Two general methods for the fabrication of thin films of conducting polymers have been studied in early exploratory activities. The radiation polymerization of aromatic heterocyclic molecules, pyrrole and thiophene, can be carried out on ultra thin condensed films in UHV. Doping studies were not attempted, since the thin films were grown on metal substrates for spectroscopic purposes. At this time these radiation-polymerized materials are only a spectroscopic curiosity, but with some potential for future applications. A chemical vapor deposition (CVD) process for the fabrication of conducting polypyrrole films has been demonstrated, using different vacuum-grades of instrumentation. Electrical conductivities of up to 10 % of those for conventional electro-chemically prepared polypyrrole films has been achieved. The CVD process has not been optimized yet, however, and even higher conductivities may be anticipated in the future.

V: Acknowledgements:

This work has been supported by grants from the Swedish Natural Science Research Council (NFR), the Swedish Board for Technical Development (STU), and Imperial Chemical Industries (I.C.I.), PLC, England.

VI: References:

1. W. R. Salaneck, J. -L. Brédas, C. R. Wu, and J. O. Nilsson: J. Synth.
    Met. (in press)
2. W. R. Salaneck, R. Erlandsson, J. Prejza, I Lundström, and O. Inganäs:
    J. Synth. Met. 5, 125 (1983)
3. C. R. Wu, J. O. Nilsson, O. Inganäs, W. R. Salaneck, and J. E.
    Österholm: J. Synth. Met. (in press)
4. J. -L. Brédas, R. L. Elsenbaumer, R. R. Chance, and R. Silbey: J. Chem.
    Phys. 78, 5656 (1983)
5. J. -L. Brédas, private communication.
6. G. Tourillon and F. Garnier: J. Electroanal. Chem. 161, 51 (1983).
7. D. Bloor: In Developments in Crystalline Polymers, ed. by D. C. Bassett
    (Applied Science, London, 1982)
8. A. Mohammadi, M. A. Hasan, B. Liedberg, I. Lundström, and W. R.
    Salaneck: J. Synth. Met. 14, 189 (1986)
9. A. Mohammadi, I. Lundström, W. R. Salaneck, and O. Inganäs: Chemtronics
    1, 171 (1986)
10. A. Mohammadi, I. Lundström, W. R. Salaneck, and O. Inganäs: J. Synth.
    Met. (in press)
11. A. Mohammadi, private communication, and to be published.

Organic Semiconducting Polymers for New Electronic Devices

F. Garnier and G. Horowitz

Laboratoire de Photochimie Solaire, C.N.R.S.,
2, rue Henry Dunant, F-94320 Thiais, France

1. Introduction

Among the many areas of applications opened to organic conjugated polymers, such as polyacetylene, polypyrrole or polythiophene, a new promising field is emerging, which concerns the elaboration of organic polymer-based electronic devices. Two main characteristics of these polymers have been used for this purpose. First the possibility to switch them in a controlled way from a highly resistive neutral state to a conducting doped state. Thus, M. WRIGHTON's group (1) has already largely shown that microelectrochemical transistors can be elaborated, in which the gate is represented by an electrochemical doping-undoping process. A small variation of the electrochemical potential leads to a large variation in the polymer conductivity, which consequently allows the current to flow between two microelectrodes, source and drain, embedded in the polymer. Even more sophisticated structures have been proposed, by the use of a solid electrolyte, leading to a solid-state device, and also by using a chemical oxidant or reductant to drive the doping-undoping process, leading thus to a chemically sensitive transistor. Although very promising, this first type of electronic devices is limited by two main factors. The channel thickness of this type of transistor being large, an important amount of charge, $\sim 10^{-7}$ mole e$^-$/cm^2, is needed for turning on the device, which limits the amplification power. Furthermore the channel material being a polymer, the doping process requires an ion transfer, whose slowness limits the operation frequency to some 10 Hz.

A second type of field effect transistors has been proposed mainly by Japanese groups (2), based on the semiconducting properties of these organic conjugated polymers when almost completely undoped. A thin polymer layer is grown on two gold microelectrodes, source and drain, deposited on a SiO$_2$-coated Si wafer. A back contact to the Si substrate forms the gate. When applying a negative voltage to this gate, a field is created in the polymer layer which attracts the remaining holes in the polymer toward the channel between the two gold microelectrodes. A current can thus flow, and the amplification ratio obtained with this polymer-based FET, $\sim 10^2$, is quite interesting. These structures are however limited by the low mobility of the ionized carriers in the semiconducting polymer. This very general problem encountered with most organic semiconductors underlines the need of a better knowledge and control of the electrical parameters of these organic semiconducting polymers. In this regard we have mainly focused our work on the characterization of an "organic-on-inorganic" diode (3), with the aim of understanding the features which control this basic device. This new diode has been obtained by the growing of a thin film of p-type semiconducting polymethylthiophene derivative, PMeT, on an inorganic semiconductor, n-GaAs, and among the many applications of this new diode, we shall describe the characteristics of a GaAs/PMeT solar cell.

2. Elaboration of the n-GaAs/p-PMeT diode

The diodes were realized using non intentionally doped (100)-oriented n-GaAs wafers. The doping level was 2×10^{16} cm^{-3}. Ohmic back contacts were made by evaporating Au-Ge alloy and annealing at 400°C under N$_2$. The n-GaAs disk was polished with 1 μm diamond paste and etched in 1 % Br$_2$-MeOH, followed by dipping in ammonia water (28 %) and rinsing in doubly distilled water. For improving the homogeneity of the polymerization of PMeT, a small amount of platinum, ca. 1.6×10^{16} atoms per cm^2, was electrochemically pre-deposited onto the photoelectrode.

The electrodeposition of PMeT was then performed in a classical three-electrode electrochemical cell with a saturated calomel reference electrode (SCE) and a platinum counter-electrode, in 0.5 M distilled 3-methylthio-phene 0.05 M (tBu$_4$N) SO$_3$CF$_3$ acetonitrile solution, under illumination with a tungsten lamp at a potential of 1.5 V/SCE. The film thickness was moni-tored by means of a digital coulometer (EG & G PAR model 179) assuming a linear variation of 1 nm per mC.cm^{-2}. The area of the polymer deposit was 0.25 cm^2. Undoping of the polymer was carried out by polarizing the elec-trode at - 0.5 V/SCE.

Finally the diode was completed by vacuum-evaporating 15 nm of gold the polymer.

3. Characterization of the n-GaAs/p-PMeT diode

The dark direct current density-voltage characteristic of this GaAs/PMeT diode (fig.1) shows a classical linear dependence of log j as a function of the applied bias, for a low voltage (V < 0.3 V). At higher voltage, the current is limited by ohmic losses through the highly resistive polymer overlayer. These characteristics follow the modified Shockley-like equa-tion (I)

$$j \quad = \quad j_S \exp \quad q(V - jR) \quad / \quad nkT \quad - \quad 1 \tag{I}$$

where j_S is the saturation current density, n the diode quality factor and R the series resistance (Ω.cm^2) of the polymer layer. The saturation cur-rent density j_S was interpreted by using the thermoionic emission theory, $j_S = A^* T^2 \exp(-\phi_B/kT)$, allowing the determination of the barrier height ϕ_B. Table 1 lists the values of the parameters j_S, n, R and ϕ_B for different thicknesses of the polymer film, zero thickness corresponding to the classi-cal GaAs/Au Schottly diode.

The current density at reverse bias measured on a diode with a 100nm PMeT layer (Fig. 2) shows that the leakage current is amazingly low, limited to less than 2×10^{-4} A.cm^{-2} at - 10 V.

TABLE I
Parameters of the dark J-V characteristics of n-GaAs/p-PMeT diodes as a function of polymer thickness.

Polymer thickness (nm)	Saturation current density (A.cm^{-2})	Quality factor n	Barrier height ϕ_B (eV)	Series resistance (ohm.cm^2)
0	7.40×10^{-9}	1.20	0.82	-
25	4.41×10^{-10}	1.64	0.90	-
60	1.34×10^{-8}	1.78	0.81	22
100	1.42×10^{-7}	1.78	0.75	86

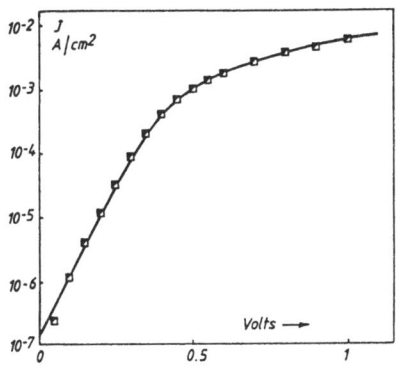

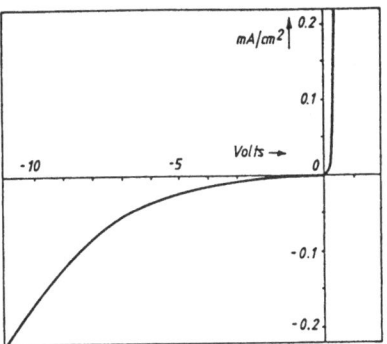

Fig. 1 Dark forward characteristic
of a n-GaAs/p-PMeT diode, with a
100 nm polymer thickness. Diode
area : 0.12 cm². Solid line is
the best fit to the Schockley-like
equation (1).

Fig. 2 Reverse bias current characteristic of the same diode as in
Fig. 1.

4. Discussion

In previous works devoted to the photovoltaic properties of PMeT (4),we have
already determined the electrical parameters of this p-type organic semiconductor : band gap, 2 eV ; conductivity, 7. 10^{-7} S.cm^{-1}. We were also able
to show that these parameters can be modified by the electrochemical control
of the doping level of PMeT. Like most organic semiconductors, the mobility of the majority carriers is very low and hence the conductivity much
lower than that of inorganic semiconductors. Moreover, the thickness of the
PMeT layer in the above diode is at the most of the same order of magnitude
as the space charge layer width. The GaAs/PMeT diode should thus behave
more like a MIS diode rather than a heterojunction. However, although the
conductivity of PMeT is much lower than that of classical inorganic semiconductors, it is still far higher than that of real insulators used in MIS
structures, such as SiO_2, Si_3N_4. This explains why current densities as
high as 10 mA.cm^{-2} can flow in GaAs/PMeT diodes at a forward polarization
of 1 V. This high current density is in fact only limited by the resistivity of the polymer layer, which can be controlled whether by the thickness of
this layer, or by the doping level of the polymer. The MIS theory is also
able to explain the increase of the quality factor n with the polymer layer
thickness, as due to the increase of the voltage drop across the diode supported by the insulator. However, MIS theory cannot explain the increase of
the saturation current density j_s with the PMeT thickness, which tends to
show that the behavior of these new n-GaAs/p-PMeT diodes stands in between
MIS and heterojunction diode.

5. Application of the diode

As a first application, a solar cell has been envisioned. In fact, organic
semiconducting polymers have already been used for photovoltaic devices :
Schottky diodes in which the undoped polymer, covered by a metallic overlayer, acts as a p-type photoactive semiconductor (5) ; heterojunctions
involving an undoped polymer and an inorganic semiconductor (6). However
in all these previous structures, the undoped polymer constituted the

photoactive layer of the device, and the power efficiencies were thus limited to very low values, as a consequence of the high resistivity, the low carrier concentration and the high recombination rates generally observed for organic semiconductors. The main difference for our diode GaAs/PMeT lies in the fact that n-GaAs is the photoactive layer, which enables thus much higher efficiencies. A photovoltaic cell has thus been realized, n-GaAs/p-PMeT (25 nm)/Au (15 nm), and analyzed under a water filtered tungsten light at an illumination of 100 mW.cm^{-2}.

The power characteristic is given in Fig. 3. The comparison with a classical Schottky cell GaAs/Au shows i- a small decrease in the photocurrent density j_{sc}, due to the absorption of incident light by the polymer, ii- a sharp increase in the open-circuit voltage V_{OC}, which is comparable to that one observed with MIS structures. When corrected for the absorption of light by the fold overlayer, the obtained power efficiencies are 14 % for the GaAs/Au, and 17.5 % for the GaAs/PMeT/Au cell.

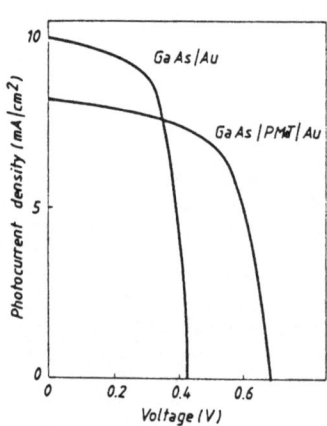

Fig. 3 Power characteristics of GaAs and GaAs/PMeT/Au cells under 100 mW.cm^2 tungsten illumination.

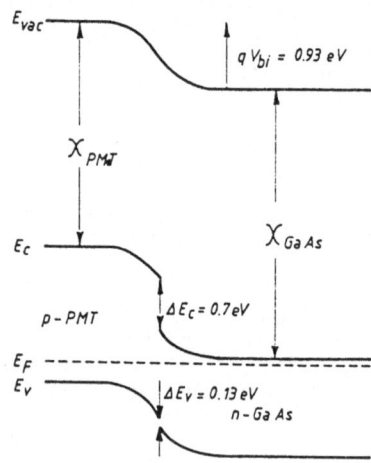

Fig. 4 Energy band diagram after Anderson's model of a p-PMeT/n-GaAs junction.

Dark JV measurements under forward bias, performed on the GaAs/Au cell, led to the determination of its quality factor, 1.18, of its barrier height, ϕ_B = 0.84eV, and of its built-in potential, V_{bi} = 0.77V. For the GaAs/PT/Au cell, the obtained quality factor is much higher, 1.53. An energy diagram for the GaAs/PMeTjunction has been established (Fig. 4), following the Anderson's model (7) and using the already determined electrical parameters for semiconducting polythiophene (4).

The determined built-in potential for this new cell, V_{bi} = 0.93V, can be directly compared to the corresponding value obtained for the GaAs/Au cell (V_{bi} = 0.77V). The increase in V_{bi} between the two cells, 0.16V, is too low for explaining the large increase observed in the open-circuit voltage, ΔV_{OC} = 0.24V. A reduction of the dark direct current must also be taken into account for explaining the large improvement of the characteristics of the n-GaAs/p-PMeT/Au diode.

This new structure differs thus from the already reported organic-on-inorganic devices in that the organic polymer front layer is very thin, allowing most of the incident light to be absorbed within the inorganic semiconductor, which acts as the photoactive substrate. The major disadvantages of organic semiconductors, namely high recombination rate and resistivity, is then avoided, and the slight reduction of the short circuit current, compared to classical Schottky cells, is overcompensated by an important improvement of the open-circuit voltage.

6. Conclusion

Many other applications can be envisioned for these organic semiconducting polymers, which present basic important advantages related to a large à-priori molecular control of their characteristics, i- easy elaboration of thin films of controlled thickness, by the use of a mild electrochemical growing technique that can be applied on fragile inorganic semiconductors and also on polycrystalline substrates, ii- precise control of their doping level, which can be electrochemically adjusted for desired resistivity, carrier concentration..., and iii- possibility of controlling their band energy levels, which can be modified by electronic structure effects on the monomer (substituted polythiophenes), and hence tuned with the band levels of the inorganic semiconductor in order to improve the characteristics of the diode. For instance, the increase of the barrier height could allow the development of MIS field effect transistors using InP substrate.

A large field of new developments can thus be expected from organic semiconducting polymers for the design of new electronic devices.

References

1 J.W. Thackeray, H.S. White and M.S. Wrighton, J. Phys. Chem., 89, 5133 (1985).
2 A. Tsumura, H. Koezuka, S. Tsunoda and T. Ando, Chem. Letters, 863 (1986).
3 C.L. Cheng, S.R. Forrest, M.L. Kaplan, P.H. Schmidt and B. Tell, Appl. Phys. Lett., 47, 1217 (1985).
4 S. Glenis, G. Horowitz, G. Tourillon and F. Garnier, Thin Solid Films, 111, 93 (1984).
  S. Glenis, G. Tourillon and F. Garnier, Thin Solid Films, 122, 9 (1984).
  G. Horowitz and F. Garnier, Solar Energy Mater., 13, 47 (1986).
5 B.R. Weinberger, M. Akhtar and S.C. Gau, Synth. Met., 4, 187 (1982).
6 M. Ozaki, D.L. Peebles, B.R. Weinberger, C.K. Chiang, S.C. Gau, A. J. Heeger and A.G. Mediarmid, Appl. Phys. Lett., 35, 83 (1979).
7 R.L. Anderson, Solid State Electron, 5, 341 (1962).

Polymer-Based Conducting Material Used for Water Vapour Detection in Air

J.P. Lukaszewicz and J. Siedlewski

Institute of Chemistry, Nicholas Copernicus University,
PL-87-100 Torun, Gagarina 7, Poland

An attempt has been made to apply an unconventional, conducting, polymer-based material for the detection of water vapour in air. The influence of the heat-treatment temperature and chemical treatment of the surface on the changes of electrical resistance observed in the course of water adsorption is presented.

1. Introduction

Polyfurfuryl alcohol obtained from the monomer by poly-condensation is a black, non-melting and non-conductive material. One possible way to convert this polymer into the conductive form is heat treatment carried out at relatively high temperatures (800-900 K). Since 1981, we have been testing thermally decomposed polyfurfuryl alcohol (TDPFA). Polyfurfuryl alcohol has the following valuable properties: the polycondensation is easy to control, it is thermosetting, it is relatively easy to prepare samples in different geometrical forms (bulk samples, powders, films), the adsorption and basic electrical properties are highly stable, fundamental electronic properties may be simply controlled by suitable heat treatment.

2. Experimental

The polycondensation of furfuryl alcohol was carried out at 373 K in the presence of oxalic acid (catalyst) previously powdered and dissolved in the monomer. The thermogravimetric data show a significant decrement of the polymer mass at about 600 K, which is related to decomposition of the original polymer structure. The heat treatment was carried out in an oxygen-free atmosphere for 1 hour

at 823 K (sample A) or at 873 K (sample B). In these
studies we used powdered TDPFA of 0.12-0.06 mm grain
diameter. A special sample holder was constructed to have
simultaneously a good contact between the water vapour
and adsorbing polymer. An ohmic contact (d.c. electrical resistance measurements) was also made on the
TDPFA-electrode interface (Table 1).

Table 1. Initial resistance values of TDPFA samples.

Sample	HT temp. $|K|$	R_o $|Mohm|$	Current carriers	Geometric form
A	823	9.253	holes	powder
B	873	1.429	holes	powder
B_1	873	2.786	holes	powder
E	1123	0.792	electrons	monolith
E_1	1123	0.746	electrons	monolith

The resistivity of TDPFA drops with an increase of the
heat-treatment temperature. The nature of majority
current carriers also depends strongly on the heat-treatment temperature (a shift from p-type to n-type semi-
conductivity). Due to a significant decrease of sensitivity with the growth of the equilibrium concentration
of current carriers, an increase of the heat-treatment
temperature causes a decrease of R_o. The samples of type A
and B are the most sensitive of all the previously investigated specimens. In this case the depth of surface layer
(conductivity changed by water adsorption) is great
enough to produce large alterations in the relative
resistance value R_m R_o even when the concentration of
water vapour in the carrier gas is low. Figures 1 and 2
(R_m/R_o = f(V) functions) confirm the rise of sensitivity
with decreasing heat-treatment temperature. R_m is the
value of the sample electrical resistance that differs
most from the initial resistance R_o due to water adsorption. V is the value of liquid water injected into the air
stream; V is directly proportional to the water concentration in the air flow.

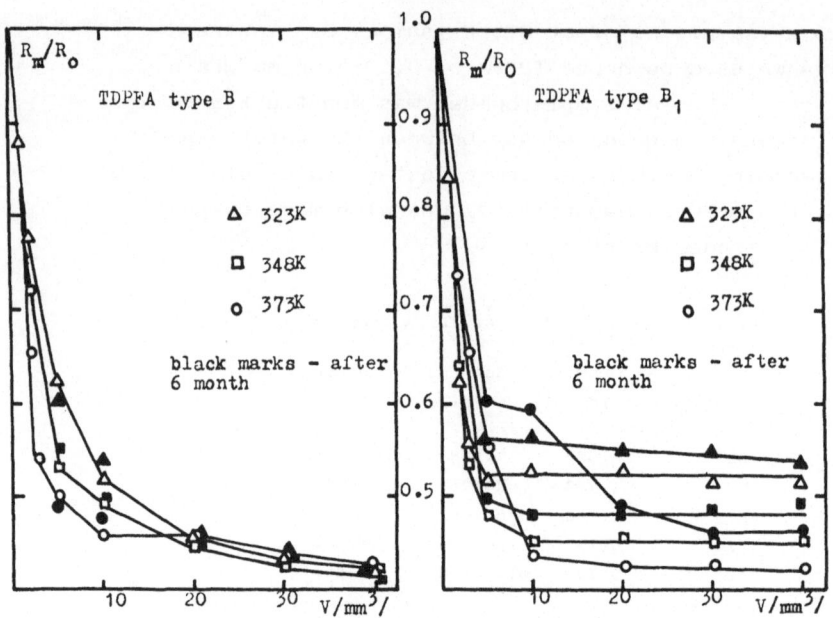

Fig. 1. Dependence of R_m/R_o on the volume of water shot.

Fig. 2. Dependence of R_m/R_o on the volume of water shot.

Table 2. Stability of R_o (sample A).

Temperature |K|	R_o |Mohm|	R_o |Mohm|
323	9.253	0.173
348	7.282	0.223
373	4.601	0.100

3. Conclusions

The analysis of experimental data is the basis for the
following conclusions:

A. Electrical resistance changes of TDPFA samples induced by water adsorption are relatively large and completely and quickly reversible (see Table 2 and Fig.3).

B. The stability of sensitivity to water vapour is satisfactory for unoxidized samples of TDPFA even after storage
for six months in room conditions. Chemical treatment,
e.g. oxidation or ion exchange on the TDPFA surface,

430

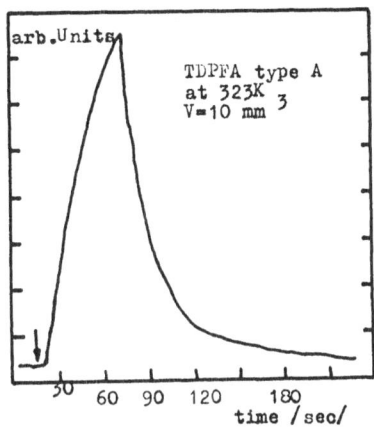

Fig. 3. Resistance of sample A Fig. 4. Dependence of R_m/R_o on
vs. time of water adsorption   the volume of water shot.

produces relatively unstable material, whose sensitivity
can change with time (Figs. 1,2,4) .

C. All TDPFA samples of type A or B show significant
sensitivity (drop of $|R_m-R_o|/R_o$ value) in the range of low
water vapour concentration (shot volumes 1-10 mm^3).

D. For high V values (10-40 mm^3), i.e. for water vapour
concentrations close to saturated vapour pressure
a "plateau" is observed.

E. The oxidation of TDPFA type B gave the sample B_1,

for which the "plateau" is wider. The sensitivity also
decreases.

F. The sensitivity of unoxidized samples A and B depends
rather weakly on temperature in the range 323-373 K.

For oxidized sample B_1 the sensitivity increases with increasing temperature in the same range.

G. The R_m/R_o=f(V) function could in general be described
as $R_m/R_o = a$ V $/(b + c$ V $)$.

For sample A and for a=1, b=0.5, c=1, n=2 the theoretical
curve fits the experimental points well.

4. References

1. J. P. Lukaszewicz: Proc. Int. Conf. Exhibition "Test &
Transducer'86", Trident Int., London 1986.

2. J. P. Lukaszewicz: Thesis, Nicholas Copernicus University, Torun 1983.

A Novel Application of Conducting Polymers: Remotely Readable Indicator Devices

R.H. Baughman, R.L. Elsenbaumer, Z. Iqbal, G.G. Miller, and H. Eckhardt

Allied-Signal Inc., Corporate Technology, Morristown, NJ 07950, USA

1. Abstract

Electron donor or electron acceptor doped polymer complexes which provide high electrical conductivities are used as ambient-responsive elements for remotely readable indicator devices. Depending upon the choice of conducting polymer, dopant, dopant-compensation agent and other details of device design, diverse indicator devices result for in-box monitoring of products: time-temperature, temperature limit, humidity, radiation dosage, mechanical abuse, and chemical release indicators. These new indicator devices are obtained from the novel combination of ambient-responsive conducting polymers and commercially available technology for radiofrequency or microwave frequency antitheft devices.

2. Introduction

It is now well known that the addition of electron donors or electron acceptors to a variety of insulating organic polymers results in metallic conductors. However, although diverse applications have been proposed for these novel materials, no major commercialization has yet occurred. The problem has been obtaining the entire desired properties profile in an economically fabricatible material. Specifically, many of the conducting polymers are insoluble and infusible and are, therefore, difficult to process economically. Also, many of those polymers which are processible, albeit usually from solution, have not yet been obtained in sufficiently high molecular weight that mechanical properties can be optimized. Finally, and probably of key importance, most of the conducting polymers are unstable in the ambient atmosphere and apparently none of the known solution or melt processible conducting polymers have long-term environmental and thermal stability (largely invariant properties for a service life longer than ten years at the extreme of normal ambient). Despite this problem, promising application possibilities are suggested by the rate at which property improvements are being made and the fact that in certain applications, such as batteries, protection from the ambient is economically feasible.

We will here describe a novel application of conducting polymers which utilizes as a positive, required feature the instability which has thwarted many other uses. This application area is remotely-readable indicator devices. Depending upon device materials and construction, indicators result for in-box measurement of various exposures: time-temperature, temperature limit (freeze and defrost), humidity, radiation dosage, mechanical abuse, or chemical release.[1] Conducting polymers have previously been examined for use as radiation dosage, gas, and biological sensors,[2-9] but not as remotely readable indicator devices.

3. Description Of Indicator Technologies

3.1 General Aspects

The new indicator technologies are based on the use of conducting polymers as ambient-responsive layers or circuit elements for devices of the type in present commercial use as antitheft targets. The antitheft targets are small antennae which operate at either radiofrequency or microwave frequencies. Detection of these targets is _via_ a transmitter-receiver, which can be small enough to fit into the palm of your hand.

The devices using conducting polymers which we have focused upon resulted from key modifications of radiofrequency antitheft targets which are commercially available from Checkpoint Systems Inc. (Thorofare, New Jersey). These targets cost only about $0.03 each and are therefore inexpensive enough to be found on diverse items in your local store. For example, we found these antitheft targets attached to items costing less than a dollar at our local pharmacy.

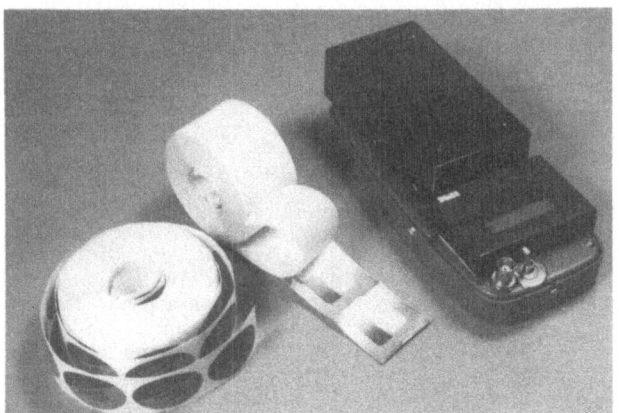

Fig. 1 A roll of deactivating stickers, a roll of rf antitheft targets, and a hand-held detector which we have modified for digital output; all are commercially available from Checkpoint Systems, Inc.

The Checkpoint Systems Inc. radiofrequency antitheft target tags, which are shown in Figures 1 and 2, are about 4 cm square.[10] Deposited on the adhesive side of each paper label is a passive radiofrequency (rf) resonance circuit consisting of a thin strip of aluminum foil configured as a squarish spiral. The detectors for these tags, which operate at a resonance frequency of about 8.2 MHz, are available in a variety of different configurations, ranging from hand-held to large dimension detectors suitable for monitoring store exits. The detector contains a transmitter which continually sweeps over a designated frequency range with a period of about a millisecond. The antitheft tag circuit, which is equivalent to an inductor and capacitor, resonates somewhere in this range. The tag absorbs and remits rf energy which varies in intensity with the transmitter's instantaneous frequency. This frequency-dependent response over the sweep frequency is monitored by a receiver and stored in a short-term memory. On each subsequent frequency sweep of the tag, the selective frequency response of the tag accumulates, while signals due to rf noise tend to average to

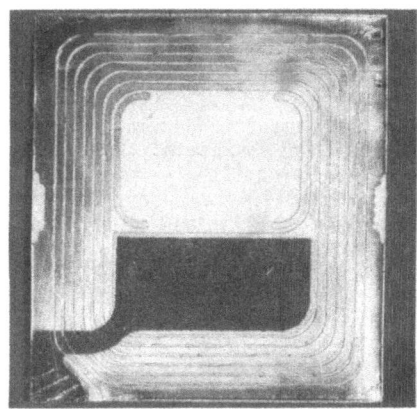

Fig. 2 Antitheft targets from Checkpoint Systems as-sold (left) and
as-modified (right) with a partial covering of an ambient-sensitive
conducting polymer.

zero. This accumulated signal provides the detector response. The
nonadhesive side of the indicator tag, which is the side which the customer
views, typically looks like a conventional bar-code label, which conceals
the function as a theft detector.

We utilize in many of our designs for rf-monitored indicators the fact
that a conductor can either shield the target or couple to it, so as to
change target response. In fact, Checkpoint Systems Inc. utilizes
adhesive-backed aluminum foil/paper composites to deactivate indicator
response. These deactivating tags appear to the customer as a paper sticker
marked "Paid, Thank you", to conceal their function. This effect of
metallic conductors on target response, which we utilize in our indicator
devices, means that the rf-based indicators cannot be in close proximity to
metallic packaging or product materials. On the other hand, the indicator
response can be easily read through a foot of water.
The contrary case exits for microwave (mw) antitheft targets, such as
produced by Sensormatic Electronics Corp. (Deerfield Beach, Florida), which
are easily shielded by water but not by metal foil.[10] Each of these
microwave tags consists of an encapsulated antenna, diode, and resonator
circuit. In this case the transmitter sends out microwaves at two
frequencies and the diode of the target generates sum and difference
frequencies that the receiver detects, thereby providing the response of the
target. Because the microwaves can conveniently be transmitted as a
directional beam, mw systems have about twice the range of the rf systems,
permitting target detection from distances up to about eight feet. Because
of increased complexity, the microwave antitheft targets are typically more
expensive than the radiofrequency targets and are therefore designed to be
re-useable. Despite this cost disadvantage, the microwave systems account
for about 75% of the surveillance systems in use in the United States. We
have devised ambient-responsive indicator devices by incorporating ambient-
responsive conducting polymers in the circuit of the microwave targets.

Knowing the way in which both indicator response and product quality
varies with a key ambient variable, measurement of indicator response
permits prediction of the effect that the shared indicator/product
environment has had on product quality. However, the signal strength from

434

an rf or mw frequency antitheft target, which corresponds to the indicator response, depends upon the orientation of the transmitter/receiver relative to the target antennae. Consequently, in order to obtain quantitative measurements on indicator devices, the measurement geometry should be fixed by a specified relationship of both indicator target and transmitter/receiver to the product container. In the case of hand-held monitoring units, this requirement is conveniently met by positioning the indicator targets in a fixed manner within the product container and by providing an outline of the hand-held monitor on the shipping container, so as to uniquely specify the position of the detector relative to the shipping container. Alternately, two targets operating at different frequencies (rf or mw) can be positioned in close proximity and with identical orientations inside each shipping container. One of these targets (the reference target) does not contain an ambient-sensitive element, while the second target contains such an element. Referencing the signal from the ambient-sensitive target with respect to the signal from the reference target relaxes constraints for obtaining quantitative measurements of ambient exposure when the geometrical relationship of transmitter/receiver and indicator target is not well defined.

The operation of our indicator devices utilizes, directly or indirectly, the effect of environment upon the electrical conductivity of conducting polymers. Conducting polymers are herein defined as polymers which undergo large changes in electrical conductivity upon the addition of either electron donors or electron acceptors, a process which is referred to as doping[11]. This increase in electrical conductivity is typically from much less than 10^{-8} S/cm for the undoped state to from 10^{-1} S/cm to well over 10^{3} S/cm for the doped state. The polymers are typically conjugated polymers having broad valence bands (or broad conduction bands) and low ionization potentials (or high electron affinities).

The reactions utilized in device operation include: (1) the dramatic increase in electrical conductivity upon the addition of an electron acceptor or an electron donor to an undoped conducting polymer to form, respectively, p-type or n-type conductors, (2) the dramatic decrease in electrical conductivity of acceptor-doped or donor-doped conducting polymers upon compensation of these dopants using, respectively, donor or acceptor dopants, and (3) thermally- or chemically-induced modification of the doped polymer, such as by disruption of the conjugation of the polymer backbone.

Fig. 1 pictures a hand-held monitor for anthitheft targets from Checkpoint Systems, which we have modified to provide a digital readout of indicator response. Also shown here is a roll of rf antitheft targets and a roll of deactivating stickers, both from Checkpoint Systems. Fig. 2 shows a closer view of the above antitheft targets. The target on the left is unmodified and the one on the right contains a film of a conducting polymer (a polyalkylthiophene) which covers part of the antenna. The effect of ambient on the conductivity of the conducting polymer determines the response of the rf indicator. The conducting polymer layer can serve as a rf shield and as a capacitive element to effect the response of the rf target. Alternately, removal of an overcoating insulating layer on the antitheft target permits direct shorting of the antenna with the conducting polymer. Such direct incorporation of the conducting polymer in the resonance circuit of the target results in an enhanced sensitivity of the indicator to changes in conductivity of the conducting polymer.

3.2 Time-Temperature Indicators

Time-temperature indicators are devices designed to respond to the accumulated effects of thermal exposure in the same way as this exposure affects product quality. The degradation kinetics observed for a particular perishable are generally fit to Arrhenius kinetics in the relevant temperature range and an indicator device is then designed which provides basically the same preexponential factor and activation energy in this range.

A variety of different types of time-temperature indicators have been proposed and several different types of visually or machine-read color indicators have been commercialized[12,13]. Of special interest to us has been the LifeLines technology which we developed at Allied-Signal.[14-16] Color transformations resulting from the solid-state polymerization of diacetylenes are used in this technology to provide machine-readable bar codes that denote time-temperature exposure. This technology has many attractive features, such as inherent simplicity. While the present indicator devices are more complicated than the LifeLines indicators, the present indicators have the advantage of being conveniently monitorable inside a closed box where the thermal environment of the indicator and the product (drugs, foodstuffs, etc.) can be identical.

We have demonstrated a variety of different designs for remotely readable time-temperature indicators which utilize conducting polymers as sensor elements[1]. The kinetics of device response is controlled by either (1) the thermal degradation of the conductivity of a doped polymer, (2) the diffusion of a dopant or dopant-compensation agent through a barrier to contact a conducting polymer, or (3) the thermal generation of a dopant or dopant compensation agent which then interacts with an undoped or doped conducting polymer. Device response is then a consequence for an undoped polymer of the conductivity increase upon polymer doping or the conductivity decrease for a doped polymer during thermal degradation or the compensation of one dopant type with another. Depending upon device design, device response kinetics can be controlled by the thickness and nature of a diffusion barrier and the choice of conducting polymer, dopant, and dopant-compensation agent.

The simplest device design consists of a layer of doped conducting polymer, such as polyaniline doped with HCl or polyacetylene doped with iodine, on the surface of an rf antitheft target, such as available from Checkpoint Systems. The high initial conductivity of the doped conducting polymer effectively shields the rf target, so this target is initially in the OFF state. Thermal degradation of the conducting polymer as a consequence of thermal exposure dramatically decreases the shielding effect of the conducting polymer, so as to gradually turn the indicator to the fully ON state, where the polymer film has no significant effect on the response of the rf target. Additives to the polymer film can be used to vary the thermal degradation rate, as can the choice of whether or not the conducting polymer is sealed against the generally accelerating effect of atmosphere on thermal degradation of polymer conductivity.

Alternately, an activatable time-temperature indicator can be constructed by microencapsulating a solution of a dopant (or a dopant-compensation agent) on top of an undoped conducting polymer (or a doped conducting polymer) which covers the rf target. Mechanical rupture of the microcapsule activates the indicator by releasing the dopant (or dopant-compensation agent), so that the state of the rf target can change from ON to OFF (or from OFF to ON) during subsequent thermal exposure. This design is generally suitable for highly labile perishables such as frozen foods. However, the kinetics of device response can be lengthened and the temperature dependence of this response can be conveniently varied at will by the use of a diffusion barrier film to separate the polymer layer and the dopant source (or the source of dopant-compensation agent).

3.3 Temperature Limit Indicators (Freeze and Defrost)

Temperature limit indicators have special value for products which undergo
substantial changes in quality as a consequence, for example, of freezing or
melting (defrost). Examples range from frozen foods, various aqueous
medical products, and water-based paints to salad dressings which phase
separate during freezing.

A freeze indicator can be constructed similarly to the above mentioned
time-temperature indicator by covering a rf target with a layer of undoped
conducting polymer, which in turn is covered with microcapsules containing a
dopant solution. The mechanical stress caused by freezing ruptures the
microcapsules, thereby releasing a dopant which dopes the conducting
polymer. Hence, freezing causes the conducting polymer to go from
insulating to metallic state, corresponding to an ON-OFF transition of the
rf indicator. Alternately, the freeze-rupturable capsules can contain a
dopant-compensation agent which causes a highly conducting polymer to become
insulating, corresponding to an OFF-ON transition of the rf device. Other
modifications of this basic design are also convenient. For example, the
freeze-ruptured microcapsules can release a solvent which transports a
solid, nonvolatile dopant or dopant-compensation agent through a barrier
(such as filter paper) to contact the conducting polymer. Note that this
design is not much different from the above mentioned design of a
time-temperature indicator, wherein dopant is released by applied mechanical
pressure. The transition is made from freeze indicator, to time-temperature
indicator for highly perishable products, and to time-temperature indicators
for high thermal stability products by design changes (dopant, dopant
compensation agent, conducting polymer and the presence and nature of
diffusion barriers) to lengthen progressively the time required between
dopant release (or compensation agent release) and the occurrence of
substantial polymer doping (or dopant compensation). If this time is long
in the operating temperature range, then the above described freeze
indicator becomes a freeze-activated time-temperature indicator.

A freeze-activated defrost indicator can be constructed by a slight
modification of the above design for a freeze indicator. The freezing
process again ruptures the microcapsules, thereby (in this case) activating
the device. However, a diffusion barrier which is permeable only by the
fluid phase separates the thereby-released solvent from the solid dopant or
solid dopant-compensation agent and the conducting polymer film (or the
thereby released solution of dopant or dopant-compensation agent from the
conducting polymer film). The barrier is sufficiently thick that complete
solvent freezing prevents complete penetration of the barrier. Hence,
contact of the polymer with dopant solution (or a solution of
dopant-compensation agent) does not occur until the device is above defrost
temperature for sufficient time after activation that this diffusion barrier
is penetrated. If the solid dopant or solid dopant compensation agent
interacts directly on contact with the conducting polymer, these solids need
be separated from the conducting polymer by a second such barrier film.

3.4 Radiation Dosage, Mechanical-Abuse, Humidity, and Chemical-Release Indicators

Radiation dosage indicators are conveniently constructed using materials
which release dopants or dopant-compensation agents for conducting polymers.
Indicators for γ-ray dosage are of increasing interest because of the
increasing commercial use of γ-rays for sterilization. One such indicator
which we have constructed consists of a film of poly(vinylchloride)
containing a dispersion of undoped polyaniline, which is deposited on the
surface of a rf target. The γ-ray radiation degrades the poly(vinyl

chloride), presumably by releasing HCl, which progressively dopes the polyaniline. The polyaniline/poly(vinyl chloride) film which is initially insulating and, therefore, transparent to the rf radiation becomes conducting and absorbing for this radiation, so that the device response progressively shifts from ON state to OFF state. Similarly, indicators for ultraviolet radiation can be constructed using chemicals such as $Ph_2I^+PF_6^-$, which photogenerate doping agents (HPF_6) for conducting polymers.

Indicators for accumulated moisture exposure can be constructed using the conductive-nonconductive transition of water-sensitive doped conducting polymers. The use of a barrier film for water vapor can be used to adjust device sensitivity.

Indicators for mechanical abuse can be constructed using mechanically fragile capsules containing fluid dopant or dopant-compensation agents. Product mechanical abuse during shipping and handling ruptures these capsules to release such a fluid, which then interacts with an undoped or doped conducting polymer which overlays the rf indicator to provide the conductivity change which generates the device response change.

Also of interest are indicators which have been constructed that provide remote, in-box monitoring of chemical release (ie., container rupture). The contact of an accidentally released oxidizing or reducing agent with a conducting polymer causes a change in the conductivity of this polymer, thereby providing a change in response of the rf target. Alternately, a released solvent for a solid, nonvolatile dopant or dopant-compensation agent transports this chemical through a permeable barrier to contact a conducting polymer, thereby providing the device response.

4.0 Conclusion

We have demonstrated the use of conducting polymers for constructing ambient-responsive layers as part of remotely readable indicator devices. Using only this class of materials, we have seen that a broad variety of indicators can be constructed. However, we have also demonstrated elsewhere the use of more general classes of organic and inorganic materials for constructing the ambient responsive layers.[1] Most generally, other classes of materials which undergo significant changes in electronic conductivity, ionic conductivity, or dielectric constant can be utilized to shield or alter the response characteristics of radiofrequency or microwave frequency target antennae. For example, molecular charge-transfer conductors can be used analogously to the conducting polymers. Going beyond these electronic conductors, it is possible to conveniently utilize the changes in electrical properties which occur during the dissolution of solid salts, such as dissolving sodium chloride in water.[1] Using the variety of conveniently available, nontoxic salts and solvents for such salts, it is possible to construct indicators for monitoring various exposures: time-temperature, temperature, temperature-limit, humidity, radiation dosage, mechanical abuse, and chemical release. The application of these salts to make remotely-readable indicator devices has the disadvantage of requiring the use of fluid components and, in some cases, a slightly more complicated design of the ambient responsive element on the antenna target. Also, the application of conducting polymers permits the fabrication of devices that can have much higher sensitivity than those fabricated using conventional salts. An example of such high sensitivity is provided by an oxygen indicator which can be constructed using alkali-metal doped polyacetylene or poly(p-phenylene). On the other hand, certain of these conventional salts can have advantages over the conducting polymers for specific application because of negligible cost, present commercial availability, and well-established nontoxicity.

References

1. R.H. Baughman, R.L. Elsenbaumer, Z. Iqbal, G.G. Miller, and H. Eckhardt, U.S. Patent No. 4,646,066 (1987).
2. K. Yoshino, S. Hayashi, G. Ishii, Y. Kohno, K. Kaneto, J. Okube, and T. Moriya, Kobunshi Ronbunshu 41, 177 (1984).
3. J.H. Lai, Macromolecules 10, 1253 (1977).
4. C. Nylander, M. Armgarth and I. Lundstrom, Analytical Chem. Sympos. Ser. Chem-Sensors 17, 203-207 (1983).
5. K. Yoshino and H. B. Gu, Japanese Journal of Applied Physics 25, 1064 (1986).
6. M. Josowicz, J. Janata, K. Ashley, and S. Pons, Anal. Chem. 59, 253 (1987).
7. H. Kita and Y. Kato, Japan Patent Office, Patent Application No. 123976 (1984).
8. G. Berthet, J.P. Blanc, J.P. Germain, A. Larbi, C. Mayeysson, and H. Robert, Synth. Met. 18, 715 (1987).
9. M.K. Malmros, U.S. Patent 4,334,880 (1982).
10. High Technology, pp. 16-17, Sept/Oct (1983).
11. For an overview of the physics and chemistry of conducting polymers see the Proceedings of the International Conference on Science and Technology of Synthetic Metals (Kyoto, Japan, 1986), which is published in Synthetic Metals, Vols. 17 and 18 (1987).
12. H.M. Schoen and C.H. Byrne, Food Technology 26, 46 (1972).
13. J.W. Farquhar, Revue Internationale du Froid 5, 50 (1982).
14. G.N. Patel, A.F. Preziosi, and R.H. Baughman, U.S. Patent 3,999,946 (1976).
15. R.H. Baughman, G.N. Patel, and G.G. Miller, U.S. Patent No. 4,389,217 (1983).
16. R.H. Baughman and R.R. Chance, Polymer Preprints 27, No. 1, 67 (1986).

Index of Contributors

Springer Series in Solid-State Sciences

Editors: M. Cardona   P. Fulde   K. von Klitzing   H.-J. Queisser

Volumes 1–39 are listed on the back inside cover